Nanotechnology in Medicine

Nanotechnology in Medicine

Toxicity and Safety

Edited by

Mahendra Rai

Sant Gadge Baba Amravati University
Maharashtra, India

Mrunali Patel

Ramanbhai Patel College of Pharmacy, Charotar University of Science and Technology
Changa, Gujarat, India

Rashmin Patel

Ramanbhai Patel College of Pharmacy, Charotar University of Science and Technology
Changa, Gujarat, India

WILEY Blackwell

This edition first published 2022

Registered Offices
John Wiley & Sons, Inc., 111 River Street, Hoboken, NJ 07030, USA
John Wiley & Sons Ltd, The Atrium, Southern Gate, Chichester, West Sussex, PO19 8SQ, UK

Editorial Office
9600 Garsington Road, Oxford, OX4 2DQ, UK

For details of our global editorial offices, customer services, and more information about Wiley products visit us at www.wiley.com.

Wiley also publishes its books in a variety of electronic formats and by print-on-demand. Some content that appears in standard print versions of this book may not be available in other formats.

Library of Congress Cataloging-in-Publication Data

Names: Rai, Mahendra, editor. | Patel, Mrunali Rashmin, 1981– editor. |
Patel, Rashmin Bharatbhai, 1979– editor.
Title: Nanotechnology in medicine : toxicity and safety / edited by Mahendra Kumar Rai,
SGB Amravati University, Maharashtra, India, Mrunali Rashmin Patel,
Charotar University of Science and Technology, Gujarat, India, Rashmin Bharatbhai Patel,
Charotar University of Science and Technology, Gujarat, India.
Description: Hoboken, NJ : Wiley-Blackwell, 2022. | Includes bibliographical references and index.
Identifiers: LCCN 2021031539 (print) | LCCN 2021031540 (ebook) | ISBN 9781119769866 (cloth) |
ISBN 9781119769873 (adobe pdf) | ISBN 9781119769880 (epub)
Subjects: LCSH: Nanotechnology. | Medical technology.
Classification: LCC R857.N34 N3617 2022 (print) | LCC R857.N34 (ebook) | DDC 610.285–dc23
LC record available at https://lccn.loc.gov/2021031539
LC ebook record available at https://lccn.loc.gov/2021031540

Cover Design: Wiley
Cover Image: © KATERYNA KON/Getty

Set in 9.5/12.5pt STIXTwoText by Straive, Pondicherry, India

10 9 8 7 6 5 4 3 2 1

Contents

Preface

Currently, an exceptional growth in research, innovation, and applications have been realized in the area of nanoscience and nanotechnology. Nanotechnology is beheld as a transformative technology. It offers ample possibilities to address the key societal challenges essential to encounter global healthcare problems. It presents engineered nanomaterials, smart innovative products incorporating these materials, and nanoenabled processes having magnificent growth potential for a large number of industrial sectors. Formulation scientists have extensively used nanoparticles for entrapment of drugs with intention of enhanced delivery to, or uptake by, target cells and/or a reduction in the toxicity of the free drug to nontarget organs. There is collective optimism that nanotechnology, as applied to medicine, has and will further lead to noteworthy developments in therapeutics, tissue engineering, active implants, nano-robotic biosensors, molecular diagnostics and imaging, bone grafting, tiny vehicles for drug delivery, nutraceuticals, genomics, and proteomics. Nanomedicine, incorporating prospective drug delivery nanocarriers, hails great promises to combat complex illnesses like cancer, neurological diseases, infectious or inflammatory diseases, and malaria due to novel physicochemical properties. However, this may damage and exert toxicological effects due to detrimental interfaces with living systems and the environment. Thus, the smart and sustainable development of these nanomaterials has created concerns about their possible undesirable effects on human health and safety as well as an environmental burden. Additionally, these necessitate contemplating its characteristic features, exposure effects, hazard mechanisms, risk assessment and management.

This book focuses on current trends to deliver therapeutic agents into targeted specific cells, cellular compartments, tissues, and organs by using nanoparticulate carriers; applications of nanotechnology in medicine with a special focus on neurodegenerative diseases, cancer, diagnostics, nano-nutraceuticals, dermatology, gene therapy, etc.; nanotoxicology with focus on nanomedicine, landscapes of *in vitro* and *in vivo* toxicological testing; regulatory framework in nanotechnology and medicine; future aspects in toxicity and safety assessment of nanotechnology, etc. Also, the book identifies the knowledge gaps related to nanomaterials' safety which tends to be the greatest hurdle in obtaining benefits of nanomedicine in healthcare. Now that we are entering a new phase from academic development to proven clinical value, it is very important for specialists in the field to find all the information on these aspects together in a book, to see what the current stage is, what the problems are and what viable solutions are being proposed for the future in this field.

This book contains valuable chapters written by professionals and experts on the diverse aspects of nanotechnology in medicine with insights into the toxicological and safety issues owing to interaction with biological systems and the environment. The chapters have been divided into four different parts. *Part I: Nanomedicine: Nanotoxicological Insights* – aims to enrich the understandings of nanotoxicological features of nanomedicine. It provides an introductory overview of nanomedicine and nanotoxicology progressing further toward recent advances in research and development dealing with microbial biopolymers and their derivatives as nanotechnological instruments for medicine. *Part II* deals with *Nanoparticles: Toxicity and Safety*. The recent technological and manufacturing innovations have led to the enhanced use of nanoparticles for diverse applications. However, this has also preceded the risks allied with its exposure. This part imparts the perceptions of toxicity and safety aspects of selenium nanoparticles, the impact of nanoparticles on protozoa, moving ahead in the direction of toxicity and safety facets of metallic, polymeric, and mesoporous silica nanoparticles. *Part III* is focused on *Nanotoxicology and Drug delivery*. Augmenting the mechanistic insights for risk assessment of employing nanomaterials in drug delivery is the need of the hour. Consequently, this part targets the toxicity and safety aspects of incorporating nanocarriers for gene delivery, brain delivery, parenteral delivery, dermatological delivery, etc. Lastly, it encompasses the current scenario and future road map for nanotoxicity and nanosafety assessment of nanocarriers. The last part (*Part IV: Nanotechnology, Ethics and Regulatory Framework*) emphasizes harmonizing the procedures for safety evaluation globally and ascertaining unambiguous regulations as well as standard protocols. The part thus delivers the perceptions of nanoethics and the regulatory framework of nanotechnology in medicine.

This book has been planned, authored, and structured for the wide spectrum of multisectoral audiences of scholars, students, practitioners, and policymakers with adequate coverage. It is intended to be a reference resource that will meet the needs of academicians, clinicians, as well as graduate and postgraduate students from nearly every field of science and engineering – toxicology, biotechnology, chemistry, physics, genetics, medical science, material science and engineering, environmental science and environmental engineering. For researchers working in the fields of nanotechnology, nanomaterials and nanocarriers, nanoengineering, nanobiotechnologies, nanomedicine and biopharmaceuticals, and other similar fields, it would be very useful.

We are highly thankful to the authors for their excellent work in providing cutting-edge information on the subject of their respective chapters. Their efforts will certainly enhance and update the readers' knowledge of the toxicity and safety aspects of nanomedicine. Further, we would like to express our sincere thanks to the publishers and authors of the chapters whose research work has been mentioned in the book. We are also thankful to Dr. Julia Squarr, Sarah Mellor, Rosie Hayden, Tom Marriott, and the Wiley team for their substantial cooperation and efforts in producing this book.

Last but not the least, we believe that this edited book would inspire engagement and intervention by concerted and collaborative action by appropriate bodies to shape the nanoworld and deliver innovative solutions to address daunting medical challenges.

25 April 2021

Mahendra Rai
Amravati, MH, India
Mrunali Patel
Anand, GJ, India
Rashmin Patel
Anand, GJ, India

List of Contributors

Nayelen Sayuri Aizawa
Department of Biotechnology
Engineering School of Lorena (EEL)
São Paulo University (USP)
Lorena, São Paulo, Brazil

Amit Alexander
National Institute of Pharmaceutical Education and Research (NIPER GUWAHATI)
Department of Pharmaceuticals
Ministry of Chemicals & Fertilizers
Near Dr. Bhupen Hazarika Regional Govt. Film & Television Institute
Sila Village, Nizsundarighopa
Changsari, Kamrup (R)
Guwahati, Assam, India

Thais Francine Ribeiro Alves
Laboratory of Biomaterials and Nanotechnology (LaBNUS)
University of Sorocaba
Sorocaba, Brazil
Sorocaba Technology Park, Innovation Agency
Sorocaba, Brazil

Mansi Athalye
Department of Pharmaceutics and Pharmaceutical Technology
L. M. College of Pharmacy
Ahmedabad, GJ, India

Fernando Batain
Laboratory of Biomaterials and Nanotechnology (LaBNUS)
University of Sorocaba
Sorocaba, Brazil

Bappaditya Chatterjee
Shobhaben Pratapbhai Patel School of Pharmacy & Technology Management (SPPSPTM)
NMIMS University
Mumbai, MH, India

Marco Vinícius Chaud
Laboratory of Biomaterials and Nanotechnology (LaBNUS)
University of Sorocaba
Sorocaba, Brazil
Sorocaba Technology Park, Innovation Agency
Sorocaba, Brazil

Iti Chauhan
Department of Pharmaceutics
I.T.S College of Pharmacy, Muradnagar
Ghaziabad, UP, India

Namdev Dhas
Department of Pharmaceutics
Institute of Pharmacy
Nirma University
Ahmedabad, GJ, India

Nataliya N. Dremina
Irkutsk Scientific Center of Surgery and Traumatology
Irkutsk, Russia

Nidhi Dubey
School of pharmacy
Devi Ahilya Vishwavidyalaya
Indore, MP, India

Nitin Dubey
IPS Academy College of Pharmacy
Indore, MP, India

Wilson Engelmann
Graduate Program in Public Law
University of Vale do Rio dos Sinos – UNISINOS
São Leopoldo, Brazil
Corporate Law and Business
University of Vale do Rio dos Sinos – UNISINOS
São Leopoldo, Brazil

Anuradha Ketan Gajjar
L. M. College of Pharmacy
Ahmedabad, GJ, India

Fernanda Gonçalves
Department of Biotechnology
Engineering School of Lorena (EEL)
São Paulo University (USP)
Lorena, São Paulo, Brazil

Kartik Hariharan
Department of Pharmaceutics
Institute of Pharmacy
Nirma University
Ahmedabad, GJ, India

Raquel Von Hohendorff
Graduate Program in Public Law
University of Vale do Rio dos Sinos – UNISINOS
São Leopoldo, Brazil

Josef Jampílek
Department of Analytical Chemistry
Faculty of Natural Sciences
Comenius University
Bratislava, Slovakia
Institute of Neuroimmunology
Slovak Academy of Sciences
Bratislava, Slovakia

Nirav Khatri
Dr. Reddy's Laboratories Ltd., Hyderabad, Telangana, India

Talita Martins Lacerda
Department of Biotechnology
Engineering School of Lorena (EEL)
São Paulo University (USP)
Lorena, São Paulo, Brazil

Daniele Weber S. Leal
CAPES fellow at the Graduate Program in Public Law
University of Vale do Rio dos Sinos – UNISINOS
São Leopoldo, Brazil

Sadhucharan Mallick
Department of Chemistry
Indira Gandhi National Tribal University (Central University)
Amarkantak, MP, India

Paulo Ricardo Franco Marcelino
Department of Biotechnology
Engineering School of Lorena (EEL)
São Paulo University (USP)
Lorena, São Paulo, Brazil

Tejal Mehta
Department of Pharmaceutics
Institute of Pharmacy
Nirma University
Ahmedabad, GJ, India

Dhaivat Parikh
Department of Pharmaceutics
Institute of Pharmacy
Nirma University
Ahmedabad, GJ, India

Mitali Patel
Maliba Pharmacy College
Uka Tarsadia University
Bardoli, GJ, India

Mrunali Patel
Department of Pharmacy
Ramanbhai Patel College of Pharmacy
Charotar University of Science and Technology (CHARUSAT)
CHARUSAT Campus
Changa, GJ, India

Priya Patel
Department of Pharmaceutical Sciences
Saurashtra University
Rajkot, GJ, India

Priyanshi Patel
Maliba Pharmacy College
Uka Tarsadia University
Bardoli, GJ, India

Rashmin Patel
Department of Pharmacy
Ramanbhai Patel College of Pharmacy
Charotar University of Science and Technology, CHARUSAT
Changa, GJ, India

Ravish J. Patel
Ramanbhai Patel College of Pharmacy
Charotar University of Science and Technology (CHARUSAT)
CHARUSAT campus
Changa, India

Viral Patel
Department of Pharmaceutics
Institute of Pharmacy
Nirma University
Ahmedabad, GJ, India

Henrique Paiva Pereira
Department of Biotechnology
Engineering School of Lorena (EEL)
São Paulo University (USP)
Lorena, São Paulo, Brazil

Daniela Plachá
Nanotechnology Centre, Energy and Environmental Technology Centre
VŠB-Technical University of Ostrava
Ostrava-Poruba, Czech Republic
ENET Centre, Energy and Environmental Technology Centre
VŠB-Technical University of Ostrava
Ostrava-Poruba, Czech Republic

Anu Puri
RNA Structure and Design Section, RNA Biology Laboratory (RBL)
Center for Cancer Research, NCI-Frederick, NIH
Frederick, MD, USA

Mahendra Rai
Department of Biotechnology
Sant Gadge Baba Amravati University
Amravati, MH, India
Department of Microbiology
Nicolaus Copernicus University
Lwowska, Torun, Poland

Mihir Raval
Department of Pharmaceutical Sciences
Saurashtra University
Rajkot, GJ, India

Piyali Sabui
Department of Zoology
The University of Burdwan
Golapbag, Bardhaman, WB, India

Amaresh Kumar Sahoo
Department of Applied Sciences
Indian Institute of Information Technology
Allahabad, UP, India

Patrícia Severino
Nanomedicine and Nanotechnology Laboratory
University of Tiradentes
Aracaju, Brazil

Shah Esha Bhavin
Babaria Institute of Pharmacy
Vadodara, GJ, India
Ramanbhai Patel College of Pharmacy
Charotar University of Science and Technology
Changa, GJ, India

Saurabh Shivalkar
Department of Applied Sciences
Indian Institute of Information Technology
Allahabad, UP, India

Michael G. Shurygin
Irkutsk Scientific Center of Surgery and Traumatology
Irkutsk, Russia

Irina A. Shurygina
Irkutsk Scientific Center of Surgery and Traumatology
Irkutsk, Russia

Silvio Silvério da Silva
Department of Biotechnology
Engineering School of Lorena (EEL)
São Paulo University (USP)
Lorena, São Paulo, Brazil

Vishal Singh
Department of Applied Sciences
Indian Institute of Information Technology
Allahabad, UP, India

Eliana Barbosa Souto
Faculty of Pharmacy
Department of Pharmaceutical Technology
University of Coimbra
Coimbra, Portugal

Irina S. Trukhan
Irkutsk Scientific Center of Surgery and Traumatology
Irkutsk, Russia

Arushi Verma
Department of Applied Sciences
Indian Institute of Information Technology
Allahabad, UP, India

Madhu Verma
Department of Pharmaceutics
I.T.S College of Pharmacy, Muradnagar
Ghaziabad, UP, India

Imran Vhora
Formulation R&D Consultant, Burnaby, British Columbia, Canada

Naimish Vyas
Department of Pharmaceutical Sciences
Saurashtra University
Rajkot, GJ, India

Mohd Yasir
Department of Pharmacy
College of Health Sciences
Arsi University
Asella, Oromia Region, Ethiopia

Aleksandra Zielińska
Institute of Human Genetics
Polish Academy of Sciences
Poznań, Poland

List of Abbreviations

AFNOR	Association Française de Normalisation
AgNP	Silver nanoparticle
AIHA	American Industrial Hygiene Association
ALP	alkaline phosphatase
ALT	alanine transaminase
ANSI	American National Standards Institute, USA
AO	Acridine orange
AST	aspartate transaminase
ASTM	ASTM International
ATF4	activating transcription factor 4
AuNC	Gold nanocube
AuNP	Gold nanoparticle
AuNR	Gold nanorod
BAD2	Bcl-2 associated agonist of cell death
BALF	Bronchoalveolar lavage fluid
Bcl-xL	B-cell lymphoma-extra large
Bcl-2	B-cell lymphoma 2
BSI	British Standards Institution
CDC	Centers for Disease Control and Prevention, USA
Cdk1	cyclin-dependent kinase 1
CEN	European Committee for Standardization
CENELEC	European Committee for Electrotechnical Standardization
CeONP	Cerium oxide nanoparticle
CEPA	Canadian Environmental Protection Act, 1999
CFR	Code of Federal Regulations
CLP	Classification, Labeling, and Packaging of Substances and Mixtures
CNST	Center for Nanoscale Science and Technology, USA
COSHH	The Control of Substances Hazardous to Health Regulations, 2002, USA
CQA	critical quality attributes
CTAB	Cetyl trimethyl ammonium bromide
CuNP	Copper nanoparticle
CuONP	Copper oxide nanoparticle
DAPI	4′,6-diamidino-2-phenylindole
DCFH-DA	2′7′-dichlorofluorescin diacetate
DHHS	Department of Health and Human Services, USA
DNA	deoxyribonucleic acid
DOC	Office of the Director, USA
DSL	Domestic Substance List, Canada
ECHA	European Chemical Agency
ED	Ethidium bromide
ELISA	Enzyme linked immunosorbent assay
ELSA	ethical, legal, and social aspects
EPA	Environmental Protection Agency
Erα	estrogen receptor alpha
ETSI	European Telecommunications Standards Institute
Fas	fatty acid synthase
FIFRA	Federal Insecticide, Fungicide, and Rodenticide Act
Fpg	Formamidopyrimidine DNA glycosylase
GB	Guobiao Standards
GDG	Guideline Development Group
GHS	Globally Harmonized System of Classification and Labeling of Chemicals

GSSG	Glutathione disulfide
IL	interleukin
IL-2	Interleukin 2
IONP	Iron oxide nanoparticle
IOMC	Inter-Organization Programme for the Sound Management of Chemicals
ISIRI	Institute of Standards & Industrial Research of *Iran*
ISO	International Organization for Standardization
HepG2	hepatoma G2
HepG2	human hepatoma cell
HSP70	70 kilodalton heat shock proteins
Ki-67	antigen Ki-67
LC_{50}	median lethal concentration
LD_{50}	median lethal dose
LPS	lipopolysaccharide
MAPK	mitogen-activated protein kinase
MCF-7	breast cancer cell line. The acronym of Michigan Cancer Foundation-7
Mcl-1	Myeloid cell leukemia 1
MDA	Malondialdehyde
MDA-MB-231	human breast cancer cells
MFDS	Ministry of Food and Drug Safety of the Republic of Korea
MMP-9	matrix metallopeptidase 9
MNPs	Metallic nanoparticles
mRNA	messenger RNA
mtTFA	mitochondrial transcription factor A
NCI	U.S. National Cancer Institute, USA
NEHI	Nanotechnology Environmental and Health Implications Working Group
NF-κB	nuclear factor kappa-light-chain-enhancer of activated B cells
NIH	National Institute of Health, USA
NIOSH	National Institute for Occupational Safety and Health, USA
NIST	National Institute of Standards and Technology, USA
NMBP	The Advisory Group for Nanotechnologies, Advanced Materials, Biotechnology, and Advanced Manufacturing and Processing
NMPA	National Medical Products Administration, China
NNCO	National Nanotechnology Coordination Office, USA
NNI	National Nanotechnology Initiative, USA
NNIN	National Nanotechnology Infrastructure Network, USA
NOAEL	No observable adverse effect level
NPs	nanoparticles
NRC	Canada National Research Council of Canada
NSET	The Nanoscale Science, Engineering, and Technology Subcommittee, USA
NSNR	New Substances Notification Regulations, Canada
NSRC	Nanoscale Science Research Centers, USA
NSTC	The National Science and Technology Council, USA
NTP	National Toxicology Program
OECD	Organization for Economic Co-operation and Development
OEL	occupational exposure limit
OSHA	Occupational Safety and Health Administration, USA
OSTP	The Office of Science and Technology Policy, USA
p53	tumor protein p53
PARP	poly(ADP-ribose)polymerase
PBMC	Peripheral blood mononuclear cells
PBT	persistent, bioaccumulative, and toxic
PCR	Polymerase chain reaction
PEG	Polyethylene glycol
PMN	premanufacture notices
PPE	personal protective equipment
PS20	Polysorbate 20
PS80	Polysorbate 80
PtD	prevention through design

REACH Regulation for the Registration, Evaluation, Authorization, and Restriction of Chemicals
REL recommended exposure limit
RITC rhodamine B isothiocyanate
ROS reactive oxygen species
SAC Standardization Administration of China
SDO standards developing organizations
SDS safety data sheet
SDS Sodium dodecyl sulphate
SOD Superoxide dismutase
SOD2 superoxide dismutase 2
SeNPs selenium nanoparticles
SNUR significant new use rules
SVHC substance of very high concern
TBARs Thiobarbituric acid reactive substances
TGF-β Transforming growth factor-β
TiO_2NP Titanium dioxide nanoparticle
TLV threshold limit value
TMAT NNN-trimethylammonium ethanethiol
TNF-α tumor necrosis factor alpha
TNSC *Taiwan* Nanotechnology Standards Council
TSCA Toxic Substances Control Act
TUNEL terminal deoxynucleotidyl transferase dUTP nick end labeling
UNCED United Nations Conference on Environment and Development
UNMZ Czech Office for Standards, Metrology, and Testing
UNP unbound engineered nanoparticles
USFDA Food and Drug Administration, USA
USPTO U.S. Patent and Trademark Office, USA
VAMAS Versailles Project on Advanced Materials and Standards
vPvB very persistent, very bioaccumulative
VEGF vascular endothelial growth factor
WEL workplace exposure limit
WPMN OECD Working Party on Manufactured Nanomaterials
WPMN Working Party on Manufactured Nanomaterials
ZnONP Zinc oxide nanoparticle

Part I

Nanomedicine: Nanotoxicological Insights

1

Nanomedicines

Applications and Toxicological Concerns

Mrunali Patel[1], *Rashmin Patel*[1], *and Mahendra Rai*[2,3]

[1] *Department of Pharmacy, Ramanbhai Patel College of Pharmacy, Charotar University of Science and Technology, (CHARUSAT), CHARUSAT Campus, Changa, GJ, India*
[2] *Department of Biotechnology, Sant Gadge Baba Amravati University, Amravati, MH, India*
[3] *Department of Microbiology, Nicolaus Copernicus University, Lwowska, Torun, Poland*

1.1 Introduction

Nanotechnology is an evolving technology that has been researched for numerous uses, from material science to life science. A widely used description by National Nanotechnology Initiative, USA, defines nanotechnology as the designing, synthesis, and characterization of materials and structures smaller than 100 nanometers (nm in size or output limit), by controlling the shapes and sizes at the nanoscale (Parappurath et al. 2018; Bayda et al. 2020). The size of 100 nm, however, is not a rigid boundary; it varies depending on the application (McDonald et al. 2015). Recently, the design and manufacture of nanotechnology have risen, including nanoscience, nanomaterials, nanoparticles, nanomedicine, and nanotoxicology. The branch of science dedicated to the study of the peculiar properties of matter at the nanoscale is known as nanoscience (Bayda et al. 2020). The structures made up of aggregated or unbound nanoparticles apply to nanomaterials. As per the International Organization of Standardization, nanoparticles (NPs) are particles whose sizes in one, two, or three dimensions are within the range from 1 to 100 nm (Jeevanandam et al. 2018). Their measurements are equivalent to those of biomolecules such as proteins, DNA, hemoglobin, viruses, and cell membranes. Some of the significant and special characteristics of NPs are higher surface-to-mass ratios than the bulk material, altered quantum properties, and their ability to adsorb and hold other substances such as drugs, probes and proteins. They are both present in nature and can be produced by traditional industrial processes and techniques of nanomanufacturing (Shubhika 2013). The nanomaterials are classified based on diverse parameters such as dimensionality, morphology, composition, uniformity, agglomeration rate, and origin (Buzea and Pacheco 2017; Jeevanandam et al. 2018; Trotta and Mele 2019; Khan 2020) as tabulated in Table 1.1.

Nanotechnology employs novel, specific, and selective medicinal products at the nanoscale in the healthcare sector to develop nanomedicines. Nanomedicine is an emerging subdivision of nanotechnology that plays an important role in disease detection, surveillance, prevention, and remediation. The term nano means minuscule, extremely

Nanotechnology in Medicine: Toxicity and Safety, First Edition. Edited by Mahendra Rai, Mrunali Patel, and Rashmin Patel.

Table 1.1 An overview of the classification of nanomaterials based on various parameters with a few examples and applications.

Parameter	Classes	Examples	Applications
Dimensionality	Zero-dimensional (0D)	Quantum dots Dendrimers Fullerenes	Cell marker, emulsifier in solution, reinforcement filler within a solid matrix
	One-dimensional (1D)	Nanowires Nanotubes Nanofibers Nanorods	Electronics, magnetism, optics, and catalysis
	Two-dimensional (2D)	Thin films Nanocoatings Nanoplates	Optoelectronics, catalysts, sensors, solar cells, energy storage facilities
	Three-dimensional (3D)	Nanocomposites (nanofillers in bulk matrix composed of ceramics, metals, or polymers) Nanostructured materials (nanoporous structures as aerogels, block polymers, nanostructured metals and alloys)	Packaging materials, shape-memory materials, coatings with nanoprotrusions, catalysts
Morphology	Low aspect ratio	Nanospheres, nanocubes, nanopyramids	Biomedical applications, plasmonic sensing platform (silver nanocube), optical applications
	High aspect ratio	Nanowires, nanotubes, nanobelts, nanofibres	Separation and isolation of specific analyte from complex mixtures, targeted delivery, magnetic resonance imaging
Chemical composition*a*	Metal and metal alloys	Silver	Medical diagnostic, antibacterial, conductive, and optical applications
		Copper	Catalyst, lubricant additive, electrical and thermal conductor, antibacterial agent
		Gold	Drug delivery, medical testing, cancer detection, electronics
		Iron	Bactericide in water treatment, superparamagnetic in drug delivery, data recording, and magnetic detection
		Aluminum–magnesium/titanium–aluminum alloys	Aerospace and high-temperature applications

	Metal oxides	Titanium dioxide	Filter for cosmetics, chemical catalyst for cleaning product and wall paint, antibacterial agents for filtration devices
		Zinc oxide	Catalytic, antibacterial, antistatic agent
		Nanosilica	Hollow carrier for drug delivery
		Iron oxides	Biomedical and electronic applications
	Semiconductors	Nanosilicon, gallium nitride, nanoflowers, nanocolumns	Photovoltaic and optoelectronic applications
	Silicates, carbonates and nitrides	Clay, calcium carbonate, Silicon nitride	Filler in composites
	Carbon nano-objects	Graphene	Biosensors,
		Carbon nanotubes	Scaffolds for bone growth,
		Carbon nanofibers	Scanning electron microscope tips,
		Fullerenes	Antiaging cosmetics, contrasting agents for medical imaging,
		Carbon black	Heat and electricity conductor, antistatic agent
	Polymers	Nanospheres Nanofibers Nanoporous membranes	Barrier membranes, membrane for fuel cells, antibacterial textiles, optical components, filtration membranes, reinforcement of structural composites
State	Uniformity	Isometric and inhomogeneous	—
	Agglomeration	Agglomerates and dispersed	—
Origin	Natural	Natural nanostructures present in microorganisms like bacteria, viruses and algae; complex organisms like plants, insects, birds, animals, and humans	Nanomaterials in insects helps them to stay alive in adverse environmental conditions, Plant-based nanomaterials acts as a source of cellular biocomposites, Nanomaterials in the human body help to perform normal physiological functions
	Incidental	Natural nanoparticles produced by photochemical reactions, volcanic eruptions, and forest fires	—

(*Continued*)

Table 1.1 (Continued)

Parameter	Classes	Examples	Applications
	Engineered	Metals, quantum dots, sunscreen pigments, nanocapsules	Cosmetic products, personal care products, nutraceuticals, health supplements, medical imaging
Applications in medicine	Solid-drug nanoparticles	Nanocrystals Nanosuspensions	Delivery of immunosuppressant drug sirolimus (Rapamune®), antiemetic aprepitant (Emend®), antihyperlipidemic fenofibrate (Tricor®)
	Polymer therapeutics	Polymer-protein conjugates Polymer-drug conjugates Polyplexes	Delivery of asparaginase in treatment of leukemia, non-Hodgkin's lymphoma, acute lymphoblastic leukemia (Oncaspar®) Delivery of glatiramer acetate for treating relapsing multiple sclerosis (Copaxone®) Delivery of methoxy polyethylene glycol-epoetin beta in treatment of anemia I associated with chronic kidney disease (Mircera®)
	Nanocarriers	Polymer nanoparticles (solid polymer nanoparticles and nanogels)	Delivery of paclitaxel as albumin polymer nanoparticle (Abraxane®)
		Metal nanoparticles (gold nanoparticles, iron nanoparticles, and silver nanoparticles)	Gold, iron, and silver nanoparticles have diagnostic (biosensors), conductive (inks and composites) and optical (metal enhanced fluorescence and surface-enhanced Raman scattering) applications (especially iron oxide nanoparticles are favored to obtain enhanced contrast in MRI imaging)
		Solid lipid nanoparticles Nanostructured lipid carriers Nanoemulsions Self-assembled structures (polymer micelle, liposome)	Delivery of estradiol hemihydrate as nanoemulsion in treatment of some symptoms of menopause (Estrasorb®) Delivery of doxorubicin as PEGylated liposomal injection (Doxil®)
	Others (nanomaterial itself act as a therapeutic agent)	Dendrimers Virusomes Silver nanoparticles	VivaGel® antibacterial and antiviral dendrimer, Delivery of vaccine (influenza vaccine Inflex V®; Hepatitis vaccine Epaxal®), Antibacterial silver nanoparticles (wound dressing)

Potential toxicity level	Rigid and biopersistent	Fiber-like nanoparticles: carbon nanotubes with and without asbestos like effects, fiber-like metal oxides	—
	Granular and biopersistent	Titanium dioxide, silica, zinc oxide, nanoclay, fullerenes, dendrimers, aluminum oxide	—
	CMAR (carcinogenic, mutagenic, asthmagenic, reproductive toxin)	Nickel, cadmium containing quantum dots, arsenic, zinc chromate	—

[a] As per chemical composition, nanomaterials are also classified as organic (liposomes, micelle, polymeric nanoparticles, dendrimer); inorganic (quantum dot, mesoporous silica nanoparticle, gold nanoparticle, silver nanoparticles); carbon-based (carbon nanotube, graphene, fullerene, graphene); composite-based (polymer/ceramic nanocomposites, metal/metal nanocomposites, carbon/metal nanocomposites).
Sources: Based on Buzea and Pacheco (2017), Jeevanandam et al. (2018), Trotta and Mele (2019) and Khan (2020).

small; so, nanomedicine is concerned with drug formulations in the nano range of sizes (nanoformulations). The nanoformulations can be utilized for delivery of an impressive range of therapeutic drugs by encapsulation or attachment on the surface for a targeted, sustained, and controlled release to almost any tissue/organ/area of the body (Mukherjee et al. 2014; Krukemeyer et al. 2015; Ventola 2017; Fadeel and Alexiou 2020). Possible applications include: nanorobots, nanoscale medicines and targeted delivery of medicines, nanocarriers for drug delivery such as nanocrystal, polymeric NPs, magnetic NPs, mesoporous silica NPs, liposomes, micelles, dendrimers, quantum dots, iron oxide, carbon nanotubes, antimicrobial medical dressings, *in vivo* imaging techniques, bone and dental prostheses, tissue regeneration, etc. Nanotechnology also enables better identification and understanding of biomarkers. Further, benefits achieved are analysis of the disease stage, three-dimensional (3D) nanomaterials for injury site, immobilized stem cells, blood glucose metering, delivery of insulin, guided implantation, etc. The financial implications of the use of nanotechnology in medicine were projected to rise to $528 billion by 2019 and will continue to grow dramatically in the coming years, according to recent industry forecasts. Further development in this field enables technology for patient-specific, targeted, and regenerative medicine (Farjadian et al. 2019).

More than 30 years of research work has been conducted and a vast number of scientific research articles published about nanomedicine, but only a handful of nanoformulations have received marketing approval or been identified to enter clinical trials for different applications, including the diagnosis and treatment of multiple cancers and the treatment of infections and other noncancerous diseases (McGoron 2020). Thus, this represents the fundamentally high degree of regression of nanomedicine's path from laboratory to market due to the unique properties of nanomaterials, lack of safety knowledge, risk control methods, and their effective management. Consequently, any nanomedicine that enters the market must be of high quality, healthy, and efficient. Without activating any undesired reaction, in particular populations, NPs and other nanomaterials in the field of nanomedicine are supposed to exhibit an acceptable response. Unfortunately, the properties which make NPs appealing for the development of nanomedicine may also prove extremely harmful in cell interaction (Mukherjee et al. 2014). Nanotoxicology is an evolving toxicology specialty which accesses the toxicological properties of nanomaterials through various *in vitro* and *in vivo* tests using cell- or animal-based models and provides evidence for the safety evaluation of this nanomaterial and its applications. The entire life cycle of nanomedicines, including the production, disposal, and environmental impact, should be considered when weighing the benefits and risks; for example, the estimation of various disposal pathways. A thorough biosafety evaluation of therapeutic nanomaterials will make a major contribution to risk control for the continuing growth of nanomedical technology, which is so urgently needed to ensure that they are produced carefully, fully exploited, and then disposed of safely. In deliberate partnership with nanomedicine, nanotoxicology will help advance the field of nanomedicine by offering information on its harmful properties and methods of preventing them (Sharma et al. 2018).

This chapter will include insights into emerging nanomedicines, numerous diagnostics, drug delivery, and tissue engineering and regenerative medicine applications of nanomedicine. Finally, it would also discuss the nanotoxicological characteristics that impede nanomedicine's clinical transformation from bench to bedside, and expect that nanomedicine will progress to the next stage, and through rationally organized and systematic methods, deliver practical and substantial benefits to human medicine and healthcare.

1.2 Nanomedicine's Revolution

In the last decade, expectations for nanomedicine have risen. All facets of medicine, ranging from therapy, cure, control, prediction, to disease prevention, are in the long run expected to be included in it. The literature became overwhelmed with publications bearing the words nanomaterials, nanoconstructs, nanoformulations, NPs, nanomedicine, and nanotoxicology early in 2000. Nanomedicine governed by nanobiotechnology explores the construction and role of cells as well as intra- and intercellular processes and contact between cells. This breakthrough was only possible at the turn of the twentieth century, when the road to the nanoworld opened with the introduction of groundbreaking microscopes. In chemistry and biology, the use of revolutionary microscopes contributed majorly to the study of cell shapes and cell constituents. The interpretation of the structure and operation of the cell membrane, diffusion processes, and hierarchical cell signaling utilizing receptors and antibodies became even more clear with the further high-resolution inventions like voltage clamp – a predecessor of the patch-clamp technique. However, only after the demonstration of nanoscale structures using scanning probe microscopy attributed to the invention of the high-resolution scanning tunnelling did the novel research disciplines suited to the nano range, including nanomedicine, appear. The physicists were able to provide an image of Si (111) – 7×7 reconstructed surface showing atomic scale resolution (Bayda et al. 2020).

Richard Feynman (American physicist, Nobel Prize laureate, Father of nanotechnology) first proposed the concept of nanotechnology in 1959, while Norio Taniguchi (Japanese scientist) invented the term "nanotechnology" in 1974. Two directions of thinking appeared after Feynman ventured out the new area of science and awakened the attention of many scientists, outlining the different possibilities for nanostructure development. The top-down approach essentially relates to the remarks of Feynman on the incremental decrease in the scale of current devices and tools. The bottom-up approach centers around the creation of nanostructures for atoms by physical and chemical approaches and through the use and regulated modulation of atomic and molecular self-organizing forces. The research seeds of nanomedicine, which at present encompasses a wider spectrum of research and development of nanometer-length scale materials and technology, were sown around 1990 (Krukemeyer et al. 2015). The definition and interpretation of DNA and RNA focused on the concept of genetic disorders and the vision of molecular-level cures personalized for patients. For the first time, NPs were updated at the beginning of the 1990s for the transport of DNA fragments and genes and were flushed into cells using antibodies (Choi et al. 2014). The nanomedicine research focused on the possibilities of targeting and delivery of nano-sized active substances for the diagnosis, treatment, and monitoring of ailments. The commencement of the twentieth century marked the initiation of the search for "magic bullets" to which medications were applied and which could be used to target viruses and eliminate all pathogens as suggested by Paul Ehrlich. The expertise gained on cells and their constituents, intra- and intercellular processes and cell connectivity, as well as developments in biochemistry and biotechnology, made it possible to create ever more sophisticated "magic bullets."

Intensive research has also been undertaken into the potential synthesis and use of different carrier systems and the physicochemical functionalization of their surface structure. Biocompatible polymers, liposomes, nanocrystals, and micelles are currently being investigated mainly as carriers of drugs, vaccines, and genomes. Nanomaterials tend to circulate in the body previous to contact their target because of their small size (usually less than 200 nm) and are not filtered out of the blood.

In their hollow interiors, active substances may be encapsulated and their surface can be modified such that they bypass natural obstacles such as cell membranes. They can also recognize certain cells and tissues with the assistance of biosensors (for example, antibodies), bind themselves to them, and release the active substances over a comparatively long period of time to the target. These pathways are of special importance for the treatment of cancer, as the controlled release of cytostatic agents solely in the tumor tissue can decrease side effects along with higher levels of the active drug than before in the affected tissue. In addition, the enhanced permeability and retention (EPR) effect has benefited cancer treatment based on the fact that NPs are accumulated in tumors to a greater degree than in healthy tissue (Choi and Han 2018). However, the advancement of nanomedicine goes beyond the idea of a "magic bullet." The idea of the creation of novel scaffolds and surfaces for biosensors or implantable systems and electronics to assist in tissue regeneration is still prompt, although professional experience has been gained by some individuals.

1.3 Potential Applications of Nanomedicine

The fabrication and utilization of nanoscale materials in nanomedicine encompasses a wide variety of fields, including drug delivery, vaccine production, antibacterial, diagnostic and imaging instruments, wearable devices, implants, and high-throughput screening platforms, using biological, nonbiological, biomimetic, or hybrid materials. Nanomedicine-based strategies possibly have several promising platforms to encapsulate vaccines and deliver them to antigen-presenting cells or to serve as antigen-presenting carriers themselves (Pelaz et al. 2017). In addition, medications may be encapsulated in such carriers to be targeted specifically at infected cells. In specific, virus imitating NPs such as self-assembled liposomes, viral proteins, and virus-like particles are capable of replicating the mechanism of infection and can be used not only as a delivery system but also to research viral infections and associated mechanisms. NP-based vaccines may also be used to make COVID-19 therapies more effective (Heinrich et al. 2020).

1.3.1 Diagnosis

In order to recommend adequate care, doctors must be able to distinguish healthy from diseased tissues that require the visualization of structures within the human body. This can be made possible by nanomaterials encompassing NPs which can be designed with distinct contrast properties offering better 3D visualization, employing which tissue types can be more readily distinguished. This is the main priority in nanomedicine for anatomical and functional imaging. However, NPs that are capable of visualizing biological tissues must be configured to be localized in individual tissues and theoretically deliver high contrast for use in various imaging techniques such as magnetic resonance imaging (MRI); computed tomography (CT); fluorescence imaging; and photoacoustic imaging. Material production therefore plays a crucial role in the design of smart NPs providing contrast in the region of interest; disclose information on the local environment after administration to the body and further aid in the imaging of organs, especially anatomical fine structures; help in predicting concentrations of molecules of interest; and, lastly, support in the direct internal examination of diseases in the human body (Pelaz et al. 2017).

Gold nanoparticles (AuNPs) are at the forefront of developing contrast agents to improve the sensitivity of noninvasive X-ray imaging, CT, and micro-CT. The high atomic number and electron density of AuNPs result in higher attenuation coefficients than iodine, a standard X-ray contrast agent. These NPs may also be coated with targeting molecules, such as folic acid, to intensify the contact between tissues and incident rays, making it easier to illuminate distinct tissue structures. Another CT contrast agent being developed is gold nanocluster (AuNC), which has shown to exhibit excellent contrast not only for CT imaging but also for molecular imaging owing to red fluorescence emission. Thus, the CT-based clinical diagnosis is completely revamped by these promising NPs, and NP-based CT imaging approaches will soon be used clinically. Besides, MRI is another noninvasive medical imaging procedure routinely used to provide anatomical details. Contrasting agents traditionally used in clinical MRI include primarily paramagnetic agents (e.g. gadolinium (Gd)-based) and superparamagnetic agents (e.g. superparamagnetic iron oxide nanoparticles [SPIONs]) (Greish et al. 2018). A substantial effort has been made to develop NPs conjugated with paramagnetic ions, such as Gd^{3+}, Mn^{2+}, and Fe^{3+}, as contrast agents for MRI in an attempt to minimize toxicity, improve chemical stability, longer circulation periods, higher contrast, more regulated functionalization, and additional imaging methods. The paramagnetic NPs offer flexibility in size and shape, boost magnetic properties, and control over pharmacokinetics, potentially leading to an improvement in blood circulation time compared to conventional coordination complexes. These NPs are created either by integration into the nanostructured matrix or the post-functionalization with the lanthanide coordination complex particles. The examples of nanoparticulate MRI contrast agents that have shown enhanced contrasting in various investigations are Gd-doped silica NPs, Gd-cerium NPs, Gd-nanodiamond conjugates, Gd_2O_3 NPs; manganese (Mn) NPs like Mn_3O_4 NPs, Mn-based double-layered hydroxide NPs, MnO nanocomposites functionalized with porous AuNCs; dysprosium(Dy)-modified NPs such as mesoporous silica NPs with Dy-DOTA(1,4,7,10-tetraazacyclododecane-1,4,7,10-tetraacetic acid) chelate in the outer pore channel, Dy_2O_3 and DyF_3 NPs, $Dy(OH)_3$ nanorods, etc. (Pellico et al. 2019).

NP-based radiolabels are now being developed rapidly to allow researchers to carry out and monitor quantitative biodistribution studies on systemic administration, and to investigate the basic *in vivo* mechanism of NPs. They are hybrid structures consisting of an organic surface coating for colloidal stabilization and targeting functionality, the operative component i.e. imaging contrast as well as the adsorbed protein layer. However, since various marking techniques and radioisotopes are used to mark NPs in order to research their fate, it is important to ensure that radioisotopes are integrated into NPs with no effect on their initial biological behavior. Picomolar amounts of nano-size radiolabeled gold nanoshells and polymeric NPs are investigated for positron emission tomography and single-photon emission tomography (Koziorowski et al. 2017). The richness of fluorophores makes fluorescence imaging ideal for different applications. The emission of excitation probes in this procedure can be visualized by the human eye or by optical microscopy at higher resolution. Quantum dots, rare earth-based nanophosphors, carbon dots, nanodiamonds, etc. are the fluorescent NPs researched as contrast labels for imaging (Pratiwi et al. 2019).

The potential of designed NPs that concurrently diagnose, administer, and even control therapeutic effectiveness has increased the hopes and aspirations of diagnostic nanomedicine. NPs have a high aspect ratio, can accommodate a high drug load and can be tuned to allow combination with drugs so that a targeted, imaging-based diagnosis can be accompanied by simultaneous therapy unique to that condition known as theranostics. It is an interesting field of research, as many diverse variations of diagnostic and therapeutic methods can be incorporated into NPs. Diagnostic application of NPs to mark cancerous cells, distinctly tumor boundaries and minor metastatic areas, can be used to direct the surgeon in the eventual removal of the tumors surgically (McDonald et al. 2015). These diagnostic NPs may also be filled with post-surgery therapeutic agents. Additional fascinating use of nano theranostics agents for tracking, and noninvasive *in vivo* imaging of transporter and drug, could allow for early evaluation of the treatment response. Carbon dots, quantum dots, Gd-DOTA, AuNPs, magnetic iron oxide NPs, ferrimagnetic nanoclusters, and nanodiamonds have high *in vivo* stability and are materialized as valuable tools for amalgamated diagnostic imaging and therapy. Multicompartment capsules as nanomedicine vehicles composed of smaller nanocapsules or NPs assembled under bigger particles are also investigated as an efficient mode of theranostic delivery. However, after major advances in this field, therapeutic NPs are still at an early stage of growth (Sangtani et al. 2017).

Nanotechnology provides a potential for early identification of diseases and genetic make-up with the assistance of modest, rapid, accessible, and affordable *in vitro* diagnostic testing of high sensitivity. In the near future, it is predicted to have instruments that are compact and decentralized, and that need only the smallest sample size for measurement and diagnosis. Such miniaturized lab-on-a-chip technique is an inexpensive boon to patients and can be used in clinics and hospitals for preventing the spread of infectious diseases (Krukemeyer et al. 2015). The overall diagnostic process is improvised using NPs having the ability to detect molecules, cells, and tissues outside the human body. For instance, modified AuNPs in combination with ligand can directly bind to a complementary protein that induces controlled agglomeration owing to the cross-linking of the NPs by the proteins. This can be identified colorimetrically by the change of color. These AuNP-based diagnostic concepts are refined, for example in fast colorimetric DNA sensing, and are now used in the clinic for testing of patients' samples (Khan et al. 2020). NPs encompassing organic and inorganic polymers are also applied as handy units for intracellular sensing purposes. The rationale behind using NPs in diagnostic applications is to identify the unique biological molecules in patients' biological liquids allied to their health. They also offer sequential detection using quantum dot-containing microbeads to produce barcodes with special optical emission. NP-based chemical nose sensors are also gaining interest in sera sensing, cancer cell genotyping, and the distinction between cell surface-based bacteria. Nanomedicine in diagnostics can generate a multiplexed platform that can exploit the ability of nanomaterials to easily detect minute changes in the cell surface that enable high-throughput screening.

1.3.2 Drug Delivery

Nanotechnology offers various nanostructures/nanoformulations in the field of nanomedicine with exceptional physical, chemical, mechanical, electrical, magnetic, and biologic properties. Nanomaterials as a key component of nanomedicine extend many benefits due to their nanoscale properties as shown in Figure 1.1. Medical developments in cancer are among the most promising treatment methods in nanomedicine. So, the additional emphasis on cardiovascular, autoimmune, psychiatric, viral, and genetic and rare diseases is a positive development within the scientific community in nanomedicine. Another new field of high promise for nanomedicine is RNA-based synthetic vaccines (De Jong and Borm 2008; Utreja et al. 2020).

In recent years, nanomedicines have been well known because nanostructures can be used as delivery agents by encapsulating or adding medicinal drugs and distributing them more specifically with a controlled release to target tissues. More than 50 nanoformulations have been approved by the FDA since 1995 and are currently on the market for a variety of indications. The commonly approved nanoformulations are liposomes, nanocrystals, iron colloids, protein-based NPs, nanoemulsions, and metal oxide NPs. The recent acceptance of three primary nanomedicine drugs (e.g. Onpattro®, Vyxeos®, and Hensify®) by the FDA has demonstrated that the field of nanomedicine is specifically capable of developing

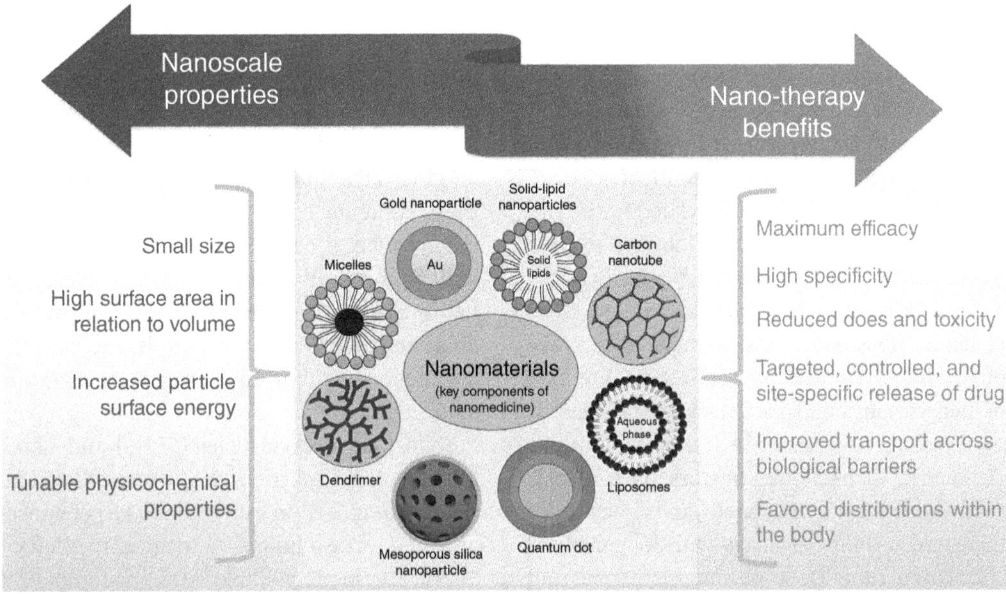

Figure 1.1 Nanoscale properties and allied benefits of nanomaterials in nanotherapy.

products that transcend crucial obstacles in conventional medicine in a special manner (Martins et al. 2020). It also offers new drug-free clinical effects through the use of pure physical modes of action within the cells, which thereby allows a difference in the lives of patients. In addition, new clinical applications are introduced commercially owing to nanomedicine formulations currently in clinical trials (above 400) unaided or jointly with foremost technologies including microfluidics, biotechnology, photonics, information and communication technology, advanced materials, biomaterials, smart systems, and robotics. Colloidal particles composed of macromolecular compounds in the 1–500 nm size range are NPs that are carriers in which either the active material is dissolved, encapsulated, or adsorbed into the matrix uniformly (Germain et al. 2020).

Liposomes are one of the most studied advanced nanoformulations, first described in 1965 for medical applications. They are biocompatible bilayered structures (single or multiple) of 50–500 nm size range composed of either or both synthetic and natural lipids that imitate cell membranes with an empty aqueous interior core. They are classified as multilamellar, oligolamellar, and unilamellar depending upon structural parameters or the number of bilayers formed and the diameter of the resultant vesicles as well as the method of preparation. They are widely used and researched nanoformulations owing to their unique properties and capacity to carry and deliver both hydrophilic as well as hydrophobic therapies in the aqueous core and lipophilic bilayer, respectively. These nanostructures are also considered to boost biodistribution and to make encapsulated medicines, such as low-molecular weight drugs, imaging agents, nucleic acids, proteins, and peptides, safe in the harsh bioenvironment of ill tissues (Vanza et al. 2020). The biological molecules like monoclonal antibodies, enzymes, and antigens can be delivered by conjugation as ligands on their surfaces (Zeb et al. 2020). Liposomes can exhibit high circulation time, sustained exposure to the site of action, strong diffusion, and penetration activity due to their unique physicochemical properties. However, clearance by RES (reticuloendothelial system), immunogenicity, and opsonization are obstacles in the use of liposomes for drug delivery. However, factors such as EPR effect aspire to improve the functioning of liposomes as drug carriers. Their simplicity of surface alteration makes them more popular in the distribution of medications (Utreja et al. 2020). PEGylated liposomes are developed by surface modification of lipid bilayer through polyethylene glycol (PEG) and are also known as long-circulating liposomes or stealth liposomes. This decreases the uptake of liposomes by RES due to steric repression of hydrophobic and electrostatic interactions with plasma proteins or cell, thus avoiding its clearance. Various investigators have shown that moderately PEGylated liposomes have increased stability of drugs with longer blood circulation time, poor plasma clearance, and low volume of distribution (Gabizon et al. 1994; Bobo et al. 2016). The liposomal delivery vehicles are expected to change drastically by the transformation of conventional liposomes into novel types such as stealth liposome, targeted liposome, and theranostic liposome. The latter is an assemblage of the previous three forms of liposomes, encompassing medicinal, imaging, and targeting molecules. The active molecule can be encapsulated in the liposome after (active approach) or during (passive approach) its formation through procedures like thin layer hydration, mechanical agitation, solvent evaporation, solvent injection, solvent dispersion, detergent removal method, and the surfactant solubilization (Eloy et al. 2014).

Accumulation of lipid vesicles at the desired location is a prerequisite for the release and absorption of the encapsulated drug besides enhanced bioavailability. The EPR effect originate passive targeting and many approved nanoliposomal formulations (e.g. Doxil®, Lipodox®, DaunoXome®, Onivyde®, etc.) have successfully increased distribution to the diseased states based on this strategy (Caster et al. 2017). However, nanoliposomes can be synthesized by incorporating antibodies, ligands, etc. on their surface for targeted and extended delivery of drugs to organs or tissues, so that the therapeutic effect is obtained only on diseased cells sparing the normal cells. Stimuli-responsive liposomes (pH-sensitive, temperature-sensitive, etc.) are also persuaded by utilizing lipids of differing fatty acid chain lengths. This allows the controlled release of their contents only on exposure to specific environmental conditions. The use of liposomal nanoformulations for drug delivery has had a major effect on anticancer, antifungal, analgesic pharmacology and is increasingly advancing to other categories as well (Patra et al. 2018).

Solid lipid nanoparticles (SLNs) and nanostructured lipid carriers (NLCs) are utilized for medicinal intention especially to target cancer and the transmission of ocular medicines. SLNs are biodegradable and biocompatible nanoscale colloidal carriers with a size range of 50–1000 nm. They are one of the nanocarrier systems to provide a highly lipophilic lipid matrix for dissolved and dispersed drugs manufactured by dispersing melted solid lipids in aqueous media or water using emulsifiers as a stabilizer. As carriers, SLNs deliver various advantages like higher drug payload, negligible *in vivo* toxicity, better bioavailability and stability of poorly soluble medicines and ease of large-scale processing than liposomes and other colloidal systems (Vanza et al. 2020). They provide a biological medium to encapsulate lipotropic cytotoxic drugs in the core, shell, or lipid matrix and can be functionalized using compounds like oligosaccharides, proteins, antibodies, or ligands for receptors. Nevertheless, they suffer from certain restrictions that comprise of high water content required to disperse, drug leakage, and low loading capacities. The above-stated problems related to SLNs can be overcome by using second-generation carriers NLCs. It comprises of two main components: one is a nanostructured solid lipid matrix (a combination of liquid and solid lipids) and another is an aqueous phase (containing surfactant). In contrast to SLNs, they facilitate elevated encapsulation of active molecules, give marginal drug leakage, exalted stability and versatility (Shidhaye et al. 2008). A wide range of solid lipids are commonly used in the formulation of lipid NPs, including free fatty acids, fatty alcohols, steroids, waxes as well as monoglycerides, diglycerides, and triglycerides. In addition, oleic acid, isopropyl myristate, vitamin E, Transcutol P, Labrafac PG, Miglyol 812 N are some of the examples of liquid lipids used to manufacture lipidic nanoparticle by any of the methods, namely microemusification, high-pressure homogenization, solvent emulsification-evaporation technique, solvent emulsification-diffusion technique, high shear homogenization or ultrasonication technique, solvent injection (or solvent displacement) technique, phase inversion temperature (PIT) method, and double emulsion technique (Utreja et al. 2020).

Polymer NPs are commonly used in nanomedical research for the delivery of various drugs. They are quickly synthesized and there is a vast volume of evidence on their effectiveness and protection. They provide many benefits over other delivery systems in terms of stability in the gastrointestinal atmosphere and the potential to shield encapsulated agents from drug efflux pumps and enzymatic degradation. They are considered promising carriers for a variety of drugs, including cancer, coronary disease, and diabetes treatments; bone-strengthening treatments; and vaccines (Wibowo et al. 2020). Implementation of different polymers to produce NPs enables simple manipulation of their properties such as surface load, hydrophilicity and particle size, along with stimulus-oriented regulated release of drugs. It is possible to change the surface of NPs by conjugating polymers with peptides and antibodies (Rahman et al. 2012). Biodegradable polymers that can be completely metabolized and eliminated from the body are of special interest in nanocarriers for drug delivery. Polymeric NPs are typically made up of polymers such as chitosan, sodium alginate, polylactic acid, poly (lactic-co-glycolic acid) (PLGA) and poly(ш-caprolactone) (Chavan et al. 2020).

Nanospheres are spherical polymeric matrix particles that are rigid with drug encapsulated or dispersed within the polymer matrix. Nanocapsules are vesicular reservoir polymeric structures that act as a reservoir in which the drug is spread or absorbed in a liquid core (oil/water) surrounded by a polymer (Allen et al. 2019). Dendrimers are another form of polymeric NPs that has a branched three-dimensional structure with ease of surface adjustment and flexibility. Various properties of dendrimers, such as high degrees of branching, size uniformity, water solubility, and the inclusion of several internal cavities, make them an effective drug delivery platform. They demonstrate the ability to enhance the solubility and bioavailability of hydrophobic drugs that may be stuck in their intramolecular cavity or conjugated to their functional surface groups. These almost monodispersed technologies represent new drug delivery systems for the treatment of various diseases and conditions of the human body. A recent emerging field of therapeutic use is the synthesis of dendrimers with bioactive ligands to promote targeted delivery and improve the effectiveness of medications with the smart use of advanced pharmaceuticals and nanomedicine. However further studies are necessary to reveal the complex structure–functional relationship of ligand–dendrimer conjugates in drug delivery processes (Lombardo et al. 2019).

Hydrophobic core–hydrophilic shell structures formed by self-assembly of amphiphilic block copolymers in the aqueous solution are nanodrugs comprising polymeric micelles. They are investigated for the delivery of various drugs (like anti-infective, anticancer molecules), genetic material (DNA and siRNA), proteins, peptides, etc. The micelles prepared by thin-film hydration, sonication, or dialysis technique can be tweaked to obtain various particle sizes (20–200 nm) with narrow size distribution, drug loading and release characteristics. They can be personalized to obtain slow controlled release circumventing prompt renal clearance, thereby allowing sustained circulation and accumulation due to the impact of EPR. They flaunt biocompatibility and stability attributable to numerous amphiphilic copolymers used like poly (vinyl alcohol) (PVA), poly(ethylenimine) (PEI), poly(ε-caprolactone)(PCL), poly-*N*-(2-hydroxypropyl) methacrylamide (HPMA), poly(D,L-lactic acid) (PLA), dextran, poly(D,L-lactic-co-glycolic acid) (PLGA), alongside additional stealth effect offered by most widely used PEG (Utreja et al. 2020). The amphiphilic character allows poorly water-soluble drugs to be loaded in the core with the shell providing aqueous solubility, colloidal stability, and essential stealth character. Although these nanoformulations offer improved penetrability, solubility, bioavailability, targeting through directing ligand complexed on the surface or combining monoclonal antibodies to the micelle corona, they suffer from poor *in vivo* stability. This effect is due to dissociation and early drug release below critical micelle concentration succeeding its administration that can also lead to drug-related toxicity. However, stimuli-responsive cross-linked micelles have exhibited micellar stability and have attracted formulation scientists for the delivery of docetaxel, camptothecin, paclitaxel, cisplatin, and oxaliplatin (Ventola 2017).

The original milled organic nanocrystal, Rapamune®, approved by FDA in 2000 opened new avenues for resourceful NPs that were capable of enhancing solubility and bioavailability. Nanocrystal-based drugs within the range of 1000 nm are peculiar because they are solely composed of drug derivatives without any carriers bound to them and are typically stabilized using polymeric steric stabilizers or surfactants. The poorly soluble organic or inorganic drugs are rendered enriched pharmacokinetic (PK)/pharmacodynamic (PD) properties by nanostructures known as nanocrystals. Nanocrystals have unique characteristics that allow them to solve problems such as increasing solubility of saturation, increased speed of dissolution, and enhanced surface/cell membrane binding. Saturation solubility increases the forces that, via biological mechanisms, such as the walls of the gastrointestinal tract, drive diffusion-based mass transfer (Farjadian et al. 2019). The oral absorption process for nanocrystal formulations, however, is not well known and their action is not completely predictable after subcutaneous injection. They are photochemically stable and exhibit a narrow, controllable, symmetric emission spectrum. They consist of an optically energetic core enclosed by a shield that creates a physical barrier to the external environment, rendering them less vulnerable to photo-oxidation or medium shifts. The methods of preparation of nanocrystals can be divided into top-down and bottom-up processes. The bottom-up method creates nanocrystals from the solution, which requires two basic stages: nucleation and crystal formation. It mainly involves high-pressure homogenization accompanied by grinding procedures. The top-down methodologies comprise of high-energy mechanical powers like milling (NanoCrystals®) or high-pressure homogenization (IDD-P®, DissoCubes® and Nanopure®), and the main benefit is that it is adaptable to the manufacturing scale (Lopalco and Denora 2018). Nevertheless, high energy, cost, and time used as well as impurity from grinding media are a downside of this technology, leading to unintended toxic undesirable results. Supercritical fluid (SCF) such as supercritical carbon dioxide exhibits superior physical characteristics, liquid solubilization, diffusivity similar to gas and minimal environmental effect. Thus, nanocrystals are lately prepared using SCF.

Inorganic nanocarriers have recently been used to create powerful nanocarriers for drug delivery applications attributable to easy alteration, high drug loading capability, and stability. A significant variety of inorganic materials can be used to produce NPs, such as silica, metal oxide, or metal. In particular, NPs of metal and metal oxide are being intensively studied for simultaneous therapeutic as well as imaging purposes. They are used and developed for an investigative picture of the diseased area because of the special magnetic and plasmonic properties. However, only a few inorganic NPs have been approved for clinical use, although others are still in the clinical testing stage. They are composed of a core containing the inorganic portion such as silica, gold, iron oxide, or quantum dots. A shell region consisting mostly of organic polymers (or metals) offers an adequate surface functionalization substrate or a way to protect from redundant physicochemical interactions with the biological microenvironment. Particle modification is usually done to strengthen the interaction with the biological membranes. In spite of these benefits, however, inorganic NPs have demonstrated only modest effectiveness in the treatment of disease tissues due to the crucial problems associated with the limited quantity of drug substances delivered and extreme toxicity (Lombardo et al. 2019). Gold NPs, silver NPs, and iron oxide NPs have been extensively studied in the biomedical field due to their special biochemical properties and high electron conductivity. With the approval of Abraxane® in 2005, which incorporates 130-nm albumin NPs conjugated with paclitaxel, a shift occurred from the use of unmodified proteins to engineered particle complexes originated to enable active targeting. Protein-based NPs include protein-conjugated medications, formulations where the protein itself is the active therapy, and complex combined platforms that utilize proteins for targeted delivery. Generally, natural proteins are preferred to reduce toxicity. In the last decade, albumin has gained the attention of research scientists as a drug carrier due to the facilitation of cellular uptake mechanisms and accumulation via EPR effect (Ventola 2017). Mesoporous silica NPs, quantum dots, and carbon nanotubes have also registered their applications in biomedical and nanomedicine fields. Moreover, nanocarriers that are equipped with complex surface functionalization of inorganic nanocarriers with organic materials or through the use of organic colloids as a template for the regulated growth of inorganic materials have recently attracted considerable global interest. For example, surface coating of mesoporous silica NPs with polyethyleneimine improves cellular uptake and enables siRNA and DNA constructs to be transported safely (Paris and Vallet-Regí 2020). Lately, lipid-coated mesoporous silica NPs have also been employed to achieve stimuli-responsive drug release, promote drug loading, prevent premature release of drug, avoid multidrug resistance, stability, and biocompatibility.

1.3.3 Tissue Engineering and Regenerative Medicine

Nanomaterials of low toxicity, contrasting agent properties, customizable characteristics, targeted/stimuli distribution ability, and accurate behavioral regulation via external stimulus are valuable tools for achieving unparalleled efficiency in driving and overcoming barriers in tissue engineering. The extraordinary mechanical strength and electrochemical properties of carbon nanostructures corresponding to graphene, carbon nanohorns, fullerenes, nanodiamonds, single/multi-walled carbon nanotubes; the fluorescence properties of quantum dots; the antimicrobial effect of silver and other metal/metal oxide NPs; and surface conjugation as well as surface conductive properties of gold NPs have made them quite successful in different tissue engineering and regenerative medicine applications (Makvandi et al. 2020) as seen in Figure 1.2.

These residing nanomaterials must unveil biocompatibility, boost the growth of cells, promote the proliferation of various types of cells, provide regulated means for delivery of bioactive and contrast agents to manage and track the engineered tissues. Furthermore, nanodiamond–polymer composites are also seen as a promising medium for repairing damaged tissues due to their unique mechanical properties, fluorescence capability, and

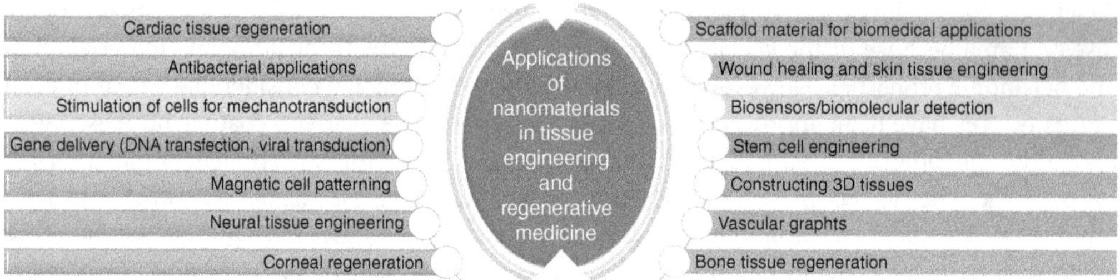

Figure 1.2 Nanomaterial applications in tissue engineering and regenerative medicine. *Source:* Based on Makvandi et al. (2020).

biocompatibility (Hasan et al. 2018). The emerging field of nanoneedles for intracellular distribution, intracellular pH estimation, probing, and cell-interfacing is currently growing promptly and displays abundant potential. Nanotechnology proposes innovative culture approaches to resolve the prevalent problem of cell-based therapy with minimal retention in target tissues. It has got the ability to generate smart surfaces that renders intermittent adhesive properties to cells. This makes detachment of cell patches from cell culture substrate feasible for further transplantation (Kubinova and Sykova 2010). Cell sheet patch therapies have paved the way for possible cell and tissue engineering therapies for patients suffering from various disorders such as cardiomyopathy, osteoarthritis, and periodontal reconstruction. Nanomedicine is now emerging as a powerful tool to mitigate organ donor shortages. The formation of *in vitro* whole organs is a present-day requirement in the field of biomedicine. This is fulfilled by nanomaterials enabling the production of artificial organs for regenerative medicine as well as organs-on-a-chip applications.

1.4 Clinical Translation of Nanomedicine

Prof. Kinam Park of Purdue University has precisely pointed out that "It is time to review the progress made in nanomedicine, and examine the sources of difficulty in clinical translation, and move forward" in the cover story "The beginning of the end of the nanomedicine hype" in Journal of Controlled Release (Park 2019a). The research world of nanomedicine has produced abundant research articles with proof of concept aiming towards the same admirable endings of countless possibilities of nanomedicine. But the enigma around the quantitative data proving its clinical efficacy still prevails. It is time to understand that what we desire is not equivalent to what we have. At this phase, it is also obligatory to shift exorbitant prejudiced focus from smart nanoformulations for treatment of tumors toward investigative endeavors for discovering remedies for other types of diseases like CVS diseases, Parkinson's disease, Alzheimer's disease to name a few. It is necessary to allow and back the clinical production of new, promising formulations for diseases that are not linked to oncology. It is the time to promote the incorporation of nanomedicines into clinical trials of new promising formulations (Park 2019b). Nevertheless, there are numerous and nuanced problems in the clinical translation of nanomedicine (Wu et al. 2020). These include lack of business management education, especially at the academic level; difficulties in carrying out early-stage preclinical characterization and safety evaluations; absence of protocols and access to characterization facilities; barriers in scale-up and GMP production; reproducibility and stability of batch-to-batch engineered nanocarriers; the absence of adequate controls and badly specified essential quality characteristics; lack of cutting-edge harmonizing assays and standard principle for toxicity estimation of nanomedicine products; the low quality and durability of the preclinical research that has been published; shortage of appropriate animal models for extrapolating the immunotoxicity data to humans; and complexity and heterogeneity within the regulatory framework (Gabizon et al. 2020; Martins et al. 2020). A key component of an effective clinical translation is the identification of the right patient and matching with the right nanoformulation.

1.5 Nanotoxicological Challenges

Nanoparticles have been commonly used in medicine in recent years, making it important to resolve human health toxicity concerns. These nanostructures generally directly enter into the human body and do not undergo the normal absorption process. The nanocarriers themselves may exhibit toxicity, penetrate the biological membrane barriers, interact with biomacromolecules, and get accumulated in organs or tissues in the human body. This stresses the fact that the toxicological evaluation of nanomaterials is an essential step in the advancement of nanomedicine (SCENIHR 2006). It has been almost more than a decade since the importance of nanotoxicology is realized in nanomedicine just as toxicology in medicine. Nanotoxicology is a developing subfield of toxicology that can be defined as the science of engineered nanodevices and nanostructures that studies the interaction between the physical and chemical properties of nanostructures and biological systems and develops means to prevent such deleterious effects (Oberdörster et al. 2005, 2009). Nanotoxicology research cannot only offer data for the safety evaluation of advanced nanostructures and materials but can also help advance the field of nanomedicine by providing awareness of their hazardous properties and methods of avoiding them. Nanotoxicological issues of nanomedicine include its physicochemical parameters at nanoscale, biological behavior, mechanisms of toxicity, and toxic effects that can be produced within the human body (Gatoo et al. 2014; Warheit and Sayes 2015; Sukhanova et al. 2018) as illustrated in Table 1.2.

Table 1.2 Diverse nanotoxicological concerns of nanomaterials.

Sr. No.	Nanotoxicological issues of nanomaterials	Associated parameters
1.	Physicochemical properties of nanocarriers	Particle size and surface area, particle shape, aspect ratio, chemical composition, hydrophilicity and hydrophobicity, surface coating, surface roughness, aggregation and concentration, degradability
2.	Biological behavior of nanomaterials	Protein corona effects, metabolism, distribution, clearance mechanisms, toxicity against cells and tissues of the reticuloendothelial system, therapeutic efficacy, safety
3.	Toxic effects of nanoparticles	Renal toxicity, spermatotoxicity, hepatotoxicity, cardiovascular toxicity, dermal toxicity, neurotoxicity, pulmonary toxicity
4.	Mechanisms of nanoparticle toxicity	Oxidative stress, cell membrane damage, disturbance of intracellular/intercellular transport, accelerate mutagenesis, cell energy imbalance, apoptosis, hindered cell division, disruption of cell, tissue and organ metabolism

Sources: Based on Gatoo et al. (2014), Warheit and Sayes (2015) and Sukhanova et al. (2018).

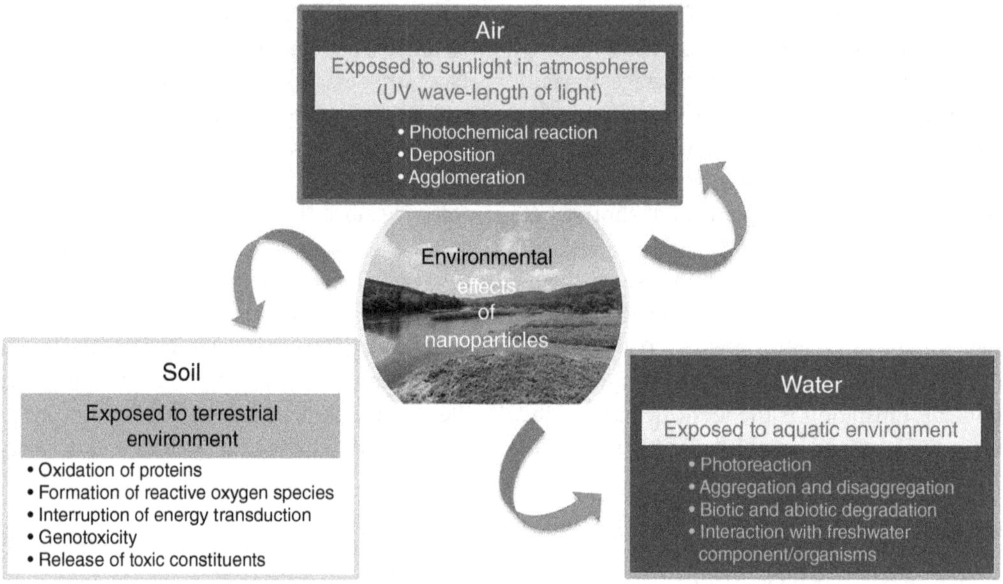

Figure 1.3 Consequences of environmental contact of nanoparticles.

Nanoecotoxicology is a subdiscipline of ecotoxicology that is primarily designed to define and forecast the impact of nano-sized materials on ecosystems. The quantitative risk assessment of NPs is inadequate owing to unsatisfactory research on the toxicity of nanomaterials to environmentally important organisms. It is indistinct that the structure, aspect ratio, morphology, and various other physicochemical properties have a greater impact on toxicity (Klaine et al. 2008). It must be mandatory to evaluate the NPs for their possible hazard to the environment before their use in products (Oberdörster et al. 2009; Viswanath and Kim 2016; Gaur et al. 2020). To accomplish this aim, to identify exposure, nanoecotoxicology requires taking into account the entry routes and consequences of environmental contact of NPs as shown in Figure 1.3.

Numerous critical toxic characteristics are closely associated with the nano-size/surface area of the nanomaterials. These include surface properties, physical absorption potential, chemical reactivity, and so on, all of which strongly dominate *in vivo* nanotoxicological behavior (Fubini et al. 2010). NPs and their biologics, such as proteins, peptides, antibody fragments, and nucleic acids, can act as bases of antigens that can induce an immune response. Its immunogenicity can be influenced by the physicochemical properties of NPs, such as size, surface area, surface charge, hydrophobicity, and solubility. An exceptional type of toxicity owing to surface modification is of distinct apprehension. The capacity for inflammatory activity and oxidative stress is shown to be largely dependent on NPs' surface chemistry and surface modifications (Azarnezhad et al. 2020). NPs' toxicity also strongly depends on their shape and chemical composition. Spherical NPs are more susceptible to endocytosis than nanotubes and nanofibers. The calcium channels are commendably blocked by single-walled carbon nanotubes than spherical fullerenes. The degradation of NPs has been shown to occur, and its magnitude depends on the conditions of the environment, such as pH or ionic strength. The toxicity also relies on the composition of the NPs' core. Leaking metal ions from the NPs' core is the most common source of the toxic impact of NPs reacting with cells. Indeed, the toxicity of NPs is determined to a large degree by their chemical composition (Gatoo et al. 2014; Viswanath and Kim 2016).

It has been reported that AuNPs possibly move through a mother's placenta to the fetus. During the neonatal phase, these NPs can act as allergens, activating the immune system. NPs can communicate with diverse immune cell networks found inside and under epithelial surfaces. The proficient gastrointestinal tract uptake of NPs also been well recorded in oral feeding and forced feeding studies. Despite the clinical approval, the dermal toxicity of nanosilver-based surgical dressings and sutures is yet a matter of concern (El-Ansary and Al-Daihan 2009). While beneficial control of wound infection is accomplished, their dermal toxicity is even of concern. In epidemiological trials, adverse cardiovascular effects attributable to exposure to NPs have been identified (De Jong and Borm 2008). Oxidative damage to DNA is exhibited by the titanium dioxide and zinc oxide NPs in *in vitro* tests and cultured human fibroblasts. Nano-sized particles inhaled can increase bloodstream access and can then be spread to other organs. There is a good likelihood that, via the lungs, skin, and gastrointestinal tract, NPs may be assimilated into the bloodstream. Besides, fluids representing the liver, blood, and airway environment are exploited to conduct experiments concerning the dissolution of NPs in artificial body fluids and classify harmful influences. NPs also have access to the brain, exhibiting toxic effects on BBB, especially high concentrations of anionic and cationic NPs, although, neutral NPs and low concentrations of anionic NPs were found not to affect the integrity of BBB (Teleanu et al. 2019). The development of reactive oxygen species and oxidative stress are caused by NPs. Further, these have shown to be involved in the development of neurodegenerative diseases such as Parkinson's and Alzheimer's diseases (Armstead and Li 2016).

The dominant role of protein–nanoparticle interactions has begun to appear in nanomedicine and nanotoxicity. The "corona" nanoparticle protein is a dynamic coating of proteins and other biomolecules that adsorbs to the surfaces of the nanoparticle (Dickinson et al. 2019). Protein corona is a nanoparticle's biological identity since it is what the cell "sees and communicates." At any given time, the structure of the corona protein can be determined by the concentrations of over 3700 plasma proteins. When exposed to a biological fluid, this corona may not achieve equilibrium automatically. The nanoparticle surface would initially be dominated by proteins with high concentrations and elevated interaction rate constants. They can also easily dissociate to be swapped by reduced concentration, leisurely exchange, and greater affinity proteins. Small NPs may cause protein malfunction, with their wide surface area as a binding interface, which may lead to pathogenesis and adverse health effects (Sukhanova et al. 2018; Singh et al. 2019).

There are two standard approaches adopted to study NPs' toxic effects on human health: *in vitro* studies on model cell lines and *in vivo* experiments on experimental animals. There are unique benefits and drawbacks of both cell culture and animal laboratory models for testing NP toxicity (Osman et al. 2020). The former offers a deeper insight into the molecular processes of toxicity and describes the key targets of NPs; however, it does not take into account the patterns of the delivery of NPs in the body and their transfer to various tissues and cells (Guadagnini et al. 2015). The study of NP toxicity in animal laboratories makes it possible to predict the delayed effects of *in vivo* NPs action. The general pattern of manifestations of toxicity, however, is so complex that it is difficult to establish which of them is the major cause of the result observed and which are the effects thereof. The long-term consequences of prolonged exposure to these NPs in humans must be analyzed. For all of the therapeutic NPs, methods must be built for detecting NPs *in situ*. Biotransformation of therapeutic NPs in the human body, their association with biological processes, and adsorption, delivery, digestion, transformation, degradation, and excretion in living systems should be studied by the characterization of NPs in terms of distribution of size, surface properties, persistence, and stabilization of initial and modified NPs (Zhao and Castranova 2011; Dickinson et al. 2019).

The latest toxicological approaches utilized for the detection of NP hazards are based on conventional toxicology approaches or complementary techniques. These tools for safety assessment of the rapidly budding list of nanomaterials have innate restrictions. There is a rising demand for strategies that could be adopted for screening by industries during the advancement of nanoproducts. These products may have to be handled on a case-by-case basis and require costly evaluations. Usually, acute and chronic toxicity is assessed in the pharmaceutical industry. The acute toxicity incorporates mitochondrial activity hemolysis, oxidative stress, inflammation, or complement activation. The chronic toxicity investigation is arduous, and it is more perplexing to scrutinize the results (Zhao and Castranova 2011). An epitome alteration has been proposed by toxicologists in using *in vitro* and small animal models with high-throughput capability. The molecular and cellular pathways extrapolative to probable pathology conditions in whole organisms are the foundation. The authentication of such analytical toxicity testing methods of nanoproducts, although, would present a challenge for the scientific society (Armstead and Li 2016). Effective methods and state-of-the-art techniques to research biological behaviors and the transformation of nanomaterials *in vivo* and *in vitro* are important for the evaluation of safety and therapeutic efficacy. Systemic and quantitative proteomics, genomics, transcriptomics, and metabolomics data can also capture improvements in metabolism, mechanisms of signaling, and biological functions (Wolfram et al. 2015; Pelaz et al. 2017).

1.6 Safety Issues and Regulations

A number of nanomedicines have been approved for cancer treatment by the FDA and the European Medicines Agency (EMA). The FDA, EMA, and other regulatory agencies have not yet explicitly adopted recommendations for drug products containing not so hard materials (Wu et al. 2020). In the absence of guidelines for the review of nanomedicine therapeutics, regulatory decisions can be taken only on the basis of an individual evaluation of benefits and risks. The regulatory process is therefore laborious and involves a sophisticated specialist in emerging technology, which could result in regulatory delays (Juillerat-Jeanneret et al. 2015; Singh et al. 2019). Also, regulatory concerns are critical for the production of cutting-edge technology to define and monitor, in addition to clinical trials and the approval process, the safety of nanomedicine products. There is an immediate need for comprehensive characterization and quality control protocols. It is important to build more sophisticated and multifunctional tools for assessing the safety of nanotherapeutics, even as the difficulty of the approval process may be increased. To date, understanding has not been reached about how to determine the safety of nanomaterials using efficient approaches and standards that are approved by regulatory bodies. There are several explanations for this: lack of established reference material for benchmarking the effects of the material being tested, lack of clarity on predictive endpoints for human toxicity, technological shortcomings of nanomaterial testing, and batch-to-batch variations due to inadequate control over the development process. Many government and nongovernment groups have been attempting to find out how to handle this topic for the past two decades (Fadeel et al. 2007; Oberdorster 2010; Pelaz et al. 2017).

In clinical trials, nanomedicine drugs should have a desirable pharmacology and toxicity profile and show safety and efficacy. In order to further link the physicochemical assets of nanomedical systems with biological consequences, successful incorporation of capital from academia, business, government organizations, and hospitals would be critical. To leverage each partner's specific strengths, it is important to cultivate collaborations between major pharmaceuticals, smaller firms, and academia early in preclinical growth (Havel et al. 2016). The definitions, quality requirements, recommendations and regulatory concerns of nanomedicine should be defined and reciprocated by health authorities (Zhao and Castranova 2011; George et al. 2015). Although any new drug conceptually needs to meet comparable challenges, the basic sophistication and multicomponent existence of nanomedicines contributes to a vast range of additional variables that may raise further concerns considerably. Nanomedicine will go on to the next step and bring practical and meaningful benefit to human medicine and healthcare by rationally planned and systematic approaches to overcome various safety issues and solve the problems.

1.7 Conclusion and Future Perspectives

Nanomedicine has raised exhilarating prospects for many medical issues. It has contributed, both basic and translational, to several major developments. Many of these developments have been made in the area of diagnosing and treating cancer. Study focuses are extending to other medical issues such as antibiotic resistance, cardiovascular problems, and artificial organs. Due to the simplicity of design, researchers believe, nanomaterial platform design is likely to be of benefit in these fields too. However, technical questions have arisen due to the lack of information about the behavior of nanomaterials inside living organisms. Conversely, relatively little is already understood about the ability of nanomaterial to induce adverse effects or humoral immune responses until they are accidentally or purposely inserted into the organism. The biggest concern with their use in medicine is the possible toxicity of NPs. The toxicity of NPs highly depends on the core and shell's nature, size, electrical charge, and chemical composition. As nanomaterial production grows, nanotoxicology and nanorisk have drawn growing interest from toxicologists and regulatory scientists. Nanotoxicology focuses on uncovering the mechanistic relationship between physicochemical characteristics of the NPs and the subsequent biological effect. Besides, it aims to refine the experimental conditions for *in vitro* and *in vivo* evaluation to identify the potential interaction of NPs with various assays and to provide evidence for the safety assessment of this nanomaterial and its applications. Though, the shortage of uniform assays and preset regulations have made it difficult to associate with the existing literature about their safety. The science community has endured comprehensive debates over two decades and still has not come to a common ground on NPs' metrics, definitions, classifications, attributes, safety/toxicity characteristics, endpoints of toxicity, target endpoints, thresholds of occupational/environmental risk, and systematic methods of assessment. In attempts to bridge international harmonization and standardization, both government and nongovernment organizations are collaborating to resolve the regulatory challenges. In addition, there are currently no specific processes and regulatory standards available for the assessment or testing of nanomaterials.

"Smart" nanomedicine demands step-by-step quality testing as an integral part of all new nanoformulations. A range of methods to design NPs with reduced toxicity are now possible relative to conventional NPs. In order to improve the chances of clinical translation, adherence to the good clinical practice standards is expected from phase I to closely supervised phase IV. However, the most significant challenge faced today in the field of nanotoxicology is the detection and evaluation of the adverse effects of a variety of engineered nanomaterials that are commonly manufactured and launched for versatile applications with their different physicochemical properties. Because of multiple reasons, for example, the precise hazard distinctions of NPs are not quite easy to discover, we are not clear about the physicochemical property of the NPs affecting the toxicity, although, the extreme improvements in molecular, chemical, and other properties that can be formulated by nanotechnology applications are so fascinating that analysis can even be valued to assess their effects on product quality, potency, or other functionality. Also, advanced techniques and approaches are required for the study of NPs' toxicity making it possible to examine, at the molecular level, multiple processes and mechanisms of toxicity and to accurately predict possible harmful effects at the level of the body. There is a need for autonomous sources like public finance organizations and foundations to back up the research needs of maturing nanodrugs. In addition to concerns relating to nanomaterials and the complexity of biological processes, there are issues relating to the putting of the substance on the market, including large-scale processing, biocompatibility, intellectual property issues, regulatory issues, and general cost-effectiveness relative to conventional treatments. Nevertheless, there is still tremendous hope and we remain optimistic that nanomedicine will certainly take its place not as a miracle cure but as a solution to unique, unmet patient needs in clinical practice.

References

Allen, S., Bobbala, S., Karabin, N., and Scott, E. (2019). On the advancement of polymeric bicontinuous nanospheres toward biomedical applications. *Nanoscale Horizons* 4 (2): 258–272. https://doi.org/10.1039/c8nh00300a.

Armstead, A. and Li, B. (2016). Nanotoxicity: emerging concerns regarding nanomaterial safety and occupational hard metal (WC-Co) nanoparticle exposure. *International Journal of Nanomedicine* 11: 6421–6433. https://doi.org/10.2147/IJN.S121238.

Azarnezhad, A., Samadian, H., Jaymand, M. et al. (2020). Toxicological profile of lipid-based nanostructures: are they considered as completely safe nanocarriers? *Critical Reviews in Toxicology* 50 (2): 148–176. https://doi.org/10.1080/10408444.2020.1719974.

Bayda, S., Adeel, M., Tuccinardi, T. et al. (2020). The history of nanoscience and nanotechnology: from chemical-physical applications to nanomedicine. *Molecules* 25 (1): 1–15. https://doi.org/10.3390/molecules25010112.

Bobo, D., Robinson, K., Islam, J. et al. (2016). Nanoparticle-based medicines: a review of FDA-approved materials and clinical trials to date. *Pharmaceutical Research* 33 (10): 2373–2387. https://doi.org/10.1007/s11095-016-1958-5.

Buzea, C. and Pacheco, I. (2017). Nanomaterials and their classification. In: *EMR/ESR/EPR Spectroscopy for Characterization of Nanomaterials* (ed. A. Shukla), 3–45. Springer (India) Pvt. Ltd.

Caster, J., Patel, A., Zhang, T., and Wang, A. (2017). Investigational nanomedicines in 2016: a review of nanotherapeutics currently undergoing clinical trials. *Wiley Interdisciplinary Reviews. Nanomedicine and Nanobiotechnology* 9 (1): e1416. https://doi.org/10.1002/wnan.1416.

Chavan, T., Muttil, P., and Kunda, N. (2020). *Introduction to Nanomedicine in Drug Delivery*. Switzerland AG: Springer Nature.

Choi, Y. and Han, H. (2018). Nanomedicines: current status and future perspectives in aspect of drug delivery and pharmacokinetics. *Journal of Pharmaceutical Investigation* 48 (1): 43–60. https://doi.org/10.1007/s40005-017-0370-4.

Choi, Y., Lee, M., David, A., and Park, Y. (2014). Nanoparticles for gene delivery: therapeutic and toxic effects. *Molecular & Cellular Toxicology* 10 (1): 1–8. https://doi.org/10.1007/s13273-014-0001-3.

De Jong, W. and Borm, P. (2008). Drug delivery and nanoparticles: applications and hazards. *International Journal of Nanomedicine* 3 (2): 133–149.

Dickinson, A., Godden, J., Lanovyk, K., and Ahmed, S. (2019). Assessing the safety of nanomedicines: a mini review. *Applied In Vitro Toxicology* 5 (3): 114–122. https://doi.org/10.1089/aivt.2019.0009.

El-Ansary, A. and Al-Daihan, S. (2009). On the toxicity of therapeutically used nanoparticles: an overview. *Journal of Toxicology* 2009: 1–9. https://doi.org/10.1155/2009/754810.

Eloy, J., Claro de Souza, M., Petrilli, R. et al. (2014). Liposomes as carriers of hydrophilic small molecule drugs: strategies to enhance encapsulation and delivery. *Colloids and Surfaces B: Biointerfaces* 123: 345–363. https://doi.org/10.1016/j.colsurfb.2014.09.029.

Fadeel, B. and Alexiou, C. (2020). Brave new world revisited: focus on nanomedicine. *Biochemical and Biophysical Research Communications* 533 (1): 36–49. https://doi.org/10.1016/j.bbrc.2020.08.046.

Fadeel, B., Kagan, V., Krug, H. et al. (2007). There's plenty of room at the forum: potential risks and safety assessment of engineered nanomaterials. *Nanotoxicology* 1 (2): 73–84. https://doi.org/10.1080/17435390701565578.

Farjadian, F., Ghasemi, A., Gohari, O. et al. (2019). Nanopharmaceuticals and nanomedicines currently on the market: challenges and opportunities. *Nanomedicine (London)* 14 (1): 93–126. https://doi.org/10.2217/nnm-2018-0120.

Fubini, B., Ghiazza, M., and Fenoglio, I. (2010). Physico-chemical features of engineered nanoparticles relevant to their toxicity. *Nanotoxicology* 4: 347–363. https://doi.org/10.3109/17435390.2010.509519.

Gabizon, A., Catane, R., Uziely, B. et al. (1994). Prolonged circulation time and enhanced accumulation in malignant exudates of doxorubicin encapsulated in polyethylene-glycol coated liposomes. *Cancer Research* 54 (4): 987–992.

Gabizon, A., de Rosales, R., and La-Beck, N. (2020). Translational considerations in nanomedicine: the oncology perspective. *Advanced Drug Delivery Reviews* 158: 140–157. https://doi.org/10.1016/j.addr.2020.05.012.

Gatoo, M., Naseem, S., Arfat, M. et al. (2014). Physicochemical properties of nanomaterials: implication in associated toxic manifestations. *BioMed Research International* 2014: 1–8. https://doi.org/10.1155/2014/498420.

Gaur, N., Sharma, N., Dahiya, A. et al. (2020). *Toxicity and Regulatory Concerns for Nanoformulations in Medicine*. Beverly, MA: Scrivener Publishing LLC.

George, S., Ho, S., Wong, E. et al. (2015). The multi-facets of sustainable nanotechnology - lessons from a nanosafety symposium. *Nanotoxicology* 9 (3): 404–406. https://doi.org/10.3109/17435390.2015.1027315.

Germain, M., Caputo, F., Metcalfe, S. et al. (2020). Delivering the power of nanomedicine to patients today. *Journal of Controlled Release* 326: 164–171. https://doi.org/10.1016/j.jconrel.2020.07.007.

Greish, K., Mathur, A., Bakhiet, M., and Taurin, S. (2018). Nanomedicine: is it lost in translation? *Therapeutic Delivery* 9 (4): 1–18. https://doi.org/10.4155/tde-2017-0118.

Guadagnini, R., Kenzaoui, B.H., Walker, L. et al. (2015). Toxicity screenings of nanomaterials: challenges due to interference with assay processes and components of classic in vitro tests. *Nanotoxicology* 9 (Suppl 1): 13–24. https://doi.org/10.3109/17435390.2013.829590.

Hasan, A., Morshed, M., Memic, A. et al. (2018). Nanoparticles in tissue engineering: applications, challenges and prospects. *Internal Journal of Nanomedicine* 13: 5637–5655. https://doi.org/10.2147/IJN.S153758.

Havel, H., Finch, G., Strode, P. et al. (2016). Nanomedicines: from bench to bedside and beyond. *The AAPS Journal* 18 (6): 1373–1378. https://doi.org/10.1208/s12248-016-9961-7.

Heinrich, M., Martina, B., and Prakash, J. (2020). Nanomedicine strategies to target coronavirus. *Nano Today* 35: 1–21. https://doi.org/10.1016/j.nantod.2020.100961.

Jeevanandam, J., Barhoum, A., Chan, Y. et al. (2018). Review on nanoparticles and nanostructured materials: history, sources, toxicity and regulations. *Beilstein Journal of Nanotechnology* 9: 1050–1074. https://doi.org/10.3762/bjnano.9.98.

Juillerat-Jeanneret, L., Dusinska, M., Fjellsbo, L. et al. (2015). Biological impact assessment of nanomaterial used in nanomedicine: introduction to the NanoTEST project. *Nanotoxicology* 9 (Suppl 1): 5–12. https://doi.org/10.3109/17435390.2013.826743.

Khan, F. (2020). *Nanomaterials: Types, Classifications, and Sources*. Singapore: Springer Nature.

Khan, S., Hasan, A., Attar, F. et al. (2020). Gold nanoparticle-based platforms for diagnosis and treatment of myocardial infarction. *ACS Biomaterials Science & Engineering* 6 (12): 6460–6477. https://doi.org/10.1021/acsbiomaterials.0c00955.

Klaine, S., Alvarez, P., Batley, G. et al. (2008). Nanomaterials in the environment: behavior, fate, bioavailability, and effects. *Environmental Toxicology and Chemistry* 27 (9): 1825–1851.

Koziorowski, J., Stanciu, A., Gomez-Vallejo, V., and Llop, J. (2017). Radiolabeled nanoparticles for cancer diagnosis and therapy. *Anti-Cancer Agents in Medicinal Chemistry* 17 (3): 333–354. https://doi.org/10.2174/1871520616666160219162902.

Krukemeyer, M., Krenn, V., and Huebner, F. (2015). History and possible uses of nanomedicine based on nanoparticles and nanotechnological progress. *Journal of Nanomedicine & Nanotechnology* 06 (06): 1–7. https://doi.org/10.4172/2157-7439.1000336.

Kubinova, S. and Sykova, E. (2010). Nanotechnologies in regenerative medicine. *Minimally Invasive Therapy* 19 (3): 144–156. https://doi.org/10.3109/13645706.2010.481398.

Lombardo, D., Kiselev, M., and Caccamo, M. (2019). Smart nanoparticles for drug delivery application: development of versatile nanocarrier platforms in biotechnology and nanomedicine. *Journal of Nanomaterials* 2019: 1–26. https://doi.org/10.1155/2019/3702518.

Lopalco, A. and Denora, N. (2018). *Nanoformulations for Drug Delivery: Safety, Toxicity, and Efficacy*, 2018/06/24e, vol. 1800. New York: Humana Press.

Makvandi, P., Ghomi, M., Padil, V. et al. (2020). Biofabricated nanostructures and their composites in regenerative medicine. *ACS Applied Nano Materials* 3 (7): 6210–6238. https://doi.org/10.1021/acsanm.0c1164.

Martins, J., das Neves, J., de la Fuente, M. et al. (2020). The solid progress of nanomedicine. *Drug Delivery and Translational Research* 10 (3): 726–729. https://doi.org/10.1007/s13346-020-00743-2.

McDonald, T., Siccardi, M., Moss, D. et al. (2015). The application of nanotechnology to drug delivery in medicine. In: *Nanoengineering* (ed. P. Dolez), 173–223. Amsterdam: Elsevier.

McGoron, A. (2020). Perspectives on the future of nanomedicine to impact patients: an analysis of US federal funding and interventional clinical trials. *Bioconjugate Chemistry* 31 (3): 436–447. https://doi.org/10.1021/acs.bioconjchem.9b00818.

Mukherjee, B., Dey, N., Maji, R. et al. (2014). Current status and future scope for nanomaterials in drug delivery. In: *Application of Nanotechnology in Drug Delivery* (ed. A. Sezer). London: IntechOpen.

Oberdorster, G. (2010). Safety assessment for nanotechnology and nanomedicine: concepts of nanotoxicology. *Journal of Internal Medicine* 267 (1): 89–105. https://doi.org/10.1111/j.1365-2796.2009.02187.x.

Oberdörster, G., Oberdörster, E., and Oberdörster, J. (2005). Nanotoxicology: an emerging discipline evolving from studies of ultrafine particles. *Environmental Health Perspectives* 113 (7): 823–839. https://doi.org/10.1289/ehp.7339.

Oberdörster, G., Stone, V., and Donaldson, K. (2009). Toxicology of nanoparticles: a historical perspective. *Nanotoxicology* 1 (1): 2–25. https://doi.org/10.1080/17435390701314761.

Osman, N.M., Sexton, D.W., and Saleem, I.Y. (2020). Toxicological assessment of nanoparticle interactions with the pulmonary system. *Nanotoxicology* 14 (1): 21–58. https://doi.org/10.1080/17435390.2019.1661043.

Parappurath, N., Kirubanandam, S., Kumar, V., and Ahmed, B. (2018). Nanomaterials history, classification, unique properties, production and market. In: *Emerging Applications of Nanoparticles and Architecture Nanostructures: Current Prospects and Future Trends* (eds. M. ASH and A. Barhoum), 341–384. United Kingdom: Elsevier.

Paris, J. and Vallet-Regí, M. (2020). Mesoporous silica nanoparticles for co-delivery of drugs and nucleic acids in oncology: a review. *Pharmaceutics* 12 (6) https://doi.org/10.3390/pharmaceutics12060526.

Park, K. (2019a). The beginning of the end of the nanomedicine hype. *Journal of Controlled Release* 305: 221–222. https://doi.org/10.1016/j.jconrel.2019.05.044.

Park, K. (2019b). Transcending nanomedicine to the next level: are we there yet? *Journal of Controlled Release* 298: 213. https://doi.org/10.1016/j.jconrel.2019.02.040.

Patra, J., Das, G., Fraceto, L. et al. (2018). Nano based drug delivery systems: recent developments and future prospects. *Journal of Nanobiotechnology* 16 (1): 1–33. https://doi.org/10.1186/s12951-018-0392-8.

Pelaz, B., Alexiou, C., Alvarez-Puebla, R. et al. (2017). Diverse applications of nanomedicine. *ACS Nano* 11 (3): 2313–2381. https://doi.org/10.1021/acsnano.6b06040.

Pellico, J., Ellis, C., and Davis, J. (2019). Nanoparticle-based paramagnetic contrast agents for magnetic resonance imaging. *Contrast Media & Molecular Imaging* 2019: 1–13. https://doi.org/10.1155/2019/1845637.

Pratiwi, F., Kuo, C., Chen, B., and Chen, P. (2019). Recent advances in the use of fluorescent nanoparticles for bioimaging. *Nanomedicine (London)* 14 (13): 1759–1769. https://doi.org/10.2217/nnm-2019-0105.

Rahman, M., Ahmad, M., Kazmi, I. et al. (2012). Advancement in multifunctional nanoparticles for the effective treatment of cancer. *Expert Opinion on Drug Delivery* 9 (4): 367–381. https://doi.org/10.1517/17425247.2012.668522.

Sangtani, A., Nag, O., Field, L. et al. (2017). Multifunctional nanoparticle composites: progress in the use of soft and hard nanoparticles for drug delivery and imaging. *Wiley Interdisciplinary Reviews. Nanomedicine and Nanobiotechnology* 9 (6) https://doi.org/10.1002/wnan.1466.

SCENIHR (2006). Modified opinion (after public consultation) on the appropriateness of existing methodologies to assess the potential risks associated with engineered and adventitious products of nanotechnologies (SCENIHR/002/05). https://ec.europa.eu/health/ph_risk/committees/04_scenihr/docs/scenihr_o_003b.pdf

Sharma, D., Sharma, N., Pathak, M. et al. (2018). Nanotechnology-based drug delivery systems. In: *Drug Targeting and Stimuli Sensitive Drug Delivery Systems* (ed. A.M. Grumezescu), 39–79. New York: William Andrew Publishing.

Shidhaye, S., Vaidya, R., Sutar, S. et al. (2008). Solid lipid nanoparticles and nanostructured lipid carriers--innovative generations of solid lipid carriers. *Current Drug Delivery* 5 (4): 324–331. https://doi.org/10.2174/156720108785915087.

Shubhika, K. (2013). Nanotechnology and medicine – the upside and the downside. *International Journal of Drug Development and Research* 5 (1): 1–10.

Singh, A., Laux, P., Luch, A. et al. (2019). Review of emerging concepts in nanotoxicology: opportunities and challenges for safer nanomaterial design. *Toxicology Mechanisms and Methods* 29 (5): 378–387. https://doi.org/10.1080/15376516.2019.1566425.

Sukhanova, A., Bozrova, S., Sokolov, P. et al. (2018). Dependence of nanoparticle toxicity on their physical and chemical properties. *Nanoscale Research Letters* 13 (1): 1–21. https://doi.org/10.1186/s11671-018-2457-x.

Teleanu, D., Negut, I., Grumezescu, V. et al. (2019). Nanomaterials for drug delivery to the central nervous system. *Nanomaterials (Basel)* 9 (3): 1–18. https://doi.org/10.3390/nano9030371.

Trotta, F. and Mele, A. (2019). *Nanomaterials: Classification and Properties*, 1e. Germany: Wiley-VCHVerlag GmbH& Co. KGaA.

Utreja, P., Verma, S., Rahman, M., and Kumar, L. (2020). Use of nanoparticles in medicine. *Current Biochemical Engineering* 6 (1): 7–24. https://doi.org/10.2174/2212711906666190724145101.

Vanza, J., Patel, R., and Patel, M. (2020). Nanocarrier centered therapeutic approaches: recent developments with insight towards the future in the management of lung cancer. *Journal of Drug Delivery Science and Technology* 60: 1–22. https://doi.org/10.1016/j.jddst.2020.102070.

Ventola, C. (2017). Progress in nanomedicine: approved and investigational nanodrugs. *Pharmacy and Therapeutics (P&T)* 42 (12): 743–755.

Viswanath, B. and Kim, S. (2016). *Influence of Nanotoxicity on Human Health and Environment: The Alternative Strategies*, 2016/10/09e, vol. 242. Cham, Switzerland: Springer.

Warheit, D. and Sayes, C. (2015). Routes of exposure to nanoparticles: hazard tests related to portal entries. In: *Nanoengineering* (ed. P. Dolez), 41–54. United Kingdom: Elsevier.

Wibowo, D., Jorritsma, S., Gonzaga, Z.J. et al. (2020). Polymeric nanoparticle vaccines to combat emerging and pandemic threats. *Biomaterials* 268: 120597. https://doi.org/10.1016/j.biomaterials.2020.120597.

Wolfram, J., Zhu, M., Yang, Y. et al. (2015). Safety of nanoparticles in medicine. *Current Drug Targets* 16 (14): 1671–1681. https://doi.org/10.2174/1389450115666140804124808.

Wu, L., Wang, D., and Li, Z. (2020). Grand challenges in nanomedicine. *Materials Science and Engineering C* 106: 1–7. https://doi.org/10.1016/j.msec.2019.110302.

Zeb, A., Rana, I., Choi, H.I. et al. (2020). Potential and applications of nanocarriers for efficient delivery of biopharmaceuticals. *Pharmaceutics* 12 (12): 1184. https://doi.org/10.3390/pharmaceutics12121184.

Zhao, J. and Castranova, V. (2011). Toxicology of nanomaterials used in nanomedicine. *Journal of Toxicology and Environmental Health, Part B Critical Reviews* 14 (8): 593–632. https://doi.org/10.1080/10937404.2011.615113.

2

Microbial Biopolymers and Their Derivatives as Nanotechnological Tools for Medicine

Applications, Advantages, Toxicity, and Safety

Paulo Ricardo Franco Marcelino, Fernanda Gonçalves, Nayelen Sayuri Aizawa, Henrique Paiva Pereira, Talita Martins Lacerda, and Silvio Silvério da Silva

Department of Biotechnology, Engineering School of Lorena (EEL), São Paulo University (USP), Lorena, São Paulo, Brazil

2.1 Introduction

In the beginning of the twentieth century, Hermann Staudinger formulated the hypothesis of the existence of very large macromolecules with high molecular weights (Mülhaupt 2004). This hypothesis was verified experimentally in the 1920s, when Theodor Svedberg and Lawrence Bragg proved that hemoglobin and cellulose consisted in macromolecules (Rånby 1995). The acceptance of the existence of macromolecules allowed the development of a myriad of polymeric materials, such as plastics, rubbers, paints, and varnishes, that are now part of our daily lives. In addition to intentional discoveries, such as nylon, polyesters, and isotactic polypropylene, there were also accidental discoveries, such as polyethylene and polytetrafluoroethylene. Today, new and interesting macromolecules are being created to obtain new mechanical, optical, and electrical properties.

The etymology of the word polymers comes from the Greek (*poly* meaning "many, several" and *mers* meaning "parts"), as the macromolecular architectures originate through the connection of several units of small molecules called monomers. Polymers are generally classified into synthetic and natural. Synthetic polymers are more frequent in our daily lives, being applied in many industrial sectors, agriculture, and services. These materials are of petrochemical origin (Figure 2.1), which represents about half of the chemical industry worldwide. The manufacture and transformation of petroleum into polymers guarantee the employment and support of millions of people, but also generate uncountable amounts of waste. It is estimated that 25 million tons of plastics are annually accumulated in the environment and due to the slow degradation rates, they remain unchanged for hundreds of years. The conversion of large pieces into smaller particles is an initial step of the degradation processes, which may cause contamination problems that directly impact the environment and health. In addition, polymers derived from fossil fuels put excessive pressure on nonrenewable energy sources (Andrady and Neal 2009; Rhodes 2018).

With sustainable development and due to the alarming pollution caused by nonbiodegradable

Nanotechnology in Medicine: Toxicity and Safety, First Edition. Edited by Mahendra Rai, Mrunali Patel, and Rashmin Patel.

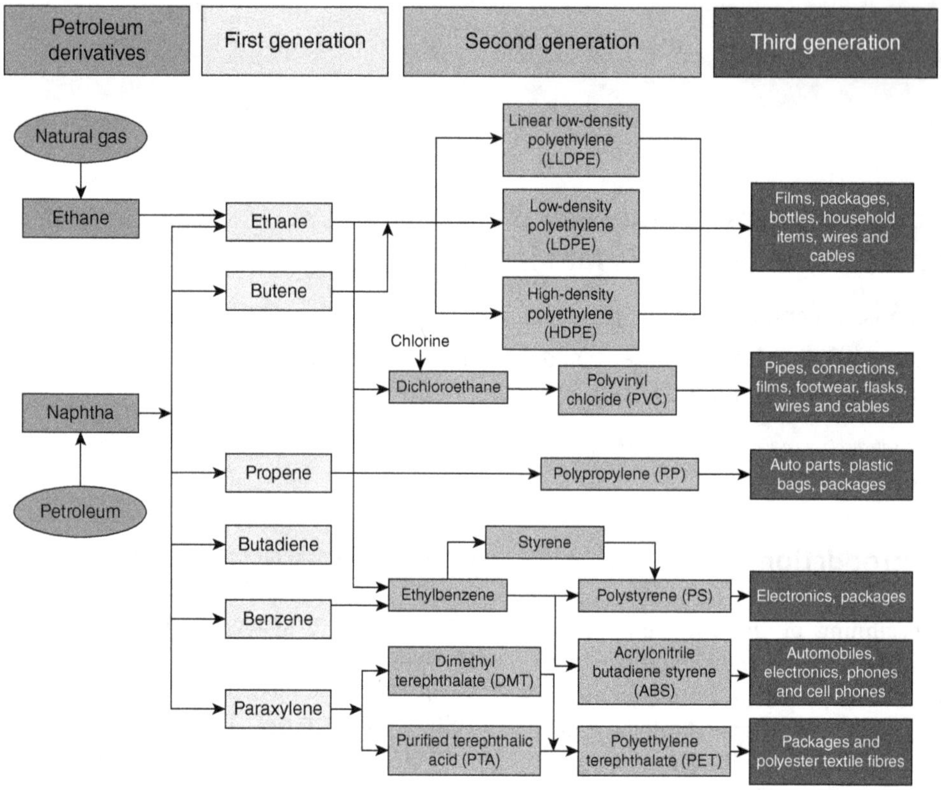

Figure 2.1 Scheme of the productive chain of fossil-based polymers. *Source:* Based on Olivatto (2017).

materials, research institutes and companies from around the world have increasingly invested in the development and application of natural polymers to reach similar performances as those of synthetic origin.

The advancement of nanotechnology for the design of several biotechnological devices such as nanoparticles, nanofilms, and liposomes based on natural polymers has gained great scientific importance due to their ecofriendly properties and biocompatibility with living systems. The nanotechnological devices based on natural polymers have been highlighted particularly in the development of pharmaceutical systems for drug delivery, tissue engineering, and bioactivation mechanisms. The low or no-toxicity and safety of natural polymers are essential characteristics for the use of these nanodevices in health, as will be discussed in this chapter.

2.2 Natural Polymers: Conceptualization, Classifications, and Physicochemical Characteristics

Natural polymers, also referred to as biopolymers, may have a natural occurrence (as is the case of chitin, cellulose, proteins, among others) or be produced by living organisms. The interest in such materials is progressively increasing due to their ecofriendly characteristics, such as low or nontoxicity, high biodegradability, and biocompatibility (Rendón-Villalobos et al. 2016). Biopolymers are generally classified into geopolymers, phytopolymers (plant polymers), zoopolymers (animal polymers), and microbial polymers. It should be noted that the concept of biopolymers does not cover all ecofriendly characteristics of natural polymers (Ogaji et al. 2012).

In recent years, vegetable and microbial biopolymers have gained attention due to their versatility, ecofriendly characteristics, and the possibility of sustainable and large-scale production. In addition to the environmental concern and the need to transition from the use of synthetic polymers to polymers from renewable resources, one must also consider geopolitical factors such as conflicts in the Middle East, Russia, and Venezuela, the main oil producers in the world (Mülhaupt 2012).

The most widely used polymers of plant and animal origins are cellulose and gelatin, respectively. Among microbial polymers, the most relevant today are the so-called exopolysaccharides, such as pullulan, curdlan, bacterial cellulose, lasiodiplodan, xanthan gum, and gellan gum, among others, and also intracellular polymers, such as microbial polyesters, for example polyhydroxyalkanoates (PHAs). Among microbial polymers, exopolysaccharides (EPSs) or extracellular polymers have been extensively studied and used for the most varied applications. EPSs are produced by some microorganisms and are found attached to the surface of cells or excreted into the extracellular medium, in the form of biofilms or slimes (Sutherland 1998). These biomolecules are usually associated with mechanisms of population cellular communication called *quorum sensing* (QS) that gives the community protection against various types of aggression, such as the lack of nutrients, or the presence of antimicrobials or biocidal chemical agents (Pyrog 2001; Gao et al. 2012; Nwodo et al. 2012; Gupta et al. 2019). In (Table 2.1) shows some examples of exopolysaccharides.

Polymers that are synthesized by classic organic routes, using bio-based molecules obtained by fermentation (building blocks) (Figure 2.2) as precursors, are also called biopolymers.

The most common examples of these polymers are poly (lactic acid) (PLA), poly (glycolic acid) (PGA), and poly (glycolic–lactic acid) (PGLA). This class of polymers has been widely used for biomedical applications, such as drug delivery in living organisms, fixators in surgeries (sutures, clips, bone pins) and special packaging. They consist of aliphatic polyesters that may be synthesized via esterification reactions (Figure 2.3), and the presence of ester moieties in the ensuing main backbone favors chemical, physical, and mainly biological degradation (Franchetti and Marconato 2006).

In addition to biodegradability, a hot topic due to the appeal for sustainable development and responsible use of polymers, biopolymers have been increasingly used due to their low toxicity and good biocompatibility. Nanotechnology has taken advantage of such properties for the development of tools for environmental, food, pharmaceutical, and medical applications. As previously mentioned, petroleum-based polymers are often chemically, physically, or biologically degraded into toxic compounds that may compromise the health of living organisms. In sequence, some examples of the toxicity of relevant synthetic polymers are presented.

- **ε-caprolactam** (ε-CAP) is a precursor of nylon-6, widely used in industry for the production of carpets, clothing, and automotive equipment, systems, components, connectors, and as additive to plastic packaging. ε-CAP waste can migrate from plastic packaging to food. Some toxicological studies indicate the possibility of ε-CAP causing eye and skin inflammation, as well as irritation in the respiratory system. Hypotension, tachycardia, palpitations, rhinorrhea, nasal dryness, genitourinary, and reproductive effects such as disorders in menstrual and ovarian functions, and complications in childbirth may also occur, in addition to neurological and hematological problems (Bomfim et al. 2009);
- **Epoxy polymers of bisphenol-A diglycidyl** ether **(DGEBA)** and aliphatic polyamine comonomers: **triethylenetetramine (TETA),**

Table 2.1 Microbial biopolymers of industrial relevance.

Biopolymer	Sources/micro-organism producers	Chemical structure	Applications	References
Xanthan gum	Bacteria from *Xanthomonas* genus.	Backbone of repeating sub-units, branched or not, consisting of 3–8 monosaccharides, such as D-glucose, D-mannose and D-glucuronic acid. Molecular weight: up to 2000 kDa.	Food Cosmetics Pharmaceutics	Jansson et al. (1975), Laws et al. (2001), Borges and Vendruscolo (2008), and Lopes et al. (2015)
Curdlan	Bacteria from *Agrobacterium* (previously taxonomically classified as *Alcaligenes*), *Rhizobium*, *Bacillus*, and *Cellulomonas* genus.	D-glucose units linked by β-(1→3) glycosidic bounds. Molecular weight: up to 2000 kDa.	Food Cosmetics Pharmaceutics	Tabernero et al. (2019)
Pullulan	Yeasts of *Aureobasidium* genus.	Maltotriose trimer made up of α-(1 → 6)-linked (1 → 4)-α-D-triglucosides. Molecular weight: up to 1000 kDa.	Foods Cosmetics Pharmaceutics	Singh et al. (2008)
Gellan gum	Bacteria from *Sphingomonas* (formerly known as *Pseudomonas elodea*).	Linear tetrasaccharide repeat unit consisting of (1→4)-L-rhamnose-(α-1→3)-D-glucose-(β-1→4)-D-glucose-(β-1→4)-D-glucose-(β-1→). Molecular weight: ≈500 kDa.	Food Cosmetics	Morris et al. (2012)
Emulsan	*Acinetobacter calcoaceticus* and *Acinetobacter venetianus*.	Polysaccharide backbone (three amino sugars in the ratio 1 : 1 : 1) with *O*-acyl and *N*-acyl fatty acid side chains (10–20 carbons representing 5–23% [w/w] of the polymer). The amino groups are either acetylated or covalently linked by an amide bond to 3-hydroxybutyric acid. Molecular weight: ≈1000 kDa.	Petrochemistry Cosmetics	Belsky et al. (1979), Gorkovenko et al. (1997), and Panilaitis et al. (2007)

Alginate	Algae (*Laminaria* and *Ascophyllum*) genus.	α-L-guluronic and β-D-mannuronic acid residues linked by β-(1–4) glycosidic bound. Molecular weight: 10–600 kDa.	Food Pharmaceutics Cosmetics	FAO (n.d.) http://www.fao.org/3/y4765e/y4765e07.htm and Liang et al. (2015)
Botriospheran	Filamentous fungus of *Botryosphaeria rhodina* species and some bacterial isolates.	D-glucose residues bonded by a β-(1 → 3;1 → 6). Molecular weight: ≈4875 kDa.	Cosmetics	Selbmann et al. (2003) and Silva et al. (2008)
Bacterial cellulose	Species of nonpathogenic aerobic bacteria, mainly belonging to the genera *Gluconacetobacter*, *Alcaligenes*, *Rhizobium*, *Agrobacterium*, and *Sarcina*.	β-D-glucopyranose units linked to each other by β-(1→4) glycosidic bonds.	Food Pharmaceutics Cosmetics Paper	Iguchi et al. (2000), Belgacem and Gandini (2008), and Klemm et al. (2005)
Chitosan	Filamentous fungi from *Mucorales* genera and some bacteria.	Molecular weight: 100–300 kDa.	Food Pharmaceutics Cosmetics	Kaur et al. (2012), Campos-Takaki et al. (2014), and Lecointe et al. (2019)

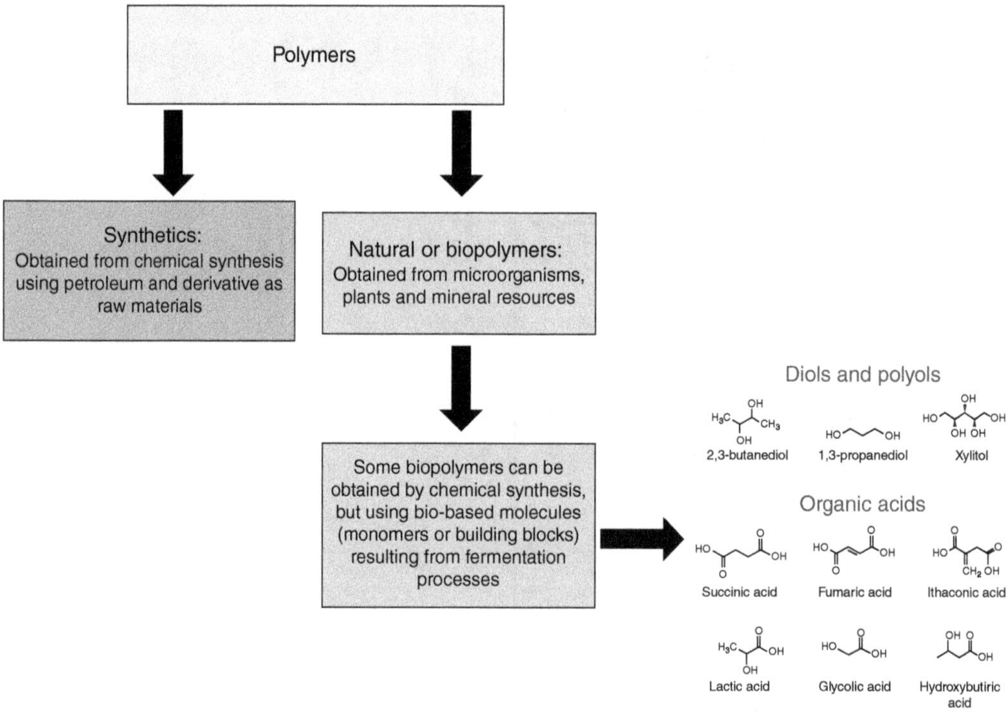

Figure 2.2 Biopolymers and building blocks obtained by fermentation.

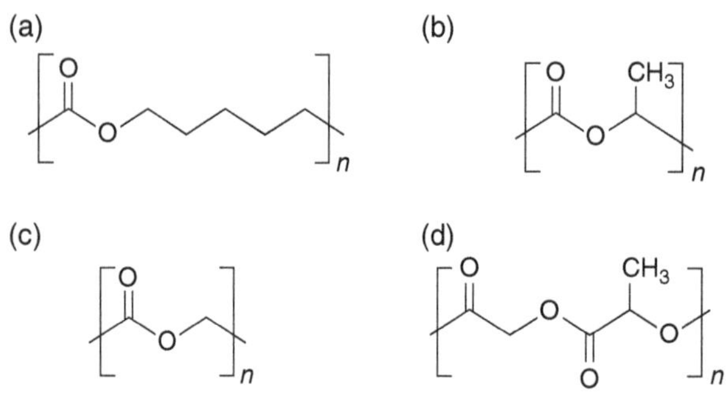

Figure 2.3 Chemical structure of (a) poly(ε-caprolactone) (PCL), (b) poly(lactic acid) (PLA), (c) poly(glycolic acid) (PGA), and (d) poly(glycolic–lactic acid) (PGLA). *Source:* Franchetti and Marconato (2006).

1-(2-aminoethyl)piperazine (AEP) and isophorone diamine (IPD) had their interactions with biological systems tested *in vitro*. Although the results show that DGEBA-IPD and DGEBA-AEP are hemocompatible and polymers based on the IPD system are not considered cytotoxic, protein adsorption tests showed that the surface of the polymers adsorbs human albumin (González Garcia et al. 2009);

- **Methyl methacrylate (MMA)**, a polymer often prepared using benzoyl peroxide as an initiator, is commonly used in the production of dental prostheses, and its toxicity cannot be neglected. When the polymerization reaction is incomplete, monomer residues in the dental prosthesis dissolve in the oral cavity and, according to reports, may cause irritations, allergies, and inflammatory and infectious processes such as stomatitis and cheilitis (Kanzaki et al. 1989; Helton and Storrs 1994; Gebhardt and Geier 1996);
- In an analysis of the acute toxicity of **polypropylene, polyethylene, and polyvinyl**

chloride (PVC), or those composed of hazardous monomers (e.g. acrylonitrile-butadiene-styrene [ABS], and epoxides) which are extensively used, it was observed that after leaching processes there is the release of hydrophobic compounds that are toxic to the environment and living organisms (Lithner et al. 2012).

2.3 Applications of Biopolymers in Nanoparticles, Nanofibers, and Drug Delivery Systems of Therapeutic Importance

Due to the toxicity of synthetic polymers derived from petroleum and the problems they can cause to the environment and the health of organisms, biopolymers have been widely used in the development of therapeutic nanotechnological tools. In addition, some biopolymers, such as chitin and its derivatives, also exhibit biological properties, such as antitumor and antimicrobial activities, which can be enhanced through nanotechnological strategies (Adhikari and Yadav 2018).

Table 2.2 presents recently reported nanotechnological devices (nanoparticles [NPs], nanofilms, nanofibers) based on biopolymers with antitumor and antimicrobial effects that have been developed for the design of tissue prostheses and in tests for the detection of various diseases.

As indicated in Table 2.2, the biopolymers in the nanocarrier systems contain hydroxyl, carboxyl, and amine groups that have good intermolecular interaction with biological surfaces and biological fluid molecules. In addition, some low-molecular weight biopolymers are necessary for the stabilization in nanosystems and even interaction with blood serum proteins which easily adhere on positively charged surfaces and cause aggregation and opsonization. These convenient physicochemical and biological characteristics are the reason why biopolymers of microbial origin are considered great alternatives for the pharmaceutical industry, mainly to the design of new drug delivery systems (Jacob et al. 2018). Some antitumor drugs have limited solubility in aqueous media, which compromises their distribution in the body and sometimes their antitumor activity. As noted in Table 2.2, the use of nanocarrier systems based on biopolymers improves absorption and distribution as well as enhances the activity of antitumor agents in the body (Eroglu et al. 2017). In the case of antibiotics, the use of nanocarrier systems improves pharmacokinetics and biodistribution aspects, decreases toxicity, enhances the antibacterial activity, and improves target selectivity (Drulis-Kawa and Dorotkiewicz-Jach 2010; Alhariri et al. 2013). The use of modified biopolymers is also very important for the development of drug delivery systems for antitumor and antimicrobial applications. Modifications of polysaccharides have followed different approaches (Efthimiadou et al. 2014):

- Association with synthetic biopolymers.
- Surface coating of micro- or nanosphere polysaccharides with biocompatible synthetic polymers.
- Cross-linking with different types of reagents.
- Increased hydrophobicity via alkylation reactions.

The main modification reactions that can be performed on polysaccharides are methylation, acetylation, phosphorylation, silylation, sulfation, carboxymethylation, and amination (Huang et al. 2016). Some studies described the conjugation of polysaccharides with molecules of low molar weight and some active ingredients. The so-called derivatization of polysaccharides is an important strategy to improve the physicochemical properties of the materials and to introduce them into microbial and tumor cells ("Trojan horses").

The monographs reported by Huang et al. (2016), Adhikari and Yadav (2018), and

Table 2.2 Antitumoral nanoparticles (NPs), nanofibers and nanofilms/antimicrobial nanoparticles (NPs), nanofibers and nanofilms/tissue engineered nanofibers and nanofilms.

Nanomaterials	Actions	References
Bicalutamide folate conjugated-chitosan functionalized PLGA NPs	NPs presented potential action against prostate cancer	Dhas et al. (2015)
Tumor-targeting delivery system of paclitaxel by PEGylated *O*-carboxymethyl-chitosan NPs grafted with cyclic Arg-Gly-Asp peptide	NPs presented potential action against Lewis lung carcinoma	Lv et al. (2012)
PEGylated chitosan nanocapsules conjugated to a monoclonal antibody anti-TMEFF-2 targeted delivery of docetaxel	Nanocapsules presented potential action against non-small cell lung carcinoma	Torrecilla et al. (2013)
Doxorubicin hyaluronic acid block copolymers NPs	NPs used as a self-targeting drug delivery in overexpressed CD44 glycoprotein cells of breast cancer	Nitta and Numata (2013)
Indocyanine green-Levan NPs	Self-assembled indocyanine green-Levan NPs used for targeted breast cancer imaging	Kim et al. (2015)
Doxorubicin-carboxymethyl xanthan gum-capped gold NPs	NPs showed *in vitro* efficacy against glioblastoma cells	Alle et al. (2019)
Doxorubicin-loaded alginic acid/poly[2-(diethylamino)ethyl methacrylate] NPs	NPs exhibited greater *in vivo* antitumoral activity compared to free doxorubicin	Cheng et al. (2012)
Cationized dextran and pullulan modified with diethyl aminoethyl methacrylate (DEAEM)	Positively charged nanocomplexes with DNA and they are cytocompatible in C6, HeLa and L929 cells. Transfection efficiency of the vectors was evaluated using p53 plasmid, which demonstrated good transfection in isolated cancer cells (C6 and HeLa)	Sherly et al. (2020)
Alginate nanogel encapsulating *Artemisia ciniformis* extract	Nanogels loaded with *A. ciniformis* extract inhibited cell proliferation and arrested the cell cycle at the G0/G1 phase. Induction of apoptosis occurred in a time- and dose-dependent manner; expression levels of pro-apoptotic genes were up-regulated; down-regulation of anti-apoptotic and metastatic genes were detected; nanogels exhibited potent anticancer activity against AGS gastric cancer cells	Rahimivand et al. (2020)
Bacterial cellulose + Fe_3O_4 NPs	Magnetic materials made from tetraaza macrocyclic Schiff base bacterial cellulose ligands with magnetite nanoparticles (Fe_3O_4 NPs) effectively inhibited the growth of the CT26 tumor models in BALB/c mice	Chaabane et al. (2020)
Clindamycin phosphate-xanthan gum-ZnO NPs	NPs applied as topical anti-inflammatory drug carrier for acne treatment	Karakuş (2019)
Chitin	Silver NPs associated with chitin/Ag nanofibers presented strong antimicrobial activity against *Escherichia coli*, *Pseudomonas aeruginosa*, and *Influenza A* virus	Park and Kim (2015)
Polycaprolactone(PCL)/gelatin + terbinafine hydrochloride (TFH)	Nanofibers presented antifungal activity against *Trichophyton mentagrophytes*, *Aspergillus fumigatus* and *Candida albicans*	Paskiabi et al. (2017)

Table 2.2 (Continued)

Nanomaterials	Actions	References
Chitosan + gentamicin loaded liposome	Liposome presented antibacterial activity against *E. coli*, *P. aeruginosa* and *Staphylococcus aureus*	Monteiro et al. (2015)
Chitosan (CS)/poly(vinyl alcohol) (PVA) + silver nanoparticles	CS/PVA nanofibers containing Ag NPs showed high antibacterial activity against *E. coli*	Nguyen et al. (2011)
Poly(D,L-lactic acid-co-glycolic acid) (PLGA) + fusidic acid (FA) and rifampicin (RIF) nanofibers	Dual-loaded nanofibers exhibited *in vitro* antimicrobial activity against two strains of methicillin-resistant *S. aureus* (MRSA) and *Staphylococcus epidermidis*	Rho et al. (2006)
Zinc ions (Zn^{2+})-loaded 2,2,6,6-tetramethylpiperidine-1-oxyl oxidized bacterial cellulose (TOBC) nanofiber-reinforced biomimetic calcium alginate hydrogel	Calcium alginate/TOBC biomimetic hydrogels loaded with Zn^{2+} exhibited good mechanical, antimicrobial, and biological properties at Zn^{2+} concentration of 0.0001 wt%.	Zhang et al. (2019)
Polyhydroxybutyrate/ poly(butyleneadipate-co-terephthalate) (PHB/PBAT)-based biodegradable antibacterial hydrophobic nanofibrous membranes	Nanofibrous membranes were tested against *E. coli* and *S. aureus* and presented good antimicrobial activity with 6.08 and 5.78 log reduction, respectively	Lin et al. (2017)
Chitosan nanofiber	Biocompatible chitosan nanofiber membranes used for bone regeneration in rabbit calvarial defects with healing effect and no evidence of inflammatory reaction	Shin et al. (2005)
Collagen/elastin nanofibers	Nanofibers coated with ECM proteins with good potential for wound dressing and scaffolds for tissue engineering	Rho et al. (2006)
Poly (L-lactide-co-glycolide) (PLGA)/ chitin nanofibers	Biodegradable electrospun nanofibers of PLGA and chitin presented cell adhesion and spreading for normal human keratinocytes, and were good matrices for normal human fibroblasts	Min et al. (2004)
Poly (L-lactide-co-glycolide) (PLGA)/ dextran nanofibers	Nanofibers based on blends of dextran and PLGA were tested in terms of interaction with dermal fibroblasts considering cell viability, proliferation, attachment, migration, extracellular matrix deposition, and cytoskeleton organization, and the functional gene expressions were characterized, scaffolds with good potential to enhance the healing of chronic or trauma wounds	Pan et al. (2006)
Poly(ε-caprolactone) (PCL)/ gelatin + metronidazole	Metronidazole was loaded in PCL/gelatine, and a sustained release was observed and significantly prevented anaerobic bacteria colonization; cytocompatibility for drug concentrations up to 30%	Xue et al. (2014)
Chitosan-alginate nanofibers + gentamicin	Chitosan-alginate nanofibers with 1–3% wt gentamicin significantly enhanced skin regeneration in mice model by stimulating the formation of a thicker dermis, increasing collagen deposition, and increasing the formation of new blood vessels and hair follicles	Bakhsheshi-Rad et al. (2019)

Gorgieva (2020) describe examples of derivatization of some microbial polysaccharides for medical and other applications. 2-phenylhydrazine or hydrazine thiosemicarbazone chitosan, chitosan–metal complexes (chitosan–Cu [II] and chitosan salicylaldehyde Schiff base–Zn [II] complexes), carboxymethyl chitosan, chitosan–thymine conjugate, sulfated chitosan and sulfated benzaldehyde chitosan, glycol-chitosan and *N*-succnyl chitosan, furanoallocolchicinoid chitosan conjugates and polypyrrole chitosan present antitumoral activity. These chitosan derivatives can act inducing apoptosis of tumor cells, affecting the cycle of tumor cells, enhancing the antioxidant activity of organism, activating the body's immune response and inhibiting the tumor angiogenesis. The chemical modifications of bacterial cellulose include oxidation, etherification, esterification, carbamation, and amidation reactions, all resulting in the formation of reactive functional and charged groups, such as sulfate, carboxyl, aldehyde, phosphate, amino, and thiol groups. Among all, the acetylated bacterial cellulose attracts significant interest as noncytotoxic material for cosmetic products, disinfectants, as well as a platform for further coupling (e.g. with drugs and other bioactives), for use in drug delivery. The studies of Zayed et al. (2017) and Tao et al. (2018) described the use of other modified microbial biopolymers to serve as carriers for antimicrobial and antitumoral drugs.

2.4 Safety of Microbial Biopolymers Used in Nanoscale-Systems for Therapeutic Applications

Several synthetic polymers are produced from aromatic monomers, mainly bearing a benzene ring. This characteristic often makes them more toxic and less biocompatible, which restrict their applications as biomaterials. The metabolism of benzene (and also some of its derivatives) in mammals and in microorganisms is a subject of study since decades, but its biotransformation and the mechanisms that lead to toxicity are still not fully understood. In mammals and in some microorganisms, the metabolism of benzene and its derivatives occurs through the family of cytochrome P450 (CYP) enzymes, responsible for the insertion of oxygen atoms in these lipophilic compounds aiming at increasing their solubility in aqueous medium (Santos et al. 2017).

In humans, benzene is metabolized mainly in the liver (primary metabolism) and subsequently in the bone marrow (secondary metabolism) (Dougherty et al. 2008; Barata-Silva et al. 2014). In the first stage of benzene metabolization, the action of a monooxygenase of the cytochrome P450 complex (CYP) is observed, leading to the formation of benzene oxides (epoxy benzene or oxepin), considered reactive. Benzene oxides can covalently bind to proteins and DNA, forming precancerous adducts, or can be metabolized to muconaldehydes and benzoquinones, which are also reactive compounds. The formation of free radicals is also possible, leading to oxidative stress (Figure 2.4) (Chaney and Carlson 1995; Ross 1996; Snyder and Hedli 1996; Monks et al. 2010; Moro et al. 2013; Barata-Silva et al. 2014).

In later stages of benzene metabolism, the formation of phenolic compounds such as phenol, catechol, and hydroquinone is observed. The toxicity of these compounds, in particular, of hydroquinones, can be highlighted, since they are precursors of myelotoxic compounds and inhibitors of the ribonucleotide reductase, essential for the DNA biosynthesis. Benzoquinones and benzene triol, formed in the bone marrow, can conjugate with sulfates and glucuronic acids, and are capable of forming adducts with DNA generating mutations favoring carcinogenesis processes (Figure 2.4) (Loureiro et al. 2002; Poirier 2004; Hartwig 2010; Monks et al. 2010; McHale et al. 2012; Snyder 2012; Barata-Silva et al. 2014).

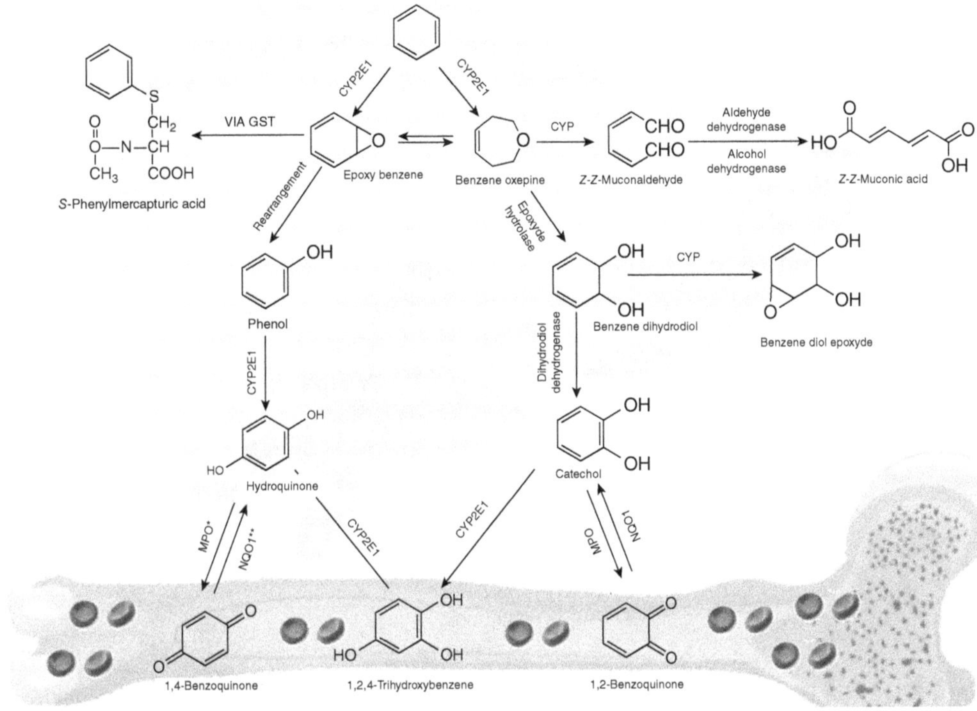

Figure 2.4 Main pathways of the metabolism of benzene and its derivatives in humans. *MPO, Myeloperoxidase; **NQO1, Quinone oxidoreductase 1 NADPH dependent. *Sources:* Based on Chaney and Carlson (1995), Ross (1996), Monks et al. (2010) and Moro et al. (2013).

The biotransformation reactions of benzene (via muconic acid and phenol) are explained by the need for obtaining water-soluble compounds for urinary excretion. Human exposure to benzene, in concentrations between 0.1 and 10 ppm, results in urinary metabolite profiles with 70–85% phenol, 5–10% hydroquinone, 5–10% trans, trans-muconic acid and catechol, and less than 1% *S*-phenylmercapturic acid (Qu et al. 2002; Kim et al. 2006; Barata-Silva et al. 2014). This is a clear indication of the toxicity of oxygenated compounds derived from benzene.

As for the microbial metabolism of benzene and its derivatives, the anaerobic pathway is a slow process and its biochemical mechanism has not been fully described (Coates et al. 1996). In general, the metabolic pathways of hydrocarbon degradation involve an aerobic metabolism carried out by bacteria, lignin-degrading fungi (lignolytics), and non-lignolytic fungi (Jacques et al. 2007).

In some fungi, the metabolism of benzene and its derivatives occurs in a similar way as humans, using the same metabolic reactions (via muconic acid and phenol) (Figure 2.5). However, the phenolic compounds originated here are conjugated with *O*-glycosides and *O*-glucoronides (Cerniglia 1984). The reactions occurring in any of these phases can be considered as detoxification measures if the products generated are less toxic than the original hydrocarbons (Cerniglia et al. 1985).

Lignolytic fungi have alternative pathways for the degradation of aromatic compounds. This degradation occurs through the action of extracellular enzymes of low specificity, such as lignin peroxidase, manganese peroxidase, and laccase (Leonowicz et al. 1999). In this process, benzene and its derivatives are

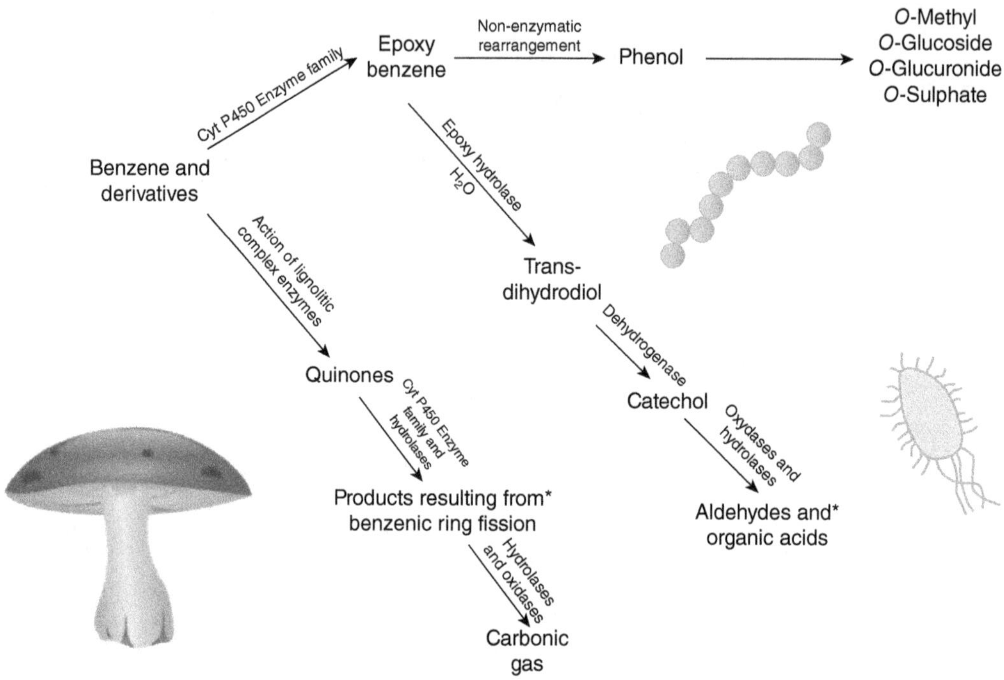

Figure 2.5 Aerobic biodegradation pathways of aromatic compounds conducted by bacteria and fungi. *Source:* Based on Cerniglia (1984, 1997). *, Products resulting from reactions of the Krebs cycle.

degraded, with the formation of quinones (Cerniglia 1997). Some lignolytic fungi have the ability to metabolize quinone PAHs by cleaving aromatic rings with subsequent breaking and formation of carbon dioxide (through the Krebs cycle reactions) (Hammel 1995) (Figure 2.5).

Understanding the microbial degradation of benzenes and other aromatic compounds present in synthetic polymers is very important considering that these materials will be disposed, and biodegradation can generate compounds that are even more toxic or that may impact other living organisms of the environment.

The metabolism of biopolymers such as exopolysaccharides (EPSs) and polyesters does not generate toxic and reactive compounds, as is the case of aromatic/synthetic polymers. The metabolization of an EPS or microbial polyesters leads to the formation of organic acids, most often found in the Krebs cycle (Eggers and Steinbuchel 2013). Both humans and capable microorganisms can degrade biopolymers by well-known metabolic pathways, such as glycolysis, respiratory chain, Lynen cycle (β-oxidation of fatty acids), among others. The metabolism of biopolymers is generally complete, leading to carbon dioxide and water, but if it is not, it generates compounds of low environmental toxicity.

2.5 Conclusions

The exploitation of microbial polymers for the design of novel therapeutic systems has evolved radically during the last decades, as they correspond to tunable, nontoxic, biodegradable, and biocompatible materials. Processing such materials in the form of nanofibers and nanoparticles is a very promising strategy that may provide more effective and convenient routes of administration, leading to sophisticated macromolecular materials for pharmaceutical and biomedical purposes. The applications of nanomaterials based on microbial polymers in biomedicine have become a large subject area, with original materials continuously being described in literature. As a consequence, commercial biomedical nanosystems based on exopolysaccharides (mainly pullulan and bacterial cellulose) and polyhydroxyalkanoates (mainly poly[hydroxybutyrate]), are expected to appear more frequently, as they may offer excellent prospects.

References

Adhikari, H.S. and Yadav, P.N. (2018). Anticancer activity of chitosan, chitosan derivatives, and their mechanism of action. *International Journal of Biomaterials* 2018: 2952085. PMID: 30693034. https://doi.org/10.1155/2018/2952085.

Alhariri, M., Azghani, A., and Omri, A. (2013). Liposomal antibiotics for the treatment of infectious diseases. *Expert Opinion on Drug Delivery* 10: 1515–1532. https://doi.org/10.1517/17425247.2013.822860.

Alle, M., Reddy, G.B., Kim, T.H. et al. (2019). Doxorubicin-carboxymethyl xanthan gum capped gold nanoparticles: microwave synthesis, characterization, and anti-cancer activity. *Carbohydrate Polymers* 115511 https://doi.org/10.1016/j.carbpol.2019.115511.

Andrady, A.L. and Neal, M.A. (2009). Applications and societal benefits of plastics. *Philosophical Transactions of the Royal Society B: Biological Sciences* 364: 1977–1984.

Bakhsheshi-Rad, H.R., Hadisi, Z., Ismail, A.F. et al. (2019). In vitro and in vivo evaluation of chitosan-alginate/gentamicin wound dressing nanofibrous with high antibacterial performance. *Polymer Testing* 106298 https://doi.org/10.1016/j.polymertesting.2019.106298.

Barata-Silva, C., Mitri, S., Pavesi, T. et al. (2014). Benzeno: reflexos sobre a saúde pública, presença ambiental e indicadores biológicos utilizados para a determinação da exposição. *Cadernos Saúde Coletiva* 22: 329–342. https://doi.org/10.1590/1414-462X201400040006.

Belgacem, M.N. and Gandini, A. (2008). *Monomers, Polymers and Composites from Renewable Resources*. Elsevier.

Belsky, I., Gutnick, D., L., and Rosenberg, E. (1979). Emulsifier of *Arthrobacter* RAG-1: determination of emulsifier-bound fatty acids. *FEBS Letters* 101: 175–178.

Bomfim, M.V.J., Abrantes, S.M.P., and Zamith, H.P.S. (2009). Estudos sobre a toxicologia da ε-caprolactama. *Brazilian Journal of Pharmaceutical Sciences* 45 (1): 21–35. https://doi.org/10.1590/S1984-82502009000100004.

Borges, C. and Vendruscolo, C. (2008). Xanthan gum: characteristics and operational conditions of production. *Semina: Ciências Biológicas e da Saúde* 29: 171–188. ISSN: 1676-5435.

Campos-Takaki, G.M., Dietrich, S.M.C., and Beakes, G.W. (2014). Cytochemistry, ultrastructure and X-ray microanalysis methods applied to cell wall characterization of *Mucoralean* fungi strains. In: *Microscopy: Advances in Scientific Research and Education* (ed. A. Mendez-Vilas), 121–127. Formatex.

Cerniglia, C.E. (1984). Microbial metabolism of polycyclic aromatic hydrocarbons. *Advances in Applied Microbiology* 30: 31–37.

Cerniglia, C.E. (1997). Fungal metabolism of polycyclic aromatic hydrocarbons: past, present and future applications in bioremediation. *Journal of Industrial Microbiology and Biotechnology* 19: 324–333.

Cerniglia, C.E., White, G.L., and Hettich, R.H. (1985). Fungal metabolism and detoxification of polycyclic aromatic hydrocarbons. *Archives of Microbiology* 143: 105–110.

Chaabane, L., Chahdoura, H., Mehdaoui, R. et al. (2020). Functionalization of developed bacterial cellulose with magnetite nanoparticles for nanobiotechnology and nanomedicine applications. *Carbohydrate Polymers* 116707 https://doi.org/10.1016/j.carbpol.2020.116707.

Chaney, A.M. and Carlson, G.P. (1995). Comparison of rat hepatic and pulmonary microsomal metabolism of benzene and the lack of benzene induced pneumotoxicity and hepatotoxicity. *Toxicology* 104: 53–62.

Cheng, Y., Yu, S., Wang, J. et al. (2012). *In vitro* and *in vivo* antitumor activity of doxorubicin-loaded alginic-acid-based nanoparticles. *Macromolecular Bioscience* 12 (10): 1326–1335. https://doi.org/10.1002/mabi.201200165.

Coates, J.D., Anderson, R.T., Woodward, J.C. et al. (1996). Anaerobic hydrocarbon degradation in petroleum contaminated harbor sediments under sulfate-reducing and artificially imposed iron-reducing conditions. *Environmental Science and Technology* 30: 2784–2789.

Dhas, N.L., Ige, P.P., and Kudarha, R.R. (2015). Design, optimization and *in-vitro* study of folic acid conjugated-chitosan functionalized PLGA nanoparticle for delivery of bicalutamide in prostate cancer. *Powder Technology* 283: 234–245.

Dougherty, D., Garte, S., Barchowsky, A. et al. (2008). NQO1, MPO, CYP2E1, GSTT1 and GSTM1 polymorphisms and biological effects of benzene exposure—a literature review. *Toxicology Letters* 182: 7–17.

Drulis-Kawa, Z. and Dorotkiewicz-Jach, A. (2010). Liposomes as delivery systems for antibiotics. *International Journal of Pharmaceutics* 387: 187–198.

Efthimiadou, E.K., Metaxa, A.F., and Kordas, G. (2014). Modified polysaccharides as drug delivery. *Polysaccharides*: 1–26. https://doi.org/10.1007/978-3-319-03751-6_23-1.

Eggers, J. and Steinbuchel, A. (2013). Poly(3-Hydroxybutyrate) degradation in *Ralstonia eutropha* H16 is mediated stereoselectively to (S)-3-hydroxybutyryl coenzyme a (CoA) via crotonyl-CoA. *Journal of Bacteriology* 195: 3213–3223. https://doi.org/10.1128/jb.00358-13.

Eroglu, M., Toksoy, O.E., Cansever, M.E., and Sennaroglu, B.M. (2017). Sugar based biopolymers in nanomedicine; new emerging era for cancer imaging and therapy. *Current Topics in Medicinal Chemistry* 17: 1507–1520.

FAO (n.d.). Seaweeds used as a source of alginate. http://www.fao.org/3/y4765e/y4765e07.htm (accessed 18 December 2020).

Franchetti, S.M.M. and Marconato, J.C. (2006). Polímeros biodegradáveis – uma solução parcial para diminuir a quantidade dos resíduos plásticos. *Química Nova* 29: 811–816.

Gao, M., Coggin, A., Yagnik, K., and Teplitski, M. (2012). Role of specific quorum-sensing signals in the regulation of exopolysaccharide II production within *Sinorhizobium meliloti* spreading colonies. *PLoS One* 7: e42611. https://doi.org/10.1371/jornal.pone.0042611.

Garcia, F.G., Leyva, M.E., Queiroz, A.A.A., and Higa, O. (2009). Novas tendências dos polímeros epoxídicos: propriedades biológicas in vitro de formulações para aplicações médicas. *Polímeros* 19: 177–182. https://doi.org/10.1590/S0104-14282009000300004.

Gebhardt, M. and Geier, J. (1996). Evaluation of patch test results with denture material series. *Contact Dermatitis* 34: 191–195.

Gorgieva, S. (2020). Bacterial cellulose as a versatile platform for research and development of biomedical materials. *Processes* 8: 624.

Gorkovenko, A., Zhang, J., Gross, R., A. et al. (1997). Bioengineering of emulsifier structure: emulsan analogs. *Canadian Journal of Microbiology* 43: 384–390.

Gupta, P., Pruthi, P.A., and Pruthi, V. (2019). Role of exopolysaccharides in biofilm formation. *Introduction to Biofilm Engineering*: 17–57. https://doi.org/10.1021/bk-2019-1323.ch002.

Hammel, K.E. (1995). Mechanism for polycyclic aromatic hydrocarbon degradation by ligninolytic fungi. *Environmental Health Perspectives* 103: 41–43.

Hartwig, A. (2010). The role of DNA repair in benzene-induced carcinogenesis. *Chemico-Biological Interactions* 184: 269–272.

Helton, J. and Storrs, F. (1994). The burning mouth syndrome: lack of a role for contact urticaria and contact dermatitis. *Journal of the American Academy of Dermatology* 31: 201–205.

Huang, G., Chen, X., and Huang, H. (2016). Chemical modifications and biological activities of polysaccharides. *Current Drug Targets* 17: 1799–1803. 27138762. https://doi.org/10.2174/1389450117666160502151004.

Iguchi, M., Yamanaka, S., and Budhiono, A. (2000). Bacterial cellulose – a masterpiece of nature's arts. *Journal of Materials Science* 35: 261–270.

Jacob, J., Haponiuk, J.T., Thomas, S., and Gopi, S. (2018). Biopolymer based nanomaterials in drug delivery systems: a review. *Materials Today Chemistry* 9: 43–55. https://doi.org/10.1016/j.mtchem.2018.05.002.

Jacques, R.J.S., Bento, F.M., Antoniolli, Z.I., and Camargo, F.A.O. (2007). Biorremediação de solos contaminados com hidrocarbonetos aromáticos policíclicos. *Ciência Rural* 37: 1192–1201.

Jansson, P., Kenne, L., and Lindberg, B. (1975). Structure of the extracellular polysaccharide from *Xanthomonas campestris*. *Carbohydrate Research* 45: 275–282. https://doi.org/10.1016/S0008-6215(00)85885-1.

Kanzaki, T., Kabasawa, Y., Jinno, T., and Isayama, K. (1989). *Contact stomatitis* due to methyl methacrylate monomer. *Contact Dermatitis* 20: 146–148. https://doi.org/10.1111/j.1600-0536.1989.tb03125.x.

Karakuş, S. (2019). A novel ZnO nanoparticle as drug nanocarrier in therapeutic applications: kinetic models and error analysis. *Journal of the Turkish Chemical Society Section A: Chemistry* 6: 119–132.

Kaur, K., Dattajirao, V., Shrivastava, V., and Bhardwaj, U. (2012). Isolation and characterization of chitosan-producing bacteria from beaches of Chennai, India. *Enzyme Research* 421683: 6.

Kim, S., Vermeulen, R., Waidyanatha, S. et al. (2006). Modeling human metabolism of benzene following occupational and environmental exposures. *Cancer Epidemiology, Biomarkers and Prevention* 15: 2246–2252.

Kim, S.J., Bae, P.K., and Chung, B.H. (2015). Self-assembled Levan nanoparticles for targeted breast cancer imaging. *Chemical Communications* 51 (1): 107–110. https://doi.org/10.1039/c4cc07679f.

Klemm, D., Heublein, B., Fink, H.P., and Bohn, A. (2005). Cellulose: fascinating biopolymer and sustainable raw material. *Angewandte Chemie International Edition* 44: 3358–3393.

Laws, A., Gu, Y., and Marshall, V. (2001). Biosynthesis, characterization, and design of bacterial exopolysaccharides from lactic acid bacteria. *Biotechnology Advances* 19: 597–625. https://doi.org/10.1016/S0734-9750(01)00084-2.

Lecointe, K., Cornu, M., Leroy, J. et al. (2019). Polysaccharides cell wall architecture of *Mucorales*. *Frontiers in Microbiology* 10: 469. https://doi.org/10.3389/fmicb.2019.00469.

Leonowicz, A., Matuszewska, M., Luterek, J. et al. (1999). Biodegradation of lignin by white rot fungi. *Fungal Genomics and Biology* 27: 173–183.

Liang, Y., Kashdan, T., Sterner, C. et al. (2015). Algal biorefineries. *Industrial Biorefineries and White Biotechnology* 1: 35–90.

Lin, X., Fan, X., Li, R. et al. (2017). Preparation and characterization of PHB/PBAT-based biodegradable antibacterial hydrophobic nanofibrous membranes. *Polymers for Advanced Technologies* 29 (1): 481–489. https://doi.org/10.1002/pat.4137.

Lithner, D., Nordensvan, I., and Dave, G. (2012). Comparative acute toxicity of leachates from plastic products made of polypropylene, polyethylene, PVC, acrylonitrile-butadiene-styrene, and epoxy to *Daphnia magna*. *Environmental Science and Pollution Research International* 19 (5): 1763–1772. https://doi.org/10.1007/s11356-011-0663-5.

Lopes, B.M., Lessa, V.L., Silva, B.M. et al. (2015). Xanthan gum: properties. Production conditions, quality and economic perspective. *Journal of Food and Nutrition Research* 54: 185–194.

Loureiro, A.P.M., Mascio, P.D., and Medeiros, M.H.G. (2002). Formação de adutos exocíclicos com bases de DNA: implicações em mutagênese e carcinogênese. *Química Nova* 25: 777–793.

Lv, P.P., Ma, Y.F., Yu, R. et al. (2012). Targeted delivery of insoluble cargo (paclitaxel) by PEGylated chitosan nanoparticles grafted with Arg-Gly-Asp (RGD). *Molecular Pharmaceutics* 4: 1736–1747. https://doi.org/10.1021/mp300051h. Epub 2012 May 15. PMID: 22559746.

McHale, C.M., Zhang, L., and Smith, M.T. (2012). Current understanding of the mechanism of benzene-induced leukemia in humans: implications for risk assessment. *Carcinogenesis* 33: 240–252.

Min, B.M., You, Y., Kim, J.M. et al. (2004). Formation of nanostructured poly(lactic-co-glycolic acid)/chitin matrix and its cellular response to normal human keratinocytes and fibroblasts. *Carbohydrate Polymers* 57: 285–292.

Monks, T.J., Butterworth, M., and Lau, S.S. (2010). The fate of benzene-oxide. *Chemico-Biological Interactions* 184: 201–206.

Monteiro, N., Martins, M., Martins, A. et al. (2015). Antibacterial activity of chitosan nanofiber meshes with liposomes immobilized releasing gentamicin. *Acta Biomaterialia* 18: 196–205. https://doi.org/10.1016/j.actbio.2015.02.018.

Moro, A.M., Charão, M.F., Brucker, N. et al. (2013). Genotoxicity and oxidative stress in gasoline station attendants. *Mutation Research* 754: 63–70.

Morris, E., R., Nishinari, K., and Rinaudo, M. (2012). Gelation of gellan – a review. *Food Hydrocolloids* 28 (2): 373–411.

Mülhaupt, R. (2004). Hermann Staudinger and the origin of macromolecular chemistry. *Angewandte Chemie International Edition* 43 (9): 1054–1063.

Mülhaupt, R. (2012). Green polymer chemistry and bio-based plastics: dreams and reality. *Macromolecular Chemistry and Physics* 214 (2): 159–174.

Nguyen, T.T.T., Tae, B., and Park, J.S. (2011). Synthesis and characterization of nanofiber webs of chitosan/poly(vinyl alcohol) blends incorporated with silver nanoparticles. *Journal of Materials Science* 46: 6528–6537. https://doi.org/10.1007/s10853-011-5599-0.

Nitta, S.K. and Numata, K. (2013). Biopolymer-based nanoparticles for drug/gene delivery and tissue engineering. *International Journal of Molecular Sciences* 14 (1): 1629–1654. https://doi.org/10.3390/ijms14011629.

Nwodo, U.U., Green, E., and Okoh, A.I. (2012). Bacterial exopolysaccharides: functionality and prospects. *International Journal of Molecular Sciences* 13 (11): 14002–14015. https://doi.org/10.3390/ijms131114002.

Ogaji, I.J., Nep, E.I., and Audu-Peter, J.D. (2012). Advances in natural polymers as pharmaceutical excipients. *Pharmaceutica Analytica Acta* 3: 146. https://doi.org/10.4172/2153-2435.1000146.

Olivatto, G.P. (2017). Estudo sobre microplásticos em águas superficiais na porção oeste da Baía de Guanabara. Dissertação de Mestrado. Pontifícia Universidade Católica do Rio de Janeiro.

Pan, H., Jiang, H., and Chen, W. (2006). Interaction of dermal fibroblasts with electrospun composite polymer scaffolds prepared from dextran and poly lactide-co-glycolide. *Biomaterials* 27 (17): 3209–3220. https://doi.org/10.1016/j.biomaterials.2006.01.032.

Panilaitis, B., Castro, G.R., Solaiman, D., and Kaplan, D.L. (2007). Biosynthesis of emulsan biopolymers from agro-based feedstocks. *Journal of Applied Microbiology* 102: 2.

Park, J.A. and Kim, S.B. (2015). Preparation and characterization of antimicrobial electrospun poly(vinyl alcohol) nanofibers containing benzyl triethylammonium chloride. *Reactive and Functional Polymers* 93: 30–37. https://doi.org/10.1016/j.reactfunctpolym.2015.05.008.

Paskiabi, F.A., Bonakdar, S., Shokrgozar, M.A. et al. (2017). Terbinafine-loaded wound dressing for chronic superficial fungal infections. *Materials Science and Engineering* 73: 130–136. https://doi.org/10.1016/j.msec.2016.12.078.

Poirier, M.C. (2004). Chemical-induced DNA damage and human cancer risk. *Nature Reviews Cancer* 4: 630–637.

Pyrog, T.P. (2001). Biologichni funktsiï mikrobnykh ekzopolisakharydiv [biological functions of microbial exopolysaccharides]. *Mikrobiolohichnyĭ Zhurnal* 63 (5): 80–101.

Qu, Q., Shore, R., Li, G. et al. (2002). Hematological changes among Chinese workers with a broad range of benzene exposures. *American Journal of Industrial Medicine* 42: 275–285.

Rahimivand, M., Tafvizi, F., and NoorBazargan, H. (2020). Synthesis and characterization of alginate nanocarrier encapsulating *Artemisia ciniformis* extract and evaluation of the cytotoxicity and apoptosis induction in AGS cell line. *International Journal of Biological Macromolecules* https://doi.org/10.1016/j.ijbiomac.2020.05.006.

Rånby, B. (1995). Svedberg – discoverer of the protein macromolecules. *Macromolecular Symposia* 98 (1): 1227–1245.

Rendón-Villalobos, R., Ortíz-Sánchez, A., Tovar- Sánchez, E., and Flores-Huicochea, E. (2016). The role of biopolymers in obtaining environmentally friendly materials. *Composites from Renewable and Sustainable Materials* https://doi.org/10.5772/65265.

Rho, K.S., Jeong, L., Lee, G. et al. (2006). Electrospinning of collagen nanofibers: effects on the behavior of normal human keratinocytes and early-stage wound healing. *Biomaterials* 27 (8): 1452–1461. https://doi.org/10.1016/j.biomaterials.2005.08.004.

Rhodes, C.J. (2018). Plastic pollution and potential solutions. *Science Progress* 101: 207–260.

Ross, D. (1996). Metabolic basis of benzene toxicity. *European Journal of Haematology* 60: 111–118.

Santos, M.V.C., Tableueiredo, V.O., Arcuri, A.S.A. et al. (2017). Aspectos toxicológicos do benzeno, biomarcadores de exposição e conflitos de interesses. *Revista Brasileira de Saúde Ocupacional* 42: e13s. https://doi.org/10.1590/2317-6369nota00017.

Selbmann, L., Stingele, F., and Petruccioli, M. (2003). Exopolysaccharide production by filamentous fungi: the example of Botryosphaeria rhodina. *Antonie Van Leeuwenhoek* 82: 135–145.

Sherly, M.C.D., Rekha, M.R., and Harikrishnan, V.S. (2020). Cationised dextran and pullulan modified with diethyl aminoethyl methacrylate for gene delivery in cancer cells. *Carbohydrate Polymers* 116426 https://doi.org/10.1016/j.carbpol.2020.116426.

Shin, S.Y., Park, H.N., Kim, K.H. et al. (2005). Biological evaluation of chitosan nanofiber membrane for guided bone regeneration. *Journal of Periodontology* 76: 1778–1784. https://doi.org/10.1902/jop.2005.76.10.1778.

Silva, I.R., Monteiro, N.K., Martinez, P.F. et al. (2008). Purificação e caracterização de uma gentioexaose obtida de botriosferana por hidrólise ácida parcial. *Química Nova* 31 (5): 1015–1019. ISSN: 1678-7064.

Singh, S.R., Sainin, G.K., and Kennedy, J.F. (2008). Pullulan: microbial sources, production and applications. *Carbohydrate Polymers* 73: 515–531.

Snyder, R. (2012). Benzene and leukemia. *International Journal of Environmental Research and Public Health* 9: 2875–2893.

Snyder, R. and Hedli, C.C. (1996). An overview of benzene metabolism. *Environmental Health Perspectives* 104: 1165–1171.

Sutherland, I.W. (1998). Novel and established applications of microbial polysaccharides. *Trends in Biotechnology* 16: 41–46.

Tabernero, A., González-Garcinuño, A., Galán, M.A., and Martín del Valle, E.M. (2019). Microbial exopolisaccharides for biomedical applications. *Materials for Biomedical Engineering* 12: 165–219.

Tao, X., Tao, T., and Wen, Y. (2018). Novel delivery of mitoxantrone with hydrophobically modified pullulan nanoparticles to inhibit bladder cancer cell and the effect of nano-drug size on inhibition efficiency. *Nanoscale Resource Letters* 13: 345.

Torrecilla, D., Lozano, M.V., Lallana, E. et al. (2013). Anti-tumor efficacy of chitosan-g-poly(ethylene glycol) nanocapsules containing docetaxel: anti-TMEFF-2 functionalized nanocapsules vs. non-functionalized nanocapsules. *European*

Journal of Pharmaceutics and Biopharmaceutics 83: 330–337.

Xue, J., He, M., Liu, H. et al. (2014). Drug loaded homogeneous electrospun PCL/gelatin hybrid nanofiber structures for anti-infective tissue regeneration membranes. *Biomaterials* 35 (34): 9395–9405. https://doi.org/10.1016/j.biomaterials.2014.07.060.

Zayed, M., Tourne-Peteilh, C., Ramonda, M. et al. (2017). Microgels of silylated HPMC as a multimodal system for drug co-encapsulation. *International Journal of Pharmaceutics* 532: 790–801.

Zhang, M., Chen, S., Zhong, L. et al. (2019). Zn^{2+}-loaded TOBC nanofiber-reinforced biomimetic calcium alginate hydrogel for antibacterial wound dressing. *International Journal of Biological Macromolecules* https://doi.org/10.1016/j.ijbiomac.2019.12.046.

Part II

Nanoparticles: Toxicity and Safety

3

Selenium Nanoparticles

Toxicity and Safety

Irina A. Shurygina, Irina S. Trukhan, Nataliya N. Dremina, and Michael G. Shurygin

Irkutsk Scientific Center of Surgery and Traumatology, Irkutsk, Russia

3.1 Introduction

Today, nanotechnology affects almost all areas of activity, including medicine. NPs are unique, biologically active objects because of their small size, relatively large area, surface modification, as well as presence of surface charges, better solubility, and multifunctionality. Their use allows us to solve many biopharmaceutical and pharmacokinetic problems. NPs often increase the therapeutic efficacy of drugs and improve the solubility of compounds in water. Also, the features of NPs allow them to be used for targeted drug delivery (Chaudhary et al. 2014; Chenthamara et al. 2019; Khan et al. 2020; Shurygina and Shurygin 2020). NPs have unique properties that are not characteristic of substances included in them in a different state (Fadeeva et al. 2015; Shurygina et al. 2016).

A special place among nanotechnologies is occupied by nanocomposites containing essential substances that are part of the mammalian enzyme systems. Often, there is a shortage of these microelements in the environment. An important role among them belongs to Se. This is an essential biologically active microelement and is part of selenoproteins (Khurana et al. 2019). So, as selenocysteine, Se is a part of antioxidant enzymes such as glutadione peroxidase, thioredoxy reductase, and selenoprotein P. In these enzymes, Se acts as a redox center and is necessary to maintain the biochemical activity of the enzymes. Se is also involved in the maintenance of redox homeostasis, inflammatory and immunological reactions by regulating the expression of inflammatory genes at the epigenetic level (Broome et al. 2004). Also, it participates in carbohydrate metabolism, shows a high reactivity to oxidizing agents (Sharifi et al. 2012; Casaril et al. 2017). Se, as a component of enzymes, affects reproductive health and the maintenance of brain function in Alzheimer's disease (Jin et al. 2017). Elemental Se in nanoform has neuroprotective properties (Amani et al. 2019), antitumor activity where SeNPs act as an apoptosis inducer of cancer cells (Sukhov et al. 2017; Trukhan et al. 2018; Xu et al. 2020), and is a highly effective prolonged-action antioxidant in regenerative medicine (Shurygina and Shurygin 2017, 2018; Tan et al. 2019; Qiao et al. 2020; Bai et al. 2020a).

Se deficiency is associated with numerous human diseases (Vinceti et al. 2017). At the same time, an excess of Se can be toxic and lead to selenosis. Over the past decades, many researchers have identified the significant

Nanotechnology in Medicine: Toxicity and Safety, First Edition. Edited by Mahendra Rai, Mrunali Patel, and Rashmin Patel.

biomedical potential of SeNPs. However, along with favorable results, it is very important to study the negative and possible side effects associated with the use of Se in nanoforms. NPs are able to enter into the human body by inhalation, oral, parenteral routes and interact with intracellular structures and macromolecules for a long time. Entered NPs can be distributed to various organs, where they are able to retain their structure or to be modified, metabolized.

Unfortunately, today there is no unambiguous information on whether NPs are absolutely safe for humans or have a certain toxicity. Thus, the study of the absorption, distribution, metabolism, and excretion of NPs also as the mechanisms of interaction with living systems studies are necessary to understand their activity, behavior in biological systems, and potential *in vivo* hazards.

3.2 Selenium Forms

In nature, Se exists mainly in the form of selenate (Se^{6+}), selenite (Se^{4+}), selenide (Se^{2-}), and elemental Se (Se^0). The latter is insoluble in aqueous media and less toxic and biologically inert. However, it was found that elemental Se in the form of nano-sized particles is not only biocompatible but also has a number of biological activities (antitumor, antimicrobial, protective). This is the reason for obtaining elemental Se in the NP form by all modern synthesis methods (physical, chemical, and biological) (Wadhwani et al. 2016; Skalickova et al. 2017). Moreover, since the biological properties of SeNPs (such as toxicity, selectivity for various cell types, biocompatibility, and biodegradability) as well as the presence of specific activities directly depend on their physical and chemical properties, special attention is paid to the ability to control the size, shape, composition, and uniformity for forming NPs (Zhang et al. 2004). NPs of the following shapes and configurations are currently known: zero-dimensional nanocrystals and quantum dots; one-dimensional nanospheres, nanorods, nanowires, nanotubes, and nanobelts; two-dimensional arrays of NPs; thin films and three-dimensional structures such as superlattices (Piacenza et al. 2018).

Conventionally, all physical, chemical, and physicochemical methods for the synthesis of SeNPs can be divided into three groups (Piacenza et al. 2018) (Figure 3.1):

1) Methods based on "upstream" approaches in which SeNPs are formed from precursors (Se salt) as a result of chemical reactions (reduction, hydrolysis, or oxidation) (Dwivedi et al. 2011; Malhotra et al. 2014; Kumar et al. 2015; Wadhwani et al. 2016; Yu et al. 2016; Skalickova et al. 2017).
2) "Downstream" processes where smaller nanostructures are formed from a larger precursor by melting or dissociation in a solvent (Triantis et al. 2009; Gusbiers et al. 2015; Sarkar et al. 2015; Badgar 2019).
3) Matrix processes involving the use of various physical or chemical matrices for the conversion of Se-containing precursors into NPs.

The third group of methods for the synthesis of SeNPs includes approaches that differ in the use of so-called matrices or substances adsorbed on the surface of NPs. These matrices play the role of stabilizing agents during the synthesis, as well as determinants for the size and shape of nanostructures. To obtain stable monodisperse NPs, block copolymers (hydroxyethylcellulose, polyvinylpyrrolidone, cetyltrimethylammonium bromide, polyvinyl alcohols, gelatin, chitosan, saccharides, proteins and alkanethiols) are preferred. These polymers can be irreversibly adsorbed on the surface of NPs, determining their size and structure (Shah et al. 2007; Malhotra et al. 2014; Wadhwani et al. 2016; Piacenza et al. 2018). It was shown that the molecular weight of saccharides (poly-, oligo-, or mono-) involved in the reaction affects the growth, morphology, and agglomeration rate of NPs, as well as that saccharides with a larger mass have better stabilizing properties (Bai

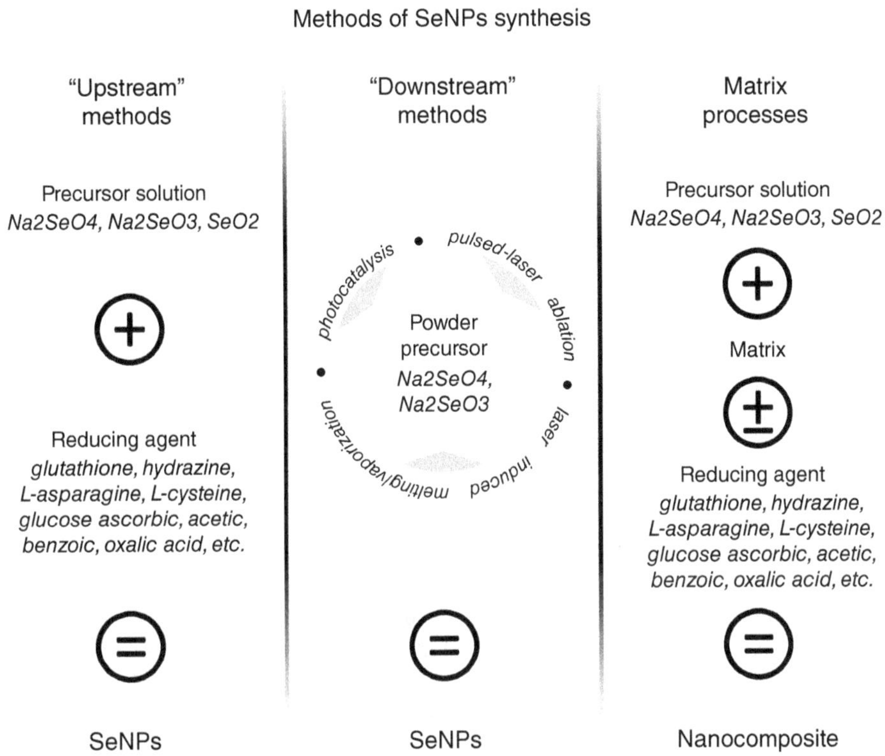

Figure 3.1 Methods for the synthesis of selenium nanoparticles. *Source:* Based on Piacenza et al. (2018).

et al. 2008). The surfactants used as a matrix (e.g. sodium-bis-[2-ethylhexyl] sulfosuccinate, hexadecyltrimethylammonium bromide), on the contrary, are able to desorb from the surface of NPs during the synthesis, which leads to the formation of larger aggregates. The morphology of the forming particles (usually one-dimensional nanostructures) is controlled using this approach, or by changing the concentration of matrix agents (Mehta et al. 2008), either by adding an acid (e.g. HCl) to the solution. This accelerates the reduction of the precursor to elemental Se and leads to the instantaneous deposition of NPs (Shah et al. 2007).

Despite the wide selection of chemical and physical methods for the synthesis of NPs, as well as the use of new reducing agents, catalysts, matrices, the choice of the most effective and least expensive synthesis method remains difficult for the researcher. In addition, chemical and physicochemical methods of synthesis often require the use of toxic and aggressive substances, as well as metal catalysts. During the synthesis process, these materials can be incorporated into the NPs as impurities.

Another group of methods associated with the "upstream" approaches is the biological synthesis of SeNPs. The basis of these methods is the use of living organisms (bacteria, fungi, plants) for the production of NPs. Over the past 20 years, the ability to reduce oxyanions as selenates (SeO_4^{2-}) and selenites (SeO_3^{2-}), to elemental Se, was found in bacteria isolated from substrates contaminated with various pollutants, including Se (*Stenotrophomonas maltophilia*, *Cupriavidus metallidurans*, *Dechloromonas* sp., *Thauera* sp.). The same possibilities were found in anaerobic microorganisms that are capable of using selenites and selenates as electron acceptors of the respiratory chain (*Desulfomicrobium sp.*, *Desulfurispirillum indicum,* and hyperthermophilic archaea), in some rhizospheric bacteria (*Rhizobium*, *Pseudomonas* sp. and *Azospirillum brasilense*), as well as in some species of Enterobacteria (*Escherichia coli* ATCC 35218, *Enterobacter cloacae* Z0206, *Pantoea agglomerans*), Lactobacilli (*Lactobacillus casei*, *Lactobacillus acidophilus*, *Lactobacillus helveticus*), Enterococci and Streptococci (*Enterococcus faecalis*, *Streptococcus thermophilus*, *Staphylococcus carnosus*), bacteria of the genus Bacillus (*Bacillus* sp., *Bacillus subtilis*, *Bacillus mycoides*, *Bacillus licheniformis*), Actinobacteria (*Streptomyces* sp., *Bifidobacterium* BB-1272), and Cyanobacteria (*Arthrospira platensis*) (Tugarova and Kamnev 2017; Badgar 2019). It was shown that the synthesis of NPs can occur both under anaerobic and aerobic conditions, and the accumulation of produced particles can be either intracellular or extracellular (Jain et al. 2014; Tugarova and Kamnev 2017).

In the case of anaerobic synthesis, recovery of selenite/selenate is the main metabolic process. This process is associated with respiration and is mediated either by specialized enzymes (periplasmic selenate reductase in *Thauera selenatis*) or nonspecific reductases, which also use sulfites and sulfates as electron acceptors (Tugarova and Kamnev 2017). Under aerobic conditions, the synthesis of SeNPs is associated with the regulation of redox homeostasis, detoxification and denitrification processes. The recovery of Se to Se^0 is carried out by systems such as glutathione/glutathione reductase, thioredoxin/thioredoxin reductase, as well as peroxiredoxins, nitrareductases (in rhizobacteria), and low-molecular weight thiols (Jain et al. 2014; Tugarova and Kamnev 2017). For extracellular synthesis, a mechanism has been proposed for the recovery of Se oxyanions with the participation of the outer membranes cytochromes (Jain et al. 2014).

SeNPs synthesized by microorganisms have a number of features. These are quite large sizes (from 50 to 500 nm, the average size usually exceeds 100 nm) and the presence of proteins, polysaccharides, or lipids associated with NPs. These biopolymers presumably perform a stabilizing function during the assembly of nanostructures. Such biogenic NPs have unique spectral properties that distinguish them not only from chemically or physically synthesized SeNPs, but also from NPs produced by other types of microorganisms (Oremland et al. 2004). In most cases, the described biogenic NPs are nanospheres; however, nanowires of the elemental Se also are found (Jain et al. 2014). The size of the formed NPs depends on the type of microorganism, the duration of the synthesis, the amount of biomaterial, and the cultivation conditions (temperature, pH, concentration of Se-containing anions and oxygen) (Jain et al. 2014; Tugarova and Kamnev 2017).

The advantages of the biological method of NPs' synthesis include low toxicity, higher biocompatibility and biodegradability of the resulting products compared to NPs synthesized by chemical and physical methods. Besides, the synthesis process is simple, of relatively low cost, does not require extreme conditions and can be carried out at ambient temperature and pressure, does not require specialized equipment and expensive chemicals, and is not a source of toxic waste. At the same time, the disadvantages of this method include the polydispersity of the obtained NPs, the rather large sizes of SeNPs, which limits their use, the need for additional purification of NPs in the case of intracellular synthesis, and the duration of the process up to several days (Oremland et al. 2004; Jain et al. 2014; Tugarova and Kamnev 2017).

Thus, the variety of modern methods of physical, chemical, and biological synthesis allows the researcher to choose a sufficiently effective and affordable method that will allow the synthesis of SeNPs with the necessary characteristics.

3.3 Toxicity of Selenium Nanoparticles

The majority of studies indicate that elemental nanoparticle Se is less toxic to mammals than inorganic or organic Se (Sun et al. 2019). For example, from experiments on Kunming mice, data showed that chemically synthesized SeNPs were 7 times less toxic than selenite (LD_{50} was 113 and 15 mg Se per kg body weight, respectively), and 3.5 times less toxic than selenomethionine (LD_{50} was 92.1 and 25.6 mg Se per kg, respectively) (Zhang et al. 2001, 2005; Wang et al. 2007). It is known that at low concentrations, SeNPs do not have a toxic effect on the animal organism (the values of the controlled parameters do not significantly differ from the control group). High concentrations above 3 mg per kg of animal weight cause changes similar to the effects of high doses of inorganic Se, although less pronounced (Jia et al. 2005; Zhang et al. 2005; Shakibaie et al. 2013; Kuršvietienė et al. 2020).

A study of the acute and subchronic toxicity of SeNPs coated with a polysaccharide–protein complex revealed low acute oral toxicity in ICR mice and Sprague–Dawley rats. Assessment of subchronic toxicity showed that the nanocomposite in the oral administration does not have a visible toxic effect up to a concentration of 200 μg Se per kg of body weight per day, which is approximately 30 times higher than the permissible upper level of human consumption of Se. With the per os exposition of this nanocomposite, there were no signs of damage to major organs, including the liver, spleen, heart, kidneys, and lungs (Zhang et al. 2019).

A classic symptom of the toxic effects of Se on the body is the inhibition of growth, which is observed when exposed to high concentrations of SeNPs. For example, animal weight loss was demonstrated in an experiment with Sprague–Dawley rats when administered orally for two weeks at concentrations exceeding 2 mg Se per kg body weight of SeNPs about 80 nm in size, obtained by reducing sodium selenite with ascorbic acid in the presence of chitosan. At the same time, the administration of these NPs in concentrations of 0.2 and 0.4 mg Se per kg of body weight stimulated the growth of animals, and a dose of 8.0 mg Se per kg of body weight inhibited the growth of animals after the first week of administration (He et al. 2014). SeNPs (size 20–60 nm) obtained by chemical reduction of selenite by glutathione also inhibited the growth of Kunming mice at a concentration of 4 and 6 mg Se per kg body weight, although less effective than selenite. In this case, the administration of SeNPs at a concentration of 6 mg Se per kg of weight completely suppressed growth during the first three days, after which the growth rate was restored (Zhang et al. 2005).

The administration SeNPs (size 20–60 nm) at a concentration of 4 and 5 ppm Se to Sprague–Dawley rats for 13 weeks led to a decrease in body weight in males from the eighth week of the experiment, and in females from the sixth and fifth weeks, respectively, although the effect was less pronounced than when exposed to selenite or a Se-enriched protein (Jia et al. 2005). However, the study indicates that during the preliminary experiment, when rats were injected with SeNPs at a concentration of 6 ppm for 13 weeks, the death of animals was observed, while there were no significant differences in mortality between the administration of SeNPs, selenite or selenium-enriched protein (Jia et al. 2005).

Other researchers found no difference between the toxic effects of SeNPs or sodium selenite. Female rats were orally administered for 28 days with either 0.05, 0.5, or 4 mg Se/kg body weight/day as SeNPs, 20 nm in size, or 0.05 or 0.5 mg Se/kg body weight/day in the form of sodium selenite. Male rats were administered 4 mg Se/kg body weight per day as SeNPs. Clear toxicity was observed at high doses of SeNPs. At all doses of SeNPs and a dose of sodium selenite 0.5 mg Se/kg body weight per day, a decrease in body weight was observed compared to the control. Relative liver mass was increased with Se at a dose of 0.5 mg Se/kg body weight per day, both as SeNPs and as sodium selenite. At the same time, no effect on brain neurotransmitters or hematological parameters was found. From the data obtained, the authors conclude that SeNPs and ionic Se have similar toxicity (Hadrup et al. 2019).

SeNPs of 80–220 nm in size were synthesized biologically using Bacillus sp. MSh-1. When these SeNPs at a concentration of 20 mg Se per kg body weight were orally administered for 14 days to Swiss mice, the animals had a weight loss and a mortality rate of 20%. The administration of lower concentrations did not cause growth retardation and did not lead to death. In this experiment, biogenic SeNPs were 26 times less toxic than Se dioxide (LD_{50} was 198.1 and 7.35 mg Se per kg body weight, respectively) (Shakibaie et al. 2013). Other researchers noted that even high doses of SeNPs do not inhibit growth and do not cause damage to the liver and kidneys, as well as changes in the hematological parameters of healthy animals (Sun et al. 2019).

Hepatotoxicity is another widely discussed toxic effect on exposure to high Se concentrations. Oral administration of SeNPs for two weeks at a concentration of 8.0 mg Se per kg of body weight in rats leads to an increase in the liver, also histopathological changes such as focal necrosis and hepatocyte degeneration. In addition, the activity of liver enzymes increased in animal blood: alanine transaminase (ALT), aspartate transaminase (AST), alkaline phosphatase (ALP), which also indicated liver damage (He et al. 2014). Similarly, an increase in the relative mass of the liver and an increase in ALT activity in the blood of males and females rats was demonstrated with long-term use of SeNPs at a concentration of 5 ppm. Exposure to concentrations of 4 and 5 ppm also showed pathological changes: mottled liver surface and vacuolar degeneration of hepatocytes (Jia et al. 2005). According to other data, no histological changes in the liver of animals were observed upon administration of 0.05, 0.5, or 4 mg Se/kg body weight/day in the form of 20 nm SeNPs to female rats orally for 28 days (Hadrup et al. 2019). In addition to liver pathology in a 14 days study with Sprague–Dawley rats, changes were found in other organs, indicating a different degree of increase in kidneys, lungs, and spleen caused by exposition of SeNPs at a concentration exceeding 2 mg Se/kg body weight body per day. At the same time, morphological changes in the heart, testicles, and thymus indicated atrophy of these organs in animals treated with SeNPs at a daily concentration of 8 mg Se/kg body weight. Histological sections of the kidneys showed signs of glomerulonephritis and necrobiosis of individual renal tubule cells. Hemorrhages were observed on lung sections with filling of the intra-alveolar and bronchial spaces with red blood cells, as well as with thickening of the epithelial septa. The thymus cortex zone in rats administered with high SeNPs concentrations was thinner than in the control samples, and the border between the cortical and medullary zones was unclear. Testicular microscopy revealed atrophy of the seminiferous tubules, impaired spermatogenesis, as well as an increase in the number of damaged seminiferous tubules. In addition, detecting DNA fragmentation by labeling the 3′-hydroxyl termini in the double-strand DNA breaks showed that the number of apoptotic cells was higher in rats administered with SeNPs at concentrations of 4 and 8 mg of Se per kg of body weight (He et al. 2014). An increase in the mass of the spleen, brain, and heart was also observed in male rats and the spleen, and heart in female rats that administered SeNPs at a concentration of 13 ppm for 13 weeks (Jia et al. 2005).

In addition to changes in the liver enzymes activity, high concentrations of SeNPs can cause changes in other blood biochemical parameters. For example, there is an increase in the activity of creatine phosphokinase (Shakibaie et al. 2013); an increase (Shakibaie et al. 2013) or decrease (Jia et al. 2005) in the number of red blood cells and hemoglobin; and an increase in the total bilirubin, creatinine, urea, triglycerides (Shakibaie et al. 2013), and cholesterol (Shakibaie et al. 2013; He et al. 2014). Also in animals, there was an increase in the number of leukocytes, platelets, and hematocrit (Shakibaie et al. 2013). After the administration of SeNPs, many indicators of the oxidative stress and the activity of the tissue detoxification enzymes may change. There is a decrease of malondialdehyde and reduced glutathione, twofold increase of the glutathione S-transferase activity compared to the control (Zhang et al. 2005), a decrease in the oxidation of lipids and proteins, an increase of glutathione peroxidase, superoxide dismutase, and catalase (Bai et al. 2020b). It indicates systemic changes caused by exposure to SeNPs and affecting all body systems. After local injection of the Se-arabinogalactan nanocomposite into the tibia fracture zone in rabbits at 9–21 days after the operation, a significant increase in skin temperature in the area of the surgical wound was noted compared with the control values. These changes may be a manifestation of the metabolic processes activation in the area of bone repair due to the presence of an additional source of Se (Rodionova et al. 2015b, 2016). At the same time, in the area of traumatic bone damage, a low intensity of mineralization of the forming bone callus is observed (Rodionova et al. 2014, 2015a; Shurygina et al. 2015). Although Se in the nanoparticle form at high concentrations causes changes in almost all organs, its effect on the mammalian body is less toxic than when exposed in the form of inorganic or organic compounds.

Fish, in contrast, have a different toxic reaction to Se compounds. SeNPs' toxicity is higher than selenite toxicity for fish *Pangasius hypophthalmus* and *Oryzias latipes*. The NPs synthesized by *Labeo rohita* most often caused the death of pangasius hypophthalmos. After 96 hours of exposure, the LC_{50} for SeNPs was 3.97 mg/l and for inorganic Se 5.82 mg/l (Kumar et al. 2018). A study of the SeNPs (size 20–60 nm) synthesized by *Latipes Oryzias* showed that SeNPs caused 100% mortality at a concentration of 3.2 mg Se per liter, while selenite at a concentration of 2 mg/l caused only to 10%, and 8 mg/l to 80% fish mortality. The calculated LC_{50} after 48 hours for SeNPs was 1.0 mg Se/l, and for selenite 4.7 mg Se/l (Li et al. 2008). Assessment of the effect of chemically synthesized SeNPs in different sizes (25–90 nm and 50–250 nm) obtained as a result of the recovery of selenite by glutathione and biologically synthesized SeNPs (100–350 nm) on fish embryos of the Danio rerio species showed a higher toxicity of chemically synthesized SeNPs. The study demonstrated that in 72 hours after fertilization, the highest mortality was observed during incubation with chemically synthesized SeNPs. They caused 56% mortality in fry even at a Se concentration of 1 mg/l, while incubation with selenite or biologically synthesized SeNPs did not result in death during this incubation period at the same concentration (Mal et al. 2017). The toxicity of chemically and biologically synthesized NPs for Danio rerio embryos was also evaluated by the decrease in the yield of embryos from eggs. In this test, chemically synthesized SeNPs with smaller particles showed the highest toxicity. The output rate slowed down even at low (0.2 mg/l) concentration of chemically synthesized SeNPs and in 72 hours after fertilization (the period with 100% yield of embryos in the control), only 75% of the embryos were released. The toxic effect of SeNPs with larger particles was observed only at concentrations above 2 mg/l Se. With an increase in the concentration of chemically synthesized SeNPs, the yield of embryos decreased to a greater extent. Only 25% of the embryos left the eggs after 72 hours at a concentration of 5 mg Se/l. The authors associated a decrease in the yield with a possible inhibition of the incubation enzyme chorionase by NPs (Mal et al. 2017).

In addition to directly affecting the viability of fish, SeNPs affect the functioning of their organs and systems. So, *Oryziaslatipes* after 10 days of incubation with SeNPs at a concentration of 100 µg Se/l had hyperaccumulation of Se in the liver. It was six times higher than during incubation with selenite in the equivalent concentration. SeNPs also caused more efficient accumulation of Se in the gills and muscles compared to selenite, with differences ranging from two to four times. When studying the clearance of Se from the fish body (after

10 days of exposure to selenite or SeNPs, the fish were placed in Se-free water for 7 days), its concentration in the muscles decreased (42.5% for selenite and 36.5% for SeNPs), whereas in the liver and gills did not change significantly (Mal et al. 2017). In elimination experiments, the rate of Se clearance for seven days was 2.34 μg Se/(kg·h) for selenite and 8.7 μg Se/(kg·h) for SeNPs. The authors attribute this to a lower incorporation of Se from NPs into selenoproteins. However, despite a higher level of clearance, SeNPs provided an active accumulation of Se in the liver and gills of fish, and therefore, the concentration of Se in these organs after seven days of elimination remained still high (Mal et al. 2017).

Hyperaccumulation of Se also occurred in the liver, gills, and brain tissue in an experiment with fish of the species *Pangasius hypophthalmus* when exposed to SeNPs in concentrations from 2.5–4.0 mg Se/l. A similar effect was demonstrated during incubation with inorganic Se, however, only at higher concentrations from 4.5 to 6.0 mg/l. At the same time, in muscles, Se from NPs accumulated less than inorganic Se and was actively excreted from the body. After exposure to SeNPs or inorganic Se, the liver of *Pangasius hypophthalmus* had several morphological changes, such as cloudy swelling, focal necrosis, interstitial edema, hemorrhage, hepatocyte hypertrophy, the presence of pyknotic nucleus and large vacuoles. Changes were also observed in the gills, such as thickening of primary lamellae epithelium, curling of secondary lamellae, epithelial hyperplasia, and fusion of secondary lamellae (Kumar et al. 2018).

For the effects of Se in the form of NPs on the fish organism of *Pangasius hypophthalmus* (Kumar et al. 2018), it was demonstrated that SeNPs affect many enzyme systems. The effect on the antioxidant system was manifested in the form of an increase in the activity of catalase by 244–514%, and inhibition of the activity of superoxide dismutase in the liver, gills, and brain tissue. The effect on the detoxification system was demonstrated by an increase in the activity of glutathione S-transferase by 153–374% in the liver, gills, and brain tissue. An increase in enzyme activity associated with tissue damage was also detected (AST and ALT activity increased in muscles, liver, gills, and brain tissue by 181–689% and by 94–315%, respectively). The change in enzyme activity occurred upon exposure to both low and high concentrations of SeNPs. A similar inhibition of superoxide dismutase activity was also observed in the liver of *Oryzias latipes*, with SeNPs showing more potent inhibitory properties than selenite. In relation to glutathione S-transferase, the effect of SeNPs caused an increase of the enzyme activity in the liver, but in contrast, selenite had no registered effect on it (Mal et al. 2017). However, SeNPs showed a full improvement of hematological and biochemical parameters more than that of sodium selenite in the elimination of Ag NPs' toxicity (Ibrahim 2020).

From the published data, we can conclude that elemental Se in the form of NPs can affect immunoregulation, reproduction, kidney function, and, to a greater extent, the liver; modulate the activity of the antioxidant and detoxification systems; as well as in high concentrations (above 2 mg Se per kg of animal weight) cause the development of Se-induced toxicity in both mammals and fish. Moreover, the toxicity of SeNPs is higher than inorganic Se and causes a more acute reaction to exposure even at low concentrations. It possibly associates with hyperaccumulation of Se in tissues, which again reminds us of the need to take into account the problems of ecotoxicity for Se nanocomposites (Shurygina et al. 2018). The toxic effects of SeNPs are as shown in Figure 3.2.

3.4 Toxicity Mechanisms

The main processes associated with intracellular toxicity are reduced to the excessive production of reactive oxygen species (ROS). ROS are both physiologically necessary and

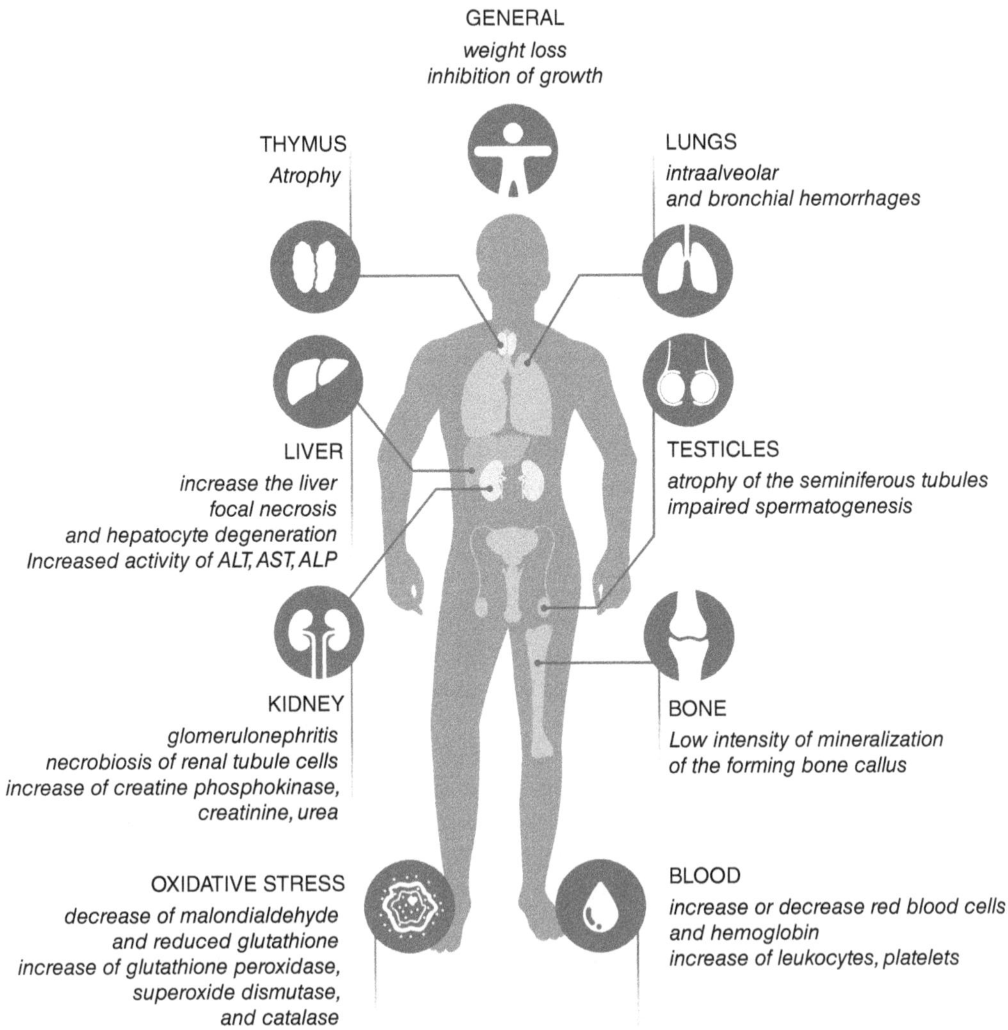

Figure 3.2 Toxic effects of selenium nanoparticles.

potentially damaging to biological objects. A moderate level of ROS plays a specific role in modulating several cellular processes, including signal transduction, proliferative response, gene expression, and regulation of redox processes. However, high ROS indicates oxidative stress and can damage cells by lipid peroxidation, protein changes, DNA breakdown, interference with signaling functions, and modulation of gene transcription (Pozhilova et al. 2015). SeNPs induce oxidative stress and apoptotic cell death in keratinocyte culture by increasing autophagy through the formation of acid lysosomes and autophagosomes (Kirwale et al. 2019). At the same time, SeNPs have a strong antioxidant effect and significantly increase the activity of antioxidant enzymes, reduce the level of lipid peroxidation markers (Hamza and Diab 2020). A study of the effect of NPs on genetic material revealed that the accumulation of single-strand breaks and oxidatively induced base damage can lead to double-strand breaks, which are considered the most unfavorable outcome of oxidative

DNA damage (Petersen and Nelson 2010). Excessive ROS production can damage mitochondrial DNA (Kirkinezos and Moraes 2001).

A serious problem is the genotoxicity of NPs, since it plays an important role in the initiation and progression of abnormalities, such as destructive genetic changes, including gene mutations, structural chromosome aberrations, and recombination (Barabadi et al. 2019). Genotoxicity is divided into primary and secondary. Primary genotoxicity implies direct or indirect interaction of NPs with genetic material (Cooke et al. 2003). The mechanism of secondary genotoxicity of NPs is associated with excessive ROS production by activated phagocytes (neutrophils and macrophages) (Singh et al. 2009). It was shown that Se nanocomposites on polysaccharide matrices from 54.35 to 123.04 nm in size with a Se content of 0.172, 0.274, and 0.305% inhibited nitric oxide (NO) production in lipopolysaccharide-induced (LPS) inflammatory macrophages RAW 264.7, and suppressed expression of mRNA TNF-α, IL-1β. Smaller SeNPs had higher anti-inflammatory effects (Hu et al. 2020). Also, oxidative stress activates specific signaling pathways, including mitogen-activated protein kinase (MAPK) (Shurygina et al. 2017) and NF-kB, which lead to the release of pro-inflammatory cytokines. The result of this signaling is a triggering of inflammation cascade. This protective reaction leads to the further release of ROS from inflammatory cells (e.g. neutrophils). Moreover, in healthy and tumor cells, the effects of SeNPs are different. SeNPs temporarily suppress the expression of p53-associated genes in healthy mice (Sun et al. 2019). Multifunctional SeNPs decorated with a polysaccharide–protein complex and loaded with tetramethylpyrazine and monosialotetrahexosylganglioside can reduce ROS overproduction to prevent mitochondrial dysfunction by inhibiting p53 and MAPK pathway activation (Rao et al. 2019).

Researchers have discovered another effect of SeNPs on tumor cells associated with features of ROS induction in tumors. Folate-conjugated SeNPs easily penetrated HepG2 hepatoma cells by endocytosis, generated the overproduction of ROS, and induced apoptosis by activating p53 and MAPK pathways (Liu et al. 2015).There were studied changes in the expression of the ATF4, Bcl-xL, BAD2, HSP70, and SOD2 genes in human dermal fibroblasts under oxidative stress, nutrient deprivation stress, and control with the admission of SeNPs. Incubation with SeNPs led to a decrease in ROS generation for all conditions tested. Under stress, the expression of ATF4 and Bcl-xL increased for cells treated with SeNPs, leading to attenuation of the cells under stress (Chung et al. 2019).

The genotoxicity of NPs, as well as general toxicity, is greatly influenced by their size, properties, composition, shape, surface properties, physicochemical characteristics (pH, temperature, etc.), solubility, and other factors, such as dose, time exposure, type of cells used. From a toxicological point of view, the size and surface area of particles are the most important characteristics, since interactions between NPs and biological objects occur, as a rule, on the surface (Hoshyar et al. 2016; Wu and Tang 2018; Keyhani et al. 2020). The smaller NPs have a larger relative surface area. Therefore, the surface of the particles becomes more reactive, and the potential catalytic surface for chemical reactions also increases (Magdolenova et al. 2014; Zijlstra and Vizio 2018). Along with the size, surface area, and charge of NPs, the chemical composition and surface coating of the particles play an important role. Thus, the toxic effects of NPs can be prevented by applying surface coatings that can stabilize particles and avoid agglomeration. Coating also has an effect for preventing the dissolution of NPs and the release of toxic ions (Kirchner et al. 2005).

The use of amino acids in the synthesis of SeNPs allows nanoselenes to be obtained with high stability and increased antitumor activity. Thus, if the production of SeNPs used valine, an initiation of cancer cell apoptosis was observed in an experiment on MCF-7 adenocarcinoma cells. A similar effect was observed with using of 5-fluorouracil. Cancer cells actively absorb such NPs by endocytosis. Also, polysaccharide–protein complexes of fungi were used as a coating for NPs. Thus, synthesized SeNPs inhibited the growth of human MCF-7 breast cancer cells by inducing apoptosis due to the cleavage of poly (ADP-ribose) polymerase (PARP) and activation of caspase (Wu et al. 2012). Other authors also indicated caspase activation in tumor cells by SeNPs (Zhang et al. 2019). Similar studies were carried out also on other cell lines. Treatment of A375 human melanoma cells with SeNPs resulted in dose-dependent cell apoptosis. In this case, DNA fragmentation was observed. Further studies of intracellular mechanisms have shown that exposure to SeNPs causes apoptotic death of A375 cells with activation of oxidative stress and the development of mitochondrial dysfunction (Chen et al. 2008; Yu et al. 2012).

SeNPs are currently considered promising for use in medicine: they are less toxic and more biocompatible agents than selenite (SeO_3^{-2}) and selenate (SeO_4^{-2}), and also have antitumor properties (Tan et al. 2019). At present, it is generally accepted that Se compounds exhibit their antitumor ability in various ways, where one of the main effects are direct or indirect antioxidant properties, which intracellularly support oxidation-reduction status and protect healthy cells from ROS (Prasad et al. 2017; Zhao et al. 2018).The cytotoxic effects of SeNPs in tumor cells are achieved by inducing apoptosis. The activation of enzymes involved in the processes of cell death occurs. This mechanism of action is similar to many chemotherapeutic agents that initiate apoptosis by affecting the regulation of intracellular ROS generation. Thus, the production of ROS induced by Se compounds is an important fact for cell cycle arrest and cell apoptosis.

Apoptosis can be activated externally or internally. The external pathway is initiated by the attachment of a proapoptotic ligand to the Fas/TNF cell death receptor (tumor necrosis factor receptor) on the cell membrane. During the interaction of the ligand, receptor, adapter, and procaspase, apoptosomes are formed. These complexes initiate cell death when autolytic activation of caspases occurs, including caspase 8 (Creagh 2014; Khan et al. 2019). The internal or mitochondrial pathway is activated by DNA damage due to excessive ROS formation or exposure to another cytotoxic factor that activates the p53 protein. This protein, in turn, increases the activity of the Bcl-2 protein, a mitochondrial membrane permeability regulator. As a result, apoptotic proteins are released, including procaspase 9, which is additionally activated in the apoptosome with the participation of cytochrome C (Wang et al. 2015). Activated caspase 9 interacts and activates effector procaspase 3. During apoptosis, caspase 3 is responsible for chromatin condensation and DNA fragmentation. The sequential activation of caspases, the so-called caspase cascade, plays a key role in cell apoptosis. In addition to the caspase pathway, PARP activity increases, and its high level is also a marker for apoptotic cells (Zhang et al. 2013).

Most chemotherapeutic drugs have low selectivity and are extremely toxic to body cells. The use of SeNPs in combination with chemotherapeutic and hormonal drugs reduces side effects. So, spherical SeNPs coated with 11-mercapto-1-undecanol have a nephroprotective and antiapoptotic effect, significantly increase mitochondrial activity and cell viability, and prevent DNA fragmentation in the cells of the proximal tubules of the kidneys. Caspase-3 activation in cells exposed to cisplatin is also blocked by this composite (Li et al. 2011). In addition, the composite penetrates the cells 10 times more actively due to enhanced endocytosis, while uncoated SeNPs easily aggregate in aqueous solutions and hardly penetrate the cell membrane (Menon et al. 2018). A possible mechanism of antitumor action is the effect of SeNPs on the expression of the estrogen receptor α (Erα). In the study (Vekariya et al. 2012), the antitumor activity of SeNPs was correlated with ERα levels in breast cancer cells both *in vivo* and *in vitro*. It has been shown that folate-conjugated SeNPs in MCF-7 cells can induce mitochondrial dysfunction through oxidative stress, followed by induction of cytoskeleton disorganization and morphological changes in the cell membrane (Pi et al. 2013).

An important factor of ruthenium-modified SeNPs is the ability to influence angiogenesis and cause its inhibition. The regulation of angiogenesis is known to be a promising area for the treatment of tumors since the uncontrolled growth of cancer cells directly depends on adequate blood supply. Ruthenium-modified SeNPs (Ru-SeNPs) can interact with proteins located in the cytoplasm. These NPs are localized mainly in the cytoplasm and perinuclear space. Then they penetrate the nucleus membrane and cause DNA fragmentation, as well as damage to the plasma membrane. Ruthenium-modified SeNPs have the ability to inhibit proliferation, endothelial cell migration, and further blood vessel formation by blocking the main fibroblast growth factor (FGFb) and its receptor (FGFR1) (Sun et al. 2013). However, Ru-SeNPs are 2–6 times more toxic than SeNPs (Chaudhary et al. 2014). Se-substituted hydroxyapatite NPs have low toxicity and reduce the expression of Ki-67 (a marker of proliferative activity of tumor cells), vascular endothelial growth factor (VEGF), and matrix metallopeptidase 9 (MMP-9) (Yanhua et al. 2016). Important for the antitumor effect is the ability of SeNPs to stop the cell cycle. Thus, Se-substituted hydroxyapatite NPs in hepatocellular carcinoma cells, in addition to damage of the cancer cell DNA, inhibit the expression of Cdk1 protein and stop the cell cycle in the S-G2/M phase (Yanhua et al. 2016).

SeNPs in MDA-MB-231 breast tumor cells delay phase S of the cell cycle, during which nuclear DNA replication occurs (Khurana et al. 2019). Due to the delay in phase S, the cell cannot go to the next phase G2, the apoptosis program is started and proliferation is inhibited (Luo et al. 2012). SeNPs have an antitumor effect by affecting the activity of individual genes. SeNPs increase expression of aldo-keto reductase family 1 member B10 and inhibitor of growth protein 3 and decrease expression of forkhead box protein P1 (Ahmed et al. 2014). To date, numerous studies with SeNPs are ongoing on cell cultures, organs, and biological organisms in general. A lot of involved molecules, systems, and pathways were revealed. However, many of the effects of SeNPs remain unclear and require further research.

3.5 Conclusion

Thus, nanoscale elemental Se is not only biocompatible, but also has a number of biological activities (antitumor, antimicrobial, protective). The biological properties of SeNPs, such as toxicity, selectivity for various cell types, biocompatibility, and biodegradability, as well as the presence of specific activities, are directly dependent on their physical and chemical properties. Particular attention in the synthesis is given to the ability to control the size, shape, composition, and uniformity of the resulting NPs. To date, data on the toxicity and safety of SeNPs are accumulating. Se in nanoscale form has a dose-dependent effect. High concentrations of SeNPs (above 2 mg Se per kg of animal weight) can cause Se-induced toxicity in mammals.

Many publications devoted the possibility of using Se nanoforms in medicine. Therefore, further studies related to the safety and mechanisms of action of SeNPs are promising.

References

Ahmed, H.H., Khalil, W.K.B., and Hamza, A.H. (2014). Molecular mechanisms of Nano-selenium in mitigating hepatocellular carcinoma induced byN-nitrosodiethylamine (NDEA) in rats. *Toxicology Mechanisms and Methods* 24 (8): 593–602.

Amani, H., Habibey, R., Shokri, F. et al. (2019). Selenium nanoparticles for targeted stroke therapy through modulation of inflammatory and metabolic signaling. *Scientific Reports* 9 (1): 6044.

Badgar, K. (2019). The synthesis of selenium nanoparticle (SeNPs). *Acta Agraria Debreceniensis* 1: 5–8.

Bai, K., Hong, B., He, J., and Huang, W. (2020a). Antioxidant capacity and hepatoprotective role of chitosan-stabilized selenium nanoparticles in concanavalin A - induced liver injury in mice. *Nutrients* 12 (3): E857.

Bai, K., Hong, B., Huang, W., and He, J. (2020b). Selenium-nanoparticles-loaded chitosan/chitooligosaccharide microparticles and their antioxidant potential: a chemical and *in vivo* investigation. *Pharmaceutics* 12 (1): E43.

Bai, Y., Wang, Y., Zhou, Y. et al. (2008). Modification and modulation of saccharides on elemental selenium nanoparticles in liquid phase. *Materials Letters* 62 (15): 2311–2314.

Barabadi, H., Najafi, M., Samadian, H. et al. (2019). A systematic review of the genotoxicity and antigenotoxicity of biologically synthesized metallic nanomaterials: are green nanoparticles safe enoughfor clinical marketing? *Medicina (Kaunas, Lithuania)* 55 (8): E439.

Broome, C.S., McArdle, F., Kyle, J.A.M. et al. (2004). An increase in selenium in take improves immune function and poliovirus handling in adults with marginal selenium status. *The American Journal of Clinical Nutrition* 80 (1): 154–162.

Casaril, A.M., Ignasiak, M.T., Chuang, C.Y. et al. (2017). Selenium-containing Indolyl compounds: kinetics of reaction with inflammation-associated oxidants and protective effect against oxidation of extracellular matrix proteins. *Free Radical Biology and Medicine* 113: 395–405.

Chaudhary, S., Umar, A., and Mehta, S.K. (2014). Surface functionalized selenium nanoparticles for biomedical applications. *Journal of Biomedical Nanotechnology* 10 (10): 3004–3042.

Chen, T., Wong, Y.S., Zheng, W. et al. (2008). Selenium nanoparticles fabricated in Undaria pinnatifida polysaccharide solutions induce mitochondria-mediated apoptosis in A375 human melanoma cells. *Colloids and Surfaces. B, Biointerfaces* 67 (1): 26–31.

Chenthamara, D., Subramaniam, S., Ramakrishnan, S.G. et al. (2019). Therapeutic efficacy of nanoparticles and routes of administration. *Biomaterials Research* 23: 20.

Chung, S., Roy, A.K., and Webster, T.J. (2019). Selenium nanoparticle protection of fibroblast stress: activation of ATF4 and Bcl-xL expression. *International Journal of Nanomedicine* 14: 9995–10007.

Cooke, M.S., Evans, M.D., Dizdaroglu, M., and Lunec, J. (2003). Oxidative DNA damage: mechanisms, mutation, and disease. *The FASEB Journal* 17 (10): 1195–1214.

Creagh, E.M. (2014). Caspase crosstalk: integration of apoptotic and innate immune signaling pathways. *Trends in Immunology* 35 (12): 631–639.

Dwivedi, C., Shah, C.P., Singh, K. et al. (2011). An organic acid-induced synthesis and characterization of selenium nanoparticles. *Journal of Nanotechnology* 2011: 1–6.

Fadeeva, T.V., Shurygina, I.A., Sukhov, B.G. et al. (2015). Relationship between the structures and antimicrobial activities of argentic nanocomposites. *Bulletin of the Russian Academy of Sciences: Physics* 79 (2): 273–275.

Gusbiers, G., Wang, Q., Khachatryan, E. et al. (2015). Anti-bacterial selenium nanoparticles produced by UV/VIS/NIR pulsed nanosecond laser ablation in liquids. *Laser Physics Letters* 12: 1–7.

Hadrup, N., Loeschner, K., Mandrup, K. et al. (2019). Subacute oral toxicity investigation of selenium nanoparticles and selenite in rats. *Drug and Chemical Toxicology* 42 (1): 76–83.

Hamza, R.Z. and Diab, A.E.A. (2020). Testicular protective and antioxidant effects of selenium nanoparticles on monosodium glutamate-induced testicular structure alterations in male mice. *Toxicology Reports* 7: 254–260.

He, Y., Chen, S., Liu, Z. et al. (2014). Toxicity of selenium nanoparticles in male Sprague-Dawley rats at supranutritional and nonlethal levels. *Life Sciences* 115 (1–2): 44–51.

Hoshyar, N., Gray, S., Han, H., and Bao, G. (2016). The effect of nanoparticle size on *in vivo* pharmacokinetics and cellular interaction. *Nanomedicine* 11 (6): 673–692.

Hu, S., Hu, W., Li, Y. et al. (2020). Construction and structure-activity mechanism of polysaccharide nano-selenium carrier. *Carbohydrate Polymers* 236: 116052.

Ibrahim, A.T.A. (2020). Toxicological impact of green synthesized silver nanoparticles and protective role of different selenium type on Oreochromis niloticus: hematological and biochemical response. *Journal of Trace Elements in Medicine and Biology* 61: 126507.

Jain, R., Gonzalez-Gil, G., Singh, V. et al. (2014). Lens biogenic selenium nanoparticles: production, characterization and challenges. In: *Biotechnology. Nanobiotechnology, Chapter: Biogenic Selenium Nanoparticles: Production, Characterization and Challenges* (ed. A. Kumar), 365–394. Studium Press LLC.

Jia, X., Li, N., and Chen, J. (2005). A subchronic toxicity study of elemental Nano-Se in Sprague-Dawley rats. *Life Sciences* 76 (17): 1989–2003.

Jin, N., Zhu, H., Liang, X. et al. (2017). Sodium selenate activated Wnt/β-catenin signaling and repressed amyloid-β formation in a triple transgenic mouse model of Alzheimer's disease. *Experimental Neurology* 297: 36–49.

Keyhani, A., Ziaali, N., Shakibaie, M. et al. (2020). Biogenic selenium nanoparticles target chronic toxoplasmosis with minimal cytotoxicity in a mouse model. *Journal of Medical Microbiology* 69 (1): 104–110.

Khan, A.U., Khan, M., Cho, M.H., and Khan, M.M. (2020). Selected nanotechnologies and nanostructures for drug delivery, nanomedicine and cure. *Bioprocess and Biosystems Engineering* https://doi.org/10.1007/s00449-020-02330-8.

Khan, S., Ullah, M.W., Siddique, R. et al. (2019). Catechins-modified selenium-doped hydroxyapatite nanomaterials for improved osteosarcoma therapy through generation of reactive oxygen species. *Frontiers in Oncology* 9: 499.

Khurana, A., Tekula, S., Saifi, M.A. et al. (2019). Therapeutic applications of selenium. *Biomedicine and Pharmacotherapy* 111: 802–812.

Kirchner, C., Liedl, T., Kudera, S. et al. (2005). Cytotoxicity of colloidal CdSe and CdSe/ZnS nanoparticles. *Nano Letters* 5 (2): 331–338.

Kirkinezos, I.G. and Moraes, C.T. (2001). Reactive oxygen species and mitochondrial diseases. *Seminars in Cell and Developmental Biology* 12 (6): 449–457.

Kirwale, S., Pooladanda, V., Thatikonda, S. et al. (2019). Selenium nanoparticles induce autophagy mediated cell death in human keratinocytes. *Nanomedicine* 14 (15): 1991–2010.

Kumar, N., Krishnani, K.K., and Singh, N.P. (2018). Comparative study of selenium and selenium nanoparticles with reference to acute toxicity, biochemical attributes, and histopathological response in fish. *Environmental Science and Pollution Research International* 25 (9): 8914–8927.

Kumar, S., Tomar, M.S., and Acharya, A. (2015). Carboxylic group-induced synthesis and characterization of selenium nanoparticles and its anti-tumor potential on Dalton's lymphoma cells. *Colloids and Surfaces. B, Biointerfaces* 126: 546–552.

Kuršvietienė, L., Mongirdienė, A., Bernatonienė, J. et al. (2020). Selenium anticancer properties and impact on cellular redox status. *Antioxidants* 9 (1): 80.

Li, H., Zhang, J., Wang, T. et al. (2008). Elemental selenium particles at nano-size (Nano-Se) are more toxic to Medaka (Oryzias latipes) as a consequence of hyper-accumulation of selenium: a comparison with sodium selenite. *Aquatic Toxicology* 89 (4): 251–256.

Li, X., Wong, Y.S., Chen, T. et al. (2011). The reversal of cisplatin-induced nephrotoxicity by selenium nanoparticles functionalized with 11-mercapto-1-undecanol by inhibition of

ROS-mediated apoptosis. *Biomaterials* 32 (34): 9068–9076.

Liu, T., Zeng, L., Jiang, W. et al. (2015). Rational design of cancer-targeted selenium nanoparticles to antagonize multidrug resistance in cancer cells. *Nanomedicine: Nanotechnology, Biology and Medicine* 11 (4): 947–958.

Luo, H., Wang, F., Bai, Y. et al. (2012). Selenium nanoparticles inhibit the growth of HeLa and MDA-MB-231 cells through induction of S phase arrest. *Colloids and Surfaces. B, Biointerfaces* 94: 304–308.

Magdolenova, Z., Collins, A., Kumar, A. et al. (2014). Mechanisms of genotoxicity. A review of *in vitro* and *in vivo* studies with engineered nanoparticles. *Nanotoxicology* 8 (3): 233–278.

Mal, J., Veneman, W.J., Nancharaiah, Y.V. et al. (2017). A comparison of fate and toxicity of selenite, biogenically, and chemically synthesized selenium nanoparticles to zebrafish (Danio rerio) embryogenesis. *Nanotoxicology* 11 (1): 87–97.

Malhotra, S., Jha, N., and Desai, K. (2014). A superficial synthesis of selenium nanospheres using wet chemical approach. *International Journal of Nanotechnology and Application* 3 (4): 7–14.

Mehta, S.K., Chaudhary, S., Kumar, S. et al. (2008). Surfactant assisted synthesis and spectroscopic characterization of selenium nanoparticles in ambient conditions. *Nanotechnology* 19 (29): 5601.

Menon, S., Devi, S., Santhiya, R. et al. (2018). Selenium nanoparticles: a potent chemotherapeutic agent and an elucidation of its mechanism. *Colloids and Surfaces. B, Biointerfaces* 170: 280–292.

Oremland, R.S., Herbel, M.J., Blum, J.S. et al. (2004). Structural and spectral features of selenium nanospheres produced by Se-respiring bacteria. *Applied and Environmental Microbiology* 70 (1): 52–60.

Petersen, E.J. and Nelson, B.C. (2010). Mechanisms and measurements of nanomaterial-induced oxidative damage to DNA. *Analytical and Bioanalytical Chemistry* 398 (2): 613–650.

Pi, J., Jin, H., Liu, R. et al. (2013). Pathway of cytotoxicity induced by folic acid modified selenium nanoparticles in MCF-7 cells. *Applied Microbiology and Biotechnology* 97 (3): 1051–1062.

Piacenza, E., Presentato, A., Zonaro, E. et al. (2018). Selenium and tellurium nanomaterials. *Physical Sciences Reviews* 3 (5): 20170100.

Pozhilova, E.V., Novikov, V.E., and Levchenkova, O.S. (2015). Reactive oxygen species in cell physiologyand pathology. *Vestnik of the Smolensk State Medical Academy* 14 (2): 13–22.

Prasad, S., Gupta, S.C., and Tyagi, A.K. (2017). Reactive oxygen species (ROS) and cancer: role of antioxidative nutraceuticals. *Cancer Letters* 387: 95–105.

Qiao, L., Dou, X., Yan, S. et al. (2020). Biogenic selenium nanoparticles synthesized by Lactobacillus casei ATCC 393 alleviate diquat-induced intestinal barrier dysfunction in C57BL/6 mice through their antioxidant activity. *Food and Function* 11 (4): 3020–3031.

Rao, S., Lin, Y., Du, Y. et al. (2019). Designing multifunctionalized selenium nanoparticles to reverse oxidative stress-induced spinal cord injury by attenuating ROS overproduction and mitochondria dysfunction. *Journal of Materials Chemistry B* 7 (16): 2648–2656.

Rodionova, L.V., Shurygina, I.A., Samoylova, L.G. et al. (2015a). Osteoresorption modelling by means of introduction of selenium preparation under conditions of reparative osteogenesis. *Siberian Medical Journal* 137 (6): 94–98.

Rodionova, L.V., Shurygina, I.A., Samoylova, L.G. et al. (2016). Effect of intraosseous introduction of selenium/arabinogalactan nanoglycoconjugate on the main indicators of primary metabolism in consolidation of bone fracture. *Acta Biomedica Scientifica* 1 (4): 104–108.

Rodionova, L.V., Shurygina, I.A., Shurygin, M.G. et al. (2014). Method for simulatung osteoresorption under osteogenic conditions. RU patent 2524128.

Rodionova, L.V., Shurygina, I.A., Sukhov, B.G. et al. (2015b). Nanobiocomposite based on selenium and arabinogalactan: sinthesis, structure, and application. *Russian Journal of General Chemistry* 85 (2): 485–487.

Sarkar, B., Bhattacharjee, S., Daware, A. et al. (2015). Selenium nanoparticles for stress-resilient fish and livestock. *Nanoscale Research Letters* 10 (1): 371.

Shah, C.P., Kumar, M., and Bajaj, P.N. (2007). Acid-induced synthesis of polyvinyl alcohol-stabilized selenium nanoparticles. *Nanotechnology* 18 (38): 385607.

Shakibaie, M., Shahverdi, A.R., Faramarzi, M.A. et al. (2013). Acute and subacute toxicity of novel biogenic selenium nanoparticles in mice. *Pharmaceutical Biology* 51 (1): 58–63.

Sharifi, S., Behzadi, S., Laurent, S. et al. (2012). Toxicity of nanomaterials. *Chemical Society Reviews* 41 (6): 2323–2343.

Shurygina, I.A., Rodionova, L.V., Shurygin, M.G. et al. (2015). Using confocal microscopy to study the effect of an original pro-enzyme Se/arabinogalactan nanocomposite on tissue regeneration in a skeletal system. *Bulletin of the Russian Academy of Sciences: Physics* 79 (2): 256–258.

Shurygina, I.A. and Shurygin, M.G. (2017). Nanoparticles in wound healing and regeneration. In: *Metal Nanoparticles in Pharma* (eds. M. Rai and R. Shegokar), 21–38. Springer.

Shurygina, I.A. and Shurygin, M.G. (2018). Perspectives of metal nanoparticles application for the purposes of regenerative medicine. *Siberian Medical Review* 4 (112): 31–37.

Shurygina, I.A. and Shurygin, M.G. (2020). Selenium nanocomposites – the prospects of application in oncology. *Journal of New Medical Technologies* 27 (1): 81–86.

Shurygina, I.A., Shurygin, M.G., and Sukhov, B.G. (2016). Nanobiocomposites of metals as antimicrobial agents. In: *Antibiotic Resistance: Mechanisms and New Antimicrobial Approaches* (eds. M. Rai and K. Kon), 167–186. Academic Press.

Shurygina, I.A., Shurygin, M.G., Zelenin, N.V., and Ayushinova, N.I. (2017). Influence on mitogen-activated protein kinases as a new direction of connective tissue growth regulation. *Bulletin of Siberian Medicine* 16 (4): 86–93.

Shurygina, I.A., Sosedova, L.M., Novikov, M.A. et al. (2018). Ecotoxicity of nanometals: the problems and solutions. In: *Nanomaterials: Ecotoxicity, Safety, and Public Perception* (eds. M. Rai and J. Biswas), 95–117. Springer.

Singh, N., Manshian, B., Jenkins, G.J.S. et al. (2009). NanoGenotoxicology: the DNA damaging potential of engineered nanomaterials. *Biomaterials* 30 (23–24): 3891–3914.

Skalickova, S., Milosavljevic, V., Cihalova, K. et al. (2017). Selenium nanoparticles as a nutritional supplement. *Nutrition* 33: 83–90.

Sukhov, B.G., Ganenko, T.V., Pogodaeva, N.N. et al. (2017). Agent with antitumor activity based on arabinogalactan nanocomposites with selenium and methods for prepariation of such nanobiocomposites. RU patent 2614363.

Sun, D., Liu, Y., Yu, Q. et al. (2013). The effects of luminescent ruthenium (II) polypyridyl functionalized selenium nanoparticles on bFGF-induced angiogenesis and AKT/ERK signaling. *Biomaterials* 34: 171–180.

Sun, F., Wang, J., Wu, X. et al. (2019). Selenium nanoparticles act as an intestinal p53 inhibitor mitigating chemotherapy-induced diarrhea in mice. *Pharmacological Research* 149: 104475.

Tan, H.W., Mo, H.Y., Lau, A.T.Y., and Xu, Y.M. (2019). Selenium species: current statusand potentials in cancer prevention and therapy. *International Journal of Molecular Sciences* 20 (1): 75.

Triantis, T., Troupis, A., Gkika, E. et al. (2009). Photocatalytic synthesis of Se nanoparticles using polyoxometalates. *Catalysis Today* 144 (1): 2–6.

Trukhan, I.S., Dremina, N.N., Lozovskaya, E.A., and Shurygina, I.A. (2018). Assessment of potential cytotoxicity during vital observation at the BioStation CT. *Acta Biomedica Scientifica* 3 (6): 48–53.

Tugarova, A.V. and Kamnev, A.A. (2017). Proteins in microbial synthesis of selenium nanoparticles. *Talanta* 174: 539–547.

Vekariya, K.K., Kaur, J., and Tikoo, K. (2012). ERα signaling imparts chemotherapeutic selectivity to selenium nanoparticles in breast cancer. *Nanomedicine: Nanotechnology, Biology and Medicine* 8 (7): 1125–1132.

Vinceti, M., Filippini, T., Cilloni, S. et al. (2017). Health risk assessment of environmental selenium: emerging evidence and challenges. *Molecular Medicine Reports* 15: 3323–3335.

Wadhwani, S.A., Shedbalkar, U.U., Singh, R., and Chopade, B.A. (2016). Biogenic selenium nanoparticles: current status and future prospects. *Applied Microbiology and Biotechnology* 100 (6): 2555–2566.

Wang, H., Zhang, J., and Yu, H. (2007). Elemental selenium at nano size possesses lower toxicity without compromising the fundamental effect on selenoenzymes: comparison with selenomethionine in mice. *Free Radical Biology and Medicine* 42: 1524–1533.

Wang, Y., Chen, P., Zhao, G. et al. (2015). Inverse relationship between elemental selenium nanoparticle size and inhibition of cancer cell growth *in vitro* and *in vivo*. *Food and Chemical Toxicology* 85: 71–77.

Wu, H., Li, X., Liu, W. et al. (2012). Surface decoration of selenium nanoparticles by mushroom polysaccharides–protein complexes to achieve enhanced cellular uptake and antiproliferative activity. *Journal of Materials Chemistry* 22: 9602.

Wu, T. and Tang, M. (2018). Review of the effects of manufactured nanoparticles on mammalian target organs. *Journal of Applied Toxicology* 38 (1): 25–40.

Xu, B., Zhang, Q., Luo, X. et al. (2020). Selenium nanoparticles reduce glucose metabolism and promote apoptosis of glioma cells through reactive oxygen species-dependent manner. *Neurorepeort* 31 (3): 226–234.

Yanhua, W., Hao, H., Li, Y., and Zhang, S. (2016). Selenium-substituted hydroxyapatite nanoparticles and their *in vivo* antitumor effect on hepatocellular carcinoma. *Colloids and Surfaces. B, Biointerfaces* 140: 297–306.

Yu, B., You, P., Song, M. et al. (2016). A facile and fast synthetic approach to create selenium nanoparticles with diverse shapes and their antioxidation ability. *New Journal of Chemistry* 40: 1118–1123.

Yu, B., Zhang, Y., Zheng, W. et al. (2012). Positive surface charge enhances selective cellular uptake and anticancer efficacy of selenium nanoparticles. *Inorganic Chemistry* 51 (16): 8956–8963.

Zhang, J., Wang, H., Bao, Y., and Zhang, L. (2004). Nano red elemental selenium has no size effect in the induction of seleno-enzymes in both cultured cells and mice. *Life Sciences* 75: 237–244.

Zhang, J., Wang, H., Yan, X., and Zhang, L. (2005). Comparison of short-term toxicity between Nano-Se and selenite in mice. *Life Sciences* 76 (10): 1099–1109.

Zhang, J.S., Gao, X.Y., Zhang, L.D., and Bao, Y.P. (2001). Biological effects of a nano red elemental selenium. *Biofactors* 15 (1): 27–38.

Zhang, Y., Li, X., Huang, Z. et al. (2013). Enhancement of cell permeabilization apoptosis-inducing activity of selenium nanoparticles by ATP surface decoration. *Nanomedicine: Nanotechnology, Biology and Medicine* 9 (1): 74–84.

Zhang, Z., Du, Y., Liu, T. et al. (2019). Systematic acute and subchronic toxicity evaluation of polysaccharide-protein complex-functionalized selenium nanoparticles with anticancer potency. *Biomaterials Science* 7 (12): 5112–5123.

Zhao, G., Wu, X., Chen, P. et al. (2018). Selenium nanoparticles are more efficient than sodium selenite in producing reactive oxygen species and hyper-accumulation of selenium nanoparticles in cancer cells generates potent therapeutic effects. *Free Radical Biology and Medicine* 126: 55–66.

Zijlstra, A. and Vizio, D.D. (2018). Size matters in nanoscale communication. *Nature Cell Biology* 20 (3): 228–230.

4

Impact of Nanoparticles on Protozoa

Daniela Plachá[1,2] *and Josef Jampílek*[3,4]

[1] *Nanotechnology Centre, Energy and Environmental Technology Centre, VŠB-Technical University of Ostrava, Ostrava-Poruba, Czech Republic*
[2] *ENET Centre, Energy and Environmental Technology Centre, VŠB-Technical University of Ostrava, Ostrava-Poruba, Czech Republic*
[3] *Department of Analytical Chemistry, Faculty of Natural Sciences, Comenius University, Bratislava, Slovakia*
[4] *Institute of Neuroimmunology, Slovak Academy of Sciences, Bratislava, Slovakia*

4.1 Introduction

Protozoa is a collective term for unicellular eukaryotic heterotrophic (or mixotrophic) organisms that were previously classified in the animal kingdom (*Animalia*) because of their mobility and their inability to photosynthesize (Hausmann et al. 2003). This differs from other protists, such as algae or fungi-like protists (mucus, oomycetes, etc.). They form colonies. The fact that they previously formed a subset of animals also implies the Latin term protozoa, or "primal animals." It should be noted that the species called "protozoa" are not closely related; they have only basic similarities (a unicellular organism formed by a eukaryotic cell with the ability to move). Therefore, the classification of protozoans has undergone extensive development in recent years. However, since this chapter discusses the impact of nanoscaled materials on these organisms and is not a biologically/taxonomically focused text, and because only the original terminology can be encountered in medicine because of the conservative nature of medical classification and the need to identify organisms based on morphology, the overall classification in this chapter is based on the original terminology (Ghosh 2017).

Probably the first scientist to observe protozoans was the Dutchman Antoni van Leeuwenhoek, who called them *Animalcula*. Until the middle of the nineteenth century, they were considered animals. In 1848, Rudolf Leuckart designated them as unicellular heterotrophs (Chatterjee 2015). The most important groups of protozoa (which are included in this empire today) are as follows: *Amoebae*, *Archamoebae*, *Arcellinida*, *Mycetozoa*, *Choanozoa*, *Microspora*, *Mesomycetozoea*, *Collodictyonidae*, *Rigifilida*, *Malawimonadida*, *Rhodelphidia*, *Stramenopila*, *Alveolata*, *Rhizaria*, *Centroheliozoa*, *Spironemida*, *Diplonema*, *Heterolobosea*, *Kinetoplastida*, *Parabasalia*, *Oxymonadea*, *Retortamonadea*, and *Diplomonadea* (Ghosh 2017).

Over 30 000 species are known. Protozoa are spread wherever it is wet. They can be found in fresh and saltwater, in soil, on the surface and inside animals and plants. They play an important role in ecology – they are important in the so-called self-purification of water, and are

Nanotechnology in Medicine: Toxicity and Safety, First Edition. Edited by Mahendra Rai, Mrunali Patel, and Rashmin Patel.

part of edaphone. They are able to survive at low and high temperatures. Some species can only live in warm seas, others require cold polar seawater. However, the most suitable temperature for most protozoa is in the range of 15–25 °C. Protozoa die at temperatures above 42 °C and when exposed to ultraviolet radiation. Rhizopoda and ciliates are found in moss and they are able to tolerate large temperature and humidity variations. Marine protozoa can be found in both the seabed and at sea surface. *Foraminifera* predominantly inhabit the seabed (benthos). *Radiolaria* live mostly on the high seas (pelagic) and are a major component of marine plankton. *Flagellates* and *Foraminifera* live in seawater in smaller numbers. Parasitic protozoa are a special environmental group (Hausmann et al. 2003; Chatterjee 2015).

Most protozoa are well observable by microscope. Most often, they range from 0.005 to 0.05 mm, but some may grow to 0.5 mm. The protozoa visible to the naked eye include, for example, *Pelomyxa palustris*, which is up to 5 mm in size. The protozoa cell contains organelles that provide food movement, digestion, secretion, and reproduction that are often unparalleled in other organisms. Protozoa have one or more equivalent or functionally different nuclei. They reproduce asexually (by division or disintegration) or sexually, sometimes the individual methods are permanently alternating in a regular life cycle. They live freely or parasitically, most are characterized by mobility and a heterotrophic way of life (Hausmann et al. 2003; Chatterjee 2015).

Protozoa are chemoorganotrophs, i.e. oxidize chemical bonds in organic compounds as their energy source (Loker and Hofkin 2015) and are also responsible for the recycling of nitrogen and phosphorus. Chemoorganotrophic organisms have three different ways of obtaining nutrients. Saprotrophic nutrition is the first method of obtaining nutrients, when they are obtained from dead organic matter by enzymatic degradation. Osmotrophic nutrition is the second way when nutrients are obtained by absorbing soluble products. Holozoic nutrition is the third method, in which solid nutrients are absorbed by phagocytosis (Hausmann et al. 2003). Some protozoa are photoautotrophic, i.e. they are capable of photosynthesis (they are strict aerobes and use photosystems I and II and produce oxygen as a result of photosynthesis). Mixotrophic protists can be photoautotrophic and chemoorganotrophic simultaneously or alternately (Hausmann et al. 2003).

The protozoans are specific by their movement. Apart from passive movement caused, e.g. by the flow of water and air, their active movement can be amoeboid (i.e. "spill-over" of the cytoplasm) or with cilium or with one or more flagellum. Based on their movement, the protozoans were traditionally divided into four groups: (i) flagellates (move by flagellates, e.g. euglena); (ii) rootworms (e.g. amoebae) – move with the help of pseudopodia; (iii) ciliate (move by cilia, e.g. paramecium); (iv) sporozoa (specialized parasites, e.g. plasmodium). Reproduction is another characteristic of protozoa. It can be divided into asexual and sexual, which can take place in different hosts. Asexual reproduction is as follows: (i) binary division (the cell is divided into two equal individuals or so-called budding, when smaller buds are separated from the parent individual, which then grow) and (ii) polytomy (breakdown into more individuals). Sexual reproduction has three variants: (i) gametogamy (confluence of two independent gametes, which may be the same [isogamy] or morphologically differentiated [anisogamy]); *ii*) gamontogamy (coupling of [nuclei] gametes precedes the connection of whole cells, or at least their larger parts); and (iii) autogamy (a process of self-fertilization of the nucleus that originated in the same individual). Frequently, metagenesis (alternation of sexual and asexual reproduction) can be found (Hausmann et al. 2003; Loker and Hofkin 2015; Ghosh 2017).

As mentioned above, some protozoa are causative agents of fatal and poorly treatable human diseases but also protozoal infections are responsible for diseases of plants, animals, and some marine life. Victims are usually infected with either an insect vector or contact with an infected agent or surface (Loker and Hofkin 2015; Ghosh 2017).

Depending on the way of life, pathogenic protozoa can be divided into ectoparasites and endoparasites. Ectoparasites live on the surface of aquatic and hygrophilous animals. They are able to cause severe diseases to the hosts and often kill them. Typical representatives of this group of protozoa are *Ichthyobodo necator*, *Chilodonella cyprini*, or *Ichthyophthirius multifiliis*. Endoparasites survive in the tissues or organs of their hosts; dangerous ones include, for example, *Entamoeba coli* or *Entamoeba histolytica*, and trypanosomes or plasmodia. Most endoparasites are harmful in host bodies – depriving them of nutrients, damaging their cells, and sometimes destroying them; they release toxins. However, it is important to note that sometimes the presence of protozoa, on the other hand, helps the host, e.g. termites that have flagellates in their intestines, or ruminants with rumen ciliates. These protozoa allow their hosts to digest food better (Chatterjee 2015; Loker and Hofkin 2015; Ghosh 2017).

Nevertheless, it can be stated that pathogenic protozoa cause many important diseases of humans and other organisms. An overview of protozoan diseases is given in Tables 4.1 and 4.2. Different types of protozoa are adapted to develop and live in different tissues. The way of infection by protozoa depends on where the protozoa live in the body. Intestinal protozoal infections result from the fecal-oral route, in which contaminated food or water is ingested or personal contact with the infected person occurs. Infection by protozoa that live in blood or tissues can be transmitted by biting infected insects (mosquitoes, flies, and sandstones). Some protozoa inhabit the genital area and can be sexually transmitted by any form of genital touching (Chatterjee 2015; Loker and Hofkin 2015; Ghosh 2017; FishBase 2020; Shah 2020).

As it follows from the above, protozoa are responsible for a number of human and other infections. Many of these infectious diseases are most common in the tropics, areas with hazardous drinking water, poor hygiene, poor housing, and little or no access to health care. It is therefore clear that protozoal infections fall into the category of so-called neglected diseases. These are diseases that affect poor regions (third-world countries of Asia, Africa, and Latin America) or the poorest in the developed "western" countries. For example, in the United States, it affects up to 1.46 million families, including 2.8 million children who live for less than $2/day. In countries like this, the burden of neglected diseases is often overshadowed by other public health issues (NIH 2020; WHO 2020a).

The World Health Organization (WHO) estimates that more than one billion people (one-sixth of the world's population) suffer from one or more neglected diseases. The list of WHO neglected diseases includes amebiasis, balantidiasis, malaria, sleeping sickness, Chagas disease, giardiasis, and leishmaniasis as protozoal infections (NIH 2020; WHO 2020a). Neglected diseases mean that illnesses are overlooked by drug developers or other drug access assistants such as government officials, public health programs, and news media. Usually, pharmaceutical companies cannot recover the costs of developing and manufacturing treatments for these diseases. Another reason why neglected diseases are not considered high priorities in prevention or treatment is that they do not usually affect people living in developed countries of the world. Neglected diseases also lack visibility, as they usually do not cause dramatic outbreaks, killing large numbers of people. Rather, these diseases exceed their toll for a longer period of time, leading to paralyzing deformities, severe disabilities, and/or relatively slow deaths (Jampílek 2016; NIH 2020; WHO 2020a).

Prevention strategies vary from disease to disease. However, the way to counter many neglected diseases is to provide people with

Table 4.1 List of selected most important protozoan diseases in humans.

Disease	Causative agent
Acanthamoeba keratitis	*Acanthamoeba*
African sleeping sickness (African trypanosomiasis)	*Trypanosoma brucei gambiense, Trypanosoma brucei rhodesiense*
Amoebiasis	*E. histolytica*
Babesiosis	*Babesia* spp.
Balantidiasis	*Balantidium coli*
Chagas disease (American trypanosomiasis)	*Trypanosoma cruzi*
Cryptosporidiosis	*Cryptosporidium* spp.
Cyclosporiasis	*Cyclospora cayetanensis*
Dientamoebiasis	*Dientamoeba fragilis*
Giardiasis	*Giardia lamblia*
Leishmaniasis	*Leishmania chagasi, Leishmania donovani, Leishmania infantum, Leishmania major, Leishmania tropica*
Malaria	*Plasmodium falciparum, Plasmodium knowlesi, Plasmodium malariae, Plasmodium ovale, Plasmodium vivax*
Toxoplasmosis	*Toxoplasma gondii*
Trichomoniasis	*Trichomonas vaginalis*

safe drinking water sources, good hygiene, adequate housing, and access to health care. For insect-borne diseases, measures such as pesticides or bed-nets can help prevent or reduce the risk of infection. For diseases transmitted by organisms that live in water or mud, people can reduce the risk of infection by not swimming or wading in water with a high risk of contamination. Other measures include administering medicines to people that prevent infection or significantly minimize the impact of the infection if it occurs. Common examples are antimalarial drugs, which many take to prevent malaria. Yet, even at a cost of less than 50 cents a day, such drugs remain too expensive for many poor people. In addition to drugs, scientists are trying to develop vaccines that would encourage the body's immune system to avert disease agents. However, the main barriers, especially economic ones, remain. Neither the inhabitants nor the governments of the affected regions have the finances in their budgets to cover the dispensation of medicines to all affected people (NIH 2020; WHO 2020a).

Therefore, many people with neglected diseases do not receive adequate treatment. One reason for this is that due to a lack of health services, they will not be diagnosed with the disease until they are in an advanced stage that is difficult to treat. Although people are diagnosed early, many cannot afford effective medications. Finally, there is no good treatment for some neglected diseases. Public health professionals are working to develop better ways to provide health reports and services to low-income people. They also educate officials in developing countries on why it makes good economic sense to provide health care and treatment to people suffering from neglected diseases. From a medical point of view, researchers are looking for cheaper medicines for neglected diseases and are trying to develop new medicines for illnesses that lack treatment. Although in recent years, public and nonprofit groups have teamed up with

Table 4.2 Examples of protozoan diseases in animals.

Disease	Causative agent
Ambiphyra infection	*Ambiphyra ameiurus*
Amoebiasis	*E. histolytica*
Anaplasmosis	*Anaplasma marginal*
Apiosoma infection	*Apiosoma* sp.
Babesiosis	*Babesia bigemina*
Balantidiasis	*B. coli*
Besnoitiosis	*Besnoitia bennetti, Besnoitia darlingi, Besnoitia jellisoni, Besnoitia saurian, Besnoitia tarandi, Besnoitia wallacei*
Chilodonelliasis	*Chilodonella cyprini, Chilodonella hexasticha*
Coccidiosis	*Eimeria acervulina, Eimeria hagani, Eimeria maxima, Eimeria necatrix, Eimeria nicatrix, Eimeria tenella, Eimeria zurnii*
Costiasis	*Ichthyobodo necatrix*
Cryptobiosis	*Cryptobia branchialis, Cryptobia iubilans*
Cryptosporidiosis	*Cryptosporidium parvum, Cryptosporidium hominis, Cryptosporidium canis, Cryptosporidium felis, Cryptosporidium meleagridis, Cryptosporidium muris*
Epistilis infection	*Epistylis* sp.
Giardiasis	*Giardia canis, G. lamblia*
Hexamitosis	*Hexamita salmonis*
Ichthyophthiriosis	*I. multifiliis*
Leishmaniasis	*Leishmania archibaldi, L. chagasi, L. donovani, L. infantum*
Malaria	*P. falciparum, P. knowlesi, P. malariae, P. ovale, P. vivax*
Microsporidiosis	*Glugea stephani, Nosema podocotyloidis, Nosema apis*
Myxozoasis	*Tetracapsuloides bryosalmonae, Myxobolus cerebralis; Enteromyxum leei*
Neosporosis	*Neospora caninum, Neospora hughesi*
Perkinsosis	*Perkinsus beihaiensis, Perkinsus chesapeaki, Perkinsus honshuensis, Perkinsus marinus, Perkinsus mediterraneus, Perkinsus olseni, Perkinsus qugwadii*
Piroplasmosis	*Theileria equi*
Piscinoodinium infection	*Piscinoodinium pillulare*
Sarcocystosis	*Sarcocystis hominis, Sarcocystis suihominis*
Tetrahymena diseases	*Tetrahymena* sp.
Theileriosis	*Theileria annulata, Theileria mutans, Theileria ovis, Theileria parva*
Toxoplasmosis	*T. gondii*
Trichodinosis	*Trichodina heterodentata, Trichodina nigra, Trichodina nobilis, Trichodina reticulata, Trichodina acuta, Trichodina kupermani, Trichodina mutabilis*
Trichophrya infection	*Trichophyra piscium*
Trypanosomiasis	*Trypanosoma brucei, Trypanosoma congolense, Trypanosoma vivax; T. cruzi*

pharmaceutical companies to produce cheaper and more effective treatments, and some progress has been made, the number of drugs developed specifically for these neglected diseases still lags behind drugs for other diseases. Behind all, we need to look for the small economic incentives that the private pharmaceutical sector receives (Jampílek 2016; NIH 2020; WHO 2020a).

4.1.1 Antiprotozoal Drugs

Antiprotozoics are a class of pharmaceuticals used in the treatment of protozoan infections that are caused by various pathogenic protozoans. Some of these infections can be found around the world (e.g. amoebiasis, giardiasis, coccidiosis, trichomoniasis, toxoplasmosis), some are in subtropic or tropic parts (e.g. malaria, trypanosomiases, leishmaniasis). Except special antiprotozoics (will be mentioned below), antibacterial chemotherapeutics are used also for treatment: (i) tetracyclines (doxycycline); (ii) aminoglycosides (paromomycin); (iii) macrolides (erythromycin, spiramycin); (iv) lincosamides (clindamycin); (v) sulfonamides (sulfadoxine, sulfamethoxypyridazine); (vi) nitroimidazoles/nitrofurans (metronidazole, tinidazole, ornidazole, nimorazole, furazolidone); and antifungals such as (i) imidazoles/triazoles, (ii) macrolides (amphotericin B); or disinfectants and antiseptics (ethacridine) (Jampílek 2014, 2016, 2017, 2018; Griffith 2020).

Classical antiprotozoals can be divided into several classes of agents: (i) organometallic compounds; (ii) halogenated quinolinol-8-ols; (iii) isoquinolines; (iv) amino acids; (v) compounds with nitro moiety; (vi) amide-like compounds (amides, ureas, amidines); (vii) naphthoquinones; and (viii) antimalarials (quinine-based derivatives, phenanthrenes, biguanides, artemisinin-based derivatives). Many of these compounds are very old; they were developed in the early twentieth century to protect colonial English and French soldiers, especially in Africa, from tropical diseases and are still in use today. Since this is a very heterogeneous group of agents, there are also different mechanisms of actions, including (i) binding to SH moieties of enzymes (arsenic-based compounds); (ii) proteosynthesis blocking (emetine-based compounds); (iii) carbohydrate metabolism (suramin); (iv) ornithine decarboxylase inhibition (eflornithine); (v) interference with DNA, RNA (quinine-based compounds, guanidines); (vi) hemoglobin metabolism, interference with Fe (quinine-based compounds, artemisinin derivatives). It should be noted that for many antiprotozoals, mechanisms of actions are not known (Jampílek 2014, 2017; Griffith 2020).

Organometallic antiprotozoal drugs include arsenic- and antimony-based agents. From the class of arsenic-based compounds with As(V) acetarsol (against amoebiasis, giardiasis, trichomoniasis) and tryparsamide (against trypanosomiases) were used. Both are prodrugs that were metabolized (oxidation of arsenobenzenes or reduction of phenylarsenic acids) to highly active arsine oxides. Less toxic arsenic As(III) agents comprise inhibitor of trypanosomial pyruvate kinase melarsoprol used for treatment of trypanosomiases. Sodium stibogluconate and meglumine antimoniate are antimony Sb(V)-based drugs used for the treatment of leishmaniosis (Griffith 2020).

Halogenated quinolinol-8-ols, such as cloroxine, clioquinol, diiodohydroxyquinoline, are typical disinfectants and being poorly absorbed drugs from the gastrointestinal tract are used as luminal amebicides acting by chelation of ferrous ions essential for metabolism. Isoquinolines emetine and synthetical dehydroemetine are used for the treatment of amoebiasis (Griffith 2020).

Eflornithine (α-difluoromethylornithine), an inhibitor of ornithine decarboxylase, is used for the treatment of African trypanosomiasis (sleeping sickness). A urea-amide derivative suramin effective against trypanosomiases is an inhibitor of enzymes involved in the oxidation of reduced nicotinamide adenine dinucleotide (NADH), which functions as a coenzyme in many reactions, such as respiration and glycolysis. Diamidines such as pentamidine, propamidine, hexamidine are used for treatment of stage I infection with *T. brucei gambiense*. Naphthoquinone – analogue of ubiquinone – atovaquone is administered for treatment of toxoplasmosis, malaria (combination with proguanil), or babesia (combination with azithromycin) (Jampílek 2014, 2017; Griffith 2020).

Among the antiprotozoal drugs, many compounds can be found whose effect is dependent on the presence of the nitro group. These compounds usually also contain amide bonds either in the chain or cyclized within the heterocycles. Nifurtimox and benznidazole are antitrypanosomal drugs for the therapy of Chagas disease. Fexinidazole can be used against African and American trypanosomiases. They form a nitroanion radical metabolite that reacts with nucleic acids of *T. cruzi* causing a breakdown of DNA. Nitazoxanide, nitrothiazolyl-salicylamide prodrug, interferes with pyruvate-ferredoxin oxidoreductase, affects electron transfer in anaerobic metabolism and is used for treatment of cryptosporidiosis and giardiasis. Another prodrug diloxanide furoate is a luminal amebicide hydrolyzing in the intestine to active 2,2-dichloro-*N*-(4-hydroxyphenyl)-*N*-methylacetamide. Triazinone derivatives such as diclazuril, letrazuril, clazuril, or ponazuril, toltrazuril are applied for the treatment of coccidiosis and cryptosporidiosis (Griffith 2020).

Historically, most drugs have been developed as malaria therapeutics. Most of the compounds used are based on the structure of quinoline and therefore they have very close mechanisms of action, which disqualifies the whole group in the fight against resistant Plasmodia. Mefloquine (8-trifluoromethylquinoline) has a specific structure. A large group of drugs is derived from 8-aminoquinoline (pamaquine, primaquine, tafenoquine) and 4-aminoquinoline (chloroquine, hydroxychloroquine, amodiaquine). Mepacrine and pyronaridine as amino-acridines also have a close structure and they are used for the treatment of malaria and giardiasis. Phenanthrenes (fluorene) can be considered as higher isosteric homologues of quinine and mefloquine and thus halofantrine and lumefantrine can be used against resistant *P. falciparum*. They inhibit conversions of toxic free heme released from hemoglobin to pigment hemozoin that is nontoxic for plasmodia. Biguanides are able to inhibit dihydrofolate reductase, prevent the biosynthesis of purines and pyrimidines, thereby halting processes of DNA replication, cell division and reproduction. Proguanil (prodrug metabolized to cycloguanil) and chloroproguanil belong to this group that can be used against *P. falciparum* and *P. vivax*. Pyrimethamine (2,4-diaminopyrimidines) close to biguanides is another agent used to inhibit dihydrofolate reductase. Artemisinin-based derivatives are the last large group of approved antiprotozoals – antimalaria drugs. All compounds are based on the structure of artemisinin used in Chinese traditional medicine isolated from the plant *Artemisia annua*, now produced using genetically engineered yeast. Artemisinin is a sesquiterpene lactone with a peroxide bridge that is crucial for efficiency. A possible mechanism of action may be in the perturbation of redox homeostasis by interfering with Fe and producing oxygen radicals. Reduction of artemisinin results in the active metabolite dihydroartemisinin, its etherification permitted preparation of artemether, artesunate, artemotil, or arterolane. These drugs form a pillar of drugs against resistant *P. falciparum* and *P. vivax* in combination with mefloquine (Jampílek 2014, 2017; Griffith 2020).

Based on the above analysis of the market for available antiprotozoal drugs, it can be stated that the development of this type of drugs is not one of the priorities of the pharmaceutical industry. On the other hand, it is also necessary to understand the attitude of private pharmaceutical manufacturers to whom the profits from the sale of the drug must cover the costs incurred in research and development. From this perspective, the development of these compounds is risky (Jampílek 2016, 2017, 2018). In recent years, many companies have begun to specialize in galenical modifications of existing drugs to expand the range of available and effective formulations, albeit based on structurally new molecules, by modifying the formulation, i.e. controlled release, targeted distribution, and possibly fixed combinations of synergistic compounds. One way to galenically improve existing molecules and to give them innovativeness is to apply nanoformulations (Jampílek and Kráľová 2017a, 2018a, 2019a, b, c, 2020a, b, c; Jampílek et al. 2019a, b, c; Mohapatra et al. 2019).

This chapter gives a comprehensive overview of recent findings related to the biological activity of various types of nanoparticles against pathogenic protozoa. The impact of various nanoformulations on the environment, including potential deleterious effects on non-target organisms, is discussed as well.

4.2 Nanosystems

Nanotechnology has begun to evolve with the new millennium, and since the second decade of the twenty-first century, we can say that it has been applied in a wide range of human activities (Wennersten et al. 2008; Drexler and Pamlin 2013). Nanotechnology takes advantage of the specific properties of materials that begin to emerge when the size of a particular material is in the range of 1–100 nm. Apart from industry and various materials, nanotechnology can be found in agriculture and the food industry and, of course, in the biomedical fields (Jampílek and Kráľová 2017a, b, c, 2018a, b, c, 2019a, b, c, d, e, 2020a, b, c, d; Jampílek et al. 2019a, b, c, 2020). Thus, the application of nanotechnology is an excellent alternative to improving the pharmacokinetic and pharmacodynamic profile of existing anti-invasive drugs. Nanomaterials are an alternative approach to the treatment and alleviation of infections caused by resistant pathogens. Although mechanisms of bacterial resistance to nanoparticles (NPs) of silver based primarily on NP agglomeration have already been entrenched (Panáček et al. 2018), it does not seem likely that such a type of resistance is widespread among pathogens. Unlike conventional anti-invasive chemotherapeutic agents, nanomaterials exhibit activity against pathogens through various mechanisms (Smith et al. 2013; Jampílek and Kráľová 2017a, 2018a, b, 2019c, 2020b, c, d; Elkodous et al. 2019; Plachá and Jampílek 2019; Jampílek et al. 2020).

Besides, the use of various nanosystems/nanoformulations can increase the bioavailability of the active agents and modify the route of administration. As mentioned, specific nanoformulations also provide a controlled release or targeted biodistribution system. Based on these facts, smaller amounts of active substances can be used, i.e. dose-dependent toxicity and thus various side/unfavorable effects can be reduced. Increased efficacy of individual medicaments can also be provided by fixed drug combinations, drug combinations with metal- or metalloid-based NPs, or by biologically active matrices that are used as stabilizers. In addition, many formulations also protect bioactive agents from degradation (Ma et al. 2013; Yah and Simate, 2015; Demetzos 2016; Aderibigbe 2017; Jampílek and Kráľová 2017a, 2018a, 2019a, b, c, 2020a, b, c; Khezerlou et al. 2018; Patra et al. 2018; Rizvi and Saleh 2018; Elkodous et al. 2019; Jain and Thareja 2019; Jampílek et al. 2019a, b; Li et al. 2019; Mohapatra et al. 2019; Rodrigues et al. 2019; Abo-Zeid and Williams 2020; Khan et al. 2020; Paunovic et al. 2020; Tavakoli and Hashemzadeh 2020). For example, Sun et al. (2019) reviewed the increased effects of various NPs, including the benefits of transport and distribution in the body. Such an overview can help to understand the trend of NPs in the treatment of parasitic diseases and to estimate new strategies in the treatment of parasitic infections in the future. Frequently used nanosystems, such as micro- and nanoemulsions, liposome drug delivery systems (DDSs), lipid-based NPs or nanospheres, which are mainly used to encapsulate natural compounds, can significantly increase the stability, bioavailability, and efficacy of these natural compounds against protozoa of the genera *Leishmania*, *Plasmodium*, and *Trypanosoma* (Pavoni et al. 2019; Souto et al. 2019).

Leishmaniasis, Chagas disease, and African trypanosomiasis cause more deaths worldwide than all other neglected diseases together. Treatment is limited to a few drugs with limited efficacy that have harmful or potentially fatal side effects at therapeutic doses. It is therefore logical that NP delivery systems have emerged as a promising area of research in neglected disease therapy and prevention. These delivery systems provide novel mechanisms for targeted drug delivery in the host, maximize therapeutic effects, and minimize systemic side effects. Simultaneously approved drugs can also be repackaged using these delivery systems, allowing their potential use in neglected diseases of protozoan origin (Santos-Valle et al. 2019; Volpedo et al. 2019). It is also important to note that a variety of metal-based nanomaterials are also used as robust and accurate diagnostics (Wang et al. 2017; Elkodous et al. 2019; Markwalter et al. 2019). For example, Benelli (2018), Jampílek and Kráľová (2019e), and Jampílek et al. (2020) described broad applications of AuNPs ranging from detection techniques to drug development, against a large number of parasites and as insecticides killing insect vectors, especially mosquitoes such as *Aedes aegypti*, *Anopheles stephensi*, and *Culex quinquefasciatus*. AuNPs show significant activity against malaria, leishmaniasis, toxoplasmosis, trypanosomiasis, and cryptosporidiosis. AuNPs can also be used very well as biomarkers, diagnostic tools (Pedrosa et al. 2015; Zhou et al. 2015), and immune-blocking agents in the study of a malaria vaccine (Kumar et al. 2015; Dykman 2020; Poon and Patel 2020). Various nanosystems are used for the development of novel vaccine adjuvants and nano-delivery targeting systems ensuring also the stability of agents. Great progress has also been made in vaccine adjuvants and DDSs based on biodegradable polymer NPs (Jin et al. 2019; Marcandalli et al. 2019; Yan et al. 2020). On the other hand, several inorganic nanomaterials, such as AuNPs, carbon nanotubes (CNs), and quantum dots (QDs), have also been found to be able to positively modulate the immune response of the body to pathogens (Teng et al. 2018; Hess et al. 2019; Sanchez-Guzman et al. 2019).

4.2.1 Preparation and Synthesis of Nanoparticles and Systems

In recent years, progress in the development of methods for the preparation of NPs of various origins has been enormous. Despite various methods of the NPs' preparation or synthesis, two main approaches are distinguished: "bottom-up" methods and "top-down" methods. It can simply be said that the difference between these two approaches lies in the starting material used for the preparation. In "top-down" methods, a macroscale bulk material is diminished and NPs are formed; in "bottom-up" methods, NPs are built from atoms or molecules (Jamkhande et al. 2019; Shrestha et al. 2020). Generally, nanomaterials are materials that possess at least one external dimension in 1–100 nm. Thus, one-dimensional nanotubes or nanowires, two-dimensional nanofilms or nanosheets and bulk nanomaterials with nanoscale structures can be distinguished. NPs have all three dimensions below 100 nm; they consist of various materials giving them specific features. Thus, metal and metal oxide NPs, carbon-based NPs, inorganic nonmetallic–ceramic NPs, semiconductor NPs with properties between metals and nonmetals, polymeric and lipid-based NPs are often studied for many prospective technological applications (Khan et al. 2019). Due to their size, they have enhanced and unique physicochemical

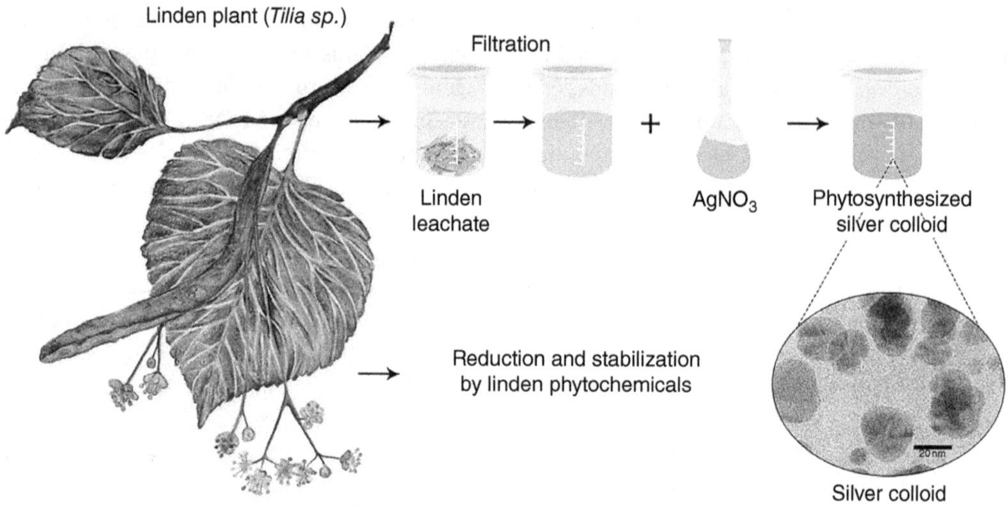

Figure 4.1 Scheme of AgNPs' biosynthesis and stabilization using linden leaves, bracts, and flowers. *Source:* Kratošová et al. (2019) with permission of Elsevier.

properties due to quantum confinement effects, such as electronic, magnetic, optical, mechanical, thermal, and structural properties. One of the main characteristics is significantly increased surface area-to-volume ratio, which influences the rate of reactions controlled by surface area or defect site occurrence (Darr et al. 2017).

Methods of preparation/synthesis are based on various mechanical, physical, chemical, or biological processes. "Top-down" syntheses use destructive and physical methods such as grinding, milling, cutting, etching, sputtering, and thermal/laser ablation such as lithography (optical, E-beam, soft and nanoimprint, block co-polymer, scanning probe) (Khan et al. 2019). The starting material is mainly a solid material; however, aerosol-based techniques (electrospraying and flame pyrolysis) and liquid-phase techniques (electrospinning) are also well described (Ayuk et al. 2017). Atomic layer deposition, sol-gel methods, molecular self-assembly, chemical or physical vapor deposition (CVD, PVD), and precipitation methods (chemical reductions, microemulsion, electrolysis, sonochemical, or solvothermal) are examples of "bottom-up" approach (Vaculíková et al. 2012, 2014, 2016; Sosna et al. 2019). The starting material is usually in the gaseous or liquid phase. The main disadvantage of physical methods is their energetic and economic loads; in the case of chemical methods, toxic solvents are often used. In addition, NPs usually have to be stabilized or shielded against aggregation. Biological methods of NP synthesis are called "green syntheses or biosyntheses." Different biomass such as algae, bacteria, fungi, yeast, or plant parts, isolated pure biomolecules, commercially available nutritional products or biowaste after mixing in aqueous solutions with precursors such as various inorganic salts reduced metal NPs from these precursors. The reduction is controlled by biomolecules such as phenolic compounds, enzymes, glycosides, flavonoids, and others; however, there are so many possibilities in choosing biomass that mechanisms have not been completely explained so far. The advantage of the biosynthesis process is that NPs are stabilized by biomolecules that are naturally present in the aqueous solutions. Thus, it is one-step of NPs formation and stabilization. Biosynthesis using extract from linden flowers, bracts, and leaves as biomass and $AgNO_3$ as precursor of AgNPs is presented in Figure 4.1 (Kratošová et al. 2019).

4.3 Nanosystems with Effect on Human Parasitic Protozoa

4.3.1 Malaria

Malaria is one of the most devastating diseases. In 2018, an estimated 228 million cases of malaria occurred worldwide. Most cases of malaria were in the African region in 2018 (213 million or 93%), followed by the Southeast Asia WHO region with 3.4% and the Eastern Mediterranean WHO region with 2.1%. In 2018, an estimated 405 000 malaria deaths were estimated worldwide. A major danger is the increasing resistance of insect vectors to insecticides of pyrethroid and organochlorine groups. There is also an increase in the number of cases caused by chloroquine and artemisinin-resistant *P. falciparum* and *P. vivax* (WHO 2019). This disease represents the challenge of the twenty-first century. In his review study, Rahman et al. (2019) summarizes the potential of nanobiotechnology for treating a patient's vector fight. Advantageously, metal NPs as well as plant extracts encapsulated in biodegradable polymers can be used to control the propagation of Plasmodia (Gnanadesigan et al. 2019). A huge number of nanoformulations of various types of insecticides are being developed, tested, and already used to kill insect plasmodia vectors (Jampílek and Kráľová 2019d, e; Jampílek et al. 2019a, 2020). However, the text below will focus on nanoformulations that are effective against their protozoa.

Chitosan tripolyphosphate (CS-TPP)-conjugated chloroquine (CQ) NPs (Tripathy et al. 2012) were administered at a dose of 250 mg/kg bodyweight for 15 days to Swiss mice infected with *Plasmodium berghei* NK65. CQ-CS-TPP NPs were found to be effective in reducing stress markers, pro-inflammatory cytokines and showing increased efficacy against *P. berghei*, thus preventing tissue damage during malaria infection (Tripathy et al. 2013, 2014). The combination of CQ and MgO NPs was tested *in vivo* on *P. berghei*. A sol-gel method was used for MgO NPs synthesis. The formed MgO NPs had the concentration of 10 mg/ml, they were of cylindrical and spherical shapes and had size under 10 nm. They were mixed with different amounts of CQ (25, 12.5, and 6.25 mg/kg) and tested *in vivo* on *P. berghei*. Mice were infected by *P. berghei* ANKA and divided into five groups which were treated by various doses of CQ and MgO NPs: the first was with no CQ and MgO, the second was only with CQ, the next four were administered with the same amount of MgO NPs and different CQ amounts. The antiplasmodial activity shows that CQ enhanced with MgO NPs was more effective (Hadiza et al. 2018).

Lipid nanoemulsions (NEs) of primaquine (PQ) had increased bioavailability and decreased uptake of PQ (25%) in *P. berghei* infection animal model (Singh and Vingkar 2008). PQ incorporated into liposomes conjugated to heparin showed *in vitro* antiplasmodial activity (Marques et al. 2014). PQ loaded in a gelatin nanopolymer had an eightfold higher concentration in the bloodstream due to significantly increased accumulation than free drug molecules (Kumar et al. 2017).

Artesunate (ATS, Artesunate 2020) is one of the frequently used therapeutics of CQ-resistant malaria. Its antimalarial activity is based on selective, irreversible, and iron-mediated damage to the parasite. Kannan et al. (2019) combined ATS with nontoxic biocompatible surface-modified iron oxide NPs that were designed and synthesized for sustained release of Fe^{2+} in a pH-dependent parasitic food vacuole to improve reactive oxygen species (ROS) efflux. Nanomaterial was evaluated in an experimental mouse malaria model (artemisinin-resistant *P. falciparum* R539T) to increase the selectivity and toxicity of ATS to the parasite. It was found that ATS enriched with NPs was highly effective as early as one-eighth of the concentration of ATS without iron NPs, there was a fatal damage to the parasite by increased ROS production.

Solid lipid NPs of artemether (ATM) formed by drug homogenization with tripalmitin and

Cremophor with size of 230.2 nm, zeta potential −31.45 mV, encapsulation efficiency 95.43%, and sustained drug release for 48 hours had 70% higher activity against *P. berghei* compared with ATM alone (Nisha et al. 2013). A 100% cure rate with the NE with peanut oil and arteether at a dose of 12.5 mg/kg/body weight in animal models of *Plasmodium yoelii* was described by Dwivedi et al. (2015). The combination of ATM and lumefantrine in a lipid NE with a lipid size of 150 nm showed 90% inhibition of parasitaemia at a concentration of 0.27 mg/kg (Ma et al. 2014). The round form of solid lipid NPs of ATM and lumefantrine prepared by a combination of glycerol monostearate and Pluronic F68 with a particle size of 145 nm, zeta potential −66 mV, encapsulation efficiency 82% for ATM and 79% for lumefantrine, and sustained drug release for 30 hours at 4 + 24 mg/kg/day in an animal model of *P. berghei* showed excellent antiparasitic activity (Parashar et al. 2016).

Probably one of the first reports of *in vitro* antimalarial activity of protoporphyrin IX-conjugated colloidal Ag against CQ-sensitive *P. faciparum* with an IC_{50} of 15.5μM (IC_{50} of CQ was 0.022 μM) was published by Begum et al. (2003). Colloidal mono- and dinuclear complexes of Ag with mono- and bis(*N*-heterocyclic carbenes) and tested against the CQ-resistant strain of *P. falciparum* were performed by Hemmert et al. (2013). Unfortunately, the most effective agents had strong hemolytic properties even at the weakest doses tested (0.5 mg/ml).

Metal NPs prepared using green biosynthesis can have a high potential to be new antiplasmodial drugs (Barabadi et al. 2019). For example, green-synthesized AgNPs using purified α-amylase or aqueous leaf extracts of Ashoka tree (*Saraca asoca*) and neem tree (*Azadirachta indica*) resulted in the preparation of triangular to hexagonal shapes of size 22–44 nm or after using extracts of spherical shapes of size 2–8 nm, and 5–20 nm, respectively. The prepared AgNPs showed antiplasmodial activity against *P. falciparum* with IC_{50} 3.75g/ml (α-amylase NPs), 8 g/ml (Ashoka NPs), and 30 g/ml (neem NPs) (Mishra et al. 2013).

As already mentioned, NPs can be biosynthesized using plant leaves or microorganisms, such as bacteria and fungi. Many research studies were focused on AgNPs, which have often spherical shapes and sizes in the range from 4 to 65 nm and they show no or little cytotoxic effect. However, the results of IC_{50} values obtained so far can be divided into three categories of NPs: (i) more active than positive controls, (ii) more active than the plant extracts, and (iii) those less active than positive controls or plant extracts (Foko et al. 2019).

Biosynthesized and stabilized TiO_2 NPs were produced using the aqueous extract of *Momordica charantia* leaves as a reducing and stabilizing agent. The highest mosquitocidal activity was confirmed against *A. stephensi* Liston larvae and pupae, the LC_{50} values were determined to be 2.50 mg/l (I instar), 2.86 mg/l (II), 3.29 mg/l (III), 3.43 mg/l (IV), and 5.04 mg/l (pupa). The antimalarial activity and the IC_{50} values were tested against CQ-sensitive (CQ-s) and CQ-resistant (CQ-r) strains of *P. falciparum*. While the IC_{50} values of *M. charantia* leaf aqueous extract were 88.14 μg/ml (CQ-r) and 83.64 μg/ml (CQ-s), the IC_{50} of TiO_2 NPs were 59.71 μg/ml (CQ-r) and 53.42 μg/ml (CQ-s). No obvious toxicity of TiO_2 NPs was observed on *Poecilia reticulata* after 24 hours of exposure (Gandhi et al. 2018).

Various metal oxide NPs, namely Al_2O_3, CeO_2, Fe_3O_4, ZrO_2, and MgO and coated with polydidecylsilane (PDDS) were tested for their antiplasmodial activity against *P. falciparum*. The NPs showed the following IC_{50} values: Al_2O_3 (71.42 ± 0.49 μg/ml), CeO_2 (82.08 ± 0.51 μg/ml), Fe_3O_4 (77.23 ± 0.42 μg/ml), ZrO_2 (81.11 ± 0.53 μg/ml), and MgO (72.33 ± 0.37 μg/ml). Coated NPs had the following activity: 69.97 ± 0.46, 67.06 ± 0.61, 48.66 ± 0.45, 79.66 ± 0.49, and 60.28 ± 0.42 μg/ml, respectively. PDDS had 49.63 ± 0.38 μg/ml. Thus, the PDDS-Fe_3O_4 formulation was the most active and can be used for further development (Inbaneson and Ravikumar 2013).

Graphene oxide (GO) has become a very popular nanomaterial for various biomedical applications, although the exact mechanism of interactions between GO-based materials and biosystems has not been elucidated (Plachá and Jampílek 2019). Kenry et al. (2017) evaluated GO nanosheets activity against *P. falciparum* and found that GO nanosheets significantly inhibit plasmodial invasion of red blood cells. Furthermore, GO nanosheets stop the plasmodial development from the ring to the trophozoite stage. Graphene quantum dots (GQDs) represent another graphene-based nanomaterial useful for combating plasmodia and one of its vectors, *A. stephensi*. Lethal doses expressed as LD_{50} of GQD against *A. stephensi* ranged from 0.157 (larva I) to 6.323 ppm (pupa). The antiplasmodial activity of GQDs was evaluated against CQ-sensitive (CQ-s) and CQ-resistant (CQ-r) strains of *P. falciparum*; IC_{50} values were 82.43 (CQ-s) and 85.17 μg/ml (CQ-r). *In vivo* tests in mice infected with *P. berghei* showed mild GQD activity compared to CQ (Murugan et al. 2017).

Polyamidoamine (PAA) NPs were conjugated to CQ and orally administered to cure mice infected by *P. yoelii*. The activity of the free drug was slightly improved and the mice immunity against malaria was induced. PAA NPs highly adhere to *P. falciparum* proteins as confirmed by liquid chromatography with tandem mass spectrometry analysis of PAA ligands. The results showed a preferential binding of PAAs to erythrocytes infected by *P. falciparum* compared to noninfected red blood cells. Some PAA NPs induce inhibition of parasite invasion and their potential for the treatment of coinfections such as *Leishmania* and *Plasmodium* can be estimated due to polymer intake by macrophages. Thus, PAA NPs were proposed to encapsulate orally administered antimalarial drugs and also to directly administrate antimalarials to mosquitoes, targeting mosquito stages of *Plasmodium* (Coma-Cros et al. 2018).

4.3.2 Trypanosomiases

The genus *Trypanosoma* can be found in tropical Africa, where its representatives *T. brucei* (*brucei*, *rhodesiense*, *gambiense*) cause sleeping sickness, as well as in South America where *T. cruzi* is the cause of Chagas disease. African trypanosome is transmitted in the saliva of tsetse fly (*Glossina spp.*), while South American trypanosome in the excrement of insects from the subfamily *Triatominae* (especially *Triatoma infestans*). *T. b. gambiense* is present in 24 countries of West and Central Africa and currently accounts for 98% of cases and causes chronic infection. *T. b. rhodesiense* is found in 13 countries of eastern and southern Africa, this form accounts for less than 2% of reported cases but causes acute infection. Other subspecies of the genus *Trypanosoma* are pathogenic to animals and cause trypanosomiasis in wild and domestic animals such as cattle called Nagana. These trypanosomiasis of livestock are a major problem for the development of agricultural areas. In addition, these other species can also be infected secondarily by humans, and domestic and wild animals also host species pathogenic to humans, making animals an important reservoir (CDC 2019; WHO 2020b, c).

While the number of people infected with African trypanosomias are stagnating or slightly decreasing (about 10 000 new cases are reported to the WHO each year; however, many cases are not diagnosed and reported), it is estimated that many more people are suffering from Chagas disease worldwide, because the epidemiological pattern has changed in recent decades as a result of population mobility. As a result, an increased number of cases have been detected in Canada, the United States and many European countries and some African, Oriental, and West Pacific countries. Due to the high number of people remaining undiagnosed or untreated, in combination with areas with remaining active transmission, the risk of infection is estimated at 75 million people (CDC 2019; WHO 2020b, c, d).

In recent years, innovative DDSs of drugs for the treatment of Chagas disease have been described by Souto et al. (2019) and by Quezada et al. (2019). Conventional treatment was based on gentian violet, nifurtimox (NFX), and

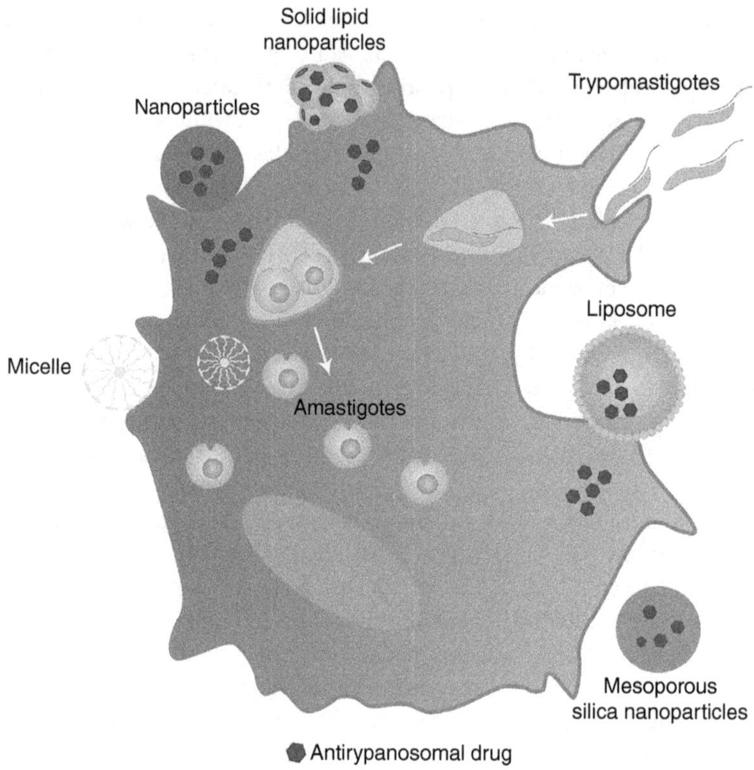

Figure 4.2 Drug delivery systems based on nanomaterials which were tested to cure Chagas disease including mesoporous silica NPs, liposomes, micelles, polymeric, and non-polymeric NPs. *Source:* Adapted from Quezada et al. (2019).

benznidazole (BZD). Even though nifurtimox has a strong activity in pathogen elimination, it was forbidden for use in several countries due to adverse effects on the peripheral nervous and central system with gastrointestinal manifestations. Also, BZD is known to have negative effects on hypersensitivity skin responses, and neurologic and bowel manifestations (Gaspar et al. 2015; Chatelain 2017). Thus, a combination therapy based on BZD and other azole derivatives, such as itraconazole, is one of the possibilities in Chagas diseases therapy (Diniz et al. 2013) or nitric oxide administered together with BZD, which prevents death from parasitemia and heart inflammation (Sesti-Costa et al. 2014). New drugs were tested in many medical researches such as posaconazole and prodrug of ravuconazole (E1224) (Diniz et al. 2013; Molina et al. 2015) or natural therapeutic molecules based on terpenes, alkaloids, heterocyclics, steroids, and others (Jones et al. 2013; Cheuka et al. 2016; Sales et al. 2017; Zulfiqar et al. 2017).

New challenges occur in regard to DDS based on nanomaterials, such as liposomes, micelles, NEs, polymeric and non-polymeric NPs, see Figure 4.2 (Quezada et al. 2019). DDS allows transferring drugs (BZD or NFX) through natural barriers in small doses, which decreases their toxic effects. Positively charged liposomes with phosphatidylcholine and stearylamine free of any drugs induced a destruction of *T. cruzi* in 30 minutes (Yoshihara and Nakae 1986; Yoshihara et al. 1987). pH-sensitive liposomes incorporated with etanidazole were able to transfer the drug to the cytoplasm, in which parasites live (Morilla et al. 2005). Unfortunately, the sesquiterpene lactone lychnofolide isolated e.g. from *Lychnophora pohlii* shows high anti-Chagas activity but high cardiotoxicity. However, it has been discovered that lychnofolide encapsulated in biodegradable polymeric nanocapsules prevents cardiac changes induced *in vivo* by repeated doses of the free form of lychnofolide (Tupinambá Branquinho et al. 2017).

Gonzalez-Martin et al. (1998) used poly(alkyl cyanoacrylate) NPs containing NFX to treat Chagas disease. The size of these NPs was below 200 nm, achieved 33.4% of NFX absorption and a drug release rate of 65.4%, after 6 hours of incubation at pH 7.4. In *in vitro* assays, the antiparasitic activity reached 98.9% at a concentration of 0.2 μg/ml compared to 40% of free NFX in standard solution. The IC_{50} values of 0.683 μg/ml for the NFX standard solution and 0.0015 μg/ml for NFX-loaded NPs were observed. The trypanocidal activity was increased also by allopurinol-loaded polyether cyclic NPs with 91.5% efficiency compared to purinol standard solution (45.9%) with 16.7 μg/ml of allopurinol (Gonzalez-Martin et al. 2000). Other polymer NPs such as poly(lactide-co-glycolide) (PLGA), poly(D,L-lactide) (PLA), polyglycolide (PGA), polycaprolactone (PCL), CS, and polyethylene glycol (PEG)-PLGA and also liposomes with various active substances were tested *in vivo/in vitro* against Chagas disease in last decades (Quezada et al. 2019).

BZD loaded into $CaCO_3$ produced a biocompatible and biodegradable formulation BZD@$CaCO_3$ NPs with low cytotoxicity against mammalian LLC-MK2 cell line. On the other hand, BZD@$CaCO_3$ NPs reduced the viability of epimastigote, trypomastigote, and amastigote (Y-strain) even at a 25-fold lower concentration of BZD compared to BNZ alone (Tessarolo et al. 2018). BZD loaded into mesoporous SiO_2 NPs and coated with CS gave a formulation with a particle size of approx. 3.3 nm and a zeta potential of -11.5 ± 0.5 mV. This formulation was found to be 30-fold more effective than free BZD in *in vitro* tests against CL-Brener epimastigotes (Nhavene et al. 2018).

In addition, it is important to underline that NPs can interact with the immune system of the host modulating the immune response to support the elimination of pathogens. They can be also used as nanobiosensors in nanodiagnostic systems for pathogen identification (Quezada et al. 2019).

Adeyemi et al. (2018) evaluated the antiparasitic action of AuNPs, AgNPs, and PtNPs stabilized with tannic acid/ethanol, alloys of the Ag–Au–Pt NPs, Ag-Pt NPs, Au-Pt NPs, and Ag-Au NPs and tryptophan-NP (Au or Ag) conjugates against *T. b. brucei*. It was found that the antitrypanosomal effect of the nanosystem was more than 200-fold selective against parasites than mammalian cells, but unfortunately, this type of NPs did not completely suppress parasite proliferation in rats in the *in vivo* study, thus failing to significantly prolong survival of infected animals compared to untreated control. Pentamidine (PTM, Pentamidine 2020) is a medicine commonly used against sleeping sickness. A nano-carrier of PTM composed of PEG-PLGA coupled to a single-domain fragment of a heavy-chain antibody that specifically recognizes the surface of *T. brucei* expressing the surface glycoprotein VSG was prepared by Arias et al. (2015). The system was sevenfold more active *in vitro* compared to the free drug. An *in vivo* test in a mouse model showed that the formulation cured all infected mice at a 10-fold lower dose than the minimal full curative dose of free PTM and 60% of mice at a 100-fold lower dose.

It is also important to note that arginine kinase, crucial enzyme for survival of trypanosome under stress, is significantly noncompetitively inhibited by AgNPs and AuNPs. A 75 and 62% decrease in activity was observed, respectively, with K_i values ranging from 1.5 nM (Ag) to 3.1 nM (Au) (Adeyemi and Whiteley 2013).

4.3.3 Leishmaniasis

After malaria, leishmaniasis, another of the neglected tropical diseases, is the most prevalent infectious disease in terms of mortality and worldwide distribution (Ahmad et al. 2016). It is a group of protozoan vector-borne infections threatening almost 350 million people in 98 countries in the world (El-khadragy et al. 2018; Saleem et al. 2019). It is caused by several intracellular protozoan parasites from 1 of over 20 *Leishmania* species that are transmitted to human bodies after the bite of an infected female phlebotomine sandfly (Valle et al. 2019; Askarizadeh et al. 2020; WHO 2020d). Diseases are classified as cutaneous, mucocutaneous, and visceral (kala-azar) leishmaniasis according to the tissue that has been infected (Valle et al. 2019). While the most serious form of this disease is visceral leishmaniasis, the most common form is its cutaneous form (GHO 2020). It occurs mainly among the poor in Latin America, Asia, and Africa, but its endemic incidence has also been confirmed in North America and Southern Europe (Leishmaniasis 2018). The visceral form, with major occurrences in Brazil, India, and East Africa, is manifested by anemia, weight loss, enlarged liver and spleen, and irregular fever. If left untreated, it is fatal in more than 95% of cases. The cutaneous form is characterized by the formation of disfiguring skin lesions on exposed parts of the body leaving lifelong scars and stigma. It occurs dominantly in both Americas, in the Middle East, Central Asia, and the Mediterranean. The last mucocutaneous form damages mucous membranes of the nose, mouth, and throat. It occurs in Bolivia, Peru, Brazil, and Ethiopia (Leishmaniasis 2020).

The current therapeutic treatment is focused on the elimination of the protozoa disease vector. However, the treatment is lengthy and demanding for the affected person especially because *Leishmania* parasites are generally resistant to conventional drugs used to treat the disease. Other complications along with resistance are only a small number of antileishmanial drugs available, undesired side effects, and a high cost of treatment that in most cases requires intravenous infusions and hospitalization (Ahmad et al. 2016; Costa et al. 2019; Sepúlveda et al. 2020).

The principal target of antileishmanial treatment is amastigotes existing within macrophage phagolysosomes. But it is not an easy process and an active agent has to go through major structural barriers (Bruni et al. 2017). There are a few antileishmanial drugs however without a prolonged action. Chemotherapeutic drugs based on pentavalent antimony, in particular PTM, already mentioned in Section 4.2, are commonly used in the treatment of leishmaniasis. However, the clinical use and administration of PTM by the parenteral route is limited due to its poor oral bioavailability. In addition, it has a number of side effects (Saleem et al. 2019; Tavakoli et al. 2019; Valle et al. 2019). The liposomal amphotericin B and miltefosine are other possibilities of treatment, however, they are also accompanied by with severe side effects, drug resistance, and high cost (Carregal et al. 2019). Clearly, research is needed to develop new effective antileishmanial therapeutics innovations in drug formulation which can overcome above-named barriers (Ahmad et al. 2016; Costa et al. 2019).

Currently, techniques and products based on nanotechnology have been designed as antileishmanial drugs, including metal and metal oxide NPs, nanotubes polymeric NPs, lipid nano-capsules, liposomes as well as nanovaccines, promising exceptional properties, such as lowered toxicity, biodegradability, bioavailability, and suitability for targeted drug delivery. It is also a promising alternative for conventional drugs which can be delivered by nano-sized carriers and it leads to their improved bioavailability and lower toxicity together with other properties successfully acting in therapeutic treatment. Colloidal carriers bearing an active agent can be uptaken by the physiological route of intravenously administered nanosystems – phagocyte system. Thus, a high drug concentration in the intracellular mononuclear phagocyte system of infected cells can be promoted (Bruni et al. 2017).

Paromomycin (PMC) was loaded into mannosylated-CS NPs using dextran (PMC-MCS-dex-NPs) and its strong antileishmanial activity was confirmed comparing with a system containing non-mannosylated-CS NPs (PMC-CS-dex-NPs). The PM-MCS-dex-NPs complex was prepared by ionic gelation and particle size analysis showed an average size of 246 nm and zeta potential value of +31 mV. Encapsulation efficiency was determined to be 83.5%. The release rate was higher in acidic media. THP-1 cell uptake level of PMC-MCS-dex-NPs after six hours was approx. three and fourfold higher compared to PMC-CS-dex-NPs. While the effect of PMC-CS-dex-NPs was more noticeable on amastigotes, both parasite stages, especially the amastigote stage, were effectively affected by PMC-MCS-dex-NPs. The novel targeted mannosylated formulation can act as a targeted delivery system with strong antileishmanial activity (Esfandiari et al. 2019).

Nano-DDSs can use existing drugs that are modified to overcome problems associated with their oral administration. An example is the DDS PTM-loaded PLGA NPs as a potential drug for an oral treatment of visceral leishmaniasis. The NPs with a size in the range from 258 to 268 nm with polydispersity index 0.17 ± 0.02 and almost neutral charge were prepared using the procedure of double emulsion. NPs contained 91.5% of captured PTM. A zero-order model was convenient to describe the release profile of PTM followed by a drug diffusion profile corresponding to the Higuchi model. The efficiency of orally given PTM-loaded PLGA NPs in *in vivo* assay in infected BALB/c mice corresponded to a dose of 0.4 mg/kg (P-value < 0.05) (Valle et al. 2019).

Polymyxin B (PlyB) adsorbed in poly(*n*-butyl cyanoacrylate) (PBCA) NPs were tested to determine a potential to kill *Leishmania amazonensis*. PBCA NPs were fabricated by anionic polymerization and coated with PlyB (PBCAnp-PlyB). They had a size from 236 to 287 nm with a low polydispersity index and they were stable at least for 45 days (1.79 ± 0.17 mV). The release profile presented that 50% of drug release from PBCAnp-PlyB was six to sevenfold slower than the free PlyB. While PBCA NPs showed an antileishmanial effect higher than the activity of PlyB, they had no antimicrobial activity against *E. coli*, *Pseudomonas*, and *Klebsiella*. The combination of PBCA NPs with PlyB led to the addition of antimicrobial properties to the NPs. Thus, the formed nano-DDS demonstrated leishmanicidal and antibacterial properties. This bactericidal ability can prevent secondary infections which cause the worst ulcers caused by cutaneous *L. amazonensis* (Costa et al. 2019).

One of the successful nanomaterials with high efficacy in leishmaniasis treatment and negligible toxicity is liposomal amphotericin B (AmB) (Saleem et al. 2019). An example of biosynthesis application in preparation of agents with antileishmanial activity is the synthesis of biogenic AgNPs that were further modified with AmB–AgNPs, see Figure 4.3. According to FTIR spectra, plant polyphenolic compounds were probably involved in Ag reduction. The aqueous extract of *I. tinctoria* was used for the preparation and stabilization of AgNPs, then AmB was adsorbed on the AgNPs surface. The formed AgNPs in the spherical shape with a particle size of 10–20 nm were observed using dynamic light scattering measurements and transmission electron microscopy. A significant activity of the biogenic AgNPs against *L. tropica* in combination with visible light irradiation was observed reaching an IC_{50} value of 4.2 μg/ml. However, the antileishmanial activity was significantly enhanced after conjugation of these NPs with AmB, when IC_{50} value dropped to 2.43 μg/ml. The preparation of this drug system is a safe, economical, and more effective alternative to the current antileishmanial therapies (Ahmad et al. 2016).

Biogenic AgNPs biosynthesized from *Fusarium oxysporium* were also used against promastigote and amastigote forms of *L. amazonensis* and action mechanisms were studied. The AgNPs effect on promastigotes results in their death leading to apoptosis-like events, which was accompanied by an increased production of ROS, mitochondrial integrity loss, phosphatidylserine exposure, and damage on

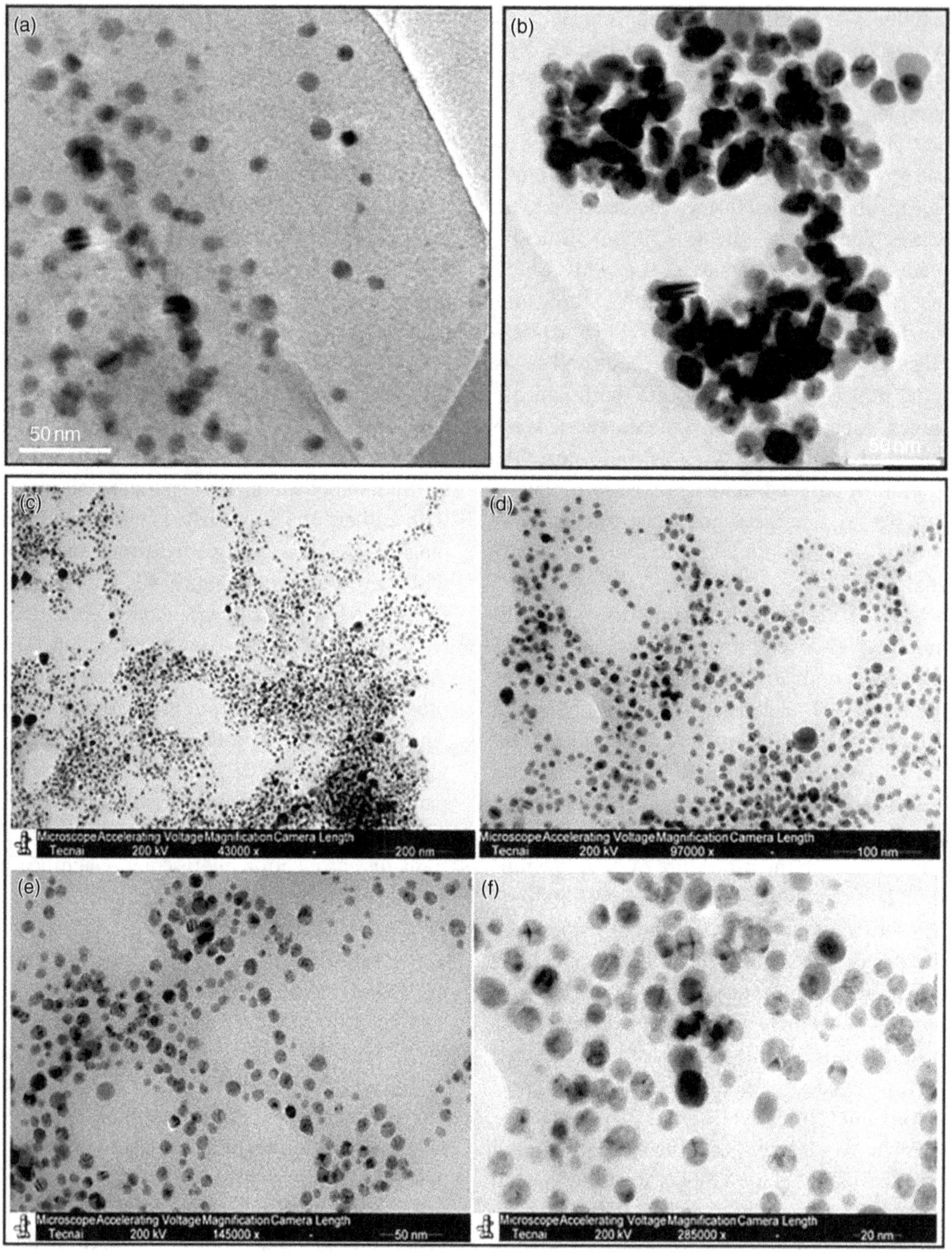

Figure 4.3 (a, b) TEM micrographs of AgNPs biosynthesized from 10 and 5 ml of *Isatis tinctoria* extract; (c–f) AgNPs were modified with amphotericin B at different magnifications. *Source:* Ahmad et al. (2016) with permission of Elsevier.

promastigotes membrane. AgNPs reduced a part of macrophages infected by *L. amazonensis* and also the corresponding amount of amastigotes per macrophage, consequently, the amount of promastigotes was recovered. The antileishmanial effect was shown together with a ROS and nitric oxide decrease. It was observed that the intracellular amastigotes were significantly damaged. It can be concluded that AgNPs have a direct effect against *L. amazonensis* and act on infected macrophages and their immunomodulatory ability, which leads to a reduction in infection and does not induce the synthesis of inflammatory mediators. It can prevent a continuous stimulation that generates and aggravates leishmaniotic lesions (Fanti et al. 2018).

AgNPs biosynthesized by *Sechium edule* were tested for their antileishmanial potential against promastigote cells of *L. donovani* in *in vitro* experiments. The IC_{50} value of $51.88 \pm 3.51\,\mu g/ml$ was determined. Furthermore, the toxicity of AgNPs was examined against normal mammalian monocyte cell line (U937) and no significant toxicity of NPs on the normal mammalian cells was found (Baranwal et al. 2018).

AgNPs biosynthesized using leaf extract of *Moringa oleifera* were tested for their activities in a murine model of *L. major* infection. Fifty mice were divided into several groups: control, infected, infected treated with Pentostam® (sodium stibogluconate), infected treated with AgNPs, and infected pretreated with AgNPs. It was concluded that therapy using AgNPs significantly reduced the average size of cutaneous lesions in comparison to untreated mice. Treatments using NPs were more efficient than common Sb(V) treatment. It seems that the presence of antioxidant enzymes in *M. oleifera* extract enhances antileishmanial activity of AgNPs (El-Khadragy et al. 2018). SbL8, amphiphilic Sb(V) complex, can form nanostructures by self-assembling in an aqueous solution. It was formed in the reaction of *N*-octanoyl-*N*-methylglucamide with Sb(V). Further SbL8/miltefosine-hexadecylphosphocholine (MiHePC) nanoformulation was prepared by incorporating MiHePC into SbL8 nanoassemblies and the therapeutic efficacy and the antileishmanial activity of the nanoformulation was tested after daily oral administration for 30 days in BALB/c mice infected by *L. amazonensis*. The treatment of cutaneous leishmaniasis was compared to the results of monotherapies (SbL8 or MiHePC) and saline control. All the treatments showed a significant reduction in the size of lesions; however, only SbL8/MiHePC nanoassemblies instigated a significant decrease of the parasite burden in the lesion. This new orally active nanoformulation showed that therapy treatment efficiency can be improved using an efficient drug combination (Carregal et al. 2019).

The potential antileishmanial effects of antimony sulfide NPs biosynthesized by *Serratia marcescens* were investigated on *L. major* in both *in vitro* and *in vivo* experiments. Various concentrations of these NPs were assessed in *in vitro* experiments using amastigotes of *L. major* confirming that an antileishmanial activity increased with the antimony sulfide NPs concentration increasing. The IC_{50} of antimony sulfide NPs against amastigotes was determined to be $62.5\,\mu g/ml$. BALB/c mice were inoculated subcutaneously with 2×10^6 *L. major* promastigotes for *in vivo* experiments and treated with antimony sulfide NPs at a concentration of $70\,\mu g/ml$, meglumine antimoniate (glucantime) as positive control and sterile phosphate-buffered saline (pH 7.4) as vehicle control. Compared with the negative control group, after administration of antimony sulfide NPs, the mean lesions size in mice was significantly reduced to $8.6 \pm 2.7\,mm^2$ ($P = 0.015$). In addition, the NPs decreased the parasite load in the test group. The greatest antileishmanial efficiency against *L. major* was found at a concentration of $62.5\,\mu g/ml$ in *in vitro* and *in vivo* experiments (Mohtasebi et al. 2019).

Antileishmanial effects of Mn_2O_3 NPs against *L. major in vitro* and *in vivo* were studied due to proven effectiveness on fungi and bacteria and they seemed to be a suitable candidate for cutaneous leishmaniasis treatment. Manganese can induce cell apoptosis by increasing FOXO3a-Bim/PUMA mRNA activation and activation of caspase-3 pathway. The antileishmanial activity of NPs was proven using light microscopic observation by calculating the number of remaining parasites. The cytotoxicity effects of Mn_2O_3 NPs against *L. major* promastigotes and macrophage cells were determined by the MTT test. The IC_{50} values of Mn_2O_3 NPs were 15 and 40 μg/ml, respectively. The flow cytometry assay showed that about 57% of the promastigotes were induced with Mn_2O_3 NPs. Additionally, in *in vivo* studies, the therapeutic effects of Mn_2O_3 NPs in BALB/c mice infected by cutaneous *L. major* were manifested by significant reduction of the ulcers' size and increase in the survival rate of the mice compared to the control group. Mn_2O_3 NPs have a beneficial effect on *L. major* promastigotes *in vitro* and *in vivo* and could be considered as a candidate for the treatment of this infection (Tavakoli et al. 2019).

The thermotherapy has a high potential for the cutaneous leishmaniasis treatment. It improves a higher compliance of patients and shows less adverse systemic effects. However, the treatment is painful, local anesthetics are required and it is not so effective against large lesions. NPs can be a very useful solution for this treatment because they can generate heat in a controlled manner in a specific localization. Iron oxide magnetic NPs were used for magnetic hyperthermia to kill *Leishmania mexicana in vitro*. The effectiveness of this therapy against the differentiated amastigote form was studied using the following viability assays: luciferase-based assay, live/dead stain, and PrestoBlue™. The findings confirmed the efficiency of magnetic hyperthermia in the treatment of host-infective amastigotes accompanied by morphological changes consistent with heat treatment. The method can be a new alternative of therapy, which can replace the expensive chemotherapy (Berry et al. 2019).

A photodynamic therapy (PDT) in combination with a photosensitizer is an alternative to AmB in the treatment of cutaneous leishmaniasis. The combination of NPs of TiO_2 doped with Zn (TiO_2/Zn) that serve as drug delivery carriers with hypericin (HPR) enhanced its photodynamic activity in the visible light region. TiO_2/Zn NPs were produced by solution combustion route and the formation of aggregates smaller than 1 μm containing particles of size smaller than 20 nm was confirmed via scanning and transmission electron microscopy. The fluorescence spectrometry measurements confirmed the complex of HPR with TiO_2/Zn (TiO_2/Zn-HPR). Further, the presence of smaller molecules in the promastigotes after 120 minutes of incubation was showed by the cellular uptake measurements. Advisable antileishmanial activity expressed as EC_{50} reached 17.5 ± 0.2 μg/ml and low cytotoxicity of CC_{50} 35.2 ± 0.3 μg/ml against murine macrophages in the visible light (22 mW/cm^2; 52.8 J/cm^2) was found. In *in vivo* analysis, the parasite load of *L. amazonensis* in BALB/c infected mice was reduced by 43–58%, after the combination of TiO_2/Zn-HPR and red/blue light with intensity of 22 mW/cm^2 and 52.8 J/cm^2 (Sepúlveda et al. 2020).

Nano-DDS for leishmaniasis treatment was prepared by a loading of betulinic acid (BLA) into CS with the aim to improve BLA therapeutic effect and reduce its side effects, The nano-DDS was tested against *L. major* in the infected BALB/c mice. The sizes of the CS and BLA-CS NPs were approx. 102 and 124 nm, respectively. The efficiency of drug loading was estimated to be 93%, and cellular uptake 97.5%. The nontoxicity of BLA-CS nano-DDS was confirmed by *in vitro* and *in vivo* tests. Dosing of 20 mg/kg can be used for wound healing and kill the parasite to a large extent ($P < 0.001$) (Mehrizi et al. 2018).

It seems that the most appropriate procedure for leishmaniasis control is the development of vaccines. No vaccine has been available yet and the main reason is that no appropriate adjuvant delivery system has been prepared. Adjuvants are important components of vaccines. Their role is to induce the immune response in the control manner. Conventional adjuvants are not adequately appropriate and for that reason, there are many efforts to develop new particulate delivery systems. It is a necessity also in the case of leishmaniasis. Adjuvants include generally mineral-, polymers-, or lipids-based delivery systems that carry antigens directly to the site of action within the organisms and enhance immune responses The potential of nano-carriers and NPs is very promising (Askarizadeh et al. 2020).

Vaccination could help in the control of leishmaniasis also in dogs. A highly immunogenic component of *Leishmania* excretory/secretory products is the promastigote surface antigen (PSA). One of a newly formed vaccine, which consists of three peptides derived from the carboxy-terminal part of *L. amazonensis* PSA and conserved among *Leishmania* species, using QA-21 as adjuvant, was tested on Beagle dogs in a preclinical trial. For the experiments, dogs were infected with *L. infantum* promastigotes four months before the vaccination. Immunization conferred immunity against experimental infection with *L. infantum* as was confirmed by macrophage nitric oxide production and by increase in specific IgG2 antibodies. Antileishmanial activity was proven by IFN-γ production in lymphocytes. The parasite burden was significantly reduced in dogs after vaccinations than in the control group. Thus, this new promising vaccine candidate ensures an effective cross-protection of a host against zoonotic leishmaniasis by inducing a Th1-type immune response joint with specific IgG2 antibodies production (Petitdidier et al. 2019).

Leishmania superoxide dismutase (SODB1) was loaded onto biodegradable and stable CS NPs by the ionotropic gelation method and the nanosystem was tested as a new nanovaccine for leishmaniasis in BALB/c mice. Phosphate buffer saline (PBS), superoxide dismutase B1 (SODB1) in PBS and NPs were implemented in mice via subcutaneous injection together with soluble leishmania antigens (SLA) and complete Freund's adjuvant (CFA). The efficiency of the vaccine was tested in blood samples collected three weeks after vaccination and evaluated using ELISA to detect IgG2a and IgG1. The analysis confirmed that IgG2a and IgG2a/IgG1 were significantly higher in mice with injected SODB1 NPs than in the other groups ($P < 0.05$) (Danesh-Bahreini et al. 2011).

4.3.4 Toxoplasmosis

Toxoplasmosis presents a global worldwide zoonotic infection caused by oblige intracellular apicomplexan protozoan parasite, foodborne pathogen, *T. gondii* (Gaafar et al. 2014; Alajmi et al. 2019). It occurs in every country in the world. The detected seropositivity ranges between 10 and 90%. *T. gondii* cause severe ophthalmic, systemic, and neurological diseases such as pneumonia, disseminated disease, or encephalitis in newborns and persons (Abdel-Malek et al. 2018). The main route of human infection is through the ingestion or usage of undercooked or raw meat contaminated by cysts or it can supervene after direct contact with cats or from water or food contaminated by oocysts excreted in feline feces (Furtado et al. 2011; Torgerson and Mastroiacovo 2013). Very often, it is without any syndromes; however, it can grow into a more severe and possibly life-threatening infection especially in immunocompromised persons. Mother-to-child transmission can occur during the prenatal period, which is called congenital toxoplasmosis. This dangerous infection can result in abortion or fetal death or it can induce syndromes of neurologic and neurocognitive deficits and chorioretinitis (Torgerson and Mastroiacovo 2013). In most cases, there is no drug treatment. If it is necessary, antibacterial chemotherapeutics, such as pyrimethamine, sulfadiazine, or spiramycin are administered. Pyrimethamine is a folic acid antagonist which is a negative aspect especially if an extra-high dose of pyrimethamine is used (Toxoplasmosis 2020).

Combination of spiramycin–metronidazole and spiramycin alone that were loaded on CS NPs were compared to the treatment of *T. gondii* using conventional spiramycin therapy with focus on tissue penetration and blood–brain barrier (BBB) passage. Mice were inoculated intraperitoneally with 2500 *T. gondii* tachyzoites RH strain. The five experimental groups were formed differing in the oral treatment: the first group received only CS NPs, the second only spiramycin, the third spiramycin–metronidazole combination, and the last two were treated with spiramycin–CS NPs in doses of 400 and 100 mg/kg. The treatment efficiency was evaluated by mice mortality rate, survival time, load of parasites in organs, and morphological study of the tachyzoites. Spiramycin–CS NPs showed the longest survival time exceeding 200 days, no mortality was reported on the sacrifice day (eighth). Unlike other applied drugs, this nano-DDS also caused the highest significant reduction by approx. 90% of tachyzoites in brain, spleen, and liver. Tachyzoites were more mutilated and some of them were ruptured in the groups with the CS NPs and spiramycin–CS NPs therapy. It can be concluded that spiramycin–CS NPs are highly effective in the acute toxoplasmosis treatment due to their antiparasitic effect and nontoxic character of CS NPs (Hagras et al. 2019).

However, CS and CS NPs are widely used as antimicrobial agents due to their strong antimicrobial potential and the CS NPs can be used as an alternative natural medicine (Rozman et al. 2019; Yilmaz 2019). The CS antimicrobial properties are influenced by its concentration and molecular weight (MW). Various MWs and concentrations of CS NPs were prepared using the ionic gelation method and applied in *in vitro* experiments against tachyzoites of *T. gondii* RH strain. The *in vivo* experiment was performed on BALB/c mice, which were first inoculated with *T. gondii* RH tachyzoites, and then CS NPs of different MWs were used for therapy. As CS NP concentrations increased and MW decreased in *in vitro* and *in vivo* experiments, antitoxoplasmosis effects increased. *In vitro* experiments showed 100% mortality of tachyzoites, which was achieved at 500, 1000, and 2000 ppm concentrations of low-molecular weight CS NPs, at 1000 and 2000 ppm of medium-molecular weight CS NPs and at 2000 ppm concentration of high-molecular weight CS NPs. Growth inhibition of tachyzoites in peritoneal mice exudates was determined to be 86, 84, and 79% for low, medium, and high MWs of CS NPs, respectively, in comparison with mice in the sulfadiazine treatment group (Teimouri et al. 2018).

Curcumin (CUR) NE was applied for the treatment of acute and chronic toxoplasmosis in mice models. NE was prepared using spontaneous emulsification to avoid the problems connected with CUR low bioavailability and poor water solubility. It consisted of 1% CUR (1% w/v), soybean as oil phase, a mixture of Tween 80 and Tween 85 as surfactant, ethanol as co-surfactant, and distilled water. The zeta potential and average particle size of CUR NE were determined as -29.46 ± 2.65 mV and 215.66 ± 16.8 nm. The NE was stable at room temperature at least for two months. 50 and 50 BALB/c mice were inoculated by virulent RH strain (type I) and avirulent Tehran strain (type II) of *T. gondii*, respectively. For the experiments, both groups were treated with CUR NE, CUR suspension (CURS, 1% w/v), and pure NE without CUR. In the acute phase, the mice survival time of mice inoculated with *T. gondii* RH strain and treated with CUR NE was prolonged only from 8 to 10 days after inoculation. The statistically significant differences in the survival time were found in mice group treated by CUR NE in comparison with negative control group ($P = 0.001$). The peritoneum tachyzoites mean counts significantly decreased from 5962.5 ± 666 in negative control group to 627.5 ± 73 in mice treated by CUR NE ($P = 0.001$). Interestingly, inhibition of peritoneum tachyzoites growth varied in groups administered with CUR NE, CURS, and NE achieving 90, 21, and 11%, respectively, compared with negative control group. In chronic phase experiment, the tissue cysts size was reduced to 17.2±15.6 and 31.5±6.26 μm, respectively, in mice inoculated with *T. gondii* Tehran strain and treated with CUR NE compared with the negative control group ($P = 0.001$). Decrease of cyst numbers was determined by BAG1 downregulation in treated groups in comparison with negative control group with a minimum relative expression in CUR NE (1.12±0.28), CURS (11.76±0.87), and NE (14.67±0.77), respectively ($P = 0.001$) (Azami et al. 2018).

Ag and CS NPs were applied to study their antitoxoplasma effects separately or in combination as pretreatment and as treatment in the mice. It was found that AgNPs (in single form or combined with CS) have a high antitoxoplasma potential. It was evident from a statistically significant reduction in the number of parasite counts in the animal livers and spleens. An instance of movement and shape deformity of the tachyzoites was observed using microscopic methods. In the animal serum, gamma interferon was increased after the administrations of NPs. In addition, the Ag detected in the selected tissues was in a safe range (Gaafar et al. 2014). AgNPs biosynthesized using extracts of plants *Phoenix dactylifera* and *Ziziphus spinachristi* were used to study the antitoxoplasmosis effects in experimental animals and were compared to the sulfadiazine drug therapy. Infection of mice by *T. gondii* was evaluated by determining the alanine transaminase (ALT) and aspartate transaminase (AST), which correspond to liver functions and which were significantly increased compared with the control group. Also, levels of hepatic nitric oxide and lipid peroxidation were increased. The activities of superoxide dismutase, catalase, and glutathione in liver homogenates were significantly reduced. AgNP pretreatment prevented liver damage, as confirmed by inhibition of enzyme activity, inhibition of hepatic nitric oxide by increase in superoxide dismutase and catalase activity. Concentrations of hepatic lipid hydroperoxides and nitric oxide as well as proinflammatory cytokines decreased by AgNPs effects but significantly enhanced activity of the antioxidant enzyme of liver homogenate. Additionally, some histological features were improved by AgNPs pretreatment as was indicated by slight infiltration and fibrosis and minimal pleomorphism (Alajmi et al. 2019). The advantage of biosynthesized AgNPs is a low toxicity and proven antiparasitic activity. They are promising therapeutic agents for toxoplasmosis treatment because they reduce the adherence, infection, and proliferation of ROS and NO and lower the production of inflammatory mediators (Machado et al. 2020).

Au, Ag, and Pt NPs in the concentration range of 0.01–1.00 μg/ml were compared for their antiparasitic activity; 90% inhibition of *T. gondii* growth as the EC_{50} values of 7, 1, and 100 μg/ml for Au, Ag, and Pt NPs, respectively, was determined. The NPs' activities are connected with redox signaling and effects on mitochondrial membranes of the parasite. NPs demonstated suitable antitoxoplasmic properties and showed no cytotoxicity in host cells even at the demanding concentrations (Adeyemi et al. 2017).

4.3.5 Cryptosporidium

Cryptosporidium is one of the most common intestinal protozoan parasites. It is the cause of unrelated diarrhea in immunocompromised people and young children. Its oocysts are considerably resistant to the water treatment processes (Peletz et al. 2013).

The effect of AgNPs on count and viability of the *C. parvum* isolated from different samples of tap water was investigated by Hassan et al. (2019). Different dosages (0.05, 0.1, and 1 ppm) and contact times (from 30 minutes to 4 hours) were used to test the AgNPs' efficiency. It was confirmed that AgNPs significantly reduced the oocyst count and viability in a dose-dependent manner. A decrease in the oocysts by 94.4 and 97.2%, respectively, was observed at the concentrations of 0.1 ppm for 1 hour and 1 ppm for 30 minutes. However, an increase in the oocyst's viability at two and four hours was found. Interactions of the *C. parvum* oocysts with AuNPs and GO were studied by Jain et al. (2019). The citrate-capped AuNPs were strongly adsorbed on the oocysts. DNA-capped AuNPs were not adsorbed on the oocysts' surface. The GO wrapped around the oocysts as a conformal shell. The integrity of oocysts' membrane was compromised by both citrate-capped AuNPs and GO, as was confirmed by the propidium iodide staining experiment. Thus, they can inactivate the oocysts and have the potential to eliminate this serious water treatment problem.

4.3.6 Acanthamoeba

The species of *Acanthamoeba* are facultative pathogens causing sight-threatening *Acanthamoeba* keratitis (occurs in approx. up to three million people, mostly those who wear contact lenses) and serious and deadly brain infection, granulomatous amoebic encephalitis. Due to the ability of *Acanthamoeba* trophozoites to form resistant cysts, most drugs are ineffective to prevent infection (Baig 2019; CDC 2020).

Prevention of amoeba adhesion to the contact lens surface can reduce the number of *Acanthamoeba* keratitis infections. At present, chlorhexidine digluconate in combination with propamidine isethionate or hexamidine can be used as antiseptic, but this is insufficient and toxic to the eyes. Niyyati et al. (2018) evaluated the efficacy of AgNPs and AuNPs against cysts of *Acanthamoeba* T4 genotype in contact lens solution. Interestingly, none of the metal types of NPs showed a cytotoxic effect on the cysts of *Acanthamoeba*. On the other hand, AgNPs and PtNPs prepared by Hendiger et al. (2020) showed activity against *Acanthamoeba* trophozoites and reduced amoeba adhesion to the contact lens surface. AgNPs showed better anti-amoeba activity against cytotoxic dependence and reduced amoeba adhesion in a wider range of contact lenses tested. It was also confirmed that ionization in addition to hydration of the contact lens material is a key parameter affecting the adhesion of *Acanthamoeba* to the contact lens surface.

AuNPs conjugated to the three antifungal drugs AmB, nystatin, and fluconazole were tested against *Acanthamoeba castellanii* genotype T4 with very good results. The activity was most pronounced in AmB–Au NPs ($P < 0.05$). In contrast, AuNPs themselves had no activity. The nanoconjugates showed reduced cytotoxicity against HeLa cells (Anwar et al. 2019a). AgNPs and AuNPs synthesized by one-phase reduction and coated with guanabenz (agonist of the α2 adrenergic receptor) used as an antihypertensive with a markedly sedative effect and having a particle size distribution of 100 nm for guanabenz–AgNPs and 60 nm for guanabenz–AuNPs showed strong anti-amoebic effects in the range of 2.5–100 μM against *A. castellanii* and *Naegleria fowleri* causing granulomatous amoebic encephalitis and primary amoebic meningoencephalitis. Conjugation of NPs increased the antiamoebic effects of guanabenz, as stronger activity was observed at lower effective concentrations (2.5 and 5 μM) compared to the drug alone. In addition, guanabenz has been found to inhibit the interconversion between trophozoite and cyst forms of *A. castellanii*. The cysticidal effects against *N. fowleri* have also been observed. The prepared nanoconjugates had insignificant cytotoxicity against human cells (Anwar et al. 2019b). Antiacanthamoebic natural substances such as quercine (QRC) and colanic acid isolated from extracts of *Polyalthia longifolia* var. *pendula* and *Caesalpinia pulcherrima* were conjugated to AgNPs. Subsequent measurements showed that the particle size distribution of NPs was approximately 45 nm. The natural compounds and the nanomaterials were tested against an isolate of *A. castellanii* genotype T4. The secondary metabolites had significant inhibition of *A. castellanii*; QRC-AgNPs showed potent antiamoebic effects, as well as disruption of the encystic and excystic activity of amoebae and at the same time showed insignificant toxic effects on human cells *in vitro* (Anwar et al. 2020). Anwar et al.

(2019c) conjugated diazepam, phenobarbital, and phenytoin (γ-aminobutyric acid receptor agonists) to AgNPs obtained by reduction of $AgNO_3$ with sodium borohydride. All conjugates were tested against *A. castellanii* genotype T4 and *N. fowleri*. The results showed strong amoebicidal effects of diazepam-, phenobarbital- and phenytoin-conjugated AgNPs compared to the bulk drugs. In addition, the conjugates had cysticidal effects. On the other hand, the drugs reduced the cytotoxicity of AgNPs on human cells.

Different cobalt NPs and their anti-acanthamoeba effects against trophozoites and cysts were tested together with parasite-mediated host cell cytotoxicity. Three different types of CoNPs were synthesized using hydrothermal and ultrasonication methods. The CoNPs' anti-acanthamoeba effects were studied as amoebicidal, encystation, excystation, and host cell cytopathogenicity assays. The $Co_3(PO_4)_2$ hexagonal microflakes and $Co(OH)_2$ nanoflakes of 100 nm showed amoebicidal activity at 10 and 100 μg/ml against *A. castellanii* in comparison with a granular Co_3O_4 of size 35–40 nm. Encystation and excystation assays presented a consistent inhibition at 100 μg/ml. Cobalt NPs inhibited amoebae-mediated host cell cytotoxicity as determined by lactate dehydrogenase release without causing significant damage to human cells when treated alone (Anwar et al. 2019d).

4.4 Nanosystems with Effect on Veterinary Parasitic Protozoa

Ciliate ectoparasites are one of the most important groups of pathogens in fish culture. The common ways of treatments can be harmful to both, the fish as well as the environment. Thus, the development of novel agents efficient at low doses and safe for fish is necessary to optimize fish treatment in aquaculture. AgNPs' efficiency against the ciliate *Tetrahymena* was studied; however, it was found that AgNPs' toxicity may vary depending on the synthesis techniques and size of NPs. The acute toxicity of two AgNPs Argovit™ and UTSA (silver NP material developed and prepared by Dr. Yacamán's group at the Department of Physics and Astronomy, University of Texas at San Antonio using microwaved-assisted techniques) was evaluated in *in vitro* test using *Tetrahymena* sp. The safety of lethal and higher doses of UTSA AgNPs for ciliates on the fish *Chirostoma estor* was evaluated. In *Tetrahymena*, the LC_{50} value for Argovit™ was 2501 ± 1717 and 796 ± 510 ng/l at 15 and 60 minutes, respectively. The LC_{50} for UTSA AgNPs was determined to be 4 ± 2 and 1 ± 0.6 ng/l at 15 and 60 minutes, respectively. A 100% effectiveness against *Tetrahymena* was found at the concentrations of 3300 ng/l for Argovit™ and 10.6 ng/l for UTSA NPs for 15 and 60 minutes, respectively. After 60 minutes of exposure to 0.25 and 0.50 ng/l UTSA AgNPs, the number of cilia was significantly reduced, small holes on the surface of cells and their membranes were damaged, see Figure 4.4 (Fuentes-Valencia et al. 2020).

Persephonella marinas is a parasite that is denoted by a World Organisation for Animal Health (OIE) and causes infections of several species of oyster, for example, *Crassostrea virginica* and *Crassostrea corteziensis* (OIE 2019). There is only limited information on how to treat this parasite. AgNPs were tested as potential agents for such a treatment. The lethal effects of the Argovir™ were described against hypnospores of *P. marinas* that are a particularly resistant parasite stage. The study was compared with $AgNO_3$ and polyvinylpyrrolidone (PVP) as a vehicle control. The hypnospores were exposed to three different concentrations of 1, 10, and 100 μg/ml of Ag contained in Argovir™, to 157.47 μg/ml of $AgNO_3$, and to 1570 μg/ml of PVP. For negative control, hypnospores in culture medium without treatment were prepared. It was observed that the efficiency is dose-dependent. The AgNPs' concentration of 0.093 mM resulted in 50% *P. marinas* mortality. The concentration

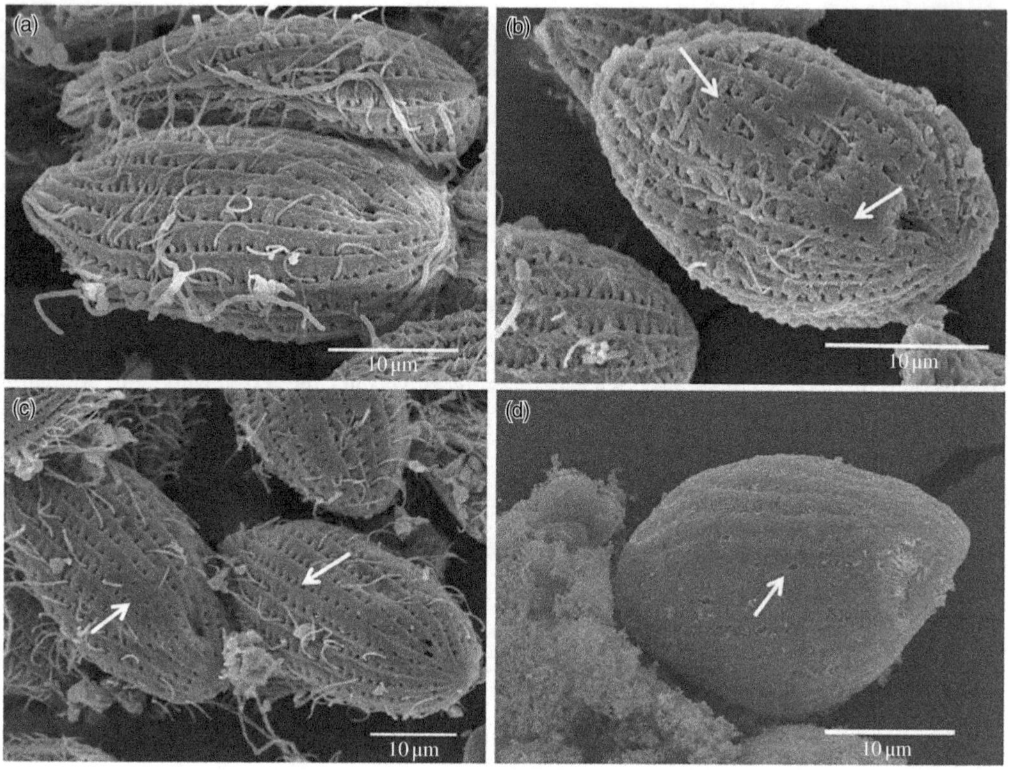

Figure 4.4 SEM images of *Tetrahymena* sp. trophozoites, after 60 minutes of exposure to UTSA AgNPs: (a) control sample without AgNPs treatment; (b) concentration of AgNPs 0.25 ng/l, significant decrease of cilia number and kinetids damaged; (c) concentration of AgNPs 0.50 ng/l; swelling and (d) damage of the cells, formation of small holes. *Source:* Fuentes-Valencia et al. (2020) with permission of Elsevier.

of 927 mM of Ag in AgNPs as well as $AgNO_3$ was highly lethal, achieving a mortality of over 90%. The hypnospores were deformed as identified by TEM, AgNPs penetrated the wall of the hypnospores and affected the degradation of lipids in the cytoplasm, see Figure 4.5 (Bravo-Guerra et al. 2020).

Trichomoniasis is a very serious cattle infection. It causes damages to females and fetus, the most serious complication is abortion. No effective treatment for trichomoniasis infection has been developed so far (BonDurant 1997; Trichomoniasis 2020). A GO and carbon nanotube oxide (GCN-O) composite was prepared. Pristine multi-walled carbon nanotubes (MWCNs) were prepared by chemical vapor deposition method and consequently, they were exfoliated and oxidized by oxygen plasma etching. The oxygen atoms' presence was determined as 20% using X-ray photoelectron spectroscopy (XPS) and energy-dispersive X-ray spectroscopy (EDX). The GCN-O composite formed a stable colloidal aqueous solution. It strongly interacted with the cell membrane of *Tritrichomonas foetus* protozoa (Zanin et al. 2014).

Trichomonas gallinae parasites cause avian trichomoniasis in bird species. The infection is treated using metronidazole especially due to economic reasons in agriculture (GWH 2017). A biocompatible nanocomposite that consists of CS nanocapsules and cellulose nanofibrils was tested for antitrichomonal activity to develop the novel potential agent. CS nanocapsules were prepared by coating Fe_3O_4 NPs by CS polymer. Fe_3O_4 NPs served as a template,

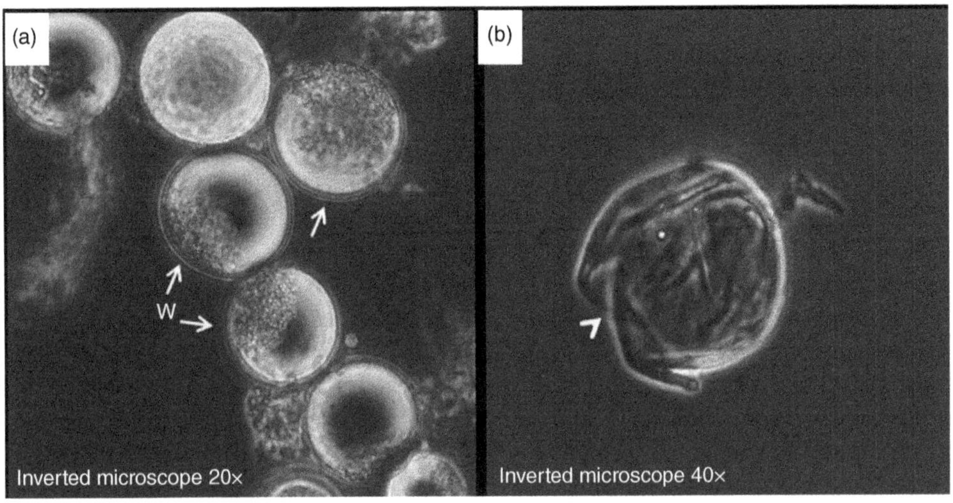

Figure 4.5 Morphology of *P. marinus* hypnospores before and after interactions with AgNPs: (a) live hypnospores posessing a spherical shape with visible cell walls and cell contents (W); (b) dead hypnospores after exposure to 10 μg/ml AgNPs for 24 hours, irregular structure, cell wall fragmentation, no cell contents. *Source:* Bravo-Guerra et al. (2020) with permission of Elsevier.

then they were removed. Furthermore, CS nanocapsules were combined with tannic acid into cellulose nanofibrils. CS nanocapsules with the highest efficiency had a diameter of 241.5 nm and the wall thickness was approx. 80.61 nm. The resulting nanoships were pH-responsive nanomaterials and their drug content and efficiency of entrapment were 24.4 and 64.56%, respectively. They had excellent antiprotozoal efficacy. *T. gallinae* was treated with CS nanocapsule-mounted cellulose nanofibrils (2 mg/ml for 60 minutes, 1 mg/ml for 120 minutes, and/or 0.5 mg/ml for 180 minutes) and complete inhibition was found. Collapse and bacterial cells' disruption was observed using FE-SEM (Yunessnia lehi et al. 2019). A pH-sensitive nano-DDS was prepared using zeolite Y NPs and tannic acid. Zeolite Y NPs were synthesized using a hydrothermal synthesis from colloidal suspensions. Prepared NPs were coated with tannic acid. Metronidazole was loaded into nano-DDS using the immersing method. This system showed pH-sensitive behavior, faster release was observed in acidic medium. NPs containing metronidazole were successfully applied in an *in vitro* test against *T. gallinae* trophozoites in trypticase-yeast extract-maltose medium. It can be concluded that NP zeolites Y containing metronidazole and tannin-modified zeolites Y are potential effective antitrichomonal agents (Rezaee et al. 2019).

4.5 Nanomaterial Toxicity on Beneficial Protozoa

NPs are defined as particles having sizes in the nanoscale, between 1 and 100 nm. They are unique in their characteristics due to specific physicochemical properties such as surface area, surface charge, surface coating, particle morphology, and degree of agglomeration (Turan et al. 2019). Due to the changed properties, extensive experiments are needed that are able to detect the ecotoxicity of nanomaterials (Miller et al. 2017; Mashjoor et al. 2019; Mayne et al. 2019). As nanosystems can suppress or kill pathogenic protozoa, they can, of course, also destroy protozoa that are desirable for a given ecosystem and serve, for example, as food for aquatic animals. For example, Du et al. (2018) in their paper discussed in detail the current knowledge gaps and safe product designs of AgNPs. The limited and existing data implied that understanding the toxicity mechanisms is crucial to the future research development of nanomaterials (Barker et al. 2018; Kheiri et al. 2019). Study of NPs' effect on phytoplankton was conducted using a simple modification of the Rosenzweig–MacArthur predator (zooplankton)–prey (phytoplankton) model. The NPs have the potential to decrease the maximum physiological *per-capita* prey growth rate. It was found that NPs induce suppression of phytoplankton population growth which probably destabilizes the system and limit cycle oscillation. If NPs and phytoplankton contact rate increases, then the equilibrium densities of phytoplankton as well as zooplankton decrease. NPs' depletion/removal from the aquatic system has an important influence on the stability of the phytoplankton and zooplankton coexistence (Sourav et al. 2015).

Metal-based NP applications have a growing trend related to their release into the environment, including water bodies. Consistent data about the NPs' biological effects for environmentally relevant species together with precise NPs characterization are limited. Ongoing research is very important for understanding the risks of NPs. Tests of 12 metal- or metalloid-based NPs including Al_2O_3, Co_3O_4, CuO, Fe_3O_4, MgO, Mn_3O_4, Sb_2O_3, SiO_2, ZnO, TiO_2, WO_3, and Pd were performed on the protozoa *Tetrahymena thermophila*, three bacterial species *Vibrio fischeri*, *Escherichia coli*, *Staphylococcus aureus*, and the alga *Pseudokirchneriella subcapitata*. The NP physicochemical properties, solubility, and ROS production were determined. The most sensitive was an algal growth inhibition assay and it was used to determine EC_{50} values for 10 NPs that ranged from 0.1 to 58 mg/l. Algal toxicity was correlated with ROS production by NPs. Most of the NPs were agglomerated. There were different sensitivities in the test, but some trends in the NP toxicity for different species were observed: the highest toxicity was found for CuO and ZnO NPs with EC_{50} values below 1 mg/l for all tested organisms except for the protozoa. Al_2O_3, SiO_2, WO_3, and Sb_2O_3 NPs were not toxic below 100 mg/l and MgO had no adverse effects below 100 mg/l to any test organisms in any test setting (Aruoja et al. 2015).

AgNPs show a high toxicity against aquatic organisms. The toxicity is caused by released Ag^+ ions or by generated ROS. AgNPs coated by proteins (Collargol) with a size of 14.6 nm were applied in viability, oxidative stress, and gene expression study in wild-type strains (CU427 and CU428) of ciliate *T. thermophila*. EC_{50} values based on viability were approx. 100 and 75 mg/l for CU427 and CU428 strains, respectively. The profiles of oxidative stress-related genes in these two strains were also different. However, even though some oxidative stress (OS)-related genes were overexpressed in ciliates exposed to AgNPs, intracellular ROS level did not increase, which could be connected with efficient antioxidant defense mechanisms of cells. Metallothionein genes were upregulated at a higher level (36- versus 5000-fold compared to OS-related genes), which leads to a conclusion that Ag^+ ions-mediated toxicity mechanism prevails over OS-related pathway. Comparison between Ag^+ ions that are released from AgNPs at EC_{50} concentration and the respective $AgNO_3$ EC_{50} values indicated that Ag^+ ions are the main factor in AgNPs' toxic action on *T. thermophila* (Juganson et al. 2017).

CeO_2 NPs in the concentrations of 1 and 10 mg/l were tested for studying effects during the development of river biofilm communities. High NPs sorption on the biofilm community and interactions with lipid molecules were observed using scanning transmission X-ray microscopy. After eight weeks, polycarbonate coupons from the reactors were removed and taken out for molecular analyses using denaturing gradient gel electrophoresis analysis (DGGE-16S rRNA) and 16S rRNA amplicon sequencing. The biofilm communities (bacterial, protozoan numbers, photosynthetic biomass, exopolymer composition, and thickness) were studied using microscopic imaging. The observations showed that no significant influence of CeO_2 NPs on bacterial and photosynthetic biomass or thickness of biofilm was confirmed. However, significant changes to the composition of the community ($P < 0.05$) due to CeO_2 were indicated using microbial diversity measures and lectin binding affinities of exopolymeric substances as of DGGE analyses results. Any significant changes were not observed in binding to residues of cyanobacterial and algal exopolymers. No significant effects were observed on protozoans present in the biofilm communities (Lawrence et al. 2020).

It is well known that Fe_3O_4 NPs have a high affinity for As. When As is associated with Fe_3O_4 NPs, it can be transferred to organisms and produce high concentrations of As locally. It can result in some toxic effects for aquatic organisms. The cultivated organism *Tetrahymena pyriformis* was used to study As/Fe_3O_4 NPs toxicity. It was found that 24 hours after As(V) exposure, the As(V) median effective concentration in *T. pyriformis* reached 1.29 mg/l. Fe_3O_4 NPs were nontoxic, they even promoted the *T. pyriformis* growth when the number of cells was increased by 32.2% 24 hours after an exposure to Fe_3O_4 NPs at the concentration of 3 mg/l. The As(V) presence influenced also the survival rate. If no Fe_3O_4 NPs were present in the test samples, the *T. pyriformis* survival rate of 60.5% was reached 24 hours after exposure to the As(V) concentration of 1.0 mg/l. If Fe_3O_4 NPs were present (in the concentrations of 13 and 19 mg/l), the survival rate was increased by 73.8 and 83.8%, respectively. However, the As(V) and Fe_3O_4 NPs' combined toxic effects in the concentrations of 1.5 mg/l As(V) and 13 mg/l Fe_3O_4 NPs were observed after 30 hours of exposition. The toxicity was significantly enhanced and the survival rates of *T. pyriformis* were reduced from 92.3% after 18 hours to 45.3% after 30 hours. If the organism was in exposure only with Fe_3O_4 NPs, it was found that after 18 hours the intracellular levels of ROS evidently increased and they were in a steady state. If the combination of As(V) and Fe_3O_4 NPs (1.0 and 19 mg/l) was used, the intracellular levels of ROS significantly increased (2.56-times) due to cumulation of As(V) in *T. pyriformis*. Interestingly, As(III) content was detected, which resulted from the saturation of the cellular arsenic methylation capability (Zou et al. 2013).

Effects of CdSe/ZnS QDs on the growth rate of four microalgae: diatom *Phaeodactylum tricornutum*, cryptophyte *Rhino-monas reticulata*, prymnesiophyte *Isochrysis galbana* and green alga *Dunaliella tertiolecta* as well as QDs effects on the copepod *Acartia tonsa were studied.* A naupliar survival was evaluated using a classical acute test with embryos in 48 hours. Fecundity as embryos per f per day and egg hatching tests were conducted via four-day chronic test with adult copepods. It was found that QDs in the concentrations from 1 to 4 nM inhibited *P. tricornutum*, *I. galbana*, *R. reticulata*, and *D. tertiolecta* growth rate. EC_{50} of 1.5, 2.4, 2.5, and 4.2 nM was determined for *P. tricornutum*, *I. galbana*, *R. reticulata*, and *D. tertiolecta*, respectively. Acute tests with *A. tonsa* using QDs in the concentration range from 0.15 to 1.5 nM showed higher naupliar mortality and a value of EC_{50} 0.7 nM. No negative effects on egg production were observed with chronic use of QDs at concentration of 2.5 nM. Also, any significant decrease in the egg hatching success percentage was confirmed during the exposure. QDs' toxicity was also studied at the molecular level by the test of the heat shock protein 70 gene expression (hsp 70). Hsp 70 was upregulated in adults exposed three days to 0.5 nM QDs. It appears that organisms that are unable to swim in the water column, such as *P. tricornutumand*, could be more vulnerable to the toxic effects of QDs, which can lead to their aggregation and deposition in water bodies (Zhou et al. 2016). Protozoa with properties of phagocytizing cells can incorporate engineered NPs. CdSe/ZnS QDs with a size of 12 nm were phagocytized in sublethal concentration into the food vacuoles of *T. thermophila* in a dose- and time-dependent manner. It was evaluated after 20 hours; however, the QDs' clearance from the cells was not complete. It could mean that QDs' phagocytosis into food vacuoles is not the only pathway of NPs uptake by *T. thermophila* and the phagocytosis inhibition does not block the QDs' internalization into protozoans (Mortimer et al. 2014).

Only limited knowledge exists about the impact of carbon nanotubes on natural ecosystems. Single-walled carbon nanotubes (SWCNs) are internalized by *T. thermophila* that allows the nanotubes to go through the food chain. The internalization also leads to the protozoa aggregation, which impedes their ability to ingest and digest their prey bacteria species (Ghafari et al. 2008). It seems based on available data that concentrations of carbon nanomaterials (CNMs) in the environment do not threaten aquatic organisms. The CNM concentrations found in aquatic environments are in ng/l or even lower, not achieving the lowest observed effect concentrations on different aquatic organisms. Toxic effects that were determined were usually reported for short-term experiments and at high concentrations. It is well known that CNM toxicity level is associated with a type of tested organisms, exposition time, and methods of CNM preparation. Another important fact, which has to be respected, is that CNMs interact with other micro-pollutants through the establishment of synergistic and/or antagonistic liaisons. CNMs can act as sorbents and carriers or as sorbents and contribute to a modification of the micro-pollutant original toxicity. This complicates the study of CNM toxicity and requires the study of the interaction of CNMs with other micropollutants at their environmental concentration levels together with their synergistic effects on biological communities in the long term (Freixa et al. 2018).

GO is a very promising material with excellent electrochemical properties. However, there is still only limited information about GO ecotoxicity to organisms, particularly to aquatic organisms. GO toxicity against protozoa *Euglena gracilis* was studied using a growth inhibition test to determine acute toxic effects. Protozoa organisms were consequently exposed to GO in the concentration range from 0.5 to 5 mg/l for 10 days. Several parameters such as the growth, protozoa ultrastructure, antioxidant enzyme activities, and photosynthetic pigment content together with GO shading effect were determined to define the toxicity effect mechanisms. GO EC_{50} value after 24 hours in *E. gracilis* was determined to be 3.76 ± 0.74 mg/l. Significant adverse effects ($P < 0.01$) on the organism were detected at the GO concentration of 2.5 mg/l. The growth inhibition of *E. gracilis* was evident. It can be concluded that oxidative stress and a shading effect can be responsible for the toxicity of GO (Hu et al. 2015).

4.6 Conclusion

This chapter aims to acquaint readers with the issue of protozoa. A distinction must be made between protozoa "beneficial" in nature for the sustainable maintenance of ecosystems and protozoa that are harmful or explicitly pathogenic. As mentioned above, because pathogens causing serious diseases in humans occur mainly in economically weaker areas, the development of modern effective drugs against protozoan pathogens is not a priority for the pharmaceutical industry. Likewise, the governments of the affected countries are not able to financially support the development of new drugs effectively enough, as, for example, in the case of the development of orphan drugs co-financed by rich countries. Therefore, in this area, the development focuses mainly on galenic (technological) modifications of existing drugs with the aim of targeting and specifying their effect, reducing dose-dependent toxicity, and potentiating efficacy in combination with other bioactive elements/molecules/macromolecules. All these essentially simple and inexpensive galenic modifications of existing therapeutics are made possible by nanotechnology (preparation of nanoformulations). Thanks to this fascinating and rapidly developing industry, agents that have completely different characteristics and, thus, properties than the parent micronized or submicron-sized compounds are created. Thanks to this approach, new therapeutics that are effective against both human and veterinary parasites have been developed. Unfortunately, a common disadvantage of active ingredients at the nanoscale is non-specific toxicity to all organisms, i.e. the ability to non-specifically damage cell membranes or vital processes in cells, resulting in non-selective toxicity against the disease carrier or other organisms including beneficial protozoa. Therefore, in addition to evaluating the efficacy of newly developed nanoformulations against putative pathogens, not only a thorough toxicity assessment against hosts of these protozoan parasites but also an ecotoxicity assessment should be performed, because even unwanted eradication of beneficial organisms can have far-reaching effects on already very disturbed ecosystems.

Acknowledgment

This study was supported by the Slovak Research and Development Agency (APVV-17-0373 and APVV-17-0318) and by MSMT project No. CZ.02.1.01/0.0/0.0/17_049/0008441 "Innovative Therapeutic Methods of Musculoskeletal System in Accident Surgery" within the Operational Programme Research, Development, and Education financed by the European Union and by the state budget of the Czech Republic.

References

Abdel-Malek, R., Wassef, R., Rizk, E. et al. (2018). Toxoplasmosis an overlooked disease: seroprevalence in cancer patients. *Asian Pacific Journal of Cancer Prevention* 19: 1987–1991.

Abo-Zeid, Y. and Williams, G.R. (2020). The potential anti-infective applications of metal oxide nanoparticles: a systematic review. *WIREs Nanomedicine and Nanobiotechnology* 12: e1592.

Aderibigbe, B.A. (2017). Metal-based nanoparticles for the treatment of infectious diseases. *Molecules* 22: 1370–1407.

Adeyemi, O.S. and Whiteley, C.G. (2013). Interaction of nanoparticles with arginine kinase from *Trypanosoma brucei*: kinetic and mechanistic evaluation. *International Journal of Biological Macromolecules* 62: 450–456.

Adeyemi, O.S., Murata, Y., Sugi, T., and Kato, K. (2017). Inorganic nanoparticles kill *Toxoplasma gondii* via changes in redox status and mitochondrial membrane potential. *International Journal of Nanomedicine* 12: 1647–1661.

Adeyemi, O.S., Molefe, N.I., Awakan, O.J. et al. (2018). Metal nanoparticles restrict the growth of protozoan parasites. *Artificial Cells, Nanomedicine, and Biotechnology* 46 (Suppl 3): S86–S94.

Ahmad, A., Wei, Y., Syed, F. et al. (2016). *Isatis tinctoria* mediated synthesis of amphotericin B-bound silver nanoparticles with enhanced photoinduced antileishmanial activity: a novel green approach. *Journal of Photochemistry and Photobiology, B: Biology* 161: 17–24. https://doi.org/10.1016/j.jphotobiol.2016.05.003.

Alajmi, R.A., AL-Megrin, W.A., Metwally, D. et al. (2019). Anti-toxoplasma activity of silver nanoparticles green synthesized with *Phoenix dactylifera* and *Ziziphus spina-christi* extracts which inhibits inflammation through liver regulation of cytokines in Balb/c mice. *Bioscience Reports* 39 (5): BSR20190379.

Anwar, A., Siddiqui, R., Shah, M.R., and Khan, N.A. (2019a). Gold nanoparticles conjugation enhances antiacanthamoebic properties of nystatin, fluconazole and amphotericin B. *Journal of Microbiolog and Biotechnology* 29 (1): 171–177.

Anwar, A., Mungroo, M.R., Anwar, A. et al. (2019b). Repositioning of guanabenz in conjugation with gold and silver nanoparticles against pathogenic Amoebae *Acanthamoeba castellanii* and *Naegleria fowleri*. *ACS Infectious Diseases* 5 (12): 2039–2046.

Anwar, A., Rajendran, K., Siddiqui, R. et al. (2019c). Clinically approved drugs against cns diseases as potential therapeutic agents to target brain-eating amoebae. *ACS Chemical Neuroscience* 10 (1): 658–666.

Anwar, A., Numan, A., Siddiqui, R. et al. (2019d). Cobalt nanoparticles as novel nanotherapeutics against *Acanthamoeba castellanii*. *Parasites Vectors* 12: 280.

Anwar, A., Ting, E.L.S., Anwar, A. et al. (2020). Antiamoebic activity of plant-based natural products and their conjugated silver nanoparticles against *Acanthamoeba castellanii* (ATCC 50492). *AMB Express* 10: 24.

Arias, J.L., Unciti-Broceta, J.D., Maceira, J. et al. (2015). Nanobody conjugated PLGA nanoparticles for active targeting of African trypanosomiasis. *Journal of Controlled Release* 197: 190–198.

Artesunate (2020). DrugBank. https://www.drugbank.ca/drugs/DB09274 (accessed 3 June 2020).

Aruoja, W., Pokhrel, S., Sihtmäe, M. et al. (2015). Toxicity of 12 metal-based nanoparticles to algae, bacteria and protozoa. *Environmental Science: Nano* 2: 630–644.

Askarizadeh, A., Badiee, A., and Khamesipour, A. (2020). Development of nano-carriers for Leishmania vaccine delivery. *Expert Opinion on Drug Delivery* 17: 167–187.

Ayuk, E., Ugwu, M., and Aronimo, S. (2017). A review on synthetic methods of nanostructured materials. *Chemistry Research Journal* 2: 97–123.

Azami, S.F., Teimouri, A., Keshavarz, H. et al. (2018). Curcumin nanoemulsion as a novel chemical for the treatment of acute and chronic toxoplasmosis in mice. *International Journal of Nanomedicine* 13: 7363–7374.

Baig, A.M. (2019). Drug targeting in *Acanthamoeba keratitis*: rational of using drugs that are already approved for ocular use in non-keratitis indications. *Eye (Lond)* 33 (3): 509–518.

Barabadi, H., Alizadeh, Z., Taghi Rahimi, M. et al. (2019). Nanobiotechnology as an emerging approach to combat malaria: a systematic review. *Nanomedicine: Nanotechnology, Biology and Medicine* 18: 221–233.

Baranwal, A., Kumar Chiranjivi, A., Kumar, A. et al. (2018). Design of commercially comparable nanotherapeutic agent against human disease-causing parasite, Leishmania. *Scientific Reports* 8: 8814.

Barker, L.K., Giska, J.R., Radniecki, T.S., and Semprini, L. (2018). Effects of short- and long-term exposure of silver nanoparticles and silver ions to *Nitrosomonas europaea* biofilms and planktonic cells. *Chemosphere* 206: 606–614.

Begum, K., Kim, H.S., Kumar, V. et al. (2003). In vitro antimalarial activity of metalloporphyrins against *Plasmodium falciparum*. *Parasitology Research* 90: 221–224.

Benelli, G. (2018). Gold nanoparticles – against parasites and insect vectors. *Acta Tropica* 178: 73–80.

Berry, S.L., Walker, K., Hoskins, C. et al. (2019). Nanoparticle-mediated magnetic hyperthermia is an effective method for killing the humaninfective protozoan parasite *Leishmania mexicana* in vitro. *Scientific Reports* 9: 1059.

BonDurant, R.H. (1997). Pathogenesis, diagnosis, and management of trichomoniasis in cattle. *Veterinary Clinics of North America: Food Animal Practice* 13: 345–361.

Bravo-Guerra, C., Cáceres-Martínez, J., Vásquez-Yeomans, R. et al. (2020). Lethal effects of silver nanoparticles on *Perkinsus marinus*, a protozoan oyster parasite. *Journal of Invertebrate Pathology* 169: 107304. https://doi.org/10.1016/j.jip.2019.107304.

Bruni, N., Stella, B., Giraudo, L. et al. (2017). Nanostructured delivery systems with improved leishmanicidal activity: a critical review. *International Journal of Nanomedicine* 12: 5289–5311.

Carregal, V.M., Lanza, J.S., Souza, D.M. et al. (2019). Combination oral therapy against *Leishmania amazonensis* infection in BALB/c mice using nanoassemblies made from amphiphilic antimony(V) complex incorporating miltefosine. *Parasitology Research* 118: 3077–3084.

CDC (2019). Global health, division of parasitic diseases – African trypanosomiasis. https://www.cdc.gov/parasites/sleepingsickness/index.html (accessed 3 June 2020).

CDC (2020). Acanthamoeba. https://www.cdc.gov/parasites/acanthamoeba/index.html (accessed 3 June 2020).

Chatelain, E. (2017). Chagas disease research and development: is there light at the end of the tunnel? *Computational and Structural Biotechnology Journal* 15: 98–103.

Chatterjee, K.D. (2015). *Parasitology: Protozoology and Helminthology*, 13e. India: CBS Publishers & Distributors.

Cheuka, P.M., Mayoka, G., Mutai, P., and Chibale, K. (2016). The role of natural products in drug discovery and development against neglected tropical diseases. *Molecules* 22: 58.

Coma-Cros, E.M., Biosca, A., Marques, J. et al. (2018). Polyamidoamine nanoparticles for the oral administration of antimalarial drugs. *Pharmaceutics* 10 (4): 225.

Costa, J.S.R., Medeiros, M., Yamashiro-Kanashiro, E.H. et al. (2019). Biodegradable nanocarriers coated with polymyxin B: evaluation of leishmanicidal and antibacterial potential. *PLOS Neglected Tropical Diseases* 13 (5): e0007388.

Danesh-Bahreini, M.A., Shokri, J., Samiei, A. et al. (2011). Nanovaccine for leishmaniasis: preparation of chitosan nanoparticles containing *Leishmania* superoxide dismutase and evaluation of its immunogenicity in BALB/c mice. *International Journal of Nanomedicine* 6: 835–842.

Darr, J.A., Zhang, J., Makwana, N.M., and Weng, X. (2017). Continuous hydrothermal synthesis of inorganic nanoparticles:applications and future directions. *Chemical Reviews* 117: 11125–11238.

Demetzos, C. (2016). *Pharmaceutical Nanotechnology: Fundamentals and Practical Applications*. Singapore: Springer.

Diniz, L.F., Urbina, J.A., de Andrade, I.M. et al. (2013). Benznidazole and posaconazole in experimental chagas disease: positive interaction in concomitant and sequential treatments. *PLOS Neglected Tropical Diseases* 7: e2367.

Drexler, E. and Pamlin, D. (2013). *Nano-solutions for the 21st Century*. Oxford, UK: Oxford Martin School, University of Oxford https://www.oxfordmartin.ox.ac.uk/downloads/academic/201310Nano_Solutions.pdf (accessed 3 June 2020).

Du, J., Tang, J., Xu, S. et al. (2018). A review on silver nanoparticles-induced ecotoxicity and the underlying toxicity mechanisms. *Regulatory Toxicology and Pharmacology* 98: 231–239.

Dwivedi, P., Khatik, R., Chaturvedi, P. et al. (2015). Arteether nanoemulsion for enhanced efficacy against *Plasmodium yoelii* nigeriensis malaria: an approach by enhanced bioavailability. *Colloids and Surfaces B: Biointerfaces* 126: 467–475.

Dykman, L.A. (2020). Gold nanoparticles for preparation of antibodies and vaccines against infectious diseases. *Expert Review of Vaccines* 19: 465–477. https://doi.org/10.1080/14760584.2020.1758070.

El-Khadragy, M., Alolayan, E.M., Metwally, D.M. et al. (2018). Clinical efficacy associated with enhanced antioxidant enzyme activities of silver nanoparticles biosynthesized using *Moringa oleifera* leaf extract, against cutaneous leishmaniasis in a murine model of *Leishmania major*. *International Journal of Environmental Research and Public Health* 15: 1037.

Elkodous, M.A., El-Sayyad, G.S., Abdelrahman, I.Y. et al. (2019). Therapeutic and diagnostic potential of nanomaterials for enhanced biomedical applications. *Colloids and Surfaces B: Biointerfaces* 180: 411–428.

Esfandiari, F., Motazedian, M.H., Asgari, Q. et al. (2019). Paromomycin-loaded mannosylated chitosan nanoparticles: synthesis, characterization and targeted drug delivery against leishmaniasis. *Acta Tropica* 197: 105045.

Fanti, J.R., Tomiotto-Pellissier, F., Menegazzo Miranda-Sapla, M. et al. (2018). Biogenic silver nanoparticles inducing *Leishmania amazonensis* promastigote and amastigote death in vitro. *Acta Tropica* 178: 46–54.

FishBase (2020). List of available fish diseases. https://www.fishbase.se/diseases/DiseaseList.php (accessed 3 June 2020).

Freixa, A., Acuña, V., Sanchís, J. et al. (2018). Ecotoxicological effects of carbon based nanomaterials in aquatic organisms. *Science of the Total Environment* 619–620: 328–337.

Foko, L.P.K., Eya'ane Meva, F., Eboumbou Moukoko, C.E. et al. (2019). A systematic review on anti-malarial drug discovery and antiplasmodial potential of green synthesis mediated metal nanoparticles: overview, challenges and future perspectives. *Malaria Journal* 18: 337.

Fuentes-Valencia, M.A., Fajer-Ávila, E.J., Chávez-Sánchez, M.C. et al. (2020). Silver nanoparticles are lethal to the ciliate model *Tetrahymena* and safe to the pike silverside *Chirostoma estor*. *Experimental Parasitology* 209: 107825. https://doi.org/10.1016/j.exppara.2019.107825.

Furtado, J.M., Smith, J.R., Belfort, R. Jr. et al. (2011). Toxoplasmosis: a global threat. *Journal of Global Infectious Diseases* 3 (3): 281–284.

Gaafar, M.R., Mady, R.F., Diab, R.G., and Shalaby, T.I. (2014). Chitosan and silver nanoparticles: promising anti-toxoplasma agents. *Experimental Parasitology* 143: 30–38.

Gandhi, P.R., Jayaseelan, C., Kamaraj, C. et al. (2018). In vitro antimalarial activity of synthesized TiO_2 nanoparticles using *Momordica charantia* leaf extract against *Plasmodium falciparum*. *Journal of Applied Biomedicine* 16 (4): 378–386.

Gaspar, L., Moraes, C.B., Freitas-Junior, L.H. et al. (2015). Current and future chemotherapy for Chagas disease. *Current Medicinal Chemistry* 22: 4293–4312.

Ghafari, P., St-Denis, C.H., Power, M.E. et al. (2008). Impact of carbon nanotubes on the ingestion and digestion of bacteria by ciliated protozoa. *Nature Nanotechnology* 3: 347–351.

GHO (2020). Leishmaniasis. Global Health Observatory data. https://www.who.int/gho/neglected_diseases/leishmaniasis/en/ (accessed 3 June 2020).

Ghosh, S. (2017). *Paniker's Textbook of Medical Parasitology*, 8e. India: Jaypee Brothers Medical Publishers.

Gnanadesigan, M., Nandagopalan, V., Kapildev, G., and Gundappa, M. (2019). Nano drugs for curing malaria: the plausibility. In: *Applications of Targeted Nano Drugs and Delivery Systems. Nanoscience and Nanotechnology in Drug Delivery* (eds. S.S. Mohapatra, S. Ranjan, N. Dasgupta, et al.), 451–467. Oxford, UK: Elsevier.

Gonzalez-Martin, G., Merino, I., Rodriguez-Cabezas, M.N. et al. (1998). Characterization and trypanocidal activity of nifurtimox-containing and empty nanoparticles of polyethylcyanoacrylates. *Journal of Pharmacy and Pharmacology* 50 (1): 29–35.

Gonzalez-Martin, G., Figueroa, C., Merino, I., and Osuna, A. (2000). Allopurinol encapsulated in polycyanoacrylate nanoparticles as potential lysosomatropic carrier: preparation and trypanocidal activity. *European Journal of Pharmaceutics and Biopharmaceutics* 49 (2): 137–142.

Griffith, R.K. (2020). Drugs used to treat parasitic infections. In: *Foye's Principles of Medicinal Chemistry*, 8e (eds. V. Roche, W.S. Zito, T.L. Lemke and D.A. Williams). USA: Wolter Kluwer, Chapter 32.

GWH (2017). Trichomonosis in Garden Birds. Date of factsheet. https://www.gardenwildlifehealth.org/wp-content/uploads/sites/12/2018/02/Garden-bird-Trichomonosis-factsheet_GWH_2017.pdf (accessed 3 June 2020).

Hadiza, A.M., Taura, D.W., and Mohammed, B. (2018). Antiplasmodial activity of nanoparticles enhanced chloroquine on *Plasmodium berghei* in vivo. *Bayero Journal of Pure and Applied Sciences* 11 (1): 27–32.

Hagras, N.A., Allam, A.F., Farag, H.F. et al. (2019). Successful treatment of acute experimental toxoplasmosis by spiramycin-loaded chitosan nanoparticles. *Experimental Parasitology* 204: 107717.

Hassan, D., Farghali, M., Eldeek, H. et al. (2019). Antiprotozoal activity of silver nanoparticles against *Cryptosporidium parvum* oocysts: new insights on their feasibility as a water disinfectant. *Journal of Microbiological Methods* 165: 105698.

Hausmann, K., Hülsmann, N., and Radek, R. (2003). *Protistology*, 3e. Germany: Schweizerbart Science Publishers.

Hemmert, C., Fabie, A., Fabre, A. et al. (2013). Synthesis, structures, and antimalarial activities of some silver(I), gold(I) and gold(III) complexes involving N-heterocyclic carbene ligands. *European Journal of Medicinal Chemistry* 60: 64–75.

Hendiger, E.B., Padzik, M., Sifaoui, I. et al. (2020). Silver nanoparticles as a novel potential preventive agent against acanthamoeba keratitis. *Pathogens* 5 (9): 350.

Hess, K.L., Medintz, I.L., and Jewell, C.M. (2019). Designing inorganic nanomaterials for vaccines and immunotherapies. *Nano Today* 27: 73–98.

Hu, C., Wang, Q., Zhao, H. et al. (2015). Ecotoxicological effects of graphene oxide on the protozoan *Euglena gracilis*. *Chemosphere* 128: 184–190.

Inbaneson, S.J. and Ravikumar, I.S. (2013). In vitro antiplasmodial activity of PDDS-coated metal oxide nanoparticles against *Plasmodium falciparum*. *Applied Nanoscience* 3: 197–201.

Jain, A.K. and Thareja, S. (2019). In vitro and in vivo characterization of pharmaceutical nanocarriers used for drug delivery. *Artificial Cells Nanomedicine Biotechnology* 47: 524–539.

Jain, S., Huang, Z., Dixon, B.R. et al. (2019). *Cryptosporidium parvum* oocyst directed assembly of gold nanoparticles and graphene oxide. *Frontiers of Chemical Science and Engineering* 13: 608–615.

Jamkhande, P.G., Ghuleb, N.W., Mohan, A.H.B., and Kalaskard, G. (2019). Metal nanoparticles synthesis: an overview on methods of preparation, advantages and disadvantages, and applications. *Journal of Drug Delivery Science and Technology* 53: 101174.

Jampílek, J. (2014). Recent advances in design of potential quinoxaline anti-infectives. *Current Medicinal Chemistry* 21: 4347–4373.

Jampílek, J. (2016). Potential of agricultural fungicides for antifungal drug discovery. *Expert Opinion on Drug Discovery* 11: 1–9.

Jampílek, J. (2017). Design of antimalarial agents based on natural products. *Current Organic Chemistry* 21: 1824–1846.

Jampílek, J. and Kráľová, K. (2017a). Nano-antimicrobials: activity, benefits and weaknesses. In: *Nanostructures in Therapeutic Medicine – Nanostructures for Antimicrobial Therapy* (eds. A. Ficai and A.M. Grumezescu), 23–54. Amsterdam, Netherlands: Elsevier.

Jampílek, J. and Kráľová, K. (2017b). Nanopesticides: preparation, targeting and controlled release. In: *Nanotechnology in the Agri-Food Industry – New Pesticides and Soil Sensors* (ed. A.M. Grumezescu), 81–127. UK: Elsevier.

Jampílek, J. and Kráľová, K. (2017c). Nanomaterials for delivery of nutrients and growth-promoting compounds to plants. In: *Nanotechnology: An Agricultural Paradigm* (eds. R. Prasad, M. Kumar and V. Kumar), 177–226. Singapore: Springer.

Jampílek, J. (2018). Design and discovery of new antibacterial agents: advances, perspectives, challenges. *Current Medicinal Chemistry* 25: 4972–5006.

Jampílek, J. and Kráľová, K. (2018a). Application of nanobioformulations for controlled release and targeted biodistribution of drugs. In: *Nanobiomaterials: Applications in Drug Delivery* (eds. A.K. Sharma, R.K. Keservani and R.K. Kesharwani), 131–208. Warentown, NJ: CRC Press.

Jampílek, J. and Kráľová, K. (2018b). Nanomaterials applicable in food protection. In: *Nanotechnology Applications in the Food*

Industry (eds. R.V. Rai and J.A. Bai), 75–96. Boca Raton, FL: Taylor & Francis Group.

Jampílek, J. and Kráľová, K. (2018c). Benefits and potential risks of nanotechnology applications in crop protection. In: *Nanobiotechnology Applications in Plant Protection* (eds. K. Abd-Elsalam and R. Prasad), 189–246. Germany: Springer.

Jampílek, J. and Kráľová, K. (2019a). Nanotechnology based formulations for drug targeting to central nervous system. In: *Nanoparticulate Drug Delivery Systems* (eds. R.K. Keservani and A.K. Sharma), 151–220. USA: Apple Academic Press & CRC Press.

Jampílek, J. and Kráľová, K. (2019b). Recent advances in lipid nanocarriers applicable in the fight against cancer. In: *Nanoarchitectonics in Biomedicine* (ed. A.M. Grumezescu), 219–294. Netherlands: Elsevier.

Jampílek, J. and Kráľová, K. (2019c). Nanoformulations – valuable tool in therapy of viral diseases attacking humans and animals. In: *Nanotheranostic – Applications and Limitations*, vol. 137–178 (eds. M. Rai and B. Jamil). Switzerland: Springer Nature.

Jampílek, J. and Kráľová, K. (2019d). Nano-biopesticides in agriculture: state of art and future opportunities. In: *Nano-Biopesticides Today and Future Perspectives* (ed. O. Koul), 397–447. Netherlands: Academic Press & Elsevier.

Jampílek, J. and Kráľová, K. (2019e). Beneficial effects of metal- and metalloid-based nanoparticles on crop production. In: *Nanotechnology for Agriculture – Advances for Sustainable Agriculture* (eds. D.G. Panpatte and Y.K. Jhala), 161–219. Singapore: Springer Nature.

Jampílek, J., Kráľová, K., Campos, E.V.R., and Fraceto, L.F. (2019a). Bio-based nanoemulsion formulations applicable in agriculture, medicine and food industry. In: *Nanobiotechnology in Bioformulations* (eds. R. Prasad, V. Kumar, M. Kumar and D.K. Choudhary), 33–84. Germany: Springer.

Jampílek, J., Kráľová, K., Novák, P., and Novák, M. (2019b). Nanobiotechnology in neurodegenerative diseases. In: *Nanobiotechnology in Neurodegenerative Diseases* (eds. M. Rai and A. Yadav), 65–138. Switzerland: Springer Nature.

Jampílek, J., Kos, J., and Kráľová, K. (2019c). Potential of nanomaterial applications in dietary supplements and foods for special medical purposes. *Nanomaterials* 9: 296.

Jampílek, J. and Kráľová, K. (2020a). 'Natural biopolymeric nanoformulations for brain drug delivery. In: *Nanocarriers for Brain Targetting: Principles and Applications* (eds. R.K. Keservani, A.K. Sharma and R.K. Kesharwani), 131–203. USA: Apple Academic Press & CRC Press.

Jampílek, J. and Kráľová, K. (2020b). Impact of nanoparticles on toxigenic fungi. In: *Nanomycotoxicology – Treating Mycotoxins in the Nano Way* (eds. M. Rai and K.A. Abd-Elsalam), 309–348. UK: Academic Press and Elsevier.

Jampílek, J. and Kráľová, K. (2020c). Nanoweapons against tuberculosis. In: *Nanoformulations in Human Health – Challenges and Approaches* (eds. S. Talegaonkar and M. Rai), 469–502. Switzerland: Springer Nature. https://www.springer.com/gp/book/9783030418571.

Jampílek, J. and Kráľová, K. (2020d). Nanocomposites: synergistic nanotools for management mycotoxigenic fungi. In: *Nanomycotoxicology – Treating Mycotoxins in the Nano Way* (eds. M. Rai and K.A. Abd-Elsalam), 349–383. UK: Academic Press and Elsevier.

Jampílek, J., Kráľová, K., and Fedor, P. (2020). Bioactivity of nanoformulated synthetic and natural insecticides and their impact on the environment. In: *Nanopesticides – From Research and Development to Mechanisms of Action and Sustainable Use in Agriculture* (eds. L.F. Fraceto, V.L. de Castro, R. Grillo, et al.), 165–225. Switzerland: Springer Nature. https://www.springer.com/gp/book/9783030448721.

Jin, Z., Gao, S., Cui, X. et al. (2019). Adjuvants and delivery systems based on polymeric nanoparticles for mucosal vaccines.

International Journal of Pharmaceutics 572: 118731.

Jones, A.J., Grkovic, T., Sykes, M.L., and Avery, V.M. (2013). Trypanocidal activity of marine natural products. *Marine Drugs* 11: 4058–4082.

Juganson, K., Mortimer, M., Ivask, A. et al. (2017). Mechanisms of toxic action of silver nanoparticles in the protozoan *Tetrahymena thermophila*: from gene expression to phenotypic events. *Environmental Pollution* 225: 481–489.

Kannan, D., Yadav, N., Ahmad, S. et al. (2019). Pre-clinical study of iron oxide nanoparticles fortified artesunate for efficient targeting of malarial parasite. *EBioMedicine* 45: 261–277.

Kenry, Lim, Y.B., Nai, M.H. et al. (2017). Graphene oxide inhibits malaria parasite invasion and delays parasitic growth in vitro. *Nanoscale* 9: 14065.

Khan, I., Saeed, K., and Khan, I. (2019). Nanoparticles: properties, applications and toxicities. *Arabian Journal of Chemistry* 12: 908–931.

Khan, A.U., Khan, M., Cho, M.H., and Khan, M.M. (2020). Selected nanotechnologies and nanostructures for drug delivery, nanomedicine and cure. *Bioprocess and Biosystems Engineering* 43: 1339–1357. https://doi.org/10.1007/s00449-020-02330-8.

Kheiri, S., Liu, X., and Thompson, M. (2019). Nanoparticles at biointerfaces: antibacterial activity and nanotoxicology. *Colloids and Surfaces B: Biointerfaces* 184: 110550.

Khezerlou, A., Alizadeh-Sani, M., Azizi-Lalabadi, M., and Ehsani, A. (2018). Nanoparticles and their antimicrobial properties against pathogens including bacteria, fungi, parasites and viruses. *Microbial Pathogenesis* 123: 505–526.

Kratošová, G., Holišová, V., Konvičková, Z. et al. (2019). From biotechnology principles to functional and low-cost metallic bionanocatalysts. *Biotechnology Advances* 37: 154–176. https://doi.org/10.1016/j.biotechadv.2018.11.012.

Kumar, R., Ray, P.C., Datta, D. et al. (2015). Nanovaccines for malaria using *Plasmodium falciparum* antigen Pfs25 attached gold nanoparticles. *Vaccine* 33: 5064–5071.

Kumar, H., Gothwal, A., Khan, I. et al. (2017). Galactose-anchored gelatin nanoparticles for primaquine delivery and improved pharmacokinetics: a biodegradable and safe approach for effective antiplasmodial activity against *P. falciparum* 3d7 and in vivo hepatocyte targeting. *Molecular Pharmaceutics* 14: 3356–3369.

Lawrence, J.R., Paule, A., Swerhone, G.D.W. et al. (2020). Microscale and molecular analyses of river biofilm communities treated with microgram levels of cerium oxide nanoparticles indicate limited but significant effects. *Environmental Pollution* 256: 113515.

Leishmaniasis (2018). Status of endemicity of cutaneous leismaniasis. https://apps.who.int/neglected_diseases/ntddata/leishmaniasis/leishmaniasis.html (accessed 3 June 2020).

Leishmaniasis (2020). https://www.who.int/news-room/fact-sheets/detail/leishmaniasis (accessed 3 June 2020).

Li, C., Wang, J., Wang, Y. et al. (2019). Recent progress in drug delivery. *Acta Pharmaceutica Sinica B* 9: 1145–1162.

Loker, E.S. and Hofkin, B. (2015). *Parasitology – A Conceptual Approach*. USA: CRC Press.

Ma, L., Kohli, M., and Smith, A. (2013). Nanoparticles for combination drug therapy. *ACS Nano* 7: 9518–9525.

Ma, Y., Lu, T., Zhao, W. et al. (2014). Enhanced antimalarial activity by a novel artemether-lumefantrine lipid emulsion for parenteral administration. *Antimicrobial Agents and Chemotherapy* 58: 5658–5665.

Machado, L.F., Sanfelice, R.A., Rodrigues Bosquia, L. et al. (2020). Biogenic silver nanoparticles reduce adherence, infection, and proliferation of *Toxoplasma gondii* RH strain in HeLa cells without inflammatory mediators induction. *Experimental Parasitology* 211: 107853.

Marcandalli, J., Fiala, B., Ols, S. et al. (2019). Induction of potent neutralizing antibody responses by a designed protein nanoparticle vaccine for respiratory syncytial virus. *Cell* 177: 1420–1431.

Markwalter, C.F., Kantor, A.G., Moore, C.P. et al. (2019). Inorganic complexes and metal-based nanomaterials for infectious disease diagnostics. *Chemical Reviews* 119: 1456–1518.

Marques, J., Moles, E., Urbán, P. et al. (2014). Application of heparin as a dual agent with antimalarial and liposome targeting activities toward Plasmodium-infected red blood cells. *Nanomedicine* 10: 1719–1728.

Mayne, R., Morgan, J., Whiting, J.G.H. et al. (2019). On measuring nanoparticle toxicity and clearance with *Paramecium caudatum*. *Scientific Reports* 9: 8957.

Mashjoor, S., Yousefzadi, M., Zolgharnein, H. et al. (2019). Phyco-linked vs chemogenic magnetite nanoparticles: route selectivity in nano-synthesis, antibacterial and acute zooplanktonic responses. *Materials Science and Engineering: C* 102: 324–340.

Mishra, A., Kaushik, N.K., Sardar, M., and Sahal, D. (2013). Evaluation of antiplasmodial activity of green synthesized silvernanoparticles. *Colloids and Surfaces B: Biointerfaces* 111: 713–718.

Mehrizi, T.Z., Ardestani, M.S., Hoseini, M.H.M. et al. (2018). Novel nanosized chitosan-betulinic acid against resistant *Leishmania major* and first clinical observation of such parasite in kidney. *Scientific Reports* 8: 11759.

Miller, R.J., Muller, E.B., Cole, B. et al. (2017). Photosynthetic efficiency predicts toxic effects of metal nanomaterials in phytoplankton. *Aquatic Toxicology* 183: 85–93.

Mohapatra, S.S., Ranjan, S., Dasgupta, N. et al. (2019). *Applications of Targeted Nano Drugs and Delivery Systems: Nanoscience and Nanotechnology in Drug Delivery*. Amsterdam: Elsevier.

Mohtasebi, S., Mohebali, M., Elikaee, S. et al. (2019). In vitro and in vivo anti-parasitic activity of biogenic antimony sulfide nanoparticles on *Leishmania major* (MRHO/IR/75/ER). *Parasitology Research* 118: 2669–2678.

Molina, I., Salvador, F., and Sánchez-Montalvá, A. (2015). The use of posaconazole against Chagas disease. *Current Opinion in Infectious Diseases* 28: 1–407.

Morilla, M.J., Montanari, J., Frank, F. et al. (2005). Etanidazole in pH-sensitive liposomes: design, characterization and *in vitro/in vivo* anti-*Trypanosoma cruzi* activity. *Journal of Controlled Release* 103: 599–607.

Mortimer, M., Kahru, A., and Slaveykova, V.I. (2014). Uptake, localization and clearance of quantum dots in ciliated protozoa *Tetrahymena thermophila*. *Environmental Pollution* 190: 58–64.

Murugan, K., Nataraj, D., Jaganathan, A. et al. (2017). Nanofabrication of graphene quantum dots with high toxicity against malaria mosquitoes, *Plasmodium falciparum* and MCF-7 cancer cells: impact on predation of non-target tadpoles, odonate nymphs and mosquito fishes. *Journal of Cluster Science* 28: 393–411.

Nhavene, E.P.F., da Silva, W.M., Trivelato, R.R. et al. (2018). Chitosan grafted into mesoporous silica nanoparticles as benznidazol carrier for Chagas diseases treatment. *Microporous and Mesoporous Materials* 272: 265–275.

NIH (2020). Neglected diseases. Therapeutics for rare and neglected diseases. https://rarediseases.info.nih.gov/files/neglected_diseases_faqs.pdf (accessed 3 June 2020).

Nisha, R., Amit, K.G., Pillai, C.R., and Goutam, R. (2013). Development and characterization of artemether loaded solid lipid nanoparticles. *Indian Journal of Pharmaceutical Education* 47: 123–128.

Niyyati, M., Sasani, R., Mohebali, M. et al. (2018). Anti-acanthamoeba effects of silver and gold nanoparticles and contact lenses disinfection solutions. *Iranian Journal of Parasitology* 13 (2): 180–185.

OIE (2019). Infection with *Perkinsus marinus*. https://www.oie.int/fileadmin/Home/eng/Health_standards/aahm/current/chapitre_perkinsus_marinus.pdf (accessed 3 June 2020).

Panáček, A., Kvítek, L., Smékalová, M. et al. (2018). Bacterial resistance to silver nanoparticles and how to overcome it. *Nature Nanotechnology* 13: 65–71.

Parashar, D., Aditya, N.P., and Murthy, R.S.R. (2016). Development of artemether and lumefantrine co-loaded nanostructured lipid carriers: physicochemical characterization and in vivo antimalarial activity. *Journal of Drug Delivery* 23: 123–129.

Patra, J.K., Das, G., Fraceto, L.F. et al. (2018). Nano based drug delivery systems: recent developments and future prospects. *Journal of Nanobiotechnology* 16: 71.

Paunovic, J., Vucevic, D., Radosavljevic, T. et al. (2020). Iron-based nanoparticles and their potential toxicity: focus on oxidative stress and apoptosis. *Chemico-Biological Interactions* 316: 108935.

Pavoni, L., Pavela, R., Cespi, M. et al. (2019). Green micro- and nanoemulsions for managing parasites, vectors and pests. *Nanomaterials* 9: 1285.

Pedrosa, P., Vinhas, R., Fernandes, A., and Baptista, P.V. (2015). Gold nanotheranostics: proof-of-concept or clinical tool? *Nanomaterials* 5: 1853–1879.

Peletz, R., Mahin, T., Elliott, M. et al. (2013). Preventing cryptosporidiosis: the need for safe drinking water. *Bulletin of the World Health Organization* 91: 238–248. https://www.who.int/bulletin/volumes/91/4/13-119990/en/ (accessed 3 June 2020).

Pentamidine (2020). DrugBank. https://www.drugbank.ca/drugs/DB00738 (accessed 3 June 2020).

Petitdidier, E., Pagniez, J., Pissarra, J. et al. (2019). Peptide-based vaccine successfully induces protective immunity against canine visceral leishmaniasis. *Vaccines* 4: 49.

Plachá, D. and Jampílek, J. (2019). Graphenic materials for biomedical applications. *Nanomaterials* 9 (12): 1758.

Poon, C. and Patel, A.A. (2020). Organic and inorganic nanoparticle vaccines for prevention of infectious diseases. *Nano Express* 1: 012001.

Quezada, C.Q., Azevedo, C.S., Charneau, S. et al. (2019). Advances in nanocarriers as drug delivery systems in Chagas disease. *International Journal of Nanomedicine* 14: 6407–6424.

Rahman, K., Khan, S.U., Fahad, S. et al. (2019). Nano-biotechnology: a new approach to treat and prevent malaria. *International Journal of Nanomedicine* 14: 1401–1410.

Rezaee, H., Ghorbani, M., Nikpay, A., and Soltani, M. (2019). Tannic acid-coated zeolite Y nanoparticles as novel drug nanocarrier with controlled release behavior and anti-protozoan activity against *Trichomonas gallinae*. *Journal of Dispersion Science and Technology* 40: 587–593.

Rizvi, S.A.A. and Saleh, A.M. (2018). Applications of nanoparticle systems in drug delivery technology. *Saudi Pharmaceutical Journal* 26: 64–70.

Rodrigues, G.R., López-Abarrateguia, C., de la Serna Gómez, I. et al. (2019). Antimicrobial magnetic nanoparticles based-therapies for controlling infectious diseases. *International Journal of Pharmaceutics* 555 (30): 356–367.

Rozman, N.A.S., Tong, W.Y., Leong, C.R. et al. (2019). Potential antimicrobial applications of chitosan nanoparticles (ChNP). *Journal of Microbiology, Biotechnology* 29: 1009–1013.

Saleem, K., Khursheed, Z., Hano, C. et al. (2019). Applications of nanomaterials in leishmaniasis: a focus on recent advances and challenges. *Nanomaterials* 9: 1749.

Sales, P.A., Molina, I., Fonseca Murta, S.M. et al. (2017). Experimental and clinical treatment of Chagas disease: a review. *American Journal of Tropical Medicine and Hygiene* 97: 1289–1303.

Sanchez-Guzman, D., Le Guen, P., Villeret, B. et al. (2019). Silver nanoparticle-adjuvanted vaccine protects against lethal influenza infection through inducing BALT and IgA-mediated mucosal immunity. *Biomaterials* 217: 119308.

Santos-Valle, A.B.C., Souza, G.R.R., Paes, C.Q. et al. (2019). Nanomedicine strategies for addressing major needs in neglected tropical diseases. *Annual Reviews in Control* 48: 423–441.

Sepúlveda, A.A.L., Arenas Velásquez, A.M., Patiño Linares, I.A. et al. (2020). Efficacy of photodynamic therapy using TiO_2 nanoparticles doped with Zn and hypericin in the treatment of cutaneous Leishmaniasis caused by *Leishmania amazonensis*. *Photodiagnosis and Photodynamic Therapy* 30: 101676.

Sesti-Costa, R., Carneiro, Z.A., Silva, M.C. et al. (2014). Ruthenium complex with benznidazole and nitric oxide as a new candidate for the treatment of Chagas disease. *PLOS Neglected Tropical Diseases* 8: 3207.

Shah, R. (2020). Protozoan diseases found in animals. http://www.biologydiscussion.com/animals-2/diseases-animals-2/protozoan-diseases-found-in-animals/25461 (accessed 3 June 2020).

Shrestha, S., Wang, B., and Dutta, P. (2020). Nanoparticle processing: understanding and controlling aggregation. *Advances in Colloid and Interface Science* 279: 102162.

Singh, K.K. and Vingkar, S.K. (2008). Formulation, antimalarial activity and biodistribution of oral lipid nanoemulsion of primaquine. *International Journal of Pharmaceutics* 347: 136–143.

Smith, D.M., Simon, J.K., and Baker, J.R. Jr. (2013). Applications of nanotechnology for immunology. *Nature Reviews Immunology* 13 (8): 592–605.

Sosna, T., Mikeska, M., Dutko, O. et al. (2019). Micronization of ibuprofen particles using supercritical fluid technology. *Journal of Nanoscience and Nanotechnology* 19: 2814–2820.

Sourav, R., Sudip, S., Sabyasachim, B. et al. (2015). The effect of nanoparticles on plankton dynamics: a mathematical model. *BioSystems* 127: 28–41.

Souto, E.B., Dias-Ferreira, J., Craveiro, S.A. et al. (2019). Therapeutic interventions for countering leishmaniasis and Chagas's disease: from traditional sources to nanotechnological systems. *Pathogens* 8: 119.

Sun, Y., Chen, D., Pan, Y. et al. (2019). Nanoparticles for antiparasitic drug delivery. *Drug Delivery* 26 (1): 1206–1221.

Tavakoli, P., Ghaffarifar, F., Delavari, H., and Shahpari, N. (2019). Efficacy of manganese oxide (Mn_2O_3) nanoparticles against *Leishmania major* in vitro and in vivo. *Journal of Trace Elements in Medicine and Biology* 56: 162–168.

Tavakoli, A. and Hashemzadeh, M.S. (2020). Inhibition of herpes simplex virus type 1 by copper oxide nanoparticles. *Journal of Virological Methods* 275: 113688.

Teimouri, A., Azami, S.J., Keshavarz, H. et al. (2018). Anti-Toxoplasma activity of various molecular weights and concentrations of chitosan nanoparticles on tachyzoites of RH strain. *International Journal of Nanomedicine* 13: 1341–1351.

Teng, Z., Sun, S., Chen, H. et al. (2018). Golden-star nanoparticles as adjuvant effectively promotes immune response to foot-and-mouth disease virus-like particles vaccine. *Vaccine* 36: 6752–6760.

Tessarolo, L.D., Bezerra de Menezes, R.R.P.P., Mello, C.P. et al. (2018). Nanoencapsulation of benznidazole in calcium carbonate increases its selectivity to *Trypanosoma cruzi*. *Parasitology* 145: 1–8.

Torgerson, P.R. and Mastroiacovo, P. (2013). The global burden of congenital toxoplasmosis: a systematic review. *Bulletin of the World Health Organization* 91: 501–508. https://www.who.int/bulletin/volumes/91/7/12-111732/en (accessed 3 June 2020).

Toxoplasmosis (2020). Mayo Foundation for Medical Education and Research. https://www.mayoclinic.org/diseases-conditions/toxoplasmosis/diagnosis-treatment/drc-20356255 (accessed 3 June 2020).

Trichomoniasis (2020). Department of Animal Sciences, University of Wisconsin-Madison. http://www.ansci.wisc.edu/jjp1/ansci_repro/lab/diseases_2002/trichomoniasis.html (accessed 3 June 2020).

Tripathy, S., Das, S., Chakraborty, S.P. et al. (2012). Synthesis, characterization of chitosan-tripolyphosphate conjugated chloroquine nanoparticle and its in vivo anti-malarial efficacy against rodent parasite: a dose and duration dependent approach. *International Journal of Pharmaceutics* 434: 292–305.

Tripathy, S., Mahapatra, S.K., Chattopadhyay, S. et al. (2013). A novel chitosan based antimalarial drug delivery against *Plasmodium berghei* infection. *Acta Tropica* 128: 494–503.

Tripathy, S., Das, S., Dash, S.K. et al. (2014). A prospective strategy to restore the tissue damage in malaria infection: approach with chitosan-trypolyphosphate conjugated nanochloroquine in Swiss mice. *European Journal of Pharmacology* 737: 11–21.

Tupinambá Branquinho, R., Roy, J., Farah, C. et al. (2017). Biodegradable polymeric nanocapsules prevent cardiotoxicity of anti-trypanosomal lychnopholide. *Scientific Reports* 7: 44998.

Turan, N.B., Erkan, H.S., Engin, G.O., and Bilgili, M.S. (2019). Nanoparticles in the aquatic environment: usage, properties,transformation and toxicity – a review. *Process Safety and Environmental Protection* 130: 238–249.

Vaculíková, E., Grünwaldová, V., Král, V. et al. (2012). Preparation of candesartan and atorvastatin nanoparticles by solvent evaporation. *Molecules* 17: 13221–13234.

Vaculíková, E., Plachá, D., Pisárčik, M. et al. (2014). Preparation of risedronate nanoparticles by solvent evaporation technique. *Molecules* 19: 17848–17861.

Vaculíková, E., Černíková, A., Plachá, D. et al. (2016). Preparation of hydrochlorothiazide nanoparticles for solubility enhancement. *Molecules* 21: 1005.

Valle, I.V., Machado, M.E., da Costa Bernardes Araújo, C. et al. (2019). Oral pentamidine-loaded poly(d,l-lactic-co-glycolic) acid nanoparticles: an alternative approach for leishmaniasis treatment. *Nanotechnology* 30 (45): 455102.

Volpedo, G., Costa, L., Ryan, N. et al. (2019). Nanoparticulate drug delivery systems for the treatment of neglected tropical protozoan diseases. *Journal of Venomous Animals and Toxins including Tropical Diseases* 25: 144118.

Wang, Y., Yu, L., Kong, X., and Sun, L. (2017). Application of nanodiagnostics in point-of-care tests for infectious diseases. *International Journal of Nanomedicine* 2: 4789–4803.

Wennersten, R., Fidler, J., and Spitsyna, A. (2008). Nanotechnology: a new technological revolution in the 21st century. In: *Handbook of Performability Engineering* (ed. K.B. Misra), 943–952. UK: Springer.

WHO (2019). World Malaria Report 2019. Geneva. https://www.who.int/publications-detail/world-malaria-report-2019 (accessed 3 June 2020).

WHO (2020a). Neglected tropical diseases. https://www.who.int/neglected_diseases/en (accessed 3 June 2020).

WHO (2020b). Trypanosomiasis. https://www.who.int/news-room/fact-sheets/detail/trypanosomiasis-human-african-(sleeping-sickness) (accessed 3 June 2020).

WHO (2020c). Human African trypanosomiasis. https://www.who.int/trypanosomiasis_african/en/ (accessed 3 June 2020).

WHO (2020d). Chagas disease – American trypanosomiasis. https://www.who.int/health-topics/chagas-disease#tab=tab_1 (accessed 3 June 2020).

WHO (2020e). Leishmaniasis. https://www.who.int/leishmaniasis/disease/en/ (accessed 3 June 2020).

Yah, C.S. and Simate, G.S. (2015). Nanoparticles as potential new generation broad spectrum antimicrobial agents. *DARU Journal of Pharmaceutical Sciences* 23: 43.

Yan, X., Zhou, M., Yu, S. et al. (2020). An overview of biodegradable nanomaterials and applications in vaccines. *Vaccine* 38: 1096–1104.

Yilmaz, A.H. (2019). Antibacterial activity of chitosan-based systems. In: *Functional Chitosan: Drug Delivery and Biomedical Applications* (eds. S. Jana and S. Jana), 457–489. Singapore: Springer Nature.

Yoshihara, E., Tachibana, H., and Nakae, T. (1987). Trypanocidal activity of the stearylamine-bearing liposome *in vitro*. *Life Sciences* 40: 2153–2159.

Yoshihara, E. and Nakae, T. (1986). Cytolytic activity of liposomes containing stearylamine. *Biochimica et Biophysica Acta – Biomembranes* 854: 93–101.

Yunessnia lehi, A., Shagholani, H., Ghorbani, M. et al. (2019). Chitosan nanocapsule-mounted cellulose nanofibrils as nanoships for smart drug delivery systems and treatment of avian trichomoniasis. *Journal of the Taiwan Institute of Chemical Engineers* 95: 290–299.

Zanin, H., Margraf-Ferreira, A., da Silva, N.S. et al. (2014). Graphene and carbon nanotube composite enabling a new prospective treatment for trichomoniasis disease. *Materials Science and Engineering C* 41: 65–69.

Zhou, W., Gao, X., Liu, D., and Chen, X. (2015). Gold nanoparticles for in vitro diagnostics. *Chemical Reviews* 115: 10575–10636.

Zhou, C., Vitiello, V., Pellegrini, D. et al. (2016). Toxicological effects of CdSe/ZnS quantum dots on marine planktonic organisms. *Ecotoxicology and Environmental Safety* 123: 26–31.

Zou, X.Y., Xu, B., Yu, C.P., and Zhang, H.W. (2013). Combined toxicity of ferroferric oxide nanoparticles and arsenic to the ciliated protozoa *Tetrahymena pyriformis*. *Aquatic Toxicology* 134–135: 66–73.

Zulfiqar, B., Jones, A.J., Sykes, M.L. et al. (2017). Screening a natural product-based library against kinetoplastid parasites. *Molecules* 22: 1715.

5

Toxicity of Metallic Nanoparticles

A Pressing Issue

Tejal Mehta, Dhaivat Parikh, Kartik Hariharan, Namdev Dhas, and Viral Patel

Department of Pharmaceutics, Institute of Pharmacy, Nirma University, Ahmedabad, GJ, India

5.1 Introduction

In this millennium, nanotechnology has flourished with a new horizon for researchers across the globe. Different nanomaterials are intensively utilized to create functional systems that work at molecular levels. Among the nanomaterials used for different purposes, greater focus is being made on the use of metallic nanoparticles (MNPs). These MNPs are explored diversely in different fields of medicine, radio-imaging, as catalyst, sensing agents, as markers, cell labelling, target drug delivery, cancer therapy, biological diagnostics, etc. (Shang et al. 2014; Ling et al. 2015). Metal-based NPs have gained popularity owing to their properties like size, shape, and surface charge, which enable their cellular uptake. Also, certain functionalities present on the surface of some MNPs. NPs have proved to enhance their interaction at cellular level (Berry 2009; Woźniak et al. 2017).

The quantum properties of these MNPs make them unpredictable when they reach biological systems, and thus they may end up with toxicity. With a decrease in size, some MNPs have revealed toxicities in contrast to their inert behavior when in bulk form. Like two-edged swords, the toxicity caused by MNPs is helpful when they are used as antibacterial; however, it is fatal when they damage the healthy cells in the human body (Aderibigbe 2017). The toxicity caused can be either genotoxicity or carcinogenicity. The key parameter that is responsible for such toxicities is its small size, due to which it can easily invade different basic biological structures and organs of the human body (Buzea et al. 2007; Leblanc et al. 2009). In addition, there are several research studies supporting that the MNPs are more toxic than their corresponding fine nanoparticles (Liang et al. 2009; Vallyathan and Gulumian 2010). The most probable toxic effects observed with different MNPs include altered redox balance, pro-inflammatory effects, enhanced oxidative stress, elevated levels of ROS, cytotoxicity, and abnormal apoptosis. MNPs interact with different proteins and enzymes at the cellular level and they impact the antioxidant defense system of the body, causing the generation of ROS which ultimately initiates pro-inflammatory response resulting in perturbation, apoptosis, or necrosis. In addition, a study suggests, the level of toxicity depends on the target cell type and possible routes of exposure (Sohaebuddin et al. 2010; Zhao and Castranova 2011). Moreover, the toxicity of such NPs is being addressed using different approaches like *in vitro*, *in vivo*, genomic and biodistribution

Nanotechnology in Medicine: Toxicity and Safety, First Edition. Edited by Mahendra Rai, Mrunali Patel, and Rashmin Patel.

studies. In addition, certain techniques like microarray, real-time reverse transcription, polymerase chain reaction (PCR) are very sensitive in measuring the changes occurring at genomic levels and are successful in providing ample data to estimate the toxicity levels (Jeng and Swanson 2006; Schrand et al. 2010).

Although lot of research has undergone for the use of MNPs, innumerable challenges exist which needs to be overcome prior to suggesting that benefits of MNPs overweigh the risk associated with its use. Recently, researchers have focused on evaluating the potential toxicities of different metallic NPs – cobalt, copper, nickel, iron, on different organs, tissues, cells, and subcellular levels owing to their physicochemical properties. Also, there is still a question that toxicity caused by MNPs originates from themselves or due to releasing metal ions. Nevertheless, profound research is still ongoing.

This chapter focuses on current knowledge of potential toxicities caused by MNPs. Also, in the further sections, aluminum nanoparticles and their toxic effects with respect to different routes of administration are discussed in detail. Physicochemical properties, their interactions with biological substances, and detrimental effects of silver nanoparticles (AgNPs), zinc oxide nanoparticles (ZnONPs), titanium dioxide nanoparticles (TiO_2NPs), copper oxide (CuONPs), gold nanoparticles (AuNPs), cerium oxide nanoparticles (CeONPs), iron oxide nanoparticles (IONPs), causing neurotoxicity, genotoxicity, immune toxicity, carcinogenicity, teratogenicity etc. are also described in detail.

5.2 Toxicity of Metal Nanoparticles and Influence of Physicochemical Properties

The use of metal nanoparticles in nanomedicine has risen over the last few years. Though the use of metals is quite old (e.g. Swarna bhasma (gold), silver utensils) application of metals has advanced a lot. Metals have found their use in healthcare and medicine either as therapeutic moiety (platinum compounds as anticancer drugs) or imaging (gadolinium complexes as MRI agents). Today, the application has advanced even more with its potential to be used as theranostics (Sharma et al. 2015), biosensors (Prakash et al. 2013), imaging and in drug delivery (Janib et al. 2010). But its myriad use is hindered by the serious concerns of toxicity (Jeng and Swanson 2006). Various metals used in biomedicine and the toxicity reported are covered in this section. Also, an attempt has been made to cover the effects of physicochemical properties like size, shape, and charge of metal nanoparticles on toxicity induced *in vitro* as well as *in vivo* models.

5.2.1 Toxicity of Copper and Copper Oxide Nanoparticles

Copper (Cu) has been used for long as an antibacterial and antimicrobial agent (Ismail et al. 2019) and today it is being explored for its combination with other metals as an anticancer agent. Several researches have been focused on understanding the metabolism and toxicity of copper nanoparticles (CuNPs) and copper ions. Like any other foreign entity, CuNPs are metabolized in the liver and the major excretory route is liver/bile (Lee et al. 2016). The oral LD50 value for nano-copper is around 413 mg/kg as compared to ion-Cu, which is around 110 mg/kg (Chen et al. 2006). In the study by Lee and his group, comparative toxicity of CuNP and Copper ions was established. It was found that CuNPs were less toxic as compared to copper ions. Interestingly, it was also found that H^+ ions were responsible for creating ionic forms of copper from copper nanoparticles, which is more toxic, and as males secreted more H^+ ions, copper produced more toxicity in them. Thus, a sex-dependent toxicity for orally administered CuNPs was established. In the same study, the analysis of dose dependence revealed that a dose of 1250 mg/kg in male rats and 2500 mg/kg in female rats caused drastic histological changes in the kidneys, liver, and spleen, wherein it was found to be accumulated as well (Lee et al. 2016). The justification for this can be found in study of Lei et al. (2015) where exposure of 200 mg/kg/day for a period of five days led to hepatoxicity and nephrotoxicity. Mitochondrial damage and oxidative stress led to hepatic necrosis and renal proximal tubule necrosis. Another study reported that nano-copper caused the rise in malondialdehyde (MDA) and decrease in superoxide dismutase (SOD) in PK-15 pig kidney cell line. It induced apoptosis in PK-15 cell line via mitochondrial pathway. The cytotoxicity of nano-copper was found to be dependent on concentration and time. It was proposed that the permeability of mitochondria increased, leading to increased osmotic potential and swelling of membrane. The disruption of membrane led to release of cytochrome C, which triggered apoptosis and cascade of pathways associated with it. Nano-copper induced oxidative stress by generating free radical species (Figure 5.1) and was evident with increased MDA and decreased SOD level. This apoptosis via mitochondrial pathway led to nephrotoxicity (Zhang et al. 2018a). Similarly, Anreddy (2018) established that copper oxide nanoparticles (CuONPs) were responsible for increased oxidative stress and toxicity to liver after their administration to Wistar albino rats. After administering for 14 days, there was a surge in levels of MDA (an oxidative stress marker) and downfall in levels of glutathione, SOD, and catalase, which led to toxicity by formation of reactive oxygen species (ROS).

Lai et al. (2018) delivered CuONPs intranasally, and studied their toxicity on lungs. To study pulmonary inflammation, C57BL/6 Mice were administered with doses of 2.5 mg/kg, 5 mg/kg, 10 mg/kg and assessed on days 7, 14, and 28 after instillation. A 10 mg/kg dose led to death of mice, 5 mg/kg was the dose estimated for inflammation and fibrosis in lungs. ELISA test revealed that TGF-β (major pro-inflammatory genes) in the bronchoalveolar lavage (BALF) was prominently higher in CuONP-treated group in contrast to control treated with phosphate buffer saline (PBS). Thus, there was evident pulmonary inflammation in mice treated with CuONP. TUNEL assay performed to assess cell apoptosis revealed that the epithelial cells of lungs were injured in a dose- and time-dependent manner. MTT assays were performed to check the cell viability on A549 and BEAS-2B. Significant rise in cytotoxicity was observed as the dose increased from 5 to 200 μg/ml. Also, on comparing the effects of CuONP and copper ions on both cell lines, it revealed that the cytotoxic effects were due to the ion form of copper. The

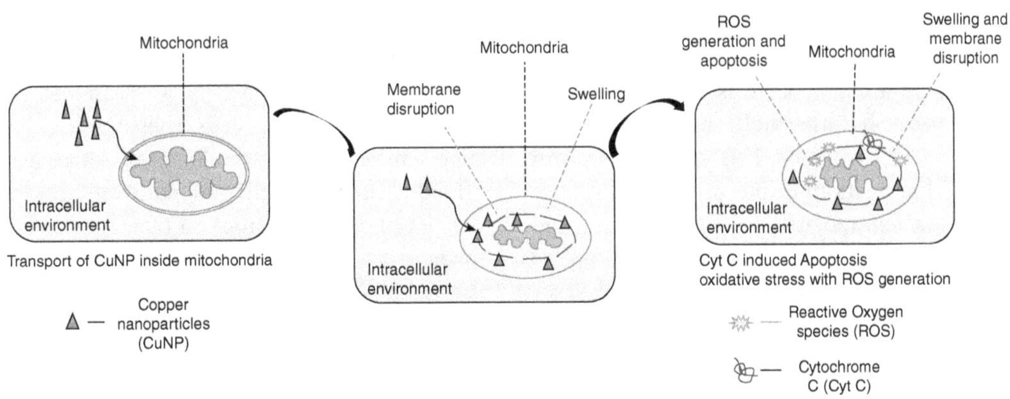

Figure 5.1 Copper nanoparticles induced cellular apoptosis via mitochondrial damage and ROS production.

treatment with dose of 5 and 10 mg/ml decreased the apoptosis inhibitor proteins BCL2 and MCL1, leading to increased apoptosis. Increased ROS levels were observed with 10 μg/ml dose in A549 and BEAS-2B cell line as the levels of SOD1 decreased. *In vivo* studies were done on C57BL/6 Mice to study the activation of myofibroblast and pulmonary fibrosis. This was evaluated with help of western blot analysis where expression of α-SMA and collagen-I was estimated. It was found that the levels of both markers increased significantly, leading to myofibroblast and pulmonary fibrosis (Lai et al. 2018). Goma et al. (2020) showed the neurotoxic effects of CuONP on rats with increased lipid and protein peroxidation levels, DNA damage in brain (Goma et al. 2020). Similarly, Fahmy et al. (2020) showed how CuNPs had the ability to induce adverse effects on brain of male Wistar rats causing increased oxidative stress and acetylcholinesterase (AChE) activity in them.

5.2.2 Toxicity of Gold Nanoparticles

Gold nanoparticles (AuNPs) belong to the noble metal class and are interesting metals as a result of their unique properties like inertness, biocompatibility, and optical properties like those of surface plasmon (Prodan et al. 2003; Ghosh and Pal 2007; Kaur et al. 2007). Due to property of surface plasmon resonance when AuNPs are exposed to NIR light it has been used for hyperthermia and photothermal therapy (Cabada et al. 2012; Kim and Lee 2017). In recent years, due to its multiple properties, its use has been extrapolated to drug delivery and theranostics. It has been vastly explored for its application in the treatment of diseases like cancer and has shown promising results. But its commercialization is often questioned with the toxicity aspect involved. The following section discusses the toxicity aspects of AuNPs.

5.2.2.1 Physicochemical Factors Influencing AuNPs' Toxicity

Physicochemical characteristics also contribute toward the toxicity of metal nanoparticles and many researchers have considered the physicochemical properties of (AuNPs) and the toxicity caused by them. Li et al. (2018) studied the toxicity produced by a series of different sizes of AuNPs ranging from 6.2 to 61.2 nm at a particular concentration (1 mM). Also, the biodistribution in various organs of different sizes over a period of 90 days was evaluated. The distribution was assessed in different organs, liver, kidney, spleen, heart, and lung at eight time points (4 hours, 24 hours, 48 hours, 6 days, 10 days, 20 days, 30 days, and 90 days). The synthesized gold nanoparticles were PEG-coated to reduce the unwanted effects of surface charge on toxicity and distribution. *In vitro* toxicity was evaluated on mammalian cell lines HepG2 and Hela. When exposing these two cell lines to different sizes and concentrations (0.25, 0.50, and 1 mM) of nanoparticles, it was found that at 1 mM concentration as the size decreased, the toxicity increased. The viability significantly decreased for 6.2 nm size gold nanoparticles of both the cell lines (Hela cells: 44.4% and HepG2 cells: 23.8%). The reactive oxygen species (ROS) generated by AuNPs was analyzed with the help of special ROS assay kits. It revealed that smaller sized AuNPs generated more ROS, causing increase in oxidative stress inside cells and damaging it (Xiaomin Li et al. 2018) (Figure 5.2).

The influence of varied sizes of AuNPs on *in vitro* and *in vivo* toxicity was assessed by Avalos et al. HepG2 cells (hepatoma cells) representing liver accumulation of AuNP and HL-60 (leukemia cells) to study effects of AuNP (30, 50, and 90 nm) on blood cells were taken as cell lines to conduct *in vitro* genotoxicity and to study the *in vivo* genotoxicity of 30 nm AuNPs, *Drosophila melanogaster* was taken. The *in vitro* comet assay on both cell lines revealed that all the nanoparticles of different sizes produced DNA strand breaks irrespective of dose range used (1–10 μg/ml).

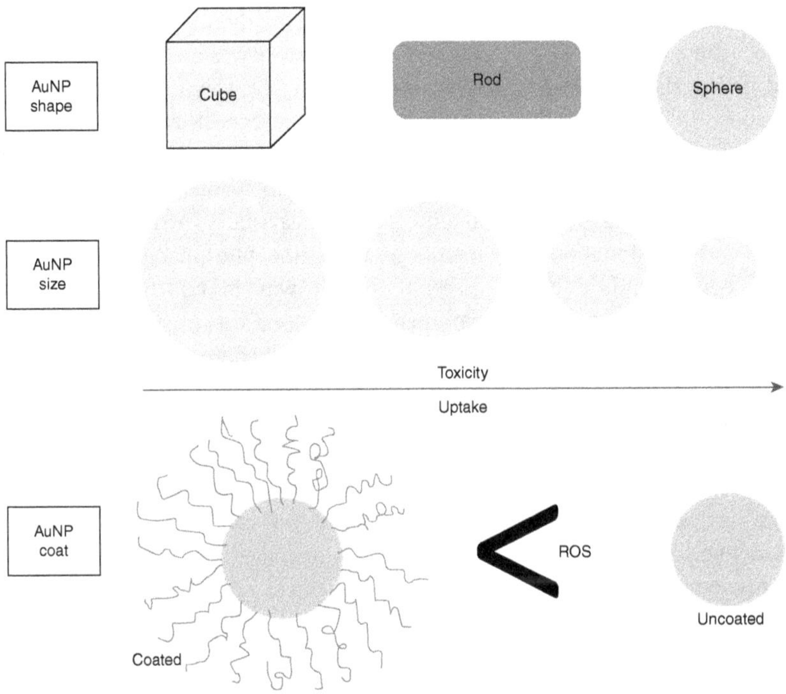

Figure 5.2 Important physicochemical factors affecting AuNP toxicity. Decrease in size causes increased uptake and higher toxicity. Sphere-shaped NP has highest uptake as compared to rod- and cube-shaped and hence more toxicity. Uncoated AuNPs have higher ROS production as compared to coated NPs. *Sources:* Based on Li et al. (2018) and Patlolla et al. (2019).

It was also established that the different sizes of AuNPs were responsible for oxidative DNA damage. There was a considerable rise in the oxidation of purines and pyrimidines which affected HepG2 cells more as compared to HL-60 cells. This DNA damage was size-dependent with 90 nm-sized AuNPs affecting more percentage of pyrimidines as compared to purines. Apart from this, the purines of HL-60 oxidized to a greater extent than pyrimidines when they were treated with 90 nm AuNPs, indicating purines as more sensitive targets for oxidation. On introducing DNA breaks by Endo III and Fpg, DNA damage was observed due to oxidized nucleotides on treating with AuNPs, indicating ROS generation as one of the prominent pathways in its damage. *In vivo* studies did not reveal any genotoxicity or mutagenicity in *D. melanogaster* (Avalos et al. 2018). The toxicity caused by AuNPs in combination with surfactants in the embryonic zebrafish model was studied. The surfactants used were polysorbate 20 (PS20), polysorbate 80 (PS80), and sodium dodecyl sulfate (SDS). The AuNPs of five different sizes (1.0–3.9 nm) and each size of five different concentrations (2.32–50 μg/ml) were given to embryonic zebrafish eight hours post fertilization (hpf). The AuNPs were delivered by two approaches, wherein one was by using a hand micropipette and the other using a digital dispenser which is BioPrinter. The synergistic toxicity produced, depended on both AuNPs' concentration and PS20 concentration. At a specific concentration of AuNP and varying PS20 concentration of 0.3 and 0.003%, significant toxicity leading to mortality was observed. Also, when PS80 and SDS were used with AuNPs, toxicity increased synergistically (Ginzburg et al. 2018).

An interesting study by Ma et al. (2018) on the influence of AuNPs of different sizes and concentrations on the developing embryo *in vitro* and *in vivo* revealed that these metallic particles can indeed affect the embryo. A systematic study involved giving the AuNPs of size 2, 15, and 50 nm at a dose of 2 μg of AuNPs per gram of body weight. The dosing on CD-1 mice was done during a period where the developmental period from blastocyst to gastrula was covered. It was found that only the 15 nm AuNPs had a considerable effect on the developing embryo. This development of the embryo was hampered by the down-regulation of expression markers at mRNA and protein levels responsible for the differentiation of embryonic germ layers (Ma et al. 2018).

Research has also been carried out to understand the effects of factors like size, shape, and surface coating on the uptake and toxicity of AuNPs. Also, the role played by serum protein in modulating the uptake and toxicity has been researched in detail. In research by Carnovale et al. (2019), the shape was found to incite toxicity. The low levels of pure AuNPs/CTAB at more than 10 μM coated nanoparticles caused significant toxicity in Prostrate cancer 3 (PC3) cells. Gold nanospheres (AuNSs) caused the highest toxicity and both gold nanorods (AuNRs) and gold nanocubes (AuNCs) (Figure 5.2) were more biocompatible even up to 100 μM. It was also found that flat surfaced nanoparticles like AuNRs and AuNCs had poor uptake as compared to AuNSs and AuNPs which had curvature, concluding that flat surfaced particles had less interaction with cellular surface proteins. Apart from this, the influence of serum proteins on uptake and toxicity was significant. The serum supplemented media formed a biological corona around the AuNPs which reduced its uptake and thus the toxicity. The effect of ligand coating was such that the AuNPs coated with tryptophan and tyrosine showed better uptake as compared to citrate and CTAB-coated AuNP. A probable justification would be that the aromatic amine groups present in the coated ligand could better interact with the cell surface receptors (Carnovale et al. 2019).

A comparative study of PEG-coated and PEG-uncoated AuNPs was made to evaluate its toxicity on Sprague–Dawley rats. After administration of different concentrations of AuNPs (12.5, 25, 50, and 100 μg/kg) body weight the rise in levels of histochemical parameters like alanine (ALN), aspartate (AST) aminotransferases and alkaline phosphatase (ALP) was assessed and it was found these chemical parameters increased in presence of PEG-uncoated AuNPs as compared to PEGylated AuNPs. ROS release as detected by DCFH-DA (2′7′-dichlorofluorescin diacetate) revealed that the effect was much higher in PEG-uncoated AuNPs as compared to coated AuNPs. The histological evaluation of rat liver after administration of coated and uncoated AuNPs revealed that uncoated AuNPs at dose of 12.5 μg/kg caused damage to the central vein. At the dose of 25 μg/kg, the uncoated AuNP caused central vein damage and vacuolation of hepatocytes and as the dose increased, further damage like liver degeneration and necrosis occurs. The PEG-coated AuNP did not increase the toxicity to that extent as compared to uncoated AuNPs (Figure 5.2) (Patlolla et al. 2019). Ginzburg et al. (2018) also studied the relationship between nanoparticle (NP) core size and toxicity of similar GNPs in the zebrafish model. The effect of different sizes on generating toxicity when the concentration, surface area, and number of particle factors changed was evaluated on the zebrafish model. Five different sizes of AuNPs (0.8–5.8 nm) were coated to produce positively charged *N,N,N*-trimethylammonium ethanethiol (TMAT) ligands. The concentration used was in the range of 0.5–50 μg/ml and it was found that higher toxicity was seen with smaller particles in case of equal mass consideration. Also, when considering the factor of the surface area of AuNPs, the toxicity was based on the core diameter of nanoparticles. When the concentration was kept constant and only surface area was evaluated for toxicity, it was observed that as the surface area increased the toxicity increased. This was because there was increase in catalytically active sites that led to more probability for ROS generation. When the number of particles was kept constant, it was found that larger-sized AuNPs had higher toxicity as compared to smaller-sized AuNPs (Figure 5.2) (Patlolla et al. 2019).

5.2.3 Toxicity of Silver Nanoparticles

Silver nanoparticles (AgNPs) like other metallic nanoparticles have an issue of toxicity. It can generate ROS when Ag^+ ions are formed, which activates NADPH oxidase. The silver ions generated have the potential to downregulate antioxidants, generate TNF-α, disrupt mitochondrial pathway, and produce excess ROS that can imbalance the oxidative status and cause toxicity to cells. Research is focused on reducing the toxicity of AgNPs and improving its applicability in biomedicine. Surface-modified AgNPs have been produced to decrease its toxicity. Das et al. (2017) prepared polyethylene glycol (PEG) and Bovine serum albumin (BSA)-coated silver nanoparticles (AgNPs) and studied their *in vitro* toxicity on peripheral blood mononuclear cells (PBMCs) and *in vivo* toxicity on mice at doses of 5, 10, 25, and 50 μg/ml. Dose-dependent toxicity was recorded following MTT assay for functionalized and non-functionalized AgNPs. BSA-coated AgNPs showed toxicity at 10 μg/ml and PEG-coated AgNPs showed toxicity at 25 μg/ml as compared to plain AgNPs, which showed toxicity at 5 μg/ml. To evaluate free radical production, superoxide radical (O_2^-) and NADPH oxidase levels were measured and the level of O_2^- marker increased to a greater extent in uncoated AgNPs as compared to PEG- and BSA-coated AgNPs (Table 5.1).

There was a significant rise in NADPH oxidase in the case of bare AgNPs as compared to PEG- and BSA-functionalized AgNPs (Table 5.1). Also, the lipid peroxidation in terms of MDA level was significantly ($p < 0.05$) increased as mentioned in Table 5.1, in

Table 5.1 *In vitro* Increase in cellular levels of O_2^-, NADPH, and MDA for different concentrations of uncoated and coated AgNPs.

	AgNPs (uncoated)	PEG-AgNPs	BSA-AgNPs
Concentration (μg/ml)	**Fold increase in O2− levels**		
5	1.186	1.187	1.236
10	2.315	1.44	1.874
25	2.655	1.975	2.88
50	3.445	2.446	3.194
Concentration(μg/ml)	**Fold increase in levels of NADPH**		
5	1.29	1.11	1.185
10	1.949	1.248	1.669
25	2.526	1.654	2.13
50	2.788	2.09	3.09
Concentration(μg/ml)	**Fold increase in levels of MDA**		
5	1.10	1.10	1.248
10	1.56	1.238	1.60
25	2.09	2.164	2.15
50	2.38	2.347	2.364

Source: Based on Das et al. (2017).

contrast to the control group. It was found that the surface functionalization caused a considerable reduction of MDA level than the surface-unmodified AgNPs-treated group. Acridine orange (AO) and ethidium bromide (ED) staining revealed that apoptosis was higher in cells treated with AgNPs uncoated as compared to coated AgNPs. DAPI staining revealed that nuclear morphology in PBMC cells was significantly affected by uncoated AgNPs to a larger extent as compared to coated AgNPs. In uncoated AgNPs, significant increase in pro-inflammatory factors (TNF-α and IL 12) and a significant decrease in anti-inflammatory factors were observed as compared to surface-modified PEG- and BSA-coated AgNPs. Similar results were obtained in *in vivo* studies. Rise in MDA, GSSG, and ROS was observed with a decrease in SOD levels (Das et al. 2017).

As reported with other metals relating to toxicity due to the influence of size, AgNPs were also found to induce toxicity based on their size. AgNPs of size 10nm, when injected intraperitonially in BALB/c mice, led to their death after 24 hours, whereas 60 and 100 nm-sized AgNPs showed no clinical effects. Smaller sized 10 nm AgNPs were also responsible for increasing the levels of AST and ALT (Cho et al. 2018). A study wherein toxicity of AgNPs was assessed in BALB/c mice revealed that there was a significant effect on the quality and quantity of sperm. It affected the growth of spermatogenic cells, led to oxidative DNA damage, affected its morphology and motility. At a 0.25–1 mg/kg dose, the permeability surged in a dose-dependent manner which caused neuronal complications in the brain of mice. Histopathology studies showed that at the lowest dose of 0.25 mg/kg, there was a significant change in the pathology of testis tissues. Higher doses caused further complications like necrosis in spermatogonial cells. After intraperitoneal (IP) administration, the tissue of the heart was affected to such extent that there was scattered nuclear enlargement of muscle fiber and formation of plump endothelial cells of intramuscular vessels. IP administration of AgNPs in Balb/c caused damage to the epithelial lining of the renal tubule with vacuolar degeneration and deposition of protein (hyaline) cast. AgNPs like other metals had the potential to cause increased MDA levels, indicating increased oxidative stress levels and lipid peroxidation as a pathway to toxicity (Moradi-Sardareh et al. 2018). The toxicological profile of stable PEG-coated AgNPs was studied by Pinzaru et al. (2018) On exposing PEG-coated AgNPs and bare AgNPs to human keratinocyte cell line (HaCat), it was found that at higher concentrations of AgNPs (10 and 50 μM), significant cytotoxic effects were observed. AgNPs have the ability to increase oxidative stress and this was assessed by Blanco et al. (2018). PVP-AgNPs were administered to Sprague–Dawley rats for 90 days orally and their effect on the liver was observed. The activity of antioxidant enzymes catalase (CAT) and SOD was analyzed. The role of caspase-3 in apoptotic cell death was checked. The levels of p53 and p21 resulting in DNA damage was also assessed. Antioxidant regulating chemical markers, SOD and CAT both increased at 50 mg/kg/day, but SOD levels decreased at a dose of 200 mg/kg/day, indicating ROS generation and consequent oxidative stress rise. A probable reason for such down-regulation at higher concentrations could be impaired nuclear translocation of the nuclear factor erythroid 2 related factor 2 (Nrf2) which could be partially responsible for the decreased antioxidant activity. The transcription factor Nrf2 is responsible for the regulation of the transcription of genes encoding for cytoprotective proteins, such as CAT and SOD. The levels of thiobarbituric acid reactive substances (TBARs) an indicative of lipid peroxidation levels also increased in a similar manner indicating a rise in ROS levels at both 100 and 200 mg/kg/day. Similarly, levels of p53 and p21 increased in a dose-dependent manner conveying damaged DNA due to exposure of AgNPs. At higher doses of 200 mg/kg/day, levels of these markers decreased, indicating excess damaged DNA could not repair cells, leading to downregulation of these markers (Blanco et al. 2018).

5.2.4 Toxicity of Zinc Oxide Nanoparticles

Zinc has long been used in cosmetic preparations (Newman et al. 2009) and recently is being explored for its application in drug delivery (Rasmussen et al. 2010), biomedical field (Jiang et al. 2018). The toxicity of zinc oxide nanoparticles (ZnONPs) owing to their wide use is essential to be considered. Shubha et al. (2019) studied the toxicological aspects of both green synthesized zinc oxide nanoparticles (Pe ZnONPs) from *Phyllanthus emblica* and marketed bulk ZnO (DZnO) *in vitro* and *in vivo*. The hemolysis study revealed significant growth inhibition by DZnO in cells as compared to PeZnONPs. The potential reasons could be the surface reactivity due to the surface coating in case of chemically synthesized ZnONPs producing more ROS as compared to green synthesized ZnONPs. *In vivo* toxicological studies were performed on silkworm *Bombyx mori* larve. It was found at lower concentrations of both PeZnONPs and DZnO 100% survival was seen and its highest concentration, 0.032 mg/ml of DZnO and PeZnO death rate was 96.67 and 26.27%, respectively. Further, at different concentrations 0.002, 0.004, 0.008, and 0.016 mg/ml of DZnO and PeZnO, the death rate of the larvae recorded were 1.67, 10.00, 28.33, and 60.00% and 0.0, 0.0, 6.67, and 15%, respectively, which is statistically significant at $p < 0.05$ (Shubha et al. 2019).

The cytotoxicity of ZnONPs is primarily a result of the formation of Zn^{2+} ions. This ion potentiates genotoxicity by damaging DNA via three possible routes *viz.*, (i) ROS generation, (ii) invading in core and associate or cross-connect with DNA strand (iii) at the time of cell division when the membrane separates and offers to ascend to open the door for chromosome distortions. Neurotoxicity in the brain is possible as the ZnONPs have the ability to cross BBB, or enter via the olfactory nerve pathway (Singh 2019). Senapati and Kumar (2018) explored the influence of pH effect on the toxicity of ZnONPs in human epidermal cell line (A-431), where it was established that ZnONPs can remain internalized inside the cells up to 72 hours at both pH 4 and pH 7, and thus, the toxicity produced due to zinc ions as well as ZnONPs can last up to 72 hours. Yousef et al. (2019) studied the toxicity of ZnONPs along with aluminum oxide nanoparticles (Al_2O_3NPs) in male Wistar rats wherein they were administered orally for 75 days. The influence of these nanoparticles either alone or in combination was evaluated for different parameters like hepatic and renal impairment, redox status, nuclear DNA fragmentation, and systemic inflammation. It was found that there was a significant increase in hepatic gene expression of peroxisome proliferator activator receptor gamma-coactivator 1α (PGC-1α) (63%) and mitochondrial transcription factor A-(mtTFA) (68%). DNA fragmentation in both liver and kidney tissues was higher for the combination of both nanoparticles. Hepatic and kidney tissues also showed a rise in levels of p53, TNF-α, and IL-6 for combination nanoparticles and similarly the levels of antioxidant markers SOD, Cat decreased. Also, ROS inducing markers, like TBARS and NO were found to be increased. Histopathological analysis revealed liver necrosis, congestion, and renal damages like hypertrophy glomeruli segmentation, hydropic degeneration in epithelial cells, and necrosis of epithelial cells in tubules for combination nanoparticles (Yousef et al. 2019). Nickel–zinc–ferrites of size less than 30 nm were tested for *in vitro* toxicity at concentrations of 50, 100, and 200 μg/ml in RBCs and peripheral blood mononuclear cells (PBMCs) of human blood and were compared with nickel iron oxide and zinc iron oxide. Hemolysis was observed at 200 μg/ml concentration of nickel–zinc–ferrite. Changes in oxidative stress levels were evident when glutathione-*S*-transferase activity changed at concentrations of zinc ferrite (50 and 100 μg/ml), nickel–zinc–ferrite (200 μg/ml), and magnetite (50 μg/ml) nanoparticles, in comparison to the negative control treated with PBS (Martínez-Rodríguez et al. 2019).

5.2.5 Toxicity of Iron Oxide Nanoparticles

Iron has one of the most diverse applications and apart from being explored as a potential MRI contrast agent, it is also being researched for use as a photothermal agent (Espinosa et al. 2016), drug carrier (Kayal and Ramanujan 2010), and as theranostic agent (Santhosh and Ulrih 2013). It has been explored for treatment in various cancers. The toxicity aspects of iron, iron oxide nanoparticles (IONPs), and superparamagnetic iron oxide nanoparticles (SPIONs) are discussed. Bare IONPs often prove to be toxic and hence surface modification is the approach considered by researchers to reduce its toxicity (Shukla et al. 2015). Green synthesis of IONPs is focused upon as there is toxicity reported when synthesized by conventional chemistry approaches. Lakshmi Pravallika et al. (2019) synthesized IONP via green method by using the ethanolic extract of *Centella asiatica* (CA). Acute oral toxicity of CAIONP was evaluated on Swiss albino mice at a dose of 2000 mg/kg b.w. After 14 days, histopathological analysis of brain, liver, heart, and adrenal gland was made. No gross physiological changes were observed in control and CAIONPs-treated tissues (Lakshmi Pravallika et al. 2019). But bare IONPs have been found to affect the human blood lymphocytes. In a study by Assadian et al. (2019), it was found that IONPs affect the blood lymphocytes majorly via ROS production. At a concentration of 10 mM, maximum decrease in cell viability of human blood lymphocytes was observed. Oxidative stress levels increased in a dose- and time-dependent manner, indicated by the rise in ROS, oxidized GSH (GSSG), lipid peroxidation levels (LPO), and decrease in GSH levels. Also, there was a collapse of mitochondrial membrane potential (MMP) at dose of 2 mM of IONPs, which could be the cause for the death of blood lymphocyte cells (Assadian et al. 2019). Thus, IONPs have the potential to damage blood cells via oxidative stress generation. Similarly, Ghosh et al. (2020) reported that surface-modified positively charged IONPs could induce genotoxicity and ROS generation in human lymphocyte cells. In mice, it was observed that IONPs were internalized in brain and spleen, causing oxidative stress and DNA damage in them (Ghosh et al. 2020). Another study on the effects of coated IONPs was carried out on (PBMCs) peripheral blood mononuclear cells (lymphocytes). The toxicity of conventional Υ-Fe_2O_3 particles was compared against modified phosphonic acid (PA)- and hydroxamic acid (HA)-terminated poly (ethylene glycol), PA-PEG@Fe_3O_4, and HA-PEG@Fe_3O_4 nanoparticles on PBMC. Blood was collected from 10 human volunteers in heparinized tubes. Modified IONPs did not show significant toxicity nor did they suppress the proliferative activity of lymphocytes. But bare Υ-Fe_2O_3 showed suppression of proliferative activity of T-lymphocytes at 75 μg/cm^2, 72 hours and also reduced the phagocytic activity of monocytes (15 μg/cm^2, 24 hours; 3–75 μg/cm^2, 72 hours) (Patsula et al. 2019). A study by Mahalakshmi and Kurian (2019), prepared green chemistry-based iron nanoparticles (IONPs) using *Desmodium gangeticum* (DG) root extract which acted as a reducing as well as a capping agent. The nanoparticles synthesized (IONPs DG) were evaluated for their toxicity in renal epithelial cells of Wistar rats at a dose of 100 mg/kg b.w. Chemically synthesized IONPs (IONP-Chem) were used to compare the toxicity with green synthesized IONPs. To evaluate the oxidative stress exerted on renal epithelial cells, the concentration of thiobarbituric acid reactive substances (TBARSs) and index for lipid peroxidation (LPO) were taken into consideration. A significant rise in levels of lipid peroxidation (30%) of TBAR, as compared to IONPs-Chem (51%), was observed. *In vitro* studies on pig kidney epithelial cell line, LLC

PK1, revealed that on exposure of IONPs DG and IONPs-Chem both at a concentration of 1 mg/ml, there was a rise in lactose dehydrogenase upto 75% in IONPs-Chem treated cells and decrease in viability upto 60% in IONPs-Chem treated cells and upto 25% decrease in IONPs DG-treated cells. A considerable reduction in the mitochondrial enzyme activities of MDH (IONPs Chem: 76%; IONPs DG: 68%), SDH (IONPs DG: 83%), and NADH dehydrogenase (IONPs-Chem: 62%; IONPs DG: 55%) respectively was observed in the renal mitochondria at the end of 15th day (Mahalakshmi and Kurian 2019). Ansari et al. (2019) studied the impact of IONPs on genotoxicity, influence on oxidative stress, and their interaction with DNA. The study involved the use of IONPs at the concentrations of 50, 100, 200, 400, and 800 μg/ml in *in vitro* studies. *In vivo* studies involved division of animals in nine groups with administration of IONPs at different concentrations 25, 50, and 100 mg/kg b.w. to different treatment groups. The cytotoxicity study revealed a significant decrease in cell viability upto 61% at dose of 200 μg/ml. ROS generation was highest at the dose of 800 μg/ml. *In vivo* studies of chromosomal aberrations increased in a dose-dependent manner with highest at 100 mg/kg b.w. Apart from this, there was an increase in catalase activity, MDA levels and a decrease in superoxide dismutase activity (SOD), GSH levels with increasing dose. Thus, IONPs induced toxicity via ROS production (Ansari et al. 2019). The study by Pereira et al. (2020) showed how IONPs affected the developmental toxicity on *Danio rerio* embryo. The sublethal effects included decreased heartbeat, accumulation of blood in the heart and pericardial edema (Pereira et al. 2020). Similarly, it has been found that IONPs in blackfish (*Capeota fusca*) caused abnormalities in gills and intestine including aneurism, hyperplasia, and indications of necrosis in gills and increased counts of lymphocytes, RBCs, in intestine (Sayadi et al. 2020).

5.2.6 Physicochemical Properties Affecting Toxicity

The toxicity of metallic nanoparticles is influenced greatly by their physicochemical properties and different factors like size, shape, charge, surface coating (Li et al. 2015). Table 5.2 covers a comprehensive list of metal nanoparticles, physicochemical properties, and factors contributing toward its toxicity.

5.3 Accumulation and Toxicity of Metal-Based Nanoparticles in Various Organs

In spite of AuNPs' colossal significant benefits in biomedical applications, there is still a restricted understanding of their tissue bioavailability and toxicity on animals. Various animal models have been explored in order to understand the toxicity and fate of AuNPs *in vivo*, including rats, mice, swine, and rabbits (Li and Chen 2011). Lopez-Chaves et al. (2018) developed AuNPs and studied them for *in vitro, in vivo* biodistribution, bioaccumulation, and toxicity in rats. The results revealed that AuNPs were unequally accumulated in all the studied organs. In fact, AuNPs with particle size 10 and 30 nm were accumulated in kidney, liver, and intestine in higher concentration as compared to AuNPs with particle size 60 nm. However, the AuNPs with particle size 60 nm were accumulated in spleen with a greater concentration of Au. Therefore, AuNPs with particle size 60 nm were accumulated mainly with greater concentration, i.e. spleen followed by intestine, liver, and kidney. Meanwhile, AuNPs with 30 nm particle size were mainly accumulated with greater concentration in spleen followed by liver, intestine, and kidney. Lastly, AuNPs with particle size 10 nm were accumulated with higher concentration in the intestine followed by spleen, liver, and kidney. Previous studies demonstrated that renal filtration cutoff is smaller

Table 5.2 Factors affecting the toxicity of metallic nanoparticles.

a. Effect of size					
Metal nanoparticles	**Size range**	***In vitro/in vivo* study**	**Toxicity concentration**	**Toxicity manifestations**	**References**
ZnONPs	8–12 nm	Bacterial growth inhibition study	4–6 mM	ROS production	Raghupathi et al. (2011)
AgNPs	i. 15 nm	C18-4 cell line i. Cell morphology ii. Mitochondrial function	i. 10 µg/ml ii. 5 and 10 ug/ml	i. Apoptosis ii. Inhibition of mitochondrial function	Braydich-Stolle et al. (2005)
(Al_2O_3NP)	30 nm	C18-4 cell line i. Cell apoptosis ii. Lactose dehydrogenase leakage	i. 5–10 ug/ml ii. 10 ug/ml	i. Necrosis ii. LDH leakage	Braydich-Stolle et al. (2005)
AgNP	i. 3 nm ii. 3 and 50 nm	Zebrafish embryo animal model i. Mortality ii. Sublethal toxicity	i. 250 uM ii. 100 uM	i. Mortality upto 80% ii. Deformities in embryo • Small head • Jaw and snout malformations • Circulatory malformations • Tail malformations • Pericardial edema • Bent spine	Bar-Ilan et al. (2009)
Gold nanoparticles AuNP	i. 2.82 and 5.52 nm ii. 2.82 nm, 5.52 nm, 38.05 nm	Cell line: i. J774 A1 macrophages ii. J774 A1 macrophages	i. 1–10 ppm 1 ppm	i. **Changes in**: cellular morphology ii. **Rise in:** pro-inflammatory genes • interlukin-1 (IL-1), interlukin-6 (IL-6), and • tumor necrosis factor (TNF-a)	Yen et al. (2009)

IONPs	i. 5 and 30 nm	Porcine aortic endothelial cells	0.5 mg/ml	Cell elongation, ROS production, changes in cell morphology	Yu et al. (2012)
ZnONP	5–16 nm	Human lung epithelium cells (A549)	EC_{50}–4.6 ug/ml	Cytotoxicity and pro-inflammatory factors like IL-8 increased	Hsiao and Huang (2011)
b. Effect of shape					
ZnONP	Rod-shaped more toxic than sphere-shaped	Human lung epithelium cells (A549)	Rod-shaped; EC_{50} 8.5 ug/ml Sphere-shaped: EC_{50} 12.1 ug/ml	Cytotoxicity	Hsiao and Huang (2011)
AgNP (10 nm)	i. Ag nanoplates > Ag nanospheres > Ag nanowires	i. *In vitro*: fish gill epithelial cell line (RT-W1)	i. At 25 ug/ml 70% cell death	i. Superoxide production ii. Crystal defects in Ag nanoplates iii. High interaction with cells leading to differential toxicity	George et al. (2012)
	ii. Ag nanoplates > Ag nanospheres > Ag nanowires	ii. Animal model: zebrafish embryo	ii. 10 ug/ml led to 100% mortality	ii. Possible oxidative stress and crystal defects led to failure of embryo to hatch or led to its death	
c. Effect of surface charge					
AgNP	Branched polyethyleneimine-coated (BPEI-AgNPs (+40 mV) > citrate AgNP (−38 mV)	PolySeed (Gram-positive bacillus species)	10–100 ug/l (highest toxicity by negative charged BPEI AgNPs)	Decrease in negative charge leading to rise in cellular interactions and thus higher toxicity	El Badawy et al. (2011)
(CeONP)	i. Poly-(acrylic acid) (PNC) (−ve)	i. Toxic to A549 lung carcinoma cell line	i. 1.0 mM	i. Localized in lysosome of the cell line and activates oxidase activity causing cytotoxicity	Asati et al. (2010)
	ii. Aminated poly(acrylic acid) (ANC) (+ve), or	ii. Toxic to H9c2, A549, and HEK 293 cell lines	ii. 1.0 mM	ii. Localized in lysosome of the cell line and activates oxidase activity causing cytotoxicity	
	iii. Dextran (0 charge)(DNC)	iii. Not toxic to any of cell lines mentioned above	iii. 1.0 mM	iii. Neutral charged and hence localized in cytoplasm	

(*Continued*)

Table 5.2 (Continued)

AuNP (2 nm)	a. Functionalized with: quaternary ammonium functionalized nanoparticle (+ve charged side chain) (MMPC 1)	a. i. Red blood cells, ii. Cos-1 cells	i. 1.1 uM ii. 1.0 uM	a. Positive charge led to better interaction with negative lipid membrane and thus induced toxicity	Goodman et al. (2004)
	b. Functionalized with: carboxylate-substituted nanoparticle (−ve charged side chain) (MMPC 2)	i. Red blood cells ii. Cos-1 cells	i. 72 uM ii. 7.37 uM	b. Negative charge resisted the uptake and thus no cytotoxicity was seen	
d. Effect of surface coating and its associated charge					
Zinc oxide quantum dots (ZnO QD)	Functional coat: poly) aminotrimethoxysilanes	*Escherichia coli* bacterial cells	1 mM	Dissolution of ZnO to release Zn^{2+} ions causing toxicity	Aboulaich et al. (2012)
IONPs (10 and 100–150 nm)	Coat: 3-aminopropyltrimethoxysilane (APTMS, +ve charged) *vs* Tetraethyl orthosilicate (TEOS) −ve charged	i. a. Human dermal fibroblasts (HDF) i. b. HT-1080	APTMS at 1000 μg/ml dose, lactose dehydrogenase leakage was 112% and for TEOS it was 109%	Damage to cell membrane integrity by positive charged APTMS due to better interaction with cell membrane as opposed to negatively charged TEOS	Yang et al. (2013)
(AgNP)	Coat: i. Mercaptoundecanoic acid (MUA)	NIH/3T3 embryonic fibroblasts	LD_{50} values [$\times 10^{-3}$] i. Ag-MUA: $0.04 \pm 2.83 \times 10^{-4}$	PEG significantly reduced the uptake and hence LD_{50} values were higher. Maximum decrease in cell viability was by Ag-MUA	Caballero-Díaz et al. (2013)
	ii. Poly(isobutylene-alt-maleic anhydride) (PMA)		ii. Ag-PMA: $0.65 \pm 8.66 \times 10^{-3}$		
	iii. Saturated polyethylene glycol (PEG)-PMA (PMA-$_{sat}$PEG)		iii. Ag-PMA-$_{sat}$PEG: $1.34 \pm 2.49 \times 10^{-2}$		

than AuNP size. Therefore AuNPs are not eliminated from kidney; however, they are accumulated in liver and spleen. Hence, spleen would be the significant organ for the metabolism of the AuNPs. Furthermore, the results showed that although AuNPs caused an increase in generation of ROS and as a result increase in the disruption of lipids and proteins, it did not impart inflammatory response and tissue damage. As far as shape is concerned, small and elongated NPs are more rapidly uptaken than large and flat NPs. In the same way, it has been reported that Au nanospheres and disc were located only at the tumor surface; however, Au nanorods were distributed throughout the tumor tissue. Additionally, Au nanorods demonstrated faster uptake and elimination over star-shaped NPs and nanospheres (Lopez-Chaves et al. 2018). Additionally, van Pomeren et al. (2019) developed AuNPs with different shapes (i.e. nanorods, nano-urchins, nanospheres, and nano-bipyramids) and studied their effect on biodistribution and immune responses. The results demonstrated that spherical AuNPs-exposed embryo showed reduced neutrophils in the whole organism scale and tail region; however, nanorods, nano-urchins, and nano-bipyramids did not impart any significant effect on whole organism scale. Additionally, when nano-urchin AuNPs were exposed to embryo, higher level of macrophages were observed; however, when nanosphere AuNPs were exposed to an embryo, the overall abundance of macrophages were decreased in both the tail region and whole organism scale (van Pomeren et al. 2019). Furthermore, Pan et al. (2007) developed various sized AuNPs and investigated for size-dependent toxicity. The results demonstrated that AuNPs with 15 nm size exhibited lower toxicity as compared to AuNPs with particle size of 1.4 nm (Pan et al. 2007). In addition to this, Jong and Paul (2008) also investigated size-dependent toxicity of spherical AuNPs. The results were similar to the abovementioned study. Briefly, the results demonstrated that the smaller particles (10–50 nm) induced greater toxicity as compared to larger particles (100–200 nm). Furthermore, Wang et al. (2008) studied the effect of shape-dependent toxicity. The results indicated that CTAB-coated Au nanorods were found to be more toxic as compared to Au nanospheres to human HaCaT keratinocytes.

IONPs are metallic carriers that had attracted lot of attention of researchers owing to their biomedical applications, however, many studies had investigated for their accumulation and toxicity in biological systems. In addition to the aforementioned statement, Kim et al. (2006) investigated the accumulation of RITC/Si-coated IONPs in various organs *viz.* kidney, liver, testes, lungs, brain, uterus, spleen, and heart followed by intra-peritoneal injection. The results indicated that IONPs@RITC/Si were able to pass easily through BBB. Moreover, the higher concentration of IONPs@RITC/Si was mostly found in spleen and liver (Kwon et al. 2008). Similarly, another case study showed that when IONPs were injected intravenously in rats, the biodistribution of iron altered over time; however, the higher concentration of iron was accumulated in spleen and liver rather than in the kidneys, brain, lungs, and heart. Additionally, magnetization measurement of the spleen and liver samples demonstrated a gradual and steady decrease over a period of three weeks, indicating degradation of NPs (Jain et al. 2008). Furthermore, as far as toxicity is concerned, IONPs are more tolerated physiologically. For example, dextran-coated IONPs did not demonstrate a measurable cytotoxicity index (Babincová et al. 2000). Nevertheless, Zhou et al. (2003) found that IONPs when inhaled, exhibited pro-inflammatory response and oxidative stress in a dose-dependent manner in the lungs. Moreover, Naqvi et al. (2010) developed tween coated IONPs and investigated their toxicity on murine macrophages cells. The results demonstrated that low concentration of IONPs, i.e. 25–200 μg/ml after 2 hours of exposure, displayed higher cytotoxicity in comparison to higher concentration, i.e. 300–500 g/ml for 6 hours. Nevertheless, dextran-coated IONPs (100–150 nm, 0.1 mg/ml) reduced cell viability in human macrophages by 20% after seven days of incubation. Furthermore, *in vivo* and *in vitro* toxicity of IONPs was investigated and reviewed. Many of the studies revealed that they exhibit long-term exposure effects or were found to be mutagenic. And some studies demonstrated that it could induce genotoxicity, developmental toxicity, and neurotoxicity. In conclusion, irrespective of the administration route, administration dose plays a significant role in inducing various types of toxicity (Naqvi et al. 2010).

TiO_2NPs are one of the metal-based NPs, which have garnered interest of the researchers in the field of biomedical applications. TiO_2NPs after administration are transported to various systems by systemic circulation and accumulated in different tissues/organs, leading to toxicity. Recently, many studies revealed that TiO_2NPs were minimally trapped by the GIT and transferred to systemic circulation in humans and rats. Additionally, after oral administration in rats, TiO_2 showed a low level of toxicity at NOAEL (no observable adverse effect level) >1000 mg/kg body weight/24 hours (Warheit and Donner 2015). Additionally, MacNicoll et al. (2015) studied the absorption of TiO_2 through GIT. The results indicated that TiO_2 dose (5 mg TiO_2 NPs/kg body weight) did not result in capturing of TiO_2NPs from the GIT into the urine, blood, or various internal organs. In addition to this, the effect of particle size of TiO_2NPs on absorption was also investigated. The results indicated that varying the particle size of TiO_2 i.e. <5000 nm (pigments), 100 nm (nanoshell), and 15 nm (nanoshell) with a dose of 5 mg/kg body weight/day did not demonstrate any effect on absorption (Jones et al. 2015; MacNicoll et al. 2015). However, Wang et al. (2007) investigated that size of TiO_2NPs had the ability to influence both accumulation and toxicity in various organs. The results showed that after oral administration of TiO_2, the smaller NPs (25 nm) were mostly accumulated in spleen, and to a smaller extent in lungs and kidney. However, larger NPs (80 nm) were mostly found in liver (Wang et al. 2007). Similarly, Ali et al. (2019) investigated the influence of dose and particle size of TiO_2NPs on toxicity. The results revealed that the toxicity of TiO_2NPs was dependent on dose, i.e. increase in the dose of TiO_2NPs lead to increase in toxicity and *vice versa*. Additionally, smaller size TiO_2NPs (21 nm) imparted considerable adverse effect in biochemical parameters, generic aberrations, and histological investigations (Ali et al. 2019). Additionally, Rosłon et al. (2019) developed noble metal (Pd, PdO, Au, Ag_2O, and Ag)-modified TiO_2NPs and investigated them for toxicity in mammalian cells. The results indicated that TiO_2NPs modified with Pd and Au showed a slight impact on cell viability at the investigated concentration (10–200 μg/ml). However, TiO_2NPs modified with Ag_2O and Ag were found to have a significantly higher toxic effect (Rosłon et al. 2019). Furthermore, many studies reported *in vitro* and *in vivo* toxic effect of TiO_2NPs on humans such as constriction of the nuclear membrane, altered cell cycle, and apoptosis. Additionally, TiO_2NPs can be responsible for DNA damage in some cases and possibly interact with the epithelium of the small intestine responsible for the absorption of nutrients. Furthermore, TiO_2NPs disturb the lipid homeostasis and glucose in rats and mice (Baranowska-Wójcik et al. 2020).

AgNPs have unique properties like electrical conductivity, optical, magnetic polarizability, microbial activity, and enhanced Raman scattering. It has been a topic of great interest for the researchers and many products of AgNPs have been commercialized till date (Pandiarajan and Krishnan 2017). However, with their wide applications, AgNPs, in biological environment, alter their transport, fate and accumulate in various organs/tissues, leading to production of certain toxicities. Ag ions have a tendency to interact with various biomolecules *viz*. sulfhydryl groups of metabolic enzymes, nucleic acids, glutathione and components of cell wall. AgNPs-based toxicity is generally prompted by Ag ion release. AgNPs when exposed to cells cause generation of ROS, which ultimately leads to toxicity (Slawson et al. 1990; Ahamed et al. 2008; Ruparelia et al. 2008; Greulich et al. 2011, 2012; Luther et al. 2011). Barbasz et al. (2017) developed three types of AgNPs (cysteamine-stabilized AgNPs [SBATE]; citrate-stabilized AgNPs [SBTC]; sodium bromohydride AgNPs [SBNMs]) and evaluated their toxicity in HL-60 and U-937 cells. The three AgNPs were differentiated from each other by surface properties. The results indicated that SBNM and SBTC when exposed to HL-60 and U-937 cells lead to significant decrease in mitochondrial functions of the cells. However, when SBATE is exposed to both cell lines, the results indicated that SBATEs were more sensitive to U-937 cells as compared to HL-60 cells. In conclusion, it can be said that SBATE showed minimal toxicity in spite of showing similar ion release profile as SBNB and SBTC. It can be concluded that induction of the apoptosis in the tumor cells depends on both, surface charge and individual features of the cells. Furthermore, Cho et al. (2018) developed various sized (10, 60, and 100 nm) AgNPs and investigated them for size-dependent toxicity in mice. The results indicated that 10 nm AgNPs when administered intraperitoneally, piloerection and reduced activity were observed after 5 hours post-administration, and lowered body temperature was observed at 6 hours post-administration, with histological changes of congestion, single cell necrosis, congestion in the spleen, vacuolation, apoptosis in the thymus cortex and focal necrosis in the liver. However, the AgNPs with particle size 60 and 100 nm when administered intraperitoneally did not show any evident histopathological changes. In conclusion, it can be said that AgNPs with smaller particle size (10 nm) showed higher acute toxicity in mice as compared to AgNPs with lager particle size (60 and 100 nm) (Cho et al. 2018).

CeONPs are also one of the metal NPs that are attracting researchers' interest in the field of pharmaceuticals. However, with great potential in biomedical applications, CeONPs impart toxicity in humans. In this regard, Park et al. (2008) developed CeONPs of various particle sizes (15, 25, 30, and 45 nm) and investigated their toxicity on cultured BEAS-2B cells. The results revealed that CeONPs impart toxicity, increase in ROS, decrease in GSH, prompt oxidative-stress related genes *viz*. catalase, heme oxygenase-1, thiorexoxin reductase and glutathione-*S*-transferase in BEAS-2B cells. It was reported that increase in ROS by CeONPs trigger, the activation of cytosolic caspase 3, and chromatin condensation lead to toxicity via the apoptosis process. Recently, Pulido-Reyes et al. (2015) developed five different types of CeONPs with difference in their percentage of surface content of Ce^{3+} sites. The results revealed that CeONPs-2 (28%), CeONPs-3 (36%), and CeONPs-4 (26%) were found to be nontoxic. However, CeONPs-1 and CeONPs-5 were found to be toxic. Additionally, blocking the sites of Ce^{3+} with PO_4 treatment reverted toxicity (Pulido-Reyes et al. 2015).

ZnOP has been widely researched for its application in biomedical applications, UV-detectors, paints, gas sensors owing to its unique properties *viz* magnetic, anti-microbial, UV-light absorption, semi-conducting and catalytic characteristics. The increased use of ZnONPs in different aspects, human exposure to ZnONPs is expected and the potential toxicity of ZnONPs in human health has been raised as a concern. Various researchers have investigated ZnONPs ZnONPs-based toxicity in various animals and *in vitro* models. ZnONPs exhibit greater toxicity as compared to other metallic oxides NPs, (Palomäki et al. 2010; Cho et al. 2013; Watson et al. 2014) due to their ion-shedding ability caused by leaching of zinc ions from the ZnONPs. ZnONPs are readily soluble in acidic environment, for example, lysosome; however, they demonstrated lower solubility at neutral environments (Donaldson et al. 2013; Jiang and Hsu-Kim 2014). In addition to this, Brunner et al. (2006) developed ZnONPs and investigated their toxicity in human mesothelioma cell and rodent fibroblast cell culture. The results showed that ZnONPs with a concentration of 49 mg/ml lead to complete cell death after exposure. Similar to this study, another *in vitro* study also revealed that, when ZnONPs are exposed to embryonic kidney cells and human hepatocytes, the alteration in mitochondrial activity, DNA damage, and alteration in morphology of cells were reported (Guan et al. 2012). Furthermore, Hong et al. developed ZnONPs and investigated their *in vitro* (RAW 264.7 murine macrophages) and *in vivo* (male ICR mice) toxicity. The results revealed a small increase in intracellular ROS level, NF-kB transcription factor expression (TF) which could be correlated with *in vivo* pathological alteration in organs. Additionally, ZnONPs can cause neural toxicity by mechanisms i.e. higher generation of ROS via the disturbance of metal–ion homeostasis due to an intracellular increase in dissolved free zinc ions (Hong et al. 2013). Recently, Liu et al. developed ZnONPs and studied their size-dependent toxicity caused in the human neuroblastoma cell line (SH-SY-5Y). Two different types of ZnONPs were developed and indicated as ZnONP(A) (irregular, rod-shaped with smooth surfaces and the average size was 47.1 nm and width 27.9 nm) and ZnONPs(B) (small and rod-shaped, average size was 18.5 nm and width 6.8 nm). The results demonstrated that ZnONPs(A) at a concentration ≤20 μg/ml did not show any significant viability change in the cell line whereas, the concentration greater than 20 μg/ml demonstrated a potent decrease in viability in a time-dependent manner. However, ZnONPs(B) when exposed to SH-Sy-5Y cells showed both concentration- and time-dependent toxic effects. Additionally, the toxic effects were higher at concentrations greater than 20 μg/ml. Furthermore, as compared to both ZnONPs, $ZnCl_2$ exhibited a greatly significant toxic effect at similar molar concentrations. Furthermore, Zn ion shedding process in cells was studied by immunocytochemistry. The results indicated that Zn ions were released intracellularly. However, ZnONPs(B) generate a significant increase in level of Zn ions in the intracellular region after one hour as compared to ZnONPs(A). Additionally, after three to five hours of both ZnONPs treatment, a significant Zn-fluorescence intensity was detected (Liu et al. 2017). Furthermore, Choi et al. (2015) developed ZnONPs and investigated their toxicity in rats followed by two different routes: intravenous and oral. For the same, blood concentration and excretion (urine and feces) of Zn was monitored for seven days and tissue distribution was determined in different organs such as thymus, lungs, testes, kidney, brain, spleen, and liver. The results showed that ZnONPs were not readily absorbed through GIT and excreted mostly through feces after oral administration. Additionally, when administered intravenously, the peak concentration was observed within 5 minutes, however, returned to normal range within 48 hours after injection. Additionally, ZnONPs were distributed in kidney, liver, spleen, and lung but not to testes, thymus, and brain. ZnONPs when administered intravenously imparted acute injuries with dark-brown pigments in the lungs; however, when administered orally, no significant damage was observed in rats at the same dose (Choi et al. 2015). Chung et al. (2013) developed ZnONPs and investigated their pharmacokinetics and toxicokinetics in male and female rats. The results demonstrated that the plasma concentration of ZnONPs significantly increased in a dose-dependent manner, however, decreased within 24 hours after single-dose administered orally upto 500 mg/kg, suggesting that ZnONPs were excreted both in case of male and female rats. But in case of repeated dose for 90 days, the oral toxicity studies demonstrated that the elevated plasma concentration of ZnONPs did not return to normal control level, indicating that the ZnONPs accumulated in systemic circulation over the time. This showed that repeated dosing led to toxicity in rats as compared to a single-dose administration (Chung et al. 2013).

Al_2O_3NPs with their great potential in biomedical applications also pose certain toxicity to humans. Many researchers concluded that Al_2O_3NPs can penetrate various biological barriers followed by accumulation in different visceral organs of the body leading to genotoxicity, immunotoxicity, neurotoxicity (Zhang et al. 2018b). Park et al. (2011) developed three different types of spherical Al_2O_3NPs (i.e. γ-aluminium oxide hydroxide NPs ($Al_2O_3(OH)$, α- Al_2O_3NPs and γ-Al_2O_3NPs) and investigated their biodistribution and toxicity. The *in vivo* study demonstrated that the Al_2O_3 (OH)NPs were accumulated more within 24 hours after oral administration in mice. Additionally, $Al_2O_3(OH)$NPs were responsible for increased ratio of neutrophils, decreased the WBC count, and enhanced the secretion of interleukin (IL)-8 in the blood of mice after the administration of dose of 10 mg/kg. Furthermore, the toxicity of three Al_2O_3 NPs was compared using six different cell lines *viz.* HaCaT (skin), HEK-293 (kidney), BEAS-2B (lung), T98G (brain), Chang (liver), H9C2 (heart). The results revealed that $Al_2O_3(OH)$NPs were greatly responsible for inducing toxicity as compared to others. In conclusion, it can be said that $Al_2O_3(OH)$ were the most toxic for biological system due to their low stability in the biological system, and presence of the hydroxyl group may be a significant factor in estimating the distribution and toxicity (Park et al. 2016). In another study, Park et al. (2011) studied the repeated dose toxicity of Al_2O_3NPs following 28 days of oral administration, particularly on gene expression in mouse brain. The results revealed that the dose of 60 mg/kg of Al_2O_3NPs for 28 days lead to significant elevation in the concentration of Al in lung, thymus, and brain. Nevertheless, there was no significant alteration in histopathology. Furthermore, the group of mice treated with 60 mg/kg dose was responsible for upregulating genes related to energy expenditure, feeding behavior, and neurodegeneration by twofold (Park et al. 2011).

CuNPs have also influenced the researchers in the pharmaceutical world owing to their antibacterial, antiviral, antifouling, antifungal activity and are also used in contraceptive devices. Nevertheless, various investigations have confirmed that CuNPs showed toxicity in biological systems most severely in kidney, spleen, and liver (Schrand et al. 2010). In addition to this, CuNPs when administered orally showed interaction with gastric fluid resulting in highly reactive copper which is then accumulated in kidney (Chen et al. 2006; Meng et al. 2007; Bahadar et al. 2016). One of the studies demonstrated that CuNPs with a particle size 23.5 nm have a LD50 of 413 mg/kg in mice, which is considered as moderate toxicity as that of zinc. However, contrast results were obtained for copper microparticles with a particle size of 17 μm that it does not show toxicity and possessed LD50 of 5000 mg/kg (Chen et al. 2006). Moreover, CuNPs demonstrated concentration- and size-dependent toxicity on somatosensory neurons in rats, probably due to enhancement in the production of ROS (Prabhu et al. 2010). Additionally, Manna et al. (2012) also investigated the dose-dependent (200–600 mg/kg/b.w.) toxicity and found that higher dose leads to hepatic dysfunction and cell death via the oxidative stress-dependent signaling cascades and mitochondrial event. CuNPs are also responsible for upregulation of ERK1/2 and p-p38 expression and enhanced NF-kB transcription activity, and caused the reciprocal regulation of Bcl-2 family proteins, caused disruption of mitochondrial membrane potential, release cytochrome c, and activated the formation of apoptosome and caspase 3 (Manna et al. 2012). Furthermore, Karlsson et al. (2008) reported that CuNPs could generate cytotoxicity and DNA damage in A549 cells. Additionally, Yokohira et al. (2009) investigated that CuONPs with high doses, chronically low doses, and with frequent dosing were responsible for inducing severe acute inflammatory alteration in the rat. Furthermore, Hu et al. (2019) investigated CuNPs-based uterine injury and gene expression patterns in female rats. For the same, various doses *viz*. 3.12, 6.25, and 12.5 mg/kg/day were considered and administered intraperitoneally for 14 consecutive days. The results demonstrated that CuNPs when exposed to biological system resulted in increased inflammatory cell infiltration, a significant decrease in the relative uterine weight coefficients, mitochondrial swelling and vacuolization, shortened and reduced endometrial epithelial cell microvilli, and apoptosis. Additionally, signal transduction mechanism studies demonstrated that CuNPs when exposed activated caspase 3, 8, and 9 and BH3 interacting domain death agonist (tBid), reduced B cell leukemia/lymphoma 2 (Bcl-2) expression, and increased the expression of apoptotic peptidase activating factor 1 (Apaf-1), BCL2-associated X, apoptosis regulator (Bax), and cytochrome c. Furthermore, microarray analysis indicated significant changes in expression of 963 genes; of these, 341 were downregulated and 622 were upregulated (Hu et al. 2019).

5.4 Conclusion and Future Perspectives

The applications of metallic nanoparticles as biosensors, for bioimaging, drug delivery, and theranostic applications are tremendous. But critical insights are essential on the toxicity aspects of metallic nanoparticles as they have the potential to generate adverse events by causing damage to DNA, cell inflammation, increased oxidative stress, etc. Various physicochemical properties influence the toxicity of metallic nanoparticles and hence there should be a thorough investigation via cell line studies to ascertain the appropriate dose of MNP. Surface coating with biocompatible agents is essential where the toxicity due to bare nanoparticles is evident. Most metallic nanoparticles are administered intravenously and many of them are primarily accumulated in liver and spleen, followed by kidney. Again, the influence of size, surface coating, charge, etc. modulates the location and extent of accumulation. The future applications of metallic nanoparticles in the pharmaceutical industry and healthcare needs a strong regulatory outlook with exhaustive guidelines in place so that it ensures its safe and effective use.

Acknowledgments

The authors would like to thank CSIR and Nirma University for supporting them with financial assistance and resources for writing this chapter.

References

Aboulaich, A., Tilmaciu, C.M., Merlin, C. et al. (2012). Physicochemical properties and cellular toxicity of (poly) aminoalkoxysilanes-functionalized ZnO quantum dots. *Nanotechnology* 23 (33): 335101. https://doi.org/10.1088/0957-4484/23/33/335101.

Aderibigbe, B.A. (2017). Metal-based nanoparticles for the treatment of infectious diseases. *Molecules* 22 (8): 1370. https://doi.org/10.3390/molecules22081370.

Ahamed, M., Karns, M., Goodson, M. et al. (2008). DNA damage response to different surface chemistry of silver nanoparticles in mammalian cells. *Toxicology and Applied Pharmacology* 233 (3): 404–410. https://doi.org/10.1016/j.taap.2008.09.015.

Ali, S.A., Rizk, M.Z., Hamed, M.A. et al. (2019). Assessment of titanium dioxide nanoparticles toxicity via oral exposure in mice: effect of dose and particle size. *Biomarkers* 24 (5): 492–498. https://doi.org/10.1080/1354750X.2019.1620336.

Anreddy, R.N.R. (2018). Copper oxide nanoparticles induces oxidative stress and liver toxicity in rats following oral exposure. *Toxicology Reports* 5 (August): 903–904. https://doi.org/10.1016/j.toxrep.2018.08.022.

Ansari, M.O., Parveen, N., Ahmad, M.F. et al. (2019). Evaluation of DNA interaction, genotoxicity and oxidative stress induced by iron oxide nanoparticles both *in vitro and in vivo*: attenuation by thymoquinone. *Scientific Reports* 9 (1): 1–14. https://doi.org/10.1038/s41598-019-43188-5.

Asati, A., Santra, S., Kaittanis, C., and Perez, J.M. (2010). Surface-charge-dependent cell localization and cytotoxicity of cerium oxide nanoparticles. *ACS Nano* 4 (9): 5321–5331. https://doi.org/10.1021/nn100816s.

Assadian, E., Dezhampanah, H., Seydi, E., and Pourahmad, J. (2019). Toxicity of Fe 2 O 3 nanoparticles on human blood lymphocytes. *Journal of Biochemical and Molecular Toxicology*, (October 2018): 4–9. https://doi.org/10.1002/jbt.22303.

Ávalos, A., Haza, A.I., Mateo, D., and Morales, P. (2018). *In vitro and in vivo* genotoxicity assessment of gold nanoparticles of different sizes comet and SMART assays. *Food and Chemical Toxicology* 120: 81–88. https://doi.org/10.1016/j.fct.2018.06.061.

Babincová, M., Leszczynska, D., Sourivong, P., and Babinec, P. (2000). Selective treatment of neoplastic cells using ferritin-mediated electromagnetic hyperthermia. *Medical Hypotheses* 54 (2): 177–179. https://doi.org/10.1054/mehy.1999.0011.

Bahadar, H., Maqbool, F., Niaz, K., and Abdollahi, M. (2016). Toxicity of nanoparticles and an overview of current experimental models. *Iranian Biomedical Journal* 20 (1): 1–11. https://doi.org/10.7508/ibj.2016.01.001.

Baranowska-Wójcik, E., Szwajgier, D., Oleszczuk, P., and Winiarska-Mieczan, A. (2020). Effects of titanium dioxide nanoparticles exposure on human health—a review. *Biological Trace Element Research* 193 (1): 118–129. https://doi.org/10.1007/s12011-019-01706-6.

Barbasz, A., Oćwieja, M., and Roman, M. (2017). Toxicity of silver nanoparticles towards tumoral human cell lines U-937 and HL-60. *Colloids and Surfaces B: Biointerfaces* 156: 397–404. https://doi.org/10.1016/j.colsurfb.2017.05.027.

Bar-Ilan, O., Albrecht, R.M., Fako, V.E., and Furgeson, D.Y. (2009). Toxicity assessments of multisized gold and silver nanoparticles in zebrafish embryos. *Small* 5 (16): 1897–1910. https://doi.org/10.1002/smll.200801716.

Berry, C.C. (2009). Progress in functionalization of magnetic nanoparticles for applications in biomedicine. *Journal of Physics D: Applied Physics* 42 (22) https://doi.org/10.1088/0022-3727/42/22/224003.

Blanco, J., Tomás-Hernández, S., García, T. et al. (2018). Oral exposure to silver nanoparticles increases oxidative stress markers in the liver of male rats and deregulates the insulin signalling pathway and p53 and cleaved caspase 3 protein expression. *Food and Chemical Toxicology* 115: 398–404. https://doi.org/10.1016/j.fct.2018.03.039.

Braydich-Stolle, L., Hussain, S., Schlager, J.J., and Hofmann, M.C. (2005). *In vitro* cytotoxicity of nanoparticles in mammalian germline stem cells. *Toxicological Sciences* 88 (2): 412–419. https://doi.org/10.1093/toxsci/kfi256.

Brunner, T.J., Wick, P., Manser, P. et al. (2006). *In vitro* cytotoxicity of oxide nanoparticles: comparison to asbestos, silica, and the effect of particle solubility. *Environmental Science and Technology* 40 (14): 4374–4381. https://doi.org/10.1021/es052069i.

Buzea, C., Pacheco, I.I., and Robbie, K. (2007). Nanomaterials and nanoparticles: sources and toxicity. *Biointerphases* 2 (4): MR17–MR71. https://doi.org/10.1116/1.2815690.

Cabada, T.F., de Pablo, C.S.L., Serrano, A.M. et al. (2012). Induction of cell death in a glioblastoma line by hyperthermic therapy based on gold nanorods. *International Journal of Nanomedicine* 7: 1511–1523. https://doi.org/10.2147/IJN.S28470.

Caballero-Díaz, E., Pfeiffer, C., Kastl, L. et al. (2013). The toxicity of silver nanoparticles depends on their uptake by cells and thus on their surface chemistry. *Particle and Particle Systems Characterization* 30 (12): 1079–1085. https://doi.org/10.1002/ppsc.201300215.

Carnovale, C., Bryant, G., Shukla, R., and Bansal, V. (2019). Identifying trends in gold nanoparticle toxicity and uptake: size, shape, capping ligand, and biological corona. *ACS Omega* 4 (1): 242–256. https://doi.org/10.1021/acsomega.8b03227.

Chen, Z., Meng, H., Xing, G. et al. (2006). Acute toxicological effects of copper nanoparticles in vivo. *Toxicology Letters* 163 (2): 109–120. https://doi.org/10.1016/j.toxlet.2005.10.003.

Cho, W.S., Kang, B.C., Lee, J.K. et al. (2013). Comparative absorption, distribution, and excretion of titanium dioxide and zinc oxide nanoparticles after repeated oral administration. *Particle and Fibre Toxicology* 10 (1): 1–9. https://doi.org/10.1186/1743-8977-10-9.

Cho, Y-M., Mizuta, Y., Akagi, J-i., Toyoda, T., Sone, M., and, Ogawa, K., (2018). Size-dependent acute toxicity of silver nanoparticles in mice. *Japanese Society of Toxicologic Pathology* 31(1): 73–80.

Choi, J., Kim, H., Kim, P. et al. (2015). Toxicity of zinc oxide nanoparticles in rats treated by two different routes: single intravenous injection and single oral administration. *Journal of Toxicology and Environmental Health – Part A: Current Issues* 78 (4): 226–243. https://doi.org/10.1080/15287394.2014.949949.

Chung, H.E., Yu, J., Baek, M. et al. (2013). Toxicokinetics of zinc oxide nanoparticles in rats. *Journal of Physics: Conference Series* 429 (1) https://doi.org/10.1088/1742-6596/429/1/012037.

Das, B., Tripathy, S., Adhikary, J. et al. (2017). Surface modification minimizes the toxicity of silver nanoparticles: an *in vitro* and *in vivo* study. *Journal of Biological Inorganic Chemistry* 22 (6): 893–918. https://doi.org/10.1007/s00775-017-1468-x.

De Jong, W.H. and Paul, J.B. (2008). Drug delivery and nanoparticles: applications and hazards. *International Journal of Nanomedicine* 3 (2): 133–149.

Donaldson, K., Schinwald, A., Murphy, F. et al. (2013). The biologically effective dose in inhalation nanotoxicology. *Accounts of Chemical Research* 46 (3): 723–732. https://doi.org/10.1021/ar300092y.

El Badawy, A.M., Silva, R.G., Morris, B. et al. (2011). Surface charge-dependent toxicity of silver nanoparticles. *Environmental Science and Technology* 45 (1): 283–287. https://doi.org/10.1021/es1034188.

Espinosa, A., Di Corato, R., Kolosnjaj-Tabi, J. et al. (2016). Duality of iron oxide nanoparticles in cancer therapy: amplification of heating efficiency by magnetic hyperthermia and photothermal bimodal treatment. *ACS Nano* 10 (2): 2436–2446. https://doi.org/10.1021/acsnano.5b07249.

Fahmy, H.M., Ali, O.A., Hassan, A.A., and Mohamed, M.A. (2020). Biodistribution and toxicity assessment of copper nanoparticles in the rat brain. *Journal of Trace Elements in Medicine and Biology* 61: 126505. https://doi.org/10.1016/j.jtemb.2020.126505.

George, S., Lin, S., Ji, Z. et al. (2012). Surface defects on plate-shaped silver nanoparticles contribute to its hazard potential in a fish gill cell line and zebrafish embryos. *ACS Nano* 6 (5): 3745–3759. https://doi.org/10.1021/nn204671v.

Ghosh, S.K. and Pal, T. (2007). Interparticle coupling effect on the surface plasmon resonance of gold nanoparticles: from theory to applications. *Chemical Reviews* 107 (11): 4797–4862. https://doi.org/10.1021/cr0680282.

Ghosh, S., Ghosh, I., Chakrabarti, M., and Mukherjee, A. (2020). Genotoxicity and biocompatibility of superparamagnetic iron oxide nanoparticles: influence of surface modification on biodistribution, retention, DNA damage and oxidative stress. *Food and Chemical Toxicology* 136: 110989. https://doi.org/10.1016/j.fct.2019.110989.

Ginzburg, A.L., Truong, L., Tanguay, R.L., and Hutchison, J.E. (2018). Synergistic toxicity produced by mixtures of biocompatible gold nanoparticles and widely used surfactants. *ACS Nano* 12 (6): 5312–5322. https://doi.org/10.1021/acsnano.8b00036.

Goma, A.A., El Okle, O.S., and Tohamy, H.G. (2020). Protective effect of methylene blue against copper oxide nanoparticle-induced neurobehavioral toxicity. *Behavioural Brain Research* 398 (May 2020): 112942. https://doi.org/10.1016/j.bbr.2020.112942.

Goodman, C.M., McCusker, C.D., Yilmaz, T., and Rotello, V.M. (2004). Toxicity of gold nanoparticles functionalized with cationic and anionic side chains. *Bioconjugate Chemistry* 15 (4): 897–900. https://doi.org/10.1021/bc049951i.

Greulich, C., Diendorf, J., Simon, T. et al. (2011). Uptake and intracellular distribution of silver nanoparticles in human mesenchymal stem cells. *Acta Biomaterialia* 7 (1): 347–354. https://doi.org/10.1016/j.actbio.2010.08.003.

Greulich, C., Braun, D., Peetsch, A. et al. (2012). The toxic effect of silver ions and silver nanoparticles towards bacteria and human cells occurs in the same concentration range. *RSC Advances* 2 (17): 6981–6987. https://doi.org/10.1039/c2ra20684f.

Guan, R., Kang, T., Lu, F. et al. (2012). Cytotoxicity, oxidative stress, and genotoxicity in human hepatocyte and embryonic kidney cells exposed to ZnO nanoparticles. *Nanoscale Research Letters* 7: 1–7. https://doi.org/10.1186/1556-276X-7-602.

Hong, T.K., Tripathy, N., Son, H.J. et al. (2013). A comprehensive *in vitro* and *in vivo* study of ZnO nanoparticles toxicity. *Journal of Materials Chemistry B* 1 (23): 2985–2992. https://doi.org/10.1039/c3tb20251h.

Hsiao, I.L. and Huang, Y.J. (2011). Effects of various physicochemical characteristics on the toxicities of ZnO and TiO2 nanoparticles toward human lung epithelial cells. *Science of the Total Environment* 409 (7): 1219–1228. https://doi.org/10.1016/j.scitotenv.2010.12.033.

Hu, S., Yang, J., Rao, M. et al. (2019). Copper nanoparticle-induced uterine injury in female rats. *Environmental Toxicology* 34 (3): 252–261. https://doi.org/10.1002/tox.22680.

Ismail, N.A., Shameli, K., Wong, M.M.T. et al. (2019). Antibacterial and cytotoxic effect of honey mediated copper nanoparticles synthesized using ultrasonic assistance. *Materials Science and Engineering C* 104 (June): 109899. https://doi.org/10.1016/j.msec.2019.109899.

Jain, T.K., Reddy, M.K., Morales, M.A. et al. (2008). Biodistribution, clearance, and biocompatibility of iron oxide magnetic nanoparticles in rats. *Molecular Pharmaceutics* 5 (2): 316–327. https://doi.org/10.1021/mp7001285.

Janib, S.M., Moses, A.S., and MacKay, J.A. (2010). Imaging and drug delivery using theranostic nanoparticles. *Advanced Drug Delivery Reviews* 62 (11): 1052–1063. https://doi.org/10.1016/j.addr.2010.08.004.

Jeng, H.A. and Swanson, J. (2006). Toxicity of metal oxide nanoparticles in mammalian cells. *Journal of Environmental Science and Health—Part A Toxic/Hazardous Substances and Environmental Engineering* 41 (12): 2699–2711. https://doi.org/10.1080/10934520600966177.

Jiang, C. and Hsu-Kim, H. (2014). Direct in situ measurement of dissolved zinc in the presence of zinc oxide nanoparticles using anodic stripping voltammetry. *Environmental Sciences: Processes and Impacts* 16 (11): 2536–2544. https://doi.org/10.1039/c4em00278d.

Jiang, J., Pi, J., and Cai, J. (2018). The advancing of zinc oxide nanoparticles for biomedical applications. *Bioinorganic Chemistry and Applications* 2018 https://doi.org/10.1155/2018/1062562.

Jones, K., Morton, J., Smith, I. et al. (2015). Human *in vivo* and *in vitro* studies on gastrointestinal absorption of titanium dioxide nanoparticles. *Toxicology Letters* 233 (2): 95–101. https://doi.org/10.1016/j.toxlet.2014.12.005.

Karlsson, H.L., Cronholm, P., Gustafsson, J., and Mo, L. (2008). Copper oxide nanoparticles are highly toxic a comparison between metal oxide nanoparticles and carbon nanotubes. *Chemical Research in Toxicology (ACS Publications)* 21: 1726–1732.

Kaur, S., Singh, S.P., Biradar, A.M. et al. (2007). Enhanced electro-optical properties in gold nanoparticles doped ferroelectric liquid crystals. *Applied Physics Letters* 91 (2): 2–4. https://doi.org/10.1063/1.2756136.

Kayal, S. and Ramanujan, R.V. (2010). Doxorubicin loaded PVA coated iron oxide nanoparticles for targeted drug delivery. *Materials Science and Engineering C* 30 (3): 484–490. https://doi.org/10.1016/j.msec.2010.01.006.

Kim, H.S. and Lee, D.Y. (2017). Photothermal therapy with gold nanoparticles as an anticancer medication. *Journal of Pharmaceutical Investigation* 47 (1): 19–26. https://doi.org/10.1007/s40005-016-0292-6.

Kim, J.S., Yoon, T.J., Yu, K.N. et al. (2006). Toxicity and tissue distribution of magnetic nanoparticles in mice. *Toxicological Sciences* 89 (1): 338–347. https://doi.org/10.1093/toxsci/kfj027.

Kwon, J.T., Hwang, S.K., Jin, H. et al. (2008). Body distribution of inhaled fluorescent magnetic nanoparticles in the mice. *Journal of Occupational Health* 50 (1): 1–6. https://doi.org/10.1539/joh.50.1.

Lai, X., Zhao, H., Zhang, Y. et al. (2018). Intranasal delivery of copper oxide nanoparticles induces pulmonary toxicity and fibrosis in C57BL/6 mice. *Scientific Reports* 8 (1): 1–12. https://doi.org/10.1038/s41598-018-22556-7.

Lakshmi Pravallika, P., Krishna Mohan, G., Venkateswara Rao, K., and Shanker, K. (2019). Biosynthesis, characterization and acute oral toxicity studies of synthesized iron oxide nanoparticles using ethanolic extract of *Centella asiatica* plant. *Materials Letters* 236: 256–259. https://doi.org/10.1016/j.matlet.2018.10.037.

Leblanc, A.J., Cumpston, J.L., Chen, B.T. et al. (2009). Nanoparticle inhalation impairs endothelium-dependent vasodilation in subepicardial arterioles. *Journal of Toxicology and Environmental Health—Part A: Current Issues* 72 (24): 1576–1584. https://doi.org/10.1080/15287390903232467.

Lee, I.C., Ko, J.W., Park, S.H. et al. (2016). Comparative toxicity and biodistribution of copper nanoparticles and cupric ions in rats. *International Journal of Nanomedicine* 11: 2883–2900. https://doi.org/10.2147/IJN.S106346.

Lei, R., Yang, B., Wu, C. et al. (2015). Mitochondrial dysfunction and oxidative damage in the liver and kidney of rats following exposure to copper nanoparticles for five consecutive days. *Toxicology Research* 4 (2): 351–364. https://doi.org/10.1039/c4tx00156g.

Li, Y.F. and Chen, C. (2011). Fate and toxicity of metallic and metal-containing nanoparticles for biomedical applications. *Small* 7 (21): 2965–2980. https://doi.org/10.1002/smll.201101059.

Li, X., Liu, W., Sun, L. et al. (2015). Effects of physicochemical properties of nanomaterials on their toxicity. *Journal of Biomedical Materials Research—Part A* 103 (7): 2499–2507. https://doi.org/10.1002/jbm.a.35384.

Li, X., Hu, Z., Ma, J. et al. (2018). The systematic evaluation of size-dependent toxicity and multi-time biodistribution of gold nanoparticles. *Colloids and Surfaces B: Biointerfaces* 167: 260–266. https://doi.org/10.1016/j.colsurfb.2018.04.005.

Liang, G., Pu, Y., Yin, L. et al. (2009). Influence of different sizes of titanium dioxide nanoparticles on hepatic and renal functions in rats with correlation to oxidative stress. *Journal of Toxicology and Environmental Health—Part A: Current Issues* 72 (11–12): 740–745. https://doi.org/10.1080/15287390902841516.

Ling, D., Lee, N., and Hyeon, T. (2015). Chemical synthesis and assembly of uniformly sized iron oxide nanoparticles for medical applications. *Accounts of Chemical Research* 48 (5): 1276–1285. https://doi.org/10.1021/acs.accounts.5b00038.

Liu, J., Kang, Y., Yin, S. et al. (2017). Zinc oxide nanoparticles induce toxic responses in human neuroblastoma SHSY5Y cells in a size-dependent manner. *International Journal of Nanomedicine* 12: 8085–8099. https://doi.org/10.2147/IJN.S149070.

Lopez-Chaves, C., Soto-Alvaredo, J., Montes-Bayon, M. et al. (2018). Gold nanoparticles: distribution, bioaccumulation and toxicity. In vitro and in vivo studies. *Nanomedicine: Nanotechnology, Biology, and Medicine* 14 (1): 1–12. https://doi.org/10.1016/j.nano.2017.08.011.

Luther, E.M., Koehler, Y., Diendorf, J. et al. (2011). Accumulation of silver nanoparticles by cultured primary brain astrocytes. *Nanotechnology* 22 (37) https://doi.org/10.1088/0957-4484/22/37/375101.

Ma, X., Yang, X., Wang, Y. et al. (2018). Gold nanoparticles cause size-dependent inhibition of embryonic development during murine pregnancy. *Nano Research* 11 (6): 3419–3433. https://doi.org/10.1007/s12274-018-1969-0.

MacNicoll, A., Kelly, M., Aksoy, H. et al. (2015). A study of the uptake and biodistribution of nano-titanium dioxide using *in vitro* and *in vivo* models of oral intake. *Journal of Nanoparticle Research* 17 (2) https://doi.org/10.1007/s11051-015-2862-3.

Mahalakshmi, A. and Kurian, G.A. (2019). Evaluation of chemical and green synthesized iron oxide nanoparticles' associated renal toxicity in different experimental models: a comparative study. *Journal of Cluster Science* 30 (2): 343–350. https://doi.org/10.1007/s10876-018-01492-6.

Manna, P., Ghosh, M., Ghosh, J. et al. (2012). Contribution of nano-copper particles to in vivo liver dysfunction and cellular damage: role of IκBα/NF-κB, MAPKs and mitochondrial signal. *Nanotoxicology* 6 (1): 1–21. https://doi.org/10.3109/17435390.2011.552124.

Martínez-Rodríguez, N.L., Tavárez, S., and González-Sánchez, Z.I. (2019). *In vitro* toxicity assessment of zinc and nickel ferrite nanoparticles in human erythrocytes and peripheral blood mononuclear cell. *Toxicology in vitro* 57 (February): 54–61. https://doi.org/10.1016/j.tiv.2019.02.011.

Meng, H., Chen, Z., Xing, G. et al. (2007). Ultrahigh reactivity provokes nanotoxicity: explanation of oral toxicity of nano-copper particles. *Toxicology Letters* 175 (1–3): 102–110. https://doi.org/10.1016/j.toxlet.2007.09.015.

Moradi-Sardareh, H., Basir, H.R.G., Hassan, Z.M. et al. (2018). Toxicity of silver nanoparticles on different tissues of Balb/C mice. *Life Sciences* 211: 81–90. https://doi.org/10.1016/j.lfs.2018.09.001.

Naqvi, S., Samim, M., Abdin, M.Z. et al. (2010). Concentration-dependent toxicity of iron oxide nanoparticles mediated by increased oxidative stress. *International Journal of Nanomedicine* 5 (1): 983–989. https://doi.org/10.2147/IJN.S13244.

Newman, M.D., Stotland, M., and Ellis, J.I. (2009). The safety of nanosized particles in titanium dioxide- and zinc oxide-based sunscreens. *Journal of the American Academy of Dermatology* 61 (4): 685–692. https://doi.org/10.1016/j.jaad.2009.02.051.

Palomäki, J., Karisola, P., Pylkkänen, L. et al. (2010). Engineered nanomaterials cause cytotoxicity and activation on mouse antigen presenting cells. *Toxicology* 267 (1–3): 125–131. https://doi.org/10.1016/j.tox.2009.10.034.

Pan, Y., Neuss, S., Leifert, A. et al. (2007). Size-dependent cytotoxicity of gold nanoparticles. *Small* 3 (11): 1941–1949. https://doi.org/10.1002/smll.200700378.

Pandiarajan, J. and Krishnan, M. (2017). Properties, synthesis and toxicity of silver nanoparticles. *Environmental Chemistry Letters* 15 (3): 387–397. https://doi.org/10.1007/s10311-017-0624-4.

Park, E.J., Choi, J., Park, Y.K., and Park, K. (2008). Oxidative stress induced by cerium oxide nanoparticles in cultured BEAS-2B cells. *Toxicology* 245 (1–2): 90–100. https://doi.org/10.1016/j.tox.2007.12.022.

Park, E.J., Kim, H., Kim, Y., and Choi, K. (2011). Repeated-dose toxicity attributed to aluminum nanoparticles following 28-day oral administration, particularly on gene expression in mouse brain. *Toxicological and Environmental Chemistry* 93 (1): 120–133. https://doi.org/10.1080/02772248.2010.495191.

Park, E.J., Lee, G.H., Yoon, C. et al. (2016). Biodistribution and toxicity of spherical aluminum oxide nanoparticles. *Journal of Applied Toxicology* 36 (3): 424–433. https://doi.org/10.1002/jat.3233.

Patlolla, A.K., Kumari, S.A., and Tchounwou, P.B. (2019). A comparison of poly-ethylene-glycol-coated and uncoated gold nanoparticle-mediated hepatotoxicity and oxidative stress in Sprague Dawley rats. *International Journal of Nanomedicine* 14: 639–647. https://doi.org/10.2147/IJN.S185574.

Patsula, V., Tulinska, J., Trachtová, Š. et al. (2019). Toxicity evaluation of monodisperse PEGylated magnetic nanoparticles for nanomedicine. *Nanotoxicology* 13 (4): 510–526. https://doi.org/10.1080/17435390.2018.1555624.

Pereira, A.C., Gonçalves, B.B., da Silva Brito, R. et al. (2020). Comparative developmental toxicity of iron oxide nanoparticles and ferric chloride to zebrafish (*Danio rerio*) after static and semi-static exposure. *Chemosphere* 254: 126792. https://doi.org/10.1016/j.chemosphere.2020.126792.

Pinzaru, I., Coricovac, D., Dehelean, C. et al. (2018). Stable PEG-coated silver nanoparticles – a comprehensive toxicological profile. *Food and Chemical Toxicology* 111 (December 2017): 546–556. https://doi.org/10.1016/j.fct.2017.11.051.

van Pomeren, M., Peijnenburg, W.J.G.M., Vlieg, R.C. et al. (2019). The biodistribution and immuno-responses of differently shaped non-modified gold particles in zebrafish embryos. *Nanotoxicology* 13 (4): 558–571. https://doi.org/10.1080/17435390.2018.1564079.

Prabhu, B.M., Ali, S.F., Murdock, R.C. et al. (2010). Copper nanoparticles exert size and concentration dependent toxicity on somatosensory neurons of rat. *Nanotoxicology* 4 (2): 150–160. https://doi.org/10.3109/17435390903337693.

Prakash, S., Chakrabarty, T., Singh, A.K., and Shahi, V.K. (2013). Polymer thin films embedded with metal nanoparticles for electrochemical biosensors applications. *Biosensors and Bioelectronics* 41 (1): 43–53. https://doi.org/10.1016/j.bios.2012.09.031.

Prodan, E., Nordlander, P., and Halas, N.J. (2003). Electronic structure and optical properties of gold nanoshells. *Nano Letters* 3 (10): 1411–1415. https://doi.org/10.1021/nl034594q.

Pulido-Reyes, G., Rodea-Palomares, I., Das, S. et al. (2015). Untangling the biological effects of cerium oxide nanoparticles: the role of surface valence states. *Scientific Reports* 5: 1–14. https://doi.org/10.1038/srep15613.

Raghupathi, K.R., Koodali, R.T., and Manna, A.C. (2011). Size-dependent bacterial growth inhibition and mechanism of antibacterial activity of zinc oxide nanoparticles. *Langmuir* 27 (7): 4020–4028. https://doi.org/10.1021/la104825u.

Rasmussen, J.W., Martinez, E., Louka, P., and Wingett, D.G. (2010). Zinc oxide nanoparticles for selective destruction of tumor cells and potential for drug delivery applications. *Expert Opinion on Drug Delivery* 7 (9): 1063–1077. https://doi.org/10.1517/17425247.2010.502560.

Rosłon, M., Jastrzębska, A., Sitarz, K. et al. (2019). The toxicity *in vitro* of titanium dioxide nanoparticles modified with noble metals on mammalian cells. *International Journal of Applied Ceramic Technology* 16 (2): 481–493. https://doi.org/10.1111/ijac.13128.

Ruparelia, J.P., Chatterjee, A.K., Duttagupta, S.P., and Mukherji, S. (2008). Strain specificity in antimicrobial activity of silver and copper nanoparticles. *Acta Biomaterialia* 4 (3): 707–716. https://doi.org/10.1016/j.actbio.2007.11.006.

Santhosh, P.B. and Ulrih, N.P. (2013). Multifunctional superparamagnetic iron oxide nanoparticles: promising tools in cancer theranostics. *Cancer Letters* 336 (1): 8–17. https://doi.org/10.1016/j.canlet.2013.04.032.

Sayadi, M.H., Mansouri, B., Shahri, E. et al. (2020). Exposure effects of iron oxide nanoparticles and iron salts in blackfish (*Capoeta fusca*): acute toxicity, bioaccumulation, depuration, and tissue histopathology. *Chemosphere* 247: 125900. https://doi.org/10.1016/j.chemosphere.2020.125900.

Schrand, A.M., Rahman, M.F., Hussain, S.M. et al. (2010). Metal-based nanoparticles and their toxicity assessment. *Wiley Interdisciplinary Reviews: Nanomedicine and Nanobiotechnology* 2 (5): 544–568. https://doi.org/10.1002/wnan.103.

Senapati, V.A. and Kumar, A. (2018). ZnO nanoparticles dissolution, penetration and toxicity in human epidermal cells. Influence of pH. *Environmental Chemistry Letters* 16 (3): 1129–1135. https://doi.org/10.1007/s10311-018-0736-5.

Shang, L., Bian, T., Zhang, B. et al. (2014). Graphene-supported ultrafine metal nanoparticles encapsulated by mesoporous silica: robust catalysts for oxidation and reduction reactions. *Angewandte Chemie—International Edition* 53 (1): 250–254. https://doi.org/10.1002/anie.201306863.

Sharma, H., Mishra, P.K., Talegaonkar, S., and Vaidya, B. (2015). Metal nanoparticles: a theranostic nanotool against cancer. *Drug Discovery Today* 20 (9): 1143–1151. https://doi.org/10.1016/j.drudis.2015.05.009.

Shubha, P., Gowda, M.L., Namratha, K. et al. (2019). *In vitro* and *in vivo* evaluation of green-hydrothermal synthesized ZnO nanoparticles. *Journal of Drug Delivery Science and Technology* 49: 692–699. https://doi.org/10.1016/j.jddst.2018.12.017.

Shukla, S., Jadaun, A., Arora, V. et al. (2015). *In vitro* toxicity assessment of chitosan oligosaccharide coated iron oxide nanoparticles. *Toxicology Reports* 2: 27–39. https://doi.org/10.1016/j.toxrep.2014.11.002.

Singh, S. (2019). Zinc oxide nanoparticles impacts: cytotoxicity, genotoxicity, developmental toxicity, and neurotoxicity. *Toxicology Mechanisms and Methods* 29 (4): 300–311. https://doi.org/10.1080/15376516.2018.1553221.

Slawson, R.M., Lee, H., and Trevors, J.T. (1990). Bacterial interactions with silver. *Biology of Metals* 3 (3–4): 151–154. https://doi.org/10.1007/BF01140573.

Sohaebuddin, S.K., Thevenot, P.T., Baker, D. et al. (2010). Nanomaterial cytotoxicity is composition, size, and cell type dependent. *Particle and Fibre Toxicology* 7: 1–17. https://doi.org/10.1186/1743-8977-7-22.

Vallyathan, V. and Gulumian, M. (2010). Nanoparticles and potential human health implications: past and future directions. Preface. *Journal of Toxicology and Environmental Health—Part A: Current Issues* 73 (5–6): 339–340. https://doi.org/10.1080/15287390903584339.

Wang, J., Zhou, G., Chen, C. et al. (2007). Acute toxicity and biodistribution of different sized titanium dioxide particles in mice after oral administration. *Toxicology Letters* 168 (2): 176–185. https://doi.org/10.1016/j.toxlet.2006.12.001.

Wang, S., Lu, W., Tovmachenko, O. et al. (2008). Challenge in understanding size and shape dependent toxicity of gold nanomaterials in human skin keratinocytes. *Chemical Physics Letters* 463 (1–3): 145–149. https://doi.org/10.1016/j.cplett.2008.08.039.

Warheit, D.B. and Donner, E.M. (2015). Risk assessment strategies for nanoscale and fine-sized titanium dioxide particles: recognizing hazard and exposure issues. *Food and Chemical Toxicology* 85: 138–147. https://doi.org/10.1016/j.fct.2015.07.001.

Watson, C., Ge, J., Cohen, J. et al. (2014). High-throughput screening platform for engineered nanoparticle-mediated genotoxicity using cometchip technology. *ACS Nano* 8 (3): 2118–2133. https://doi.org/10.1021/nn404871p.

Woźniak, A., Malankowska, A., Nowaczyk, G. et al. (2017). Size and shape-dependent cytotoxicity profile of gold nanoparticles for biomedical applications. *Journal of Materials Science: Materials in Medicine* 28 (6): 1–11. https://doi.org/10.1007/s10856-017-5902-y.

Yang, W.J., Lee, J.H., Hong, S.C. et al. (2013). Difference between toxicities of iron oxide magnetic nanoparticles with various surface-functional groups against human normal fibroblasts and fibrosarcoma cells. *Materials* 6 (10): 4689–4706. https://doi.org/10.3390/ma6104689.

Yen, H.J., Hsu, S.H., and Tsai, C.L. (2009). Cytotoxicity and immunological response of gold and silver nanoparticles of different sizes. *Small* 5 (13): 1553–1561. https://doi.org/10.1002/smll.200900126.

Yokohira, M., Hashimoto, N., Yamakawa, K. et al. (2009). Lung carcinogenic bioassay of CuO and TiO_2 nanoparticles with intratracheal instillation using F344 male rats. *Journal of Toxicologic Pathology* 22 (1): 71–78. https://doi.org/10.1293/tox.22.71.

Yousef, M.I., Mutar, T.F., and Kamel, M.A.E.N. (2019). Hepato-renal toxicity of oral sub-chronic exposure to aluminum oxide and/or zinc oxide nanoparticles in rats. *Toxicology Reports* 6 (March): 336–346. https://doi.org/10.1016/j.toxrep.2019.04.003.

Yu, M., Huang, S., Yu, K.J., and Clyne, A.M. (2012). Dextran and polymer polyethylene glycol (PEG) coating reduce both 5 and 30 nm iron oxide nanoparticle cytotoxicity in 2D and 3D cell culture. *International Journal of Molecular Sciences* 13 (5): 5554–5570. https://doi.org/10.3390/ijms13055554.

Zhang, H., Chang, Z., Mehmood, K. et al. (2018a). Nano copper induces apoptosis in PK-15 cells via a mitochondria-mediated pathway. *Biological Trace Element Research* 181 (1): 62–70. https://doi.org/10.1007/s12011-017-1024-0.

Zhang, Q., Ding, Y., He, K. et al. (2018b). Exposure to alumina nanoparticles in female mice during pregnancy induces neurodevelopmental toxicity in the offspring. *Frontiers in Pharmacology* 9 (MAR): 1–12. https://doi.org/10.3389/fphar.2018.00253.

Zhao, J. and Castranova, V. (2011). Toxicology of nanomaterials used in nanomedicine. *Journal of Toxicology and Environmental Health—Part B: Critical Reviews* 14 (8): 593–632. https://doi.org/10.1080/10937404.2011.615113.

Zhou, Y.M., Zhong, C.Y., Kennedy, I.M., and Pinkerton, K.E. (2003). Pulmonary responses of acute exposure to ultrafine iron particles in healthy adult rats. *Environmental Toxicology* 18 (4): 227–235. https://doi.org/10.1002/tox.10119.

6

Toxicity, Safety, and Biodistribution of Multifunctional Mesoporous Silica Nanoparticles

Mansi Athalye[1], *Rashmin Patel*[2], *and Mrunali Patel*[2]

[1] *Department of Pharmaceutics and Pharmaceutical Technology, L. M. College of Pharmacy, Ahmedabad, GJ, India*
[2] *Department of Pharmacy, Ramanbhai Patel College of Pharmacy, Charotar University of Science and Technology, CHARUSAT, Changa, GJ, India*

6.1 Introduction

The use of nanoparticles as a drug delivery carrier has received much attention in recent decades (Douroumis et al. 2013). Organic materials such as liposomes, micelles, dendrimers, and nanoparticles are widely used as nanocarriers though they are not much clinically translocated, which is attributed to their premature drug release, early abstraction before reaching the site of action, uncontrollable drug release rates for low local concentration, ineffective cellular drug uptake, and endosomal escape. Thus, there is a greater need for controlled drug delivery systems which release the drug at the required rate and time and that indicate "zero release" effect in blood circulation so as to prevent the decomposition of drugs and protect the healthy tissues from adverse effects of toxic drugs (Yang et al. 2014). Recently, gold nanoparticles, magnetic nanoparticles, and silica nanoparticles as novel systems have been developed, which have shown their potential in the manufacture of controlled drug delivery systems. Among these delivery systems, mesoporous silica nanoparticles (MSNs) have provoked significant interest and have been proven as a significant candidate for nanomedical applications (Yang et al. 2014). In 2001 for the first time, the use of MCM-41 (Mobile Crystalline Material) type of mesoporous silica material was reported as a drug delivery system. MSNs are solid materials that consist of numerous empty channels called mesopores and are arranged in a 2D network of a honeycomb-like porous structure. They possess unique modifiable properties such as high surface area (>700 m^2/g), pore volume (>0.9 cm^3/g), and tunable pore diameter (2–10 nm) with dual-functional surfaces (exterior and interior) (Lu et al. 2010; Douroumis et al. 2013). They also exhibit a rigid and stable framework having exceptional mechanical, chemical, and thermal stability; also, their fabrication is easy, cost-effective, and controllable (Douroumis et al. 2013; Li et al. 2015). Owing to their unique properties, MSNs were proposed as drug delivery carriers for various pharmacological drugs, which have shown different biotechnological and nanomedicinal applications (Vallet-Regi et al. 2001; Lu et al. 2010; Douroumis et al. 2013).

The size, shape, and controllable pores of MSNs allow them to load pharmaceutical drugs and prevent their degradation as well as

Nanotechnology in Medicine: Toxicity and Safety, First Edition. Edited by Mahendra Rai, Mrunali Patel, and Rashmin Patel.

premature release before reaching the required site of action (Lu et al. 2010). This is because the drug loaded in the mesopores of mesoporous silica nanoparticles may safeguard the drug against enzymatic degradation until it gets released and thus, it can improve the pharmacokinetics of the drug (Linden 2018). Moreover, due to its high surface area and larger pore volume, it can incorporate a high dose of drugs and it is also a suitable candidate for targeted delivery so that lower nanoparticle doses can be used. This is beneficial from the perspective of toxicity as other nanoparticulate drug delivery systems indicate lower drug loading levels (Li et al. 2015; Linden 2018). The physicochemical properties such as surface characteristics of the particles also impart dispersion stability in the physiological system and open up the probability to develop further selective functionalization regimes, and thus broaden the applicability of the nanoparticles (Desai et al. 2014).

Moreover, MSNs can be functionalized on the surface with stimuli-responsive agents for controlled release, which serves as an innovative formulation for different biomedical and biotechnological applications and for future research as all the properties of MSNs determine their *in vitro* and *in vivo* behavior (Lu et al. 2010). MSNs loaded with cargos can also be closed with various gatekeepers in which these cargos will not be released from the nanoparticles until triggered by stimuli that open up an opportunity to design stimuli-responsive drug delivery systems for controlled release. The stimulus is of two types, namely endogenous and exogenous stimulus, wherein endogenous stimuli arise from redox potential, intercellular or intracellular pH, and from the increased level of certain enzymes. However, exogenous stimuli are based on extracorporeal physical alterations, including temperature changes, ultrasounds, magnetic fields, along with light and electric fields (Yang et al. 2014).

Above all, the distinctive applications of MSNs as compared to other formulations are that they are well-suited for the incorporation of theranostic agents into a single nanoparticle, having separate domains for three components: first, for the incorporation of the contrast agent/optical label which aids in traceable imaging of the target and second domain for loading of the drug/therapeutic agents. The third domain facilitates incorporating the biomolecular ligand in the case of targeted drug delivery (Desai et al. 2014). Thus, all these exceptional features of MSNs unwrap the new dimension of drug delivery systems in the medical field.

6.2 Bioapplicability of Mesoporous Silica Nanoparticles

The use of silica nanoparticles for various biomedical and industrial applications has increased in the last few years. They are used in pharmaceutical applications, cosmetics, coating materials, printer toners as well as in the form of additives. The diverse applicability of the silica nanoparticles is due to their exclusive properties such as hydrophilicity, ease of surface modification or functionalization, ease of synthesis, and low toxicity (Lee et al. 2014). MSNs have exhibited their potential for oral drug delivery consisting of water-insoluble drugs through significant enhancement in oral bioavailability (Li et al. 2015). MSNs loaded with chemotherapeutic drugs can also be formulated so that the use of solvents that are toxic for healthy tissues can be avoided as opposed to other formulations (Lu et al. 2010). Thus, silica nanoparticles exhibit broad applications in the biomedicinal field and are used as drug delivery vehicles, for drug or DNA delivery, carriers of molecules, gene-transfection reagents, for cancer therapy as well as biomarkers, cell markers, and biotracers (Lu et al. 2010; Lee et al. 2014). They can also be explored for the delivery of bioactives such as small-molecule drugs, proteins, enzymes, and peptides (Linden 2018).

The ability to be conjugated with specific ligands or antibodies makes MSNs more

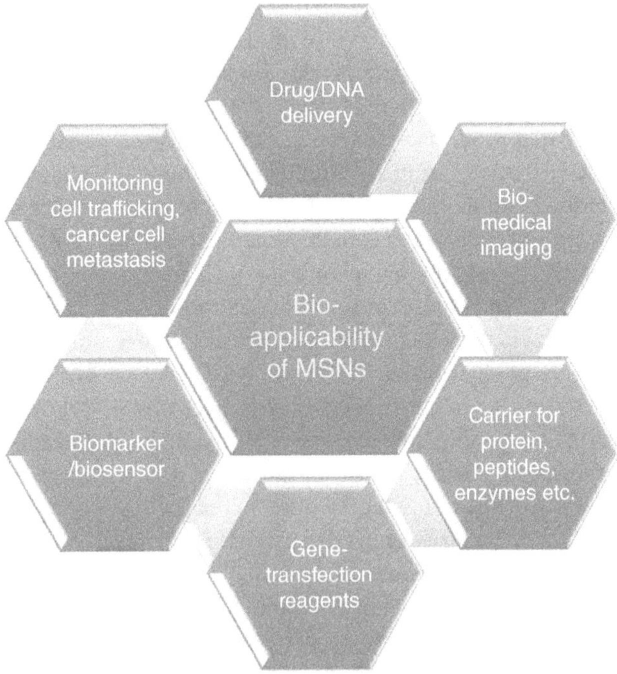

Figure 6.1 Bioapplicability of mesoporous silica nanoparticles (MSNs). *Source:* Based on Huang et al. (2011).

valuable to be used as targeted drug delivery systems especially in cancer. They can also incorporate various molecules into the pores or outside the pores which can respond to external stimuli, and thus generate infinite opportunities for different controllable delivery systems. Moreover, the incorporation of superparamagnetic iron oxide nanocrystals in the core of MSNs explores its application for biomedical imaging (Lu et al. 2010). Few reports have also suggested the application of nonspherical MSNs in the biomedical field such as drug or DNA delivery, for cancer cell metastasis, and monitoring cell trafficking as illustrated in Figure 6.1 (Huang et al. 2011).

6.3 Biodistribution, Toxicity, and Safety of MSN

Silica contains silanol groups on its surface which can interact with the various components of the membrane and cause cell lysis and consequently leakage of components of a cell. This is believed to be the major reason for toxicity caused by the use of silica in the body (Al-Salam et al. 2011; Jafari et al. 2019). On the contrary, it is also observed that MSNs possess a lesser hemolytic effect than nonporous silica particles, which is ascribed to the low density of silanol groups present over the surface of MSNs. Further, it is shown that the surface characteristics of MSNs also have a significant effect on their biocompatibility and biodistribution (Jafari et al. 2019).

To date, the biodistribution reports of only small animals are conveyed and the knowledge about the influencing factors is relatively devastated due to the reasonably limited number of studies. Furthermore, the available literature data are partly indecisive as there is a great difference in experimental design performed among the different biodistribution studies. The broad variation is believed due to differences in particle size, shape, size of mesopore, different ways of detecting particles, surface functionalization, different dissolution rates of particles, protein corona composition, different administered doses, and incorporated animal models (Linden 2018). There are many underlying reasons and challenges for the insufficient data about the biodistribution of MSNs. Generally, most of the studies have focused on the increased activity of MSNs as compared to the free drug rather than *in vivo* biodistribution studies. Additionally, these studies could not conclude on the dissolution kinetics of particles in serum, protein adsorption, etc. Moreover, few parameters related to MSNs are such that it makes it difficult to correlate their physicochemical properties to the long term-biodistribution. These parameters are the kinetic nature of the particle's mechanical stability, the influence of flow, possible changes in polarity of a particle as well as charge with time owing to different particle dissolution inclusive of the detachment of originally covalently attached functional moieties as well as time-dependent evolution of the protein corona composition (Linden 2018). The lack of suitable analytical methods for detecting silica nanoparticles in biological fluids further enhances the difficulty (Malfatti et al. 2012; Lee et al. 2014). While several *in vivo* studies have been published over time, these difficulties make it more difficult to identify conditions for particle design to achieve biodistribution, clearance, and the optimal therapeutic profile of MSNs for a specific application (Linden 2018).

In this context, much of the research work has recently focused on few critical issues such as the biological effects of silica nanoparticles at various levels including their cytotoxicity, biokinetics, compatibility with blood, toxicity on acute and repeated doses, etc. Although concentrating on these parameters, there is still a lack of insight into the distribution, clearance, and pharmacokinetics of silica nanoparticles. These data are important to determine absorption amounts, residence times, and target organs so that their short-term and long-term potential adverse effects can be predicted. The major approaches adopted to determine kinetics of MSNs *in vivo* include the use of dye-conjugated nanoparticles or dye-embedded particles followed by fluorescence detection by microscopy or some other imaging modality. This method does not, however, interpret the exact findings, as the MSN particles are structurally configured such that the surface charge and molecular weight might have changed. These improvements can affect kinetic behavior and biological interactions at the systemic level (Lee et al. 2014).

Investigators have also pointed out that various biological systems are likely to show different responses for MSNs, so that biological research should consist of effective synthetic routes/assay methods and sufficient biological systems so that biocompatibility studies can be carried out (Asefa and Tao 2012). Moreover, it is very essential that the production protocols of synthesis of MSNs and surface functionalization process if any get standardized so that the reproducibility can be assured. The optimized MSNs should also indicate the proper dispersion and stability of the formulation (Manzano and Vallet-Regí 2020).

6.3.1 Factors Affecting Biodistribution of Mesoporous Silica Nanoparticles

6.3.1.1 Particle Size

The particle size of MSNs has great influence in predicting their biodistribution and fate in the biological system as it can influence biological factors such as blood-circulation time, *in vivo* distribution, and clearance.

Waegeneers et al. (2018) had examined the biodistribution and excretion of SiO_2 nanoparticles (SiO_2-NP) (200 nm) in female Sprague–Dawley rats at a dose of 20 mg/kg body weight, and performed autopsy after 6 and 24 hours. The results indicated that the silica nanoparticles were primarily accumulated in the spleen and liver wherein there was a significant reduction in silicon concentration in the spleen between 6 and 24 hours while in the liver the same pattern was found but the effect was insignificant. This is attributed to the clearance of the spleen than the liver via the splenic vein. The clearance from the liver takes more time due to hepatic processing and biliary excretion as well as due to the reabsorption of a fraction of the silica nanoparticles from the gastrointestinal tract. Moreover, a significant rise of macrophages was observed and within the first 24 hours, silicon was found to be excreted primarily through urine.

Lee et al. (2014) reported the effects of colloidal silica nanoparticles having a particle size of 20 and 100 nm on the tissue distribution and excretion after a single oral administration to male and female rats. The authors observed that the nanoparticles were mainly targeted to the liver, kidneys, spleen, and lungs regardless of particle size or sex. The biological fate of nanoparticles was in the particulate form in tissues of the liver while decomposed morphologies of particulates were observed in kidneys representing the *in vivo* degradation of nanoparticles. Further, it was found that the elimination rate of 20 nm nanoparticles was higher than that of 100 nm nanoparticles in urine and feces, indicating size-dependent excretion kinetics which can be attributed to the rapid decomposition of small-sized particles (Lee et al. 2014). Lin et al. (2006) stated that silica nanoparticles of 15 and 46 nm at greater than 10 μg/ml dose for 42–72 hours decreased the viability of cultured bronchoalveolar carcinoma-derived cell. Moreover, the other group of investigators researched that the differences in toxicities of silica particles are because of differences in reactive oxygen species production, which is measured as decreased level of cellular glutathione, lipid peroxidation, and membrane damage (Konecny et al. 2001; Lin et al. 2006). Thus, the effect of particle size on the biodistribution of MSNs might predict the potential toxicological effects on the organs.

Chan et al. (2017) mentioned that smaller particles of silica nanoparticles (SiNPs) might exhibit toxic effects to some cell lines, and larger particles (100 nm size) were genotoxic to the cocultured cells. They concluded that silica nanoparticles of 150 nm sizes were safe for *in vivo* use in mice as they were biocompatible when administered intravascularly (Chan et al. 2017). However, the investigators have also indicated that as the particle size increases, its accumulation in the spleen and liver after intravenous administration also increases (Hudson et al. 2008). This may increase the interaction and toxicity to biological tissues. Lu et al. (2010) believed that MSNs having 100–130 nm size might be able to accumulate in a solid tumor and thus may deliver a higher number of chemotherapeutic drugs to tumors as that of normal tissues for which they used human tumor xenografts in mice. They reported that MSNs demonstrated biocompatibility at the effective dosages, got accumulated in tumor xenografts with or without targeting, and could reduce the toxicity of anticancer drugs. They also get excreted quickly, effectively deliver an anticancer drug to tumors, and exhibit an excellent tumor-suppressing effect. This creates concern about the successful delivery of drugs through MSNs as vehicle and reproducibility of *in vitro* data in an *in vivo* animal study (Lu et al. 2010).

Jafari et al. (2019) reported that spherical MSNs having a particle size of 1220 nm at greater than 25 mg/ml concentration exhibited a degree of toxicity and also hemolytic activity dependent on the size of the particle. They observed that the particles having a size greater than 100 nm indicated low cytotoxicity whereas sub-50 nm particles showed significantly induced necrotic cell death. The MSNs having smaller sizes undergo slow degradation comparatively as they can escape the degradation by the liver and spleen. They also observed that the particle uptake of greater than 1 μm involves phagocytosis wherein specific cells such as macrophages, monocytes, dendritic cells, and neutrophils engulf the particles while the uptake of smaller particles with the size range of <200–300 nm occurs via endocytosis pathway in most of the cases. This uptake mechanism is mediated through various pathways such as receptor-mediated, clathrin, and caveolin-dependent as well as clathrin and caveolin-independent endocytosis based on the cell types and their culture conditions. It is also dependent upon a few structural characteristics of MSNs such as particle size, shape, and surface charge. They confirmed that MSNs having positive charge show less cytotoxicity as compared to bare mesoporous silica nanoparticles (Jafari et al. 2019). Asefa and Tao (2012) also revealed that appropriate surface modification of MSNs might improve the drug uptake by endocytosis or exocytosis while also pacifying their intracellular trafficking (Asefa and Tao 2012). As a result, the smaller particle size implies a longer duration of blood circulation and, with the rise in particle size, the excretion also increases, which may influence biocompatibility due to a difference in its rate of degradation.

6.3.1.2 Particle Shape

In addition to the other particle characteristics, particle shape was also found to affect the *in vivo* biodistribution, cellular uptake, the flow, and adhesion in the vasculature and angiogenesis (Li et al. 2015). Harini et al. (2019) reported that the MSNs indicating non-spherical shape influence drug delivery and cytotoxicity, which further requires *in vivo* data to predict their biodistribution and bioavailability. Li et al. (2015) synthesized MSNs and evaluated the *in vivo* biodistribution and toxicity post oral administration with the different aspect ratios (ARs) of 1, 1.75, and 5. When the acute toxicity of the MSNs with different shapes was evaluated, they observed that as the aspect ratio decreased, the systemic absorption through the small intestine and other organs increased and there was a reduction in urinary excretion. They also cited that nanorods exhibit decreased phagocytosis and extended *in vivo* circulation time which proves that the study of particle shape on biocompatibility and *in vivo* kinetics is essential to improve the performance of nanoparticles (Li et al. 2015).

Huang et al. (2005) have reported based on the animal study that short rod-shaped MSNs distribute mostly in the liver while long rod-shaped MSNs get easily trapped in the spleen and exhibit a slower clearance rate than short rod-shaped MSNs. They observed that the rate of endocytosis is independent of the shape of MSNs as per *in vitro* data. The greater aspect ratio of MSNs can result in a more extended circulation time and thus there may be different biodegradation, biocompatibility, and toxicity profile of nanoparticles (Huang et al. 2005). It is also reported that diverse shaped MSNs having similarities in particle diameter, surface charge, and chemical composition but different aspect ratios (ARs) indicated a significant effect on cellular uptake of nanoparticles (Huang et al. 2011). A similar report has been observed by Niculescu (2020) wherein they also suggested that particle shape does not affect the rate of endocytosis.

The Janus-type hybrid composites based on MSNs having irregular shapes are also reported, which are basically anisotropic nanoarchitectures exhibiting physicochemical characteristics, intrinsic functionalities, and stability attributes which cause improved performances of the overall construct over conventional MSNs. They also overcome the undesired aggregation of MSNs owing to the altered charge densities by incorporation of multiple species either on the surface of the particle or within the mesoporous frameworks (Kankala et al. 2020).

6.3.1.3 Functionalization

The biodistribution of silica nanoparticles is additionally affected by nanoparticle surface properties, porosity, and to a lesser extent by geometry. The investigators have proved that the accumulation of MSNs was higher as compared to the nonporous silica nanoparticles having a similar size. They suggested that primary amine modification could reduce the accumulation which was not possible in high-aspect ratio amine-modified nanorods. The amine modification can also affect the tissue affinity as indicated by MSNs in which they had a high affinity to the lung, while nonporous particles had a high affinity to the liver.

This affinity of silica nanoparticles to the lung and kidneys was reduced by amine modification (Yu et al. 2012).

Linden (2018) reported the effect of the functionalization of MSNs using hydrophilic polymers like polyethylene glycol (PEG) on the biodistribution and clearance of the system. They observed that the longer circulation time of PEGylated MSNs resulted in an increased spleen-to-liver ratio and therefore clearly different circulation times were obtained due to the variable size of these two organs despite comparable particle sizes. The adsorption of protein onto a PEGylated nanoparticle surface is dependent on their surface concentration, the conformation of the PEG-chains, and its packing density. Thus, hydrolytically more stable MSNs having an optimal surface functionalization and particle size of 100–150 nm are expected to indicate extended blood residence times as compared to the corresponding nanoparticles which biodegrade with faster kinetics. Conversely, for effective internalization of MSNs through the target cells, the drug should be retained inside the MSNs which in the case of long-circulating MSNs would take a longer time. Moreover, particles exhibiting a slower disintegration or dissolution kinetics might delay the clearance kinetics and lead to negative side effects. This may further limit the dosing frequency of MSNs. Thus, to enhance the concentration of drug in the target cells, it is preferable to utilize MSNs having short blood circulation time but faster cellular uptake kinetics. The particles of MSNs should show stability in serum which after internalization would undergo fast degradation. The MSNs consisting of redox cleavable S–S bonds as a part of their network structure are preferable from this point of view as this permits the combination of extended blood circulation time and faster particle clearance kinetics from the particular defense organs (Linden 2018).

Desai et al. (2014) reported that surface functionalization had a distinct impact on the magnitude of cellular uptake and degree of serum protein adsorption, whereas the route of uptake or toxicity did not seem to be affected by the nature or charge of the surface coating of MSNs. They also confirmed that regardless of the coating procedure (surface grafting/electrostatic adsorption), the well-known cytotoxicity of large-molecular weight polyethyleneimine (PEI) could be mitigated when used as a component in a hybrid material together with silica, even at very high particle concentrations (Desai et al. 2014). Li et al. (2012) reported that silica nanoparticles or silica-coated nanoparticles permeate the cell without affecting the survival of cells and thus exhibit a good degree of biocompatibility. Kong et al. (2017) prepared lipid-coated biodegradable hollow MSNs for co-delivery of drugs and to study the chemo-immunotherapeutic antitumor effects of nanoparticles. They found that the MSNs indicated exceptional encapsulation efficiency, acceptable stability, favorable biodistribution, and less systemic toxicity. Jafari et al. (2019) stated that if the MSNs are modified on the surface by PEG functionalization, it would help the nanoparticles to escape from degradation by spleen, liver, and lung tissues. This is because PEG-MSNs would have longer circulation time and so they can easily escape from the endocytosis pathway. During the study of uptake of MSNs into Hela cancer cells, the investigators observed that uptake mechanism and efficiency can be controlled by the surface functionalization of nanoparticles. The functionalization of MSNs by functional groups that cells can express related to specific receptors such as folic acid may significantly enhance the cellular uptake of nanoparticles. Further, the relationship between surface charge and internalization of MSNs was also established (Jafari et al. 2019).

The extent of surface adsorptive properties of silica nanoparticles can also affect the toxicity, which can be confirmed by the report suggesting that surface-coated silicas such as PEGylated or PEI-modified crystalline silicas exhibit less toxicity as compared to the uncoated silica nanoparticles (Al-Salam et al. 2011). Di Pasqua et al. (2008) performed the comparison of cytotoxicity of native MSN with MSNs functionalized with thiol or amino groups. They reported that unmodified MSNs were more cytotoxic as compared to the modified MSNs. Moreover, the PEGylation of MSNs results in a significant reduction in their nonspecific attachment to serum proteins and also decreased hemolysis of human red blood cells than native MSNs, which improves the blood compatibility required for intravenous administration of dosage form. The MSNs were functionalized with Fe_3O_4, PEG polymers, or lipid bilayers and administered at a dose of 40 mg/kg to mice intravenously to compare their biodistribution and toxicity to bare Fe_3O_4-MSNs via inductively coupled plasma–mass spectrometry (ICP–MS) analysis. All the prepared MSNs irrespective of their functionalization accumulated in the spleen and liver in which lipid bilayer-coated MSNs accumulated more in the liver and less in the lungs. These nanoparticles indicated faster clearance from the blood than the bare and PEG-coated MSNs (found in the blood after 24 hours of injection). At the administered doses, no MSNs showed toxicity to the kidney, liver, and spleen tissues (Rascol et al. 2018).

The surface functionalization of MSNs also facilitated the delivery of nucleic acid by decorating the nanoparticle surface with positively charged amino groups such as 3-aminopropyl triethoxysilane (APTES). The entry of nucleic acid cargo inside the cells was facilitated by interaction between amino groups on the surface of nanoparticle and negatively charged phosphate groups of nucleic acids (Paris and Vallet-Regí 2020).

6.3.1.4 Dose

The dosage of MSNs affects the biological responses and thus, the appropriate dosage selection is very essential to decrease or eradicate its toxicity. Li et al. (2012) formulated rhodamine B-conjugated MSNs and performed the blood compatibility study through investigation of hemolysis and coagulation behavior at a wide concentration range from 50 to 500 mg/ml under *in vitro* conditions. The authors experimented that rhodamine B-conjugated MSNs indicated good blood compatibility. He et al. (2010) performed the blood compatibility study of SBA-15 (Santa Barbara Amorphous-15) type MSNs and rhodamine B-conjugated MSNs having negative and positive surface potentials respectively to study the hemolysis and coagulation behavior. The results suggested that both nanoparticles exhibited no effect on normal coagulation or anticoagulation characteristics of plasma and hence it was biocompatible. Moreover, even at a higher interaction ratio of 10 mg/ml, it did not indicate any aggregation of blood cells on incubation (He et al. 2010).

Li et al. (2012) disclosed that silica nanoparticles when administered in moderate doses exhibited low toxicity. Though, silica nanoparticles when administered at a dose of 25 μg/ml *in vitro* tend to agglomerate and undergo protein aggregation. This might be attributed to increased reactive oxygen species, lipid peroxidation, and decreased cellular glutathione or GSH level. They further investigated that there was no similarity concerning dose-response, and particle size did not show a significant difference in cytotoxicity. Lu et al. (2010) concluded that at concentrations less than 100 mg/ml, MSNs do not induce any cytotoxicity in different cell lines while at the concentration greater than 200 mg/ml, it shows some growth inhibition. This is in contrast to the report that it does not induce any toxicity if used for a short period during *in vivo* imaging. On the contrary, in one study, when MSNs were administered at a dose of 1.2 g/kg through intravenous or intraperitoneal route, they were lethal to Sv129 mice strain and were safe at a decreased dose of 40 mg/kg, suggesting conducting the *in vivo* study with an optimal dose of the drug (Lu et al. 2010). Repeated doses or high doses of the medication resulted in toxicity and accumulation in a limited number of organs in which no tissue adverse effects were seen when used as a single intravenous injection while serious tissue damage was found at long-term and repeated doses (20 mg/kg thrice). This also includes dysfunction of glomerular filtration, biliary excretion, thrombocytopenia, hepatotoxicity, and animal death but DNA damage was not induced in the spleen, lung, or liver (Waegeneers et al. 2018). The increase in dose of MSNs is also required according to the age group of the people to maintain the same concentration of drug as reported by one study. The investigators indicated that when biodistribution analysis of MSNs was performed, the greater uptake of MSNs was found in elder mice than in young mice (Dos Reis et al. 2020).

These findings suggest that many surface properties of MSNs are responsible for biocompatibility with blood. In addition, though MSNs can incorporate a high dose of drugs, it must be biocompatible with the body tissues and should deliver the drug at a specific target without imparting any kind of toxicity.

6.3.1.5 Route of Administration

After the administration of drug-loaded MSNs, it is very essential to have information about its *in vivo* biodistribution. The MSNs typically get accumulated in reticuloendothelial system organs such as the spleen, liver, and lung after i.v. (intravenous) administration (Linden 2018). The magnetic MSNs exhibited hepatic biocompatibility (Pisani et al. 2017), efficient uptake, and biocompatibility in murine pneumocytes and mesothelioma cells (Blumen et al. 2007). However, MCM-41 and SBA-15 type of MSNs exhibited toxicity in mesothelial cells after intraperitoneal and intravenous injections, causing pulmonary thromboembolic events (Hudson et al. 2008). As opposed to this, Al-Salam et al. (2011) reported that the interactions of calcinated MCM41 and calcinated SBA15 with cell components resulted in no adverse effect on alveolar morphology. Further, despite their relatively large sizes, they appeared in endothelial cells, pneumocytes, fibroblasts, macrophages, and interstitium as well as in different areas of the cytoplasm abundantly. This suggests their intracellular movements without affecting any kind of micro-organelles or cellular configuration and respiration rate (Al-Salam et al. 2011). The biodistribution of MSNs after administration through various routes is to be studied in-depth as it is not identical across the different animal species utilized for *in vivo* studies. After intravenous administration, silica nanoparticles have been shown to reside in the liver and spleen for up to eight weeks despite their excretion through the urinary tract and biliary route (Malfatti et al. 2012; Waegeneers et al. 2018).

It was observed in a research investigation that MSNs coated with PEI polymer (25 kDa) are cytotoxic *in vitro*. The i.v. administration neither induced any adverse effect nor damaged the red cell membrane. The particles do not reside in tissue but they get degraded and excreted (Clemens et al. 2012). Saroj and Rajput (2018) reported the biocompatibility, effectiveness, and biosafety of MSNs for prostate cancer cells by indicating extended blood circulation time and insignificant toxic effect after i.v. administration to major organs. MSNs are also reported to be efficient for breast cancer treatment owing to their higher residence and circulation time in the bloodstream without damaging any healthy tissues, and the capability to release the drug at the target site once reached (Khaleghian et al. 2018). The silica NPs have also shown biocompatibility in the ischemic cardiac muscle for the treatment of coronary artery disease (Galagudza et al. 2010). The inhalation of microscale silica might cause pulmonary disease silicosis in humans but the exceptional characteristics of silica nanoparticles might elicit different biological effects than that of microscale particles. This can only be proved by the *in vivo* investigation of these nanoparticles (Malfatti et al. 2012).

6.4 Safety Evaluation of Mesoporous Silica Nanoparticles

MSNs promise to exhibit good biodistribution and biocompatibility owing to low toxicity. But, the evaluation of toxicity and safety of MSNs is a critical parameter that needs to be investigated systematically before proceeding to clinical use. The critical factors such as protein adsorption to particles, aggregation of particles,

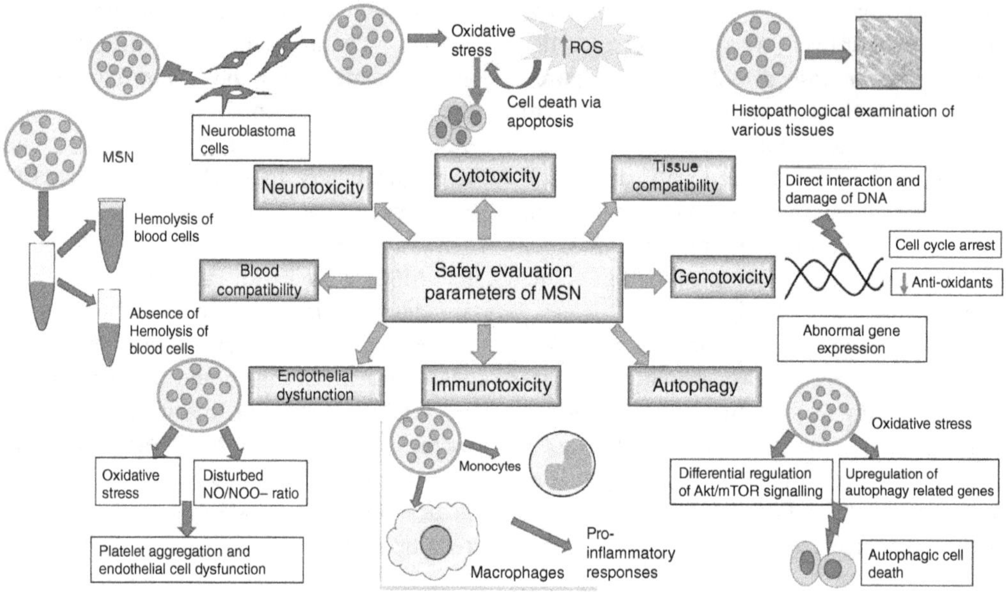

Figure 6.2 Safety evaluation parameters of mesoporous silica nanoparticles (MSNs).

nano biointerface interactions, cellular toxicity, biological behavior, and intracellular particle trafficking are affected by the surface property, and physicochemical properties of MSNs.

The various tests for safety evaluation of MSN are as follows and indicated in Figure 6.2:

1) Cytotoxicity through measurement of oxidative stress.
2) Tissue distribution and compatibility by histological examination.
3) Genotoxicity.
4) Immunotoxicity.
5) Autophagy.
6) Endothelial dysfunction and toxic effects on blood cells.
7) Blood compatibility by hemolysis test.
8) Neurotoxicity.

6.4.1 Cytotoxicity

The nanoparticles induce oxidative stress through increased production of reactive oxygen species, which can damage the cell components and ultimately cause cell death via apoptosis. Thus, it is an important parameter to be considered for nanoparticle toxicity. Murugadoss et al. (2017) investigated the cytotoxicity of silica nanoparticles in different cell lines at different incubation times in which they observed significant cytotoxic effects equal to or greater than 25 μg/ml concentration as compared to untreated cells. They also observed that silica nanoparticles-induced oxidative stress is dependent upon the size and dose of nanoparticles. He and Shi (2011) mentioned that MSNs get internalized easily into most of the normal cells and cancerous cells without apparent toxic effects on growth, differentiation, and proliferation of cells. Though MSNs exhibit negligible cytotoxicity, they have residual toxic surfactants (e.g. cetyl trimethyl ammonium bromide (CTAB)) that remain in the pore channels and show significantly exaggerated cytotoxicity. Therefore, before loading the drug in MSNs, complete removal of surfactant must be assured to lessen the toxicity caused by MSNs (He and Shi 2011).

Kim et al. (2015) also indicated that cytotoxicity of silica nanoparticles occurs because of the generation of reactive oxygen species and changes in membrane integrity due to cellular uptake. The intensity of the toxic effects exerted by these nanoparticles also depends on the type of cell and cell line. Thus, the selection of silica nanoparticles as a drug delivery carrier should be done considering interactions between the organ/site and the nanoparticles. To determine the interaction, an assessment of cytotoxicity should be performed using multiple cell lines from the target organ (Kim et al. 2015).

It was also reported that MSNs synthesized with commercial sodium silicate were found to be compatible with the biological system. This was proved from the absence of cytotoxicity to different cells such as Caco-2, HepG2 or 3T3, HEK-293, and the absence of hemolysis of chicken RBCs and other pathological changes in tissues. The reason suggested was that the MSNs synthesized using sodium silicate comprise a small amount of inorganic salt in their structure, which proved as safe up to 40 mg/kg dose for multiple administration. Further, it was found that these nanoparticles degrade as silicic acid and get excreted within three to four days from the body, suggesting the effectiveness and safety of the drug carrier system (Bhavsar et al. 2019).

A similar finding was reported by the authors that MSNs prepared from sodium silicate as silica source, such as nanoshell-based Fe–SiO_2, C_{16}-L-tryptophan template-based silica, dipalmitoyl phosphatidylcholine template-based silica, were nontoxic and indicated effective disintegration in biological fluid. Further, silica nanoparticle synthesized using cetylpyridinium bromide as template and C_{16}-L-histidine and C16-L-poline as heterocyclic amino acid-based templates were reported to be toxic. The method of synthesis also influenced the toxicity as evident from the report that MSNs prepared by Stober method using cetyltrimethylammonium bromide as template was toxic, while its absence led to nontoxic nanoparticles (Ravinayagam and Jermy 2020).

6.4.2 Tissue Compatibility

It is yet another significant parameter to be measured for MSNs' biosafety, since it is intended for sustained and controlled release, and will therefore be stored at the target site for a longer period of time. Many investigations assess the biosafety of MSNs through histopathological examination in which no abnormality of the tissues such as spleen, liver, heart, kidney, and lung is reported during gross and microscopic examinations after intravenous and intraperitoneal injection. Moreover, it was found that when a high dose of 50 mg/kg/day MCM-41-type or SBA-15-type MSNs consisting of five doses twice per week for 14 days was administered, the acute toxicity of MSNs was not as significant and almost negligible as that of blank control. Further, at an extremely higher dose of 1200 mg/kg given through intravenous or intraperitoneal injection, no mouse survived, which was not observed in the case of subcutaneous injection. Thus, it was concluded that up to 200 mg/kg dose, it was safe to be administered by intraperitoneal or intravenous injection (He and Shi 2011).

6.4.3 Genotoxicity

The toxicity of MSNs in body system through oxidative DNA damage, depletion of antioxidants, direct interaction with DNA, abnormal expression of genes, and cell cycle arrest can be correlated through genotoxicity studies. The genotoxicity induced by colloidal silica nanoparticles and Stober (method of preparation of silica nanoparticles) silica nanoparticles in human tumor cell lines such as lung, skin, kidney, and gastrointestinal system as well as the amplitude of the effect are negatively correlated with the nanoparticle size and generally attributed to the induction of oxidative stress. Such data are very limited for other types of silica nanoparticles such as pyrogenic and precipitated nanoparticles. There might exist other mechanisms for direct DNA damage but such a genotoxic effect is either direct or indirect, and cannot be easily predicted. This is due to insufficient data about the cell uptake and subcellular localization of silica nanoparticles. Moreover, many parameters such as properties of silica nanoparticles, type of cell, and exposure conditions such as assays, endpoints, and concentrations may affect the outcomes that make the comparison of studies difficult. These limitations suggest the need for the standardization of studies related to genotoxicity (Murugadoss et al. 2017).

Yazdimamaghani et al. (2019) suggested the importance of two major classes of parameters that change the genotoxic output of silica nanoparticles. One class includes physicochemical parameters of particles like size, surface area, charge, porosity, agglomeration state, geometry, density, dissolution, and degradability. The second class of parameters include different treatment conditions such as medium, concentration, cell type, exposure time, animal model, dosimetry, and single or consecutive treatments. The authors recommended a few leveraging technologies named organs-on-a-chip cultures, mass-cytometry, and dynamic simulation analysis. Furthermore, the application of next generation RNA sequencing as an emerging technology for high-throughput gene expression quantification was emphasized to study the genotoxicity of nanoparticles, which can reveal the actual mechanism and pathways of genotoxicity and cytotoxicity. To elucidate the mechanism of toxicity, a whole-genome analysis might be useful as it explores transcriptionally controlled processes associated with nanoparticle exposure (Yazdimamaghani et al. 2019).

6.4.4 Immunotoxicity

During the circulation and after the internalization of NPs, they might interact with immune cells such as macrophages and monocytes which are called “innate” cells. Though very limited data is available regarding the immune response of MSNs, the performed *in vitro* study focuses on pro-inflammatory responses and potential phagocytosis specifically which is affected by many parameters of MSNs such as shape and porosity besides particle size (Murugadoss et al. 2017).

Immunotoxicity is induced by the interaction of silica nanoparticles with immunocompetent cells. Chen et al. (2018) observed that the immunotoxicity of silica nanoparticles is dependent upon the physicochemical characteristics of nanoparticles and the type of the cell. The parameters to assess immunotoxicity consist of the determination of cytotoxicity, genotoxicity, and cell dysfunctions. They also reported that the primary mechanisms of immunotoxicity were oxidative stress, autophagy, and pro-inflammatory responses. Further, it can be mitigated by making the surface and shape modifications of nanoparticles. It was also reported that pro-inflammatory responses are an important component of silica nanoparticle-induced immunotoxicity wherein toll-like receptors and NLRP3 (NLR family pyrin domain containing protein 3) inflammasome are believed to be the main signal. Also, the underlying mechanism of immunotoxicity is oxidative stress where the MAPK (mitogen-activated protein kinase) pathway and NADPH (nicotinamide adenine dinucleotide phosphate) oxidase pathways play an important role.

6.4.5 Autophagy

Autophagy is not a much explored and investigated parameter of NP toxicity. However, it is very crucial to be performed for assessing safety as it is associated with mechanisms of cytotoxicity and genotoxicity. Autophagy is believed to be a cellular defense mechanism against the toxicity of nanoparticles as it significantly removes the misfolded proteins and clears the damaged organelles. The Stober silica nanoparticles-induced autophagy occurs through oxidative stress-mediated upregulation of autophagy-related genes and differential regulation of Akt (protein kinase B)/mTOR (mammalian target of rapamycin) signaling. As observed in cytotoxicity, the lowest exposure concentration at which silica nanoparticles showed remarkable effects was 25 μg/ml. Moreover, induced autophagy resulted in severe damage to cells and autophagic cell death when exposed to silica nanoparticles. In addition to the induction of autophagy, lysosomal dysfunction could be a probable mechanism of toxicity of silica nanoparticles, which has, however, not been investigated for silica nanoparticles (Murugadoss et al. 2017). This finding is also supported by the reports of Chen et al. (2018) that silica nanoparticles when entering hemocytes initiate autophagy and apoptosis and so autophagy dysfunction is considered as a potential toxic mechanism caused by silica nanoparticles.

6.4.6 Endothelial Dysfunction and Toxic Effects on Blood Cells

Murugadoss et al. (2017) suggested that when nanoparticles are ingested or inhaled, they can translocate across barriers such as the air–blood of the body, and once they enter into the circulation, they may interact with the cardiovascular system. In this way, endothelial cells and platelets together perform a significant function to maintain the vascular homeostasis. It was also reported that oxidative stress induced by colloidal silica nanoparticles and disturbed NO/NOO^- (nitric oxide/peroxynitrite) ratio might result in platelet aggregation as well as endothelial dysfunction. This eventually may lead to pro-inflammatory signals expressed through the secretion of adhesion molecules and cytokines and, thus, results in atherosclerosis and vascular thrombosis. Moreover, hemolysis of RBCs caused by silica nanoparticles occurs in a size-, porosity-, and charge-dependent way. Such types of study must be performed for other types of NPs for determining their safety.

6.4.7 Blood Compatibility

The MSNs are mainly administered by the intravenous route for the effective targeting of drugs to the targeted sites; so, they must be proved safe and compatible with blood and blood components and also should indicate compatibility with various parameters such as blood platelet number and function, hemolysis, blood corpuscle number, coagulation, nonspecific protein binding, immunoreaction, phagocytosis, leucocytic function, and enzymatic activity. It was also reported that at a dose of 50–500 mg/ml of SBA-15 type MSNs, they cannot activate the extrinsic, intrinsic, and common coagulation pathways and hemolysis activity was almost imperceptible at the concentrations of less than 100 mg/ml. However, MSNs exhibited significant hemolysis at high concentrations of 500 mg/ml while PEGylation of MSNs having an optimal molecular weight of PEG (10 k) and the corresponding optimal PEG chain density (0.75 wt%) significantly reduced the hemolysis of cells (He and Shi 2011).

6.4.8 Neurotoxicity

Given the limited evidence available, the value of neurotoxicity cannot be overlooked as small NPs prepared for targeting CNS cells are capable of translocation across the blood–brain barrier. It was reported that *in vitro* studies involving neuroblastoma cells indicate possible adverse effects of silica nanoparticles (Murugadoss et al. 2017).

The safety evaluation tests carried out by a few researchers are listed in Table 6.1. This indicates that most of the work is focused on carrying out cell line and hemolysis studies. Besides these, other related findings should also be carried out including an *in vivo* study for establishing the safety profile of MSNs.

6.5 Conclusion and Future Directions

The various literature reports discussed suggest that MSNs among the various established drug delivery systems unwrap the new aspect of drug delivery in the biomedical field. Owing to its exceptional properties, its applications not only confine to drug delivery but also extend to the field of bioimaging and theranostics possessed by very few drug delivery systems. Much of the research work has been carried to demonstrate the biosafety and to determine the biodistribution of MSNs in the body system, though extensive research is still required to bring about the concrete results by overcoming the various differentiating factors. This can be accomplished by conducting multiple and elaborative toxicity studies that

Table 6.1 Safety evaluation tests for mesoporous silica nanoparticles (MSNs).

Sr. No.	Safety evaluation tests of MSN	Type of MSNs	Drug	Relevant outcome of the study	References
1	Cytotoxicity	Chitosan-capped pH responsive mesoporous silica nanoparticles	Raloxifene hydrochloride	Indicated concentration-dependent and time-dependent toxicity toward MCF-7 breast carcinoma cells with least off-site side effects	Shah and Rajput (2018)
2		MSN for increasing the oral bioavailability and permeation of poorly water-soluble drugs	Telmisartan	MSNs were nontoxic even at a high concentration of 500 μg/ml when studied on Caco-2 cell line	Zhang et al. (2012)
3		PEG-PEI-coated spherical and rod-shaped MSNs across the blood–brain barrier	[a]Not applicable	No considerable toxic effect was observed on MDCK II and RBE4 cells by MSNs at the concentration less than or equal to 50 μg/ml	Baghirov et al. (2016)
4		Redox-responsive MSN for targeted breast cancer therapy	Curcumin	No toxicity to MDA-MB-231 cells was observed at the concentrations of 5–400 μg/ml after treatment for 48 h	Li et al. (2018)
5		Sterically stabilized polymeric MSN	Doxorubicin	There were no cytotoxic responses induced with treatments of DOX-loaded MSNs in normal HEK293 cells, suggesting potential selective targeting to cancer cells	Moodley and Singh (2020)
6		Calcinated MSNs synthesized from liquid crystal	Methotrexate and camptothecin	MSNs caused no significant change compared to the control group in human lung fibroblast (WI-38) and human ovarian cancer (ES-2) cells	Lin et al. (2013)
7		Biodegradable hollow MSN for regulating tumor microenvironment and enhancing antitumor efficiency	All-trans retinoic acid (ATRA), doxorubicin (DOX), and interleukin-2 (IL-2)	ATRA/DOX/IL-2 co-loaded dHMSN coated with lipid bilayer demonstrated significant tumor growth and metastasis inhibition when studied on L929 fibrosarcoma cells, and also exhibited favorable biodegradability and safety	Kong et al. (2017)
8	Tissue distribution and compatibility	Effect of larger silica nanoparticles	[a]Not applicable	The brain, heart, lungs, liver, spleen, and kidneys, were harvested and no abnormality was observed	Chan et al. (2017)
9		MSN as an oral drug delivery system	Curcumin	No significant macroscopic and microscopic changes were observed by curcumin-loaded MSNs	Hadisoewignyo et al. (2018)
10		Antitumor effect of 131 I-Labeled Anti-VEGFR2 Targeted MSNs in Anaplastic Thyroid Cancer	Penicillin G, streptomycin, and nystatin	131 I-Labeled Anti-VEGFR2 targeted nanoparticles showed no significant pathologic changes in the vital organs in the tumor-bearing mice	Zhang et al. (2019)

11		Redox-responsive MSN for targeted breast cancer therapy	Curcumin	The MSNs modified with hyaluronan (HA) and polyethyleneimine-folic acid (PEI-FA) exhibited its distribution in the kidneys and liver along with their high accumulation in the tumor	Li et al. (2018)
12		Biodegradable hollow MSNs for regulating tumor microenvironment and enhancing antitumor efficiency	All-trans retinoic acid (ATRA), doxorubicin (DOX), and interleukin-2 (IL-2)	The dHMSN coated with lipid bilayer indicated the major accumulation of MSN in liver at 6, 24, or 48 h, and only less than 3% of the total injected dose was presented in the heart, spleen, lung, and kidney	Kong et al. (2017)
13		Nonporous silica nanoparticles for ocular topical administration and oral administration	[a]Not applicable	The histopathological examinations revealed no target organ damage	Kim et al. (2017)
14	Immunotoxicity	Mesoporous silica particles for pulmonary drug delivery	Dexamethasone	Empty MSNs had no apparent effect on inflammation in MEL (melphalan)-induced inflammation model while, in the LPS (lipopolysaccharide)-induced inflammation model, inhaled particles significantly down-modulated the inflammatory response regardless of the presence of drug	Gulin-Sarfraz et al. (2019)
15		Functionalized mesoporous silica nanoparticles for targeted drug delivery	[a]Not applicable	MSN-NH_2 nanoparticles were nontoxic to primary murine leukocytes and provoked only trace immune activation. In addition, surface functionalized MSN-PVP served as a pH-triggered drug delivery tool for the synthetic TLR7/8 ligand R848 to induce potent immune activation in responder cells	Heidegger et al. (2016)
16		Calcinated MSN synthesized from liquid crystal	Methotrexate and camptothecin	After exposure to MSNs at 1–10 μg/ml, there were no significant differences compared to the blank in lactate dehydrogenase release as an indicator of inflammation	Lin et al. (2013)
17		Biodegradable hollow MSN for regulating tumor microenvironment and enhancing antitumor efficiency	All-trans retinoic acid (ATRA), doxorubicin (DOX), and interleukin-2 (IL-2)	Indicated enhanced antitumor immunity by synergistic effect of drug-loaded and lipid-coated MSN with immunosuppressive cells depletion and immune effector cells activation in tumor microenvironment and thus dHMSN worked as an excellent vehicle as well as a potential immunoregulatory agent	Kong et al. (2017)

(*Continued*)

Table 6.1 (Continued)

Sr. No.	Safety evaluation tests of MSN	Type of MSNs	Drug	Relevant outcome of the study	References
18	Toxic effects on blood cells and endothelial dysfunction	Nonporous silica nanoparticles for ocular topical administration and oral administration	[a]Not applicable	Blood cell count as well as hemoglobin and hematocrit values, differential count of white blood cells etc. were measured which indicated no toxicity.	Kim et al. (2017)
19		Larger silica nanoparticles	[a]Not applicable	Complete blood counts were measured and no obvious difference was found as compared to control group which indicated the absence of toxicity	Chan et al. (2017)
20	Blood compatibility by hemolysis test	Chitosan-capped pH responsive mesoporous silica nanoparticles	Raloxifene hydrochloride	The formulation indicated <5% hemolysis which proves it to be hemocompatible	Shah and Rajput (2018)
21		Redox-responsive MSN for targeted breast cancer therapy	Curcumin	The hemolysis ratio of MSN was observed <5%, indicating good hemocompatibility of MSN	Li et al. (2018)
22		Calcinated MSNs synthesized from liquid crystal	Methotrexate and camptothecin	The hemolysis of RBCs was detected at about 3–4% at the dose of 5 and 10 μg/ml of MSNs suggesting its hemocompatibility	Lin et al. (2013)
23		Biodegradable hollow MSN for regulating tumor microenvironment and enhancing antitumor efficiency	All-trans retinoic acid (ATRA), doxorubicin (DOX), and interleukin-2 (IL-2)	Hemolysis assay demonstrated that dHMSN and dHMLB were not likely to induce hemolysis	Kong et al. (2017)

[a] Not applicable: The toxicity study was performed on mesoporous silica nanoparticles without the incorporation of drug.

dHMLB, lipid-coated biodegradable hollow mesoporous silica nanoparticle; dHMSN, biodegradable hollow mesoporous silica nanoparticle; PEG, polyethylene glycol; PEI, polyethyleneimine; RBCs, red blood cells; TLR7/8, toll-like receptors (TLRs) 7 and 8; VEGFR2, vascular endothelial growth factor receptor 2.

generate the reproducible results of the safety of MSNs and thus establishes the reliance on its biomedical applications. Moreover, all the studies are lacking in one or other aspect of establishing its IVIVC (*in vitro–in vivo* correlation), and hence there is a strong need for the research that establishes the impeccable information about the biodistribution, toxicity, and safety of MSNs so as to lead it to the clinical applications.

References

Al-Salam, S., Balhaj, G., Al-Hammadi, S. et al. (2011). *In vitro* study and biocompatibility of calcined mesoporous silica microparticles in mouse lung. *Toxicological Sciences* 122 (1): 86–99.

Asefa, T. and Tao, Z. (2012). Biocompatibility of mesoporous silica nanoparticles. *Chemical Research in Toxicology* 25 (11): 2265–2284.

Baghirov, H., Karaman, D., Viitala, T. et al. (2016). Feasibility study of the permeability and uptake of mesoporous silica nanoparticles across the blood-brain barrier. *PLoS One* 11 (8): e0160705.

Bhavsar, D., Patel, V., and Sawant, K. (2019). Systematic investigation of in vitro and in vivo safety, toxicity and degradation of mesoporous silica nanoparticles synthesized using commercial sodium silicate. *Microporous and Mesoporous Materials* 284: 343–352.

Blumen, S.R., Cheng, K., Ramos-Nino, M.E. et al. (2007). Unique uptake of acid-prepared mesoporous spheres by lung epithelial and mesothelioma cells. *American Journal of Respiratory Cell and Molecular Biology* 36 (3): 333–342.

Chan, W.T., Liu, C.C., Chiau, J.S.C. et al. (2017). In vivo toxicologic study of larger silica nanoparticles in mice. *International Journal of Nanomedicine* 12: 3421.

Chen, L., Liu, J., Zhang, Y. et al. (2018). The toxicity of silica nanoparticles to the immune system. *Nanomedicine* 13 (15): 1939–1962.

Clemens, D.L., Lee, B.Y., Xue, M. et al. (2012). Targeted intracellular delivery of antituberculosis drugs to Mycobacterium tuberculosis-infected macrophages via functionalized mesoporous silica nanoparticles. *Antimicrobial Agents and Chemotherapy* 56 (5): 2535–2545.

Desai, D., Karaman, D.S., Prabhakar, N. et al. (2014). Design considerations for mesoporous silica nanoparticulate systems in facilitating biomedical applications. *Open Material Sciences* 1 (open-issue): 16–43.

Di Pasqua, A.J., Sharma, K.K., Shi, Y.L. et al. (2008). Cytotoxicity of mesoporous silica nanomaterials. *Journal of Inorganic Biochemistry* 102 (7): 1416–1423.

Dos Reis, S.R.R., Pinto, S.R., de Menezes, F.D. et al. (2020). Senescence and the impact on biodistribution of different nanosystems: the discrepancy on tissue deposition of graphene quantum dots, polycaprolactone nanoparticle and magnetic mesoporous silica nanoparticles in young and elder animals. *Pharmaceutical Research* 37 (3): 40.

Douroumis, D., Onyesom, I., Maniruzzaman, M., and Mitchell, J. (2013). Mesoporous silica nanoparticles in nanotechnology. *Critical Reviews in Biotechnology* 33 (3): 229–245.

Galagudza, M.M., Korolev, D.V., Sonin, D.L. et al. (2010). Targeted drug delivery into reversibly injured myocardium with silica nanoparticles: surface functionalization, natural biodistribution, and acute toxicity. *International Journal of Nanomedicine* 5: 231.

Gulin-Sarfraz, T., Jonasson, S., Wigenstam, E. et al. (2019). Feasibility study of mesoporous silica particles for pulmonary drug delivery: therapeutic treatment with dexamethasone in a mouse model of airway inflammation. *Pharmaceutics* 11 (4): 149.

Hadisoewignyo, L., Hartono, S.B., Kresnamurti, A. et al. (2018). Evaluation of anti-inflammatory activity and biocompatibility of curcumin loaded mesoporous silica nanoparticles as an oral drug delivery system. *Advances in Natural Sciences: Nanoscience and Nanotechnology* 9 (3): 035007.

Harini, L., Srivastava, S., Gnanakumar, G.P. et al. (2019). An ingenious non-spherical mesoporous silica nanoparticle cargo with curcumin induces mitochondria-mediated apoptosis in breast cancer (MCF-7) cells. *Oncotarget* 10 (11): 1193.

He, Q. and Shi, J. (2011). Mesoporous silica nanoparticle based nano drug delivery systems: synthesis, controlled drug release and delivery, pharmacokinetics and biocompatibility. *Journal of Materials Chemistry* 21 (16): 5845–5855.

He, Q., Zhang, J., Chen, F. et al. (2010). An anti-ROS/hepatic fibrosis drug delivery system based on salvianolic acid B loaded mesoporous silica nanoparticles. *Biomaterials* 31 (30): 7785–7796.

Heidegger, S., Gößl, D., Schmidt, A. et al. (2016). Immune response to functionalized mesoporous silica nanoparticles for targeted drug delivery. *Nanoscale* 8 (2): 938–948.

Huang, D.M., Hung, Y., Ko, B.S. et al. (2005). Highly efficient cellular labeling of mesoporous nanoparticles in human mesenchymal stem cells: implication for stem cell tracking. *The FASEB Journal* 19 (14): 2014–2016.

Huang, X., Li, L., Liu, T. et al. (2011). The shape effect of mesoporous silica nanoparticles on biodistribution, clearance, and biocompatibility in vivo. *ACS Nano* 5 (7): 5390–5399.

Hudson, S.P., Padera, R.F., Langer, R., and Kohane, D.S. (2008). The biocompatibility of mesoporous silicates. *Biomaterials* 29 (30): 4045–4055.

Jafari, S., Derakhshankhah, H., Alaei, L. et al. (2019). Mesoporous silica nanoparticles for therapeutic/diagnostic applications. *Biomedicine and Pharmacotherapy* 109: 1100–1111.

Kankala, R.K., Han, Y.H., Na, J. et al. (2020). Nanoarchitectured structure and surface biofunctionality of mesoporous silica nanoparticles. *Advanced Materials* 32 (23): 1907035.

Khaleghian, A., Hasanzadeh, H., Moshtaghian, A. et al. (2018). Inhibitory effects of silica nanoparticles loaded with hematoporphyrin on breast cancer cell line. *Middle East Journal of Rehabilitation and Health Studies* 5 (3): 1–6.

Kim, I.Y., Joachim, E., Choi, H., and Kim, K. (2015). Toxicity of silica nanoparticles depends on size, dose, and cell type. *Nanomedicine: Nanotechnology, Biology and Medicine, vol.* 11 (6): 1407–1416.

Kim, M., Park, J.H., Jeong, H. et al. (2017). An evaluation of the in vivo safety of nonporous silica nanoparticles: ocular topical administration versus oral administration. *Scientific Reports* 7 (1): 1–8.

Konecny, R., Leonard, S., Shi, X. et al. (2001). Reactivity of free radicals on hydroxylated quartz surface and its implications for pathogenicity of silicas: experimental and quantum mechanical study. *Journal of Environmental Pathology, Toxicology and Oncology* 20 (Suppl. 1): 119–132.

Kong, M., Tang, J., Qiao, Q. et al. (2017). Biodegradable hollow mesoporous silica nanoparticles for regulating tumor microenvironment and enhancing antitumor efficiency. *Theranostics* 7 (13): 3276.

Lee, J.A., Kim, M.K., Paek, H.J. et al. (2014). Tissue distribution and excretion kinetics of orally administered silica nanoparticles in rats. *International Journal of Nanomedicine* 9 (Suppl 2): 251.

Li, X., Wang, L., Fan, Y. et al. (2012). Biocompatibility and toxicity of nanoparticles and nanotubes. *Journal of Nanomaterials* 2012: 1–19.

Li, L., Liu, T., Fu, C. et al. (2015). Biodistribution, excretion, and toxicity of mesoporous silica nanoparticles after oral administration depend on their shape. *Nanomedicine: Nanotechnology, Biology and Medicine, vol.* 11 (8): 1915–1924.

Li, N., Wang, Z., Zhang, Y. et al. (2018). Curcumin-loaded redox-responsive mesoporous silica nanoparticles for targeted breast cancer therapy. *Artificial Cells, Nanomedicine, and Biotechnology* 46 (Suppl 2): 921–935.

Lin, W., Huang, Y.W., Zhou, X.D., and Ma, Y. (2006). In vitro toxicity of silica nanoparticles in human lung cancer cells. *Toxicology and Applied Pharmacology* 217 (3): 252–259.

Lin, Y.C., Lin, L.Y., Gao, M.Y., and Fang, Y.P. (2013). Mesoporous silica nanoparticles synthesized from liquid crystal display manufacturing extracts as a potential candidate for a drug delivery carrier: evaluation of their safety and biocompatibility. *International Journal of Nanomedicine* 8: 3833.

Linden, M. (2018). Biodistribution and excretion of intravenously injected mesoporous silica nanoparticles: implications for drug delivery efficiency and safety. *In The Enzymes* 43: 155–180.

Lu, J., Liong, M., Li, Z. et al. (2010). Biocompatibility, biodistribution, and drug delivery efficiency of mesoporous silica nanoparticles for cancer therapy in animals. *Small* 6 (16): 1794–1805.

Malfatti, M.A., Palko, H.A., Kuhn, E.A., and Turteltaub, K.W. (2012). Determining the pharmacokinetics and long-term biodistribution of SiO_2 nanoparticles in vivo using accelerator mass spectrometry. *Nano Letters* 12 (11): 5532–5538.

Manzano, M. and Vallet-Regí, M. (2020). Mesoporous silica nanoparticles for drug delivery. *Advanced Functional Materials* 30 (2): 1902634.

Moodley, T. and Singh, M. (2020). Sterically stabilised polymeric mesoporous silica nanoparticles improve doxorubicin efficiency: tailored cancer therapy. *Molecules* 25 (3): 742.

Murugadoss, S., Lison, D., Godderis, L. et al. (2017). Toxicology of silica nanoparticles: an update. *Archives of Toxicology* 91 (9): 2967–3010.

Niculescu, V.C. (2020). Mesoporous silica nanoparticles for bio-applications. *Frontiers in Materials* 7: 36.

Paris, J.L. and Vallet-Regí, M. (2020). Mesoporous silica nanoparticles for co-delivery of drugs and nucleic acids in oncology: a review. *Pharmaceutics* 12 (6): 526.

Pisani, C., Rascol, E., Dorandeu, C. et al. (2017). Biocompatibility assessment of functionalized magnetic mesoporous silica nanoparticles in human HepaRG cells. *Nanotoxicology* 11 (7): 871–890.

Rascol, E., Pisani, C., Dorandeu, C. et al. (2018). Biosafety of mesoporous silica nanoparticles. *Biomimetics* 3 (3): 22.

Ravinayagam, V. and Jermy, B.R. (2020). Nanomaterials and their negative effects on human health. *Applications of Nanomaterials in Human Health*: 249–273.

Saroj, S. and Rajput, S.J. (2018). Etoposide encased folic acid adorned mesoporous silica nanoparticles as potent nanovehicles for enhanced prostate cancer therapy: synthesis, characterization, cellular uptake and biodistribution. *Artificial Cells, Nanomedicine, and Biotechnology* 46 (Supp 3): S1115–S1130.

Shah, P.V. and Rajput, S.J. (2018). Facile synthesis of chitosan capped mesoporous silica nanoparticles: a pH responsive smart delivery platform for raloxifene hydrochloride. *AAPS PharmSciTech* 19 (3): 1344–1357.

Vallet-Regi, M., Ramila, A., Del Real, R.P., and Pérez-Pariente, J. (2001). A new property of MCM-41: drug delivery system. *Chemistry of Materials* 13 (2): 308–311.

Waegeneers, N., Brasseur, A., Van Doren, E. et al. (2018). Short-term biodistribution and clearance of intravenously administered silica nanoparticles. *Toxicology Reports* 5: 632–638.

Yang, K.N., Zhang, C.Q., Wang, W. et al. (2014). pH-responsive mesoporous silica nanoparticles employed in controlled drug delivery systems for cancer treatment. *Cancer Biology and Medicine* 11 (1): 34.

Yazdimamaghani, M., Moos, P.J., Dobrovolskaia, M.A., and Ghandehari, H. (2019). Genotoxicity of amorphous silica nanoparticles: status and prospects. *Nanomedicine: Nanotechnology, Biology and Medicine* 16: 106–125.

Yu, T., Hubbard, D., Ray, A., and Ghandehari, H. (2012). In vivo biodistribution and pharmacokinetics of silica nanoparticles as a function of geometry, porosity and surface characteristics. *Journal of Controlled Release* 163 (1): 46–54.

Zhang, Y., Wang, J., Bai, X. et al. (2012). Mesoporous silica nanoparticles for increasing the oral bioavailability and permeation of poorly water soluble drugs. *Molecular Pharmaceutics* 9 (3): 505–513.

Zhang, R., Zhang, Y., Tan, J. et al. (2019). Antitumor effect of 131 I-labeled anti-VEGFR2 targeted mesoporous silica nanoparticles in anaplastic thyroid cancer. *Nanoscale Research Letters* 14 (1): 96.

7

Safety and Toxicity issues of Polymeric Nanoparticles

A Serious Concern

Priya Patel, Naimish Vyas, and Mihir Raval

Department of Pharmaceutical Sciences, Saurashtra University, Rajkot, GJ, India

7.1 Introduction

Nanotechnology has been the most popular field of research in the last decades. Since "nanotechnology" was presented by Nobel laureate Richard P.Feynman throughout his popular lecture in 1959 "There's Plenty of Room at the bottom," there have been made various revolutionary growth in the field of nanotechnology. Nanotechnology produces materials of various types at the nanoscale. Nanoparticles (NPs) are a wide class of materials that comprise particulate substances, which have at least one dimension less than 100 nm (Khan et al. 2017). European and other global committees have defined a nanoparticle as a separate entity that has three dimensions in the order of 100 nm or less (Moreno Vega et al. 2012). Nanomaterials have been mostly used in biomedical approaches due to their specific chemical and physical properties, which change the normal biological action, as compared to bulk substances (Hajipour et al. 2012).

Nanosystems have a much more specific exterior or interface area, i.e. a larger area-to-mass ratio, than coarser materials. For round-shaped nanoparticles, the specific surface area increases inversely with the diameter. From purely geometrical considerations, as an example, 10 g of silver in the form of spherical nanoparticles with a diameter of 10 nm exhibits a total surface area of almost 600 m^2. This can be compared to a single solid silver sphere with the same mass which has a surface area of nearly 5 cm^2, giving an increase in total surface area for the nanomaterial form of a factor of about 1 200 000. A powder with grains of a diameter of 50 μm (similar to cement powder) would only have a surface area increase of a factor of less than 250 concerning the single sphere. Since biological and chemical reactions often take place at the surface of materials, one expects nanomaterials to be much more reactive than the same mass of material made up of coarser structures. In addition, as a consequence of their small size, nanoparticles may migrate more easily in biological systems like the human body, and be able to cross biological barriers in the lung, gut, or the brain, and therefore cause unexpected and unusual exposure. Furthermore, there are intrinsic nanoscale properties that result from the confinement of atoms and electrons within the boundaries of a few nanometres. These effects are most dominant at sizes below a few tens of nanometers (less than about 30 nm). They can considerably change fundamental physical material characteristics like the optical,

Nanotechnology in Medicine: Toxicity and Safety, First Edition. Edited by Mahendra Rai, Mrunali Patel, and Rashmin Patel.

electrical, and magnetic properties of the nanomaterial (Lovestam et al. 2010).

Nanocarriers can protect the payload from degradation and allow sustained and controlled drug release. Furthermore, nanoparticles have the ability to decrease the clearance and increase the accumulation of drugs in diseased tissue, thus increasing therapeutic effectiveness and minimizing side effects. Many hurdles in the body must be crossed in order for a drug to reach its intended location (Ferrari 2010). These barriers include enzymatic and mechanical degradation, immunological clearance, renal clearance, vascular endothelium, extracellular matrix, cell membrane, lysosome, and membrane pumps. The properties of nanoparticles can, in turn, be designed to overcome such barriers. In particular, nanoplatforms allow a multifaceted approach, where it is possible to combine different materials and compartments to effectively combat the challenges of targeted distribution. For instance, the multistage vector that releases cargo sequentially is well prepared to handle the different *in vivo* conditions and compartments encountered by systemic injection (Godin et al. 2011).

Taken together, these benefits signify that nanoparticles may be used effectively to battle multiple diseases. Since the field of nanomedicine shows great promise, the production of safety tests that can reliably predict the possible toxicity of nanotherapeutics is also imperative. In addition, since nanoparticles exhibit distinct and specific properties that cannot be predicted from the bulk material analysis, adequate assays should be taken for the assessment of nanoparticle toxicity. Nevertheless, it may not be necessary to establish stricter guidelines for the approval of nanoparticles, as compared to small-molecule drugs. Instead, the strategies for determining safety could be different in some situations. Moreover, since nanoparticles are solely characterized by size parameters and contain a large number of particles of different compositions and morphology, it is difficult to make general claims as to the protection or toxicity of nano-sized objects (Wolfram et al. 2005).

7.2 Application of Nanomaterials

By using nanosystems, gradually consumer products may be made cleaner, stronger, lighter, less expensive, more systematic, more valuable, or pure. That means the quality of life can be improved by using nanomaterials that are better-organized target-driven pharmaceuticals, better medical diagnosis tools, for cleaner energy production, etc. Several of the consumer end products available nowadays that utilize nanomaterials have been developed from existing products, for example by the incorporation of nanomaterials into solid, viscous, or liquid materials. Examples of areas and end products comprising nanomaterials (Lovestam et al. 2010) are:

- Intracellular targeting
- Ocular drug delivery
- Treatment of leprosy
- Brain-targeted drug delivery
- Lymph targeting
- Lipoidal carrier targeting
- Cosmetics and personal care products
- Paints and coatings
- Household products
- Catalysts and lubricants
- Sports products
- Textiles
- Food and nutritional ingredients
- Food packaging
- Agrochemicals
- Veterinary medicines
- Construction materials
- Weapons and explosives
- Consumer electronics

7.3 Classification of Nanoparticles (NPs)

1) **Polymeric nanoparticles:** These are usually organic-based nanoparticles. A particular term polymer-based polymeric nanoparticle (PNP) is collectively used for

it. They are nanosphere- or nanocapsule-shaped mostly. The nanospheres are matrix molecules whose overall bulk is normally solid and the further particles are adsorbed at the exterior boundary of the spherical surface. In the case of nanocapsular particles, the solid bulk is encapsulated within the particle completely.

2) **Lipid-based nanoparticles:** These nanoparticles comprise lipid moieties and are effectively utilized in many biomedical approaches. Generally, lipid-based nanoparticles are characteristically round-shaped with a diameter from 10 to 1000 nm. Like polymeric nanoparticles, lipid nanoparticles have a solid core made of lipid and a matrix contains soluble lipophilic molecules. Surfactants or emulsifiers stabilize the outer core of these nanoparticles. Lipid nanotechnology is a special field, which focuses on the synthesis and design of lipid nanoparticles for different approaches such as drug carriers and delivery and RNA release in cancer treatment (Khan et al. 2017).
3) **Carbon-based nanoparticles:** Fullerenes and carbon nanotubes (CNTs) constitute two major classes of carbon-based NPs. Fullerenes contain nanoparticles that are made of globular hollow cages such as allotropic forms of carbon. They have generated noteworthy trading interest also in nanocomposites for numerous commercial applications such as fillers, efficient gas adsorbents for environmental remediation, and as a support medium for specific organic and inorganic catalysts.
4) **Metal-based nanoparticles:** Metal-based nanoparticles are entirely prepared with metal precursors. Due to popular localized surface plasmon resonance (LSPR) characteristics, these nanomaterials possess ideal optoelectrical properties. Nanoparticles of the alkali and noble metals, for example, Cu, Ag, and Au have a wide absorption band in the visible area of the electromagnetic solar spectrum. The surface-, size-, and shape-controlled synthesis of metal nanoparticles is important in present-day cutting-edge materials. Due to their advanced optical properties, metal-based nanoparticles find applications in numerous research areas. Gold nanoparticles coating is mostly used for the sampling SEM, to increase the electronic stream, which assists in obtaining high-quality SEM photographs.

7.3.1 Polymeric Nanoparticles

Polymers are composed of macromolecules which are a substantial number of repeating units organized in a chain-like molecular architecture exhibiting a multiplicity of compositions, properties, and structures. It is because of this variety of compositions, structures, and properties that polymers are utilized in nanosystems to create nanoparticles suited for each specific biomedical approach. The principal use of polymeric nanoparticles is in drug delivery, although they are also utilized in biosensing and bioimaging assays (Jain et al. 2013). The use of nanosystems in drug delivery has received a lot of attention due to the emerging signs of targeted delivery in medicine. Thus, enormous research studies are focused on preparing polymeric nanosystems that are efficient, tissue-specific, and most importantly, nontoxic. For the preparation of nanomaterial for drug delivery, there are a variety of procedures that depend on the drug loading onto the nanosystem. The resulting nanoparticle-drug compounds may have the structure of capsules (polymeric nanomaterial or polymeric nanoconjugates), amphiphilic shell/core (polymeric micelles), or hyperbranched macromolecules of nanometer dimensions (dendrimers) (Cho et al. 2008). Polymeric nanomaterials are solid colloidal systems with a size in the range of 10–1000 nm and are made of biocompatible and biodegradable polymers or copolymers, in which the drug can be encapsulated or entrapped within the carrier, physically adsorbed on the surface of the carrier, or chemically linked to the surface. These nanocarriers are known for their features such as biodegradability, small size, nontoxicity, water solubility, long shelf life, and stability during storage. These characteristics make them attractive for the delivery of drugs, proteins, and DNA or genes to specific targeted tissues or organs. Thus, they are applied in anticancer therapy, crossing the BBB, vaccine and gene therapy, diagnosis, and a lot of other uses in the medical field. The major function of polymeric nanoparticles is to deliver the pharmaceutical agent to a specific site of action, achieving a higher concentration of drug and increasing the stability of volatile drugs, effectiveness, and efficiency of the drug (Yadav et al. 2019). Polymeric nanoparticles are promising vehicles for drug delivery by easy manipulation to produce carriers with the objective of delivering the drugs to a specific target; such a benefit improves drug safety (Ai et al. 2011). Polymer-based nanosystems effectively carry drugs, proteins, and DNA to target organs and cells. Their nanometer size promotes effective permeation through cell membranes and stability in the bloodstream. Polymers are very convenient materials for the manufacture of countless and varied molecular designs that can be integrated into unique nanoparticle constructs with many potential medical applications (Peer et al. 2007).

7.3.1.1 Advantages of Polymeric Nanoparticles

The advantages exhibited by polymeric nanoparticles (Nagavarma et al. 2012) are as stated below:

1) They increase the stability of any volatile pharmaceutical substances, are easily and cheaply fabricated in large quantities by a multitude of techniques.
2) They offer a notable improvement over conventional oral and intravenous techniques of administration in way of effectiveness and efficiency.
3) They deliver a higher concentration of pharmaceutical substances to a specific location.
4) The choice of polymer and the ability to modify drug release from polymeric nanoparticles have made them perfect candidates for cancer treatment, delivery of vaccines, contraceptives, and delivery of targeted antibiotics.
5) Polymeric nanomaterials can be easily incorporated into other activities corresponding to drug delivery, such as tissue engineering technology.

7.3.1.2 Polymers Used in the Preparation of Polymeric Nanoparticles

The polymers should be compatible with the body in the terms of nontoxicity and nonantigenicity and should be biocompatible and biodegradable. The various polymers utilized are as follows:

Natural polymers: The most commonly used natural polymers in preparation of polymeric nanoparticles are:

Chitosan: Chitosan is utilized in several pharmaceutical formulations and specifically in cosmetic preparations. The performance and suitability of chitosan as a component of pharmaceutical dosage forms for drug delivery applications have been investigated in numerous researches (Nagavarma et al. 2012).

Gelatin: Gelatin is mostly utilized in various pharmaceutical dosage forms, including its use as biodegradable matrix substances in an implantable delivery system, although it is commonly used to form either hard or soft gelatin capsules in various dosage forms.

Sodium alginate: Sodium alginate is used in the preparation of sustained-release oral dosage forms since it can delay the dissolution of a drug from tablets, capsules, and aqueous suspensions. Sodium alginate is utilized in various topical and oral pharmaceutical dosage forms as a binder, disintegrating agent, and also as a diluent in capsule formulations.

Albumin: Albumin is mainly used as an excipient in parenteral pharmaceutical dosage forms, where it is used as a stabilizing agent for dosage forms containing proteins and enzymes. It is also used in the preparation of microcapsules and microspheres for experimental drug-delivery systems.

Synthetic polymers: There are many synthetic polymers utilized such as:

Polyorthoesters: Polyorthoesters are utilized as hydrophobic implant substances for drug depots for continuous drug delivery by surface erosion.

Polycaprolactone: Polycaprolactone (PCL) is mostly used in long-term controlled drug release approaches and implants. However, when it comes to tissue engineering techniques, it suffers from shortcomings such as slow degradation rate process, low cell adhesion, and poor mechanical properties.

Poly glutamic acid: Poly-γ-Glutamic acid (PGA) is a naturally occurring anionic polymer composed of extensively viscous homo-polyamide of D and L-Glutamic acid units. Its multifunctionality, biodegradability, nontoxicity, compatibility, and edibility have made it a promising biopolymer for the food, cosmetics, and pharmaceutical industries.

Poly malic acid: Poly(β-L-malic acid) (PMA) is a biodegradable polymer with many potential biomedical applications. PMA can be readily hydrolyzed to malic acid (MA), which is widely used as an acidulant in foods and pharmaceuticals.

The polymeric drug carrier systems deliver the drug at the tissue site by any one of the three general physicochemical mechanisms: (i) by the swelling of the polymer nanomaterials due to hydration followed by release through diffusion system; (ii) by an enzymatic reaction resulting in rupture or cleavage or degradation of the polymer at the site of delivery, thereby releasing the drug from the entrapped inner core; (iii) dissociation of the drug from the polymer and its de-adsorption/release from the swelled nanoparticles (Gaur et al. 2000).

7.3.1.3 Methods of Preparation of Polymeric Nanoparticles

1) **Dialysis method**: In this methodology, a dialysis bag is used to incorporate the mixture of the drug and polymer in an organic solvent system (Figure 7.1). The dialysis methodology requires 36 hours for efficient and accurate drug loading. To overcome this problem, mainly utilize a more cost-effective technology, which will dissolve the polymer and the drug, followed by a redispersion in a proper solvent system. For example, the preparation of a morin hydrate (MH) polymeric nanomaterial system is done by the dialysis method. The preparation begins by using 15 mg of hyaluronic acid poly(butyl cyanoacrylate) block copolymer, which dissolved in 3 ml of phosphate-buffered saline at a specific pH 7.4; then, a mixture of morin hydrate (MH) in 0.5 ml dimethyl sulfoxide is added to the previous polymer suspension with repetitive stirring at 150 rpm at 25 °C. After this, a probe-type ultrasonicator is used to ultrasonicate the final mixture for 30 minutes in an ice bath. Then, the solution is dialyzed against an excess amount of DW (distilled water) for 12 hours by dialysis bag followed by lyophilization and filtration (Abbad et al. 2015).
2) **Solvent evaporation method**: In this type of method, drug and polymer are properly mixed in a volatile solvent and form an emulsion system (Figure 7.2). In this method, ethyl acetate is mostly used as a polymer, due to its lower toxicological profile as compared to other solvents. Then, by evaporating, the solvent alters the emulsion system to a suspension system. Solid particles can be collected by ultracentrifugation followed by washing to eliminate additives and impurities. Lyophilization was performed to obtained dried nanoparticles. Specifically, the polymer in acetone solution is added to a stirred aqueous solution of poly (vinyl alcohol). This mixture is kept under continuous stirring until the

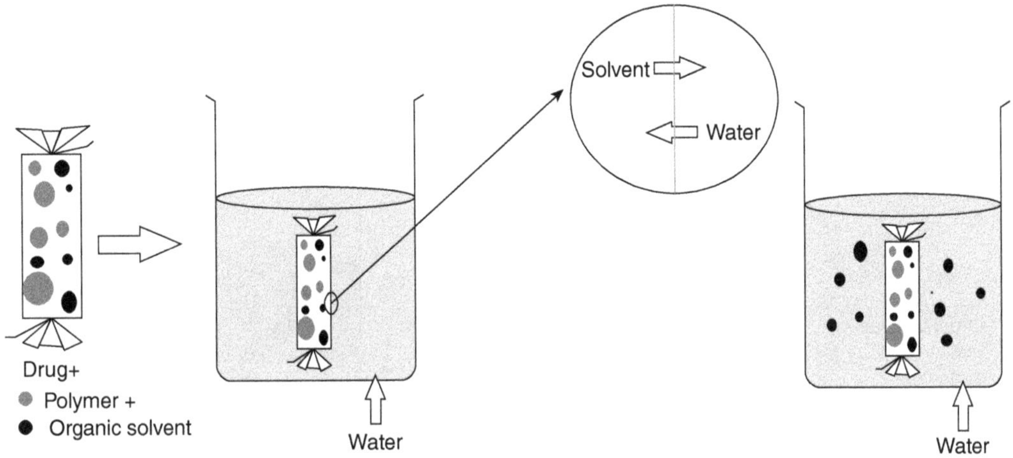

Figure 7.1 Dialysis method.

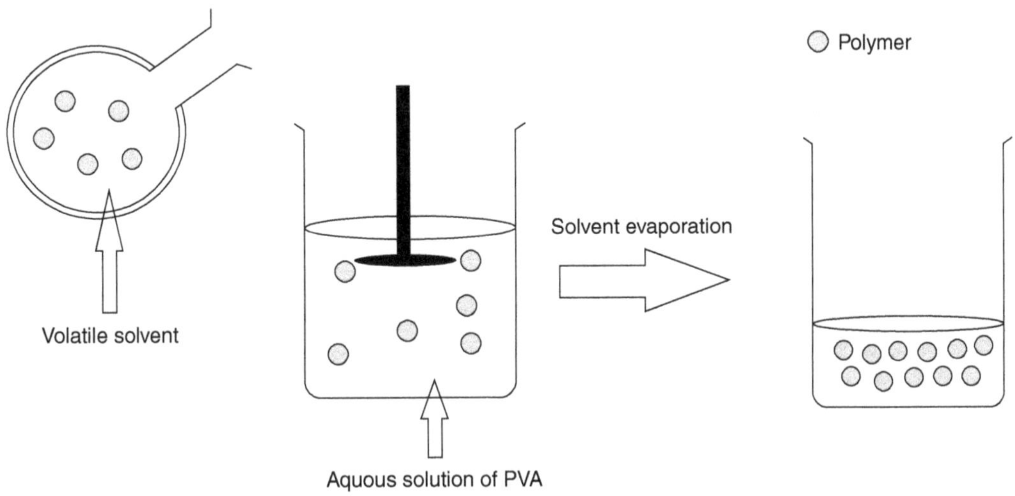

Figure 7.2 Solvent evaporation method.

evaporation of acetone and nanoparticles dispersion formed is compressed in a rotary evaporator and filtered. At the end of the process, the nanomaterials are collected by centrifugation for 25 minutes and washed two times with water before lyophilization (Ge et al. 2012a).

3) **Nanoprecipitation**: Solvent displacement is the synonym of nanoprecipitation, which is the precipitation of the polymer in an organic solvent and diffusion of the organic solvent in an aqueous medium, this technique appears with or without the presence of a surfactant (Figure 7.3). Poly(lactic acid), poly(ε-caprolactone), and PLGA are a few polymers that are utilized in this procedure for preparing polymeric nanoparticles, by dissolving them in a water-miscible solvent system. The drawbacks of this simple method are related to the water-miscible solvent, leading to solution instability due to the rapid rate of diffusion. The formed droplets having a high coalescence rate to avoid spontaneous emulsification. Acetone

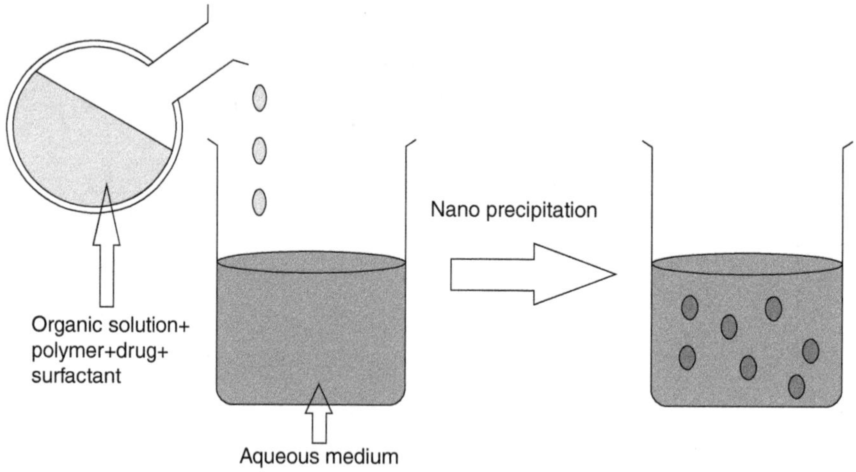

Figure 7.3 Nanoprecipitation method.

or dichloromethane is used in this method mainly when preparing lipophilic drugs, as they increase the entrapment and the particle size of the drug. For example, Bifonazole and Clotrimazole are antifungal drugs that are prepared using the nanoprecipitation method (Nagavarma et al. 2012).

4) **Interfacial polymerization**: This is one of the key methods for the formation of polymeric nanoparticles. Two reactive materials are available, which were dispersed in two different phases, such as dispersed and continuous phase, whereby they become polymerized. The process occurs at the interface or between the two different phases. A major use of this method is the formulation of oil-containing nanocapsules system by polymerization. In interfacial polymerization, some aprotic solvents are used, such as ethanol, acetonitrile, and *n*-butanol (Nagavarma et al. 2012).

7.3.1.4 Polymeric Nanoparticles in Drug Delivery

Polymeric nanoparticles represent one of the most innovative and noninvasive approaches for drug delivery. Surface-modified and Stealth nanosystems can be used for passive and active delivery of the drugs. Diseases of the CNS (central nervous system), like Alzheimer's disease and brain tumors, are very hard to diagnose, due to the BBB, which locate between the brain and the blood, hence prevents 100% of macromolecules and 98% of small particles from entering the brain. To solve this problem Researchers developed a several kind of strategies to temporarily open the BBB, including penetrating the BBB and also prepare drug molecules, which are capable of passing through the BBB to the brain. Nanoparticles can be used as a carrier system to directly deliver the therapeutic agent to the diseased cells in the brain. In brain tumors, the BBB is concentrated in the tumor; so, nanomaterials carrying the drug can reach the tumor simply and directly from the blood circulation system. There are major concerns that should be kept in mind when utilizing brain-targeted delivery systems, such as potential neurotoxicity of the nanosystems; the homogeneity of brain-targeted drug delivery; API loss along the way to the target; the protein corona, which may disturb the targeted delivery; and the off-target potential of brain-targeted delivery. Various medidinal application of polymeric nanoparticles are shown in Figure 7.4.

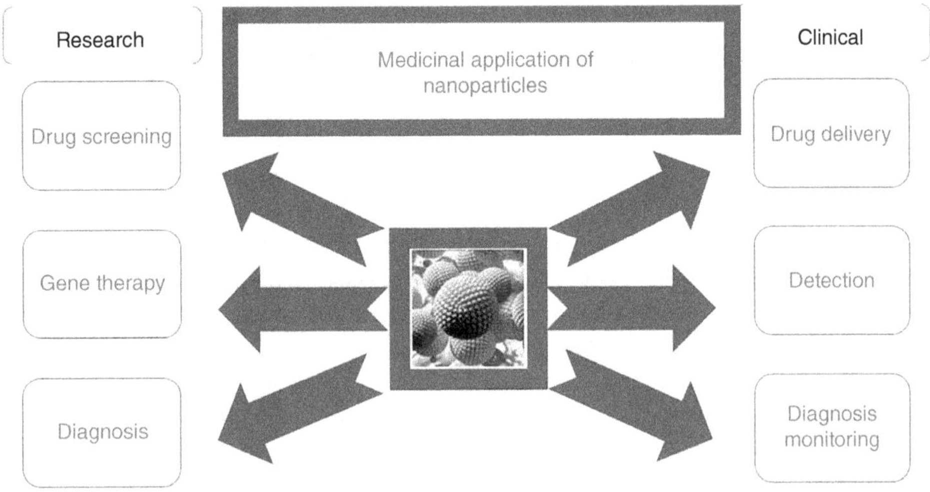

Figure 7.4 Application of polymeric nanoparticles.

7.4 Nanotoxicology

Nanotoxicology is a branch of bio-nano science, which deals with the research and application of the toxicity of nanosystems. Nanomaterials, even when made of inert elements such as gold, become highly active at the nanometer dimensions. Nanotoxicological studies are used to determine whether and to what extent these properties may pose a threat to the environment and to human health. Nanomaterials play a remarkable role in toxicity study, which is important for toxicologists, especially in respiratory-related diseases. Their size is an important element in the happening of disease. Some researches worked on the different sizes of titanium oxide and carbon and showed that reduction in nanomaterial size increases its toxicity in the lungs. Also noticeable is that the mixing of some metals with each other causes complex toxicity, which does not occur with single metals. In 1975, research showed the effect of oxidative stress caused by asbestos as the main part of asbestosis and also in disrupting cell structure. In 1998, Zhang presented work on the effects of nanomaterials on respiratory inflammation and toxicity. Some of the particle features such as surface chemistry, size, and oxidative stress functions play major roles in nanotoxicity. By gaining control over dangerous particle system, it can increase the utilization of nanomaterials by decreasing their harmfulness, thus allowing them to be used in the curing of various diseases (Bogart et al. 2014).

7.4.1 Toxicity of Nanoparticles

Knowledge of the toxicity of these small materials is limited but is growing rapidly. Many researches have shown that some nanomaterials display toxicological effects in biological systems. Thus, research in the external and internal environment is needed. Some researchers have shown that most of the nanosystems can release active oxygen and cause oxidative stress and inflammation by the reticuloendothelial system. In one study, acute toxicological effect from nanomaterials has been explored in the mouth of rats. The results showed that the toxicity depends on the coating, size, and chemical components of the nanomaterials. Also, the systemic effects of nanoparticles have been indicated in different tissue and organs. The effects on immunological and inflammatory systems may include pro-inflammatory cytotoxic or oxidative stress activity in the liver, lung, brain, and heart. The effects on the circulatory system can include prethrombosis effects and paradox effects on heart function. Carcinogenicity, genotoxicity, and teratogenicity may occur as a result on the effects of nanosystems (Akbarzadeh et al. 2013).

Due to the high loading of nanomaterials, macromolecule absorption will increase, so that they can cross the digestive tract. Because, for example, lectin is such an immunologic substance for coating, it can be toxigenic and also cause inflammatory reactions or digestive stimulation (Jafar et al. 2011). The harmful properties of certain nanomaterials do not necessarily hinder them from being utilized for medical uses. Since nanomaterials that have the ability and biocompatibility to deliver drugs to a target tissue are generally considered to be the optimal. Namely, the toxic properties of nanomaterials could be directly harnessed to ablate diseased tissue, thereby eliminating the need for a drug component. Nevertheless, this approach would need a selective targeting of nanomaterials, in order to spare healthy tissues from damage. Additionally, the harmful effects of pristine nanomaterials can be reduced using several approaches, such as surface modification (Kumari et al. 2010).

7.4.2 Tissue Toxicity of Nanomedicine

It is foreseeable that nanoparticle-induced damage elicited upon cells and endogenous biomolecules could affect the function of entire organs. There are some examples in the literature of incidents where nanoparticles have caused harm to tissues. Such damage can be traced back to molecular and cellular defects triggered by nanoparticle exposure. In general, the organs that are affected the most are the ones that have the highest levels of nanoparticle accumulation, e.g. the lungs following intratracheal instillation and the liver following intravenous injection. For instance, the deposition of polymers in the lungs resulted in lung inflammation. Moreover, positively charged lipid nanoparticles injected into the blood were found to cause hepatotoxicity, which was evident from the increased presence of liver enzymes in the blood. The suggested mechanism for cytotoxicity was nanoparticle-mediated activation of the immune system through toll-like receptor signaling (Rubin 1975; Wolfram et al. 2015).

Nanoparticles have the potential to revolutionize medical imaging, diagnostics, thera peutics as well as perform functional biological processes because of their small size and physical similarity to physiological molecules such as proteins. But their toxicity can underlie these features. Also, toxicity can vary in severity depending on the mode of administration and sites of deposition. Therefore, to maintain clinical relevance, information on toxicity is presented using a system-based approach focusing on experimental lung and dermal target (Table 7.1)

7.4.3 Mechanisms of Nanoparticle Toxicity

The toxicity of certain nanoparticles can be manifested at the molecular, cellular, and tissue level. Indeed, as nanoparticles move through the body, they can be exposed to different biological microenvironments, including the blood, the extracellular matrix, the cytoplasm, and cellular organelles. Consequently, the interactions that occur at the nano-bio interface may impact the function of biomolecules, cellular components, and tissue structures (Wolfram et al. 2015).

7.4.4 Toxicity of Polymeric Nanoparticles

Nondegradable nanoparticles create toxicity in the system because they cannot degrade in the body and remain unchanged, causing toxicological effects in the body.

Polybutylcyanoacrylate nanoparticles (PBCA NPs) are candidates for a drug delivery system, which can cross the blood–brain

Table 7.1 Summary of *in vivo* evaluations of nanoparticle toxicity based on organ.

Target	Nanoparticle	Conjugation	Concentration (time/size)/route of administration	Cellular target	Major outcomes	References
Lung	Poly(lactic-co glycolic acid) nanoparticles	Chitosan	300–5000 μg/ml (4 h)	A549 human lung cancer cells	Nontoxic even at highest concentrations	Tahara et al. (2009)
Dermal	Silica nanoparticles	Polyethylene glycol	30–300 μg/ml	Cultured human keratinocytes	Reduced cell viability and no irritation	Park et al. (2010)

barrier (BBB). PBCA nanoparticles were fabricated with different surfactant systems according to the mini-emulsion process. Viabilities of the HeLa and HEK293 cells after nanomaterial incubation were quantified by analyzing cellular metabolic activity (MTT-test). Then, intravenous rhodamine-labeled PBCA nanoparticles were repeatedly injected into the rats through intravenous route and monitored for survival as well as morphology of retrogradely labeled neurons by *in vivo* confocal neuroimaging (ICON) for five weeks.

Polymers such as Abraxane are utilized in the clinical treatment of metastatic breast cancer and the treatment of many other types of cancers, such as non-small-cell lung cancer and used in brain targeting when functionalized for specific or nonspecific binding mechanisms. Poly(L-glutamic acid) is a polymer that has been used in conjugate synthesis and also tested for representing chemotherapeutics that are used mostly in the clinic, which showed amazing abilities and potential to circumvent the shortcomings of their free drug counterparts. *N*-(2-hydroxypropyl) methacrylamide and PEG are widely used polymeric nanomaterials and are used as anticancer agents. They basically have shown huge success in tests of chemotherapy and/or radiation (Yadav et al. 2019).

There are certain categories of nanoparticles that have frequently been reported to have cytotoxic effects. The hazardous properties of certain nanoparticles do not necessarily hinder them from being used for medical purposes. Although a nanoparticle that simultaneously displays biocompatibility and the ability to deliver drugs to a tissue of interest is usually considered optimal, harmful nanoparticles could also be utilized. Namely, the toxic properties of nanoparticles could be directly harnessed to ablate diseased tissue, thereby eliminating the need for a drug component. Nevertheless, this approach would require selective targeting of nanoparticles, in order to spare healthy tissue from damage. Additionally, the harmful effects of pristine nanoparticles can be reduced using several approaches, such as surface modification. As an illustration, the addition of hydroxyl groups to gadolinium fullerene particles prevented the generation of ROS, thereby reducing toxicity. Similarly, a polymer coating on the surface of iron oxide nanoparticles dramatically improved cell viability (Jafar et al. 2011). Various other toxicity studies of polymeric nanoparticles are depicted in Table 7.2.

7.4.5 Nanoformulations Showing Toxicity

Polyglutamic acid-conjugated (polyglumex) paclitaxel is marketed as Xyotax® (Opaxio) and there has been fairly large clinical practice with this drug. Results in NSCLC (non-small-cell lung carcinoma) were disappointing. Single-agent Xyotax® only marginally reduced toxicity and did not significantly increase

Table 7.2 Review of toxicity study of polymeric nanoparticles.

Nanomaterial	Polymer characterization	Testing method	Model	Dose	Results	References
Chitosan nanoparticles	Chitosan hydrochloride salt	Acute toxicity *in vivo* method	New Zealand Rabbit	30 μl of the 0.5 mg/ml nanoparticles formulations in the right eye at every 30 min for 6 h	No sign of discomfort after the administration and no changes seen in histopathological study	De Salamanca et al. (2006)
Eudragit® S100 alginate enclosed chitosan-calcium phosphate-loaded lactoferrin nanocapsules	Eudragit, Chitosan	*in vivo* exposure (acute toxicity: 24 h)	Artemia salina (brine shrimp)	20–5000 μg/ml	No toxicity	Leng et al. (2018)
Chitosan-coated polylactide glycolic acid nanoparticles	75 : 25 Resomer® RG756 and Protasan® UP CL113, 75–90% deacetylation, 50–150 kDa	*in vivo* exposure	Balb/cJ mice	250 μg/50 μl in 5% glucose	Coated polylactide glycolic acid nanoparticles did not induce an inflammatory response in mice, with no alterations of cellular population, protein quantity, or expression of cytokines in BAL	Santiago et al. (2015)

survival compared to Taxol. The combination with carboplatin resulted in more rates of grade III/VI hematologic and neurologic toxicity with no improvements in survival or disease control. In contrast, early-stage study trials in the setting of fallopian and ovarian tube cancers have been completed and show promising therapeutic profiles. In phase III study, Taxol was compared with Xyotax® as maintenance therapy for patients with ovarian cancer who had complete responses after platinum therapy and taxane (Richards et al. 2005).

Currently, research is also ongoing into the use of polyglumex paclitaxel for radiosensitizer. Chemotherapies such as cisplatinum and 5-FU (5-Fluorouracil) are mostly combined with radiation to sensitize tumor cells to the toxic effects of radiation. Unfortunately, with standard drug delivery, normal tissues are also sensitized to the effects of ionizing radiation, resulting in significant improvement in toxicities. Nanoformulation has the potential to significantly increase the therapeutic index of chemoradiotherapy by increasing tumor-specific delivery, which should improve efficacy and concomitantly decrease normal tissue toxicity. Polyglumex paclitaxel has shown potential as a radiosensitizer with cisplatin for esophageal cancer and in combination with temozolomide for high-grade gliomas (Caster et al. 2017).

Amikacin is a strong aminoglycoside antibiotic which is useful for the treatment of multidrug-resistant Gram-negative bacteria as well as resistant strains of *Mycobacterium tuberculosis*. Amikacin is also limited by a relatively short half-life of two to three hours. It is specifically reserved for the treatment of severe infections and requires frequent blood level monitoring. Arikayce is a liposomal preparation of amikacin designed as an inhaled medication for the treatment of drug-resistant pseudomonal infections in patients with cystic fibrosis. Arikayce shows a remarkable increasing the drug half-life and in a phase II trial, there was no significant difference in toxicity between the placebo and liposomal drug treatment (Clancy et al. 2013).

Vivagel (Starpharma Inc.) is a dendrimeric antiviral/antibiotic compound which is currently under clinical trials. This novel dendrimer shows naphthalene disulphate surface groups and displays potent *in vitro* activity against herpes simplex virus (HSV) and human immunodeficiency virus (HIV). Several phase I studies have been done and indicated that vaginal use is well tolerated and that cervicovaginal fluids retain antiviral activity up to 24 hours after administration in the majority of patients. Vivagel is available as a lubricant in contraceptives in Australian markets, and additional studies as an antiviral agent are ongoing. Vivagel has also displayed activity against the common polymicrobial infection bacterial vaginosis (BV). Phase II study suggested high rates of pathogenic cure (cleared infection on laboratory testing) and clinical cure (symptomatic improvement). However, phase III studies were less clear, as they observed high rates of symptomatic improvement but lower rates of clinical cure (Caster et al. 2017).

7.5 Safety Assessment of Nanomedicines by Methodological Considerations

When testing the cytotoxicity of nanomaterials *in vitro*, it is important to select assays that are acceptable for such approaches. The majority of available safety tests have been developed for conventional therapeutic agents. However, due to their distinct size and surface properties, nanomaterials behave differently than small materials. In particular, nano-sized objects can sediment, aggregate, and show different diffusion dynamics. Therefore, nanomaterials may not be suitable for test conditions or they may interfere with test results (Wolfram et al. 2015).

The earliest nanotherapeutics were approved based on similar efficacy, but lower toxicity than their free-drug counterparts. Doxil, which is a liposomal doxorubicin formulation, was the first nanomedicine to obtain clinical approval. In 1995, Doxil was approved for AIDS-related Kaposi sarcoma by the US Food and Drug Administration (FDA) and has since been approved for other cancers, such as multiple myeloma (Barenholz 2012). The main advantage of Doxil in comparison to free-doxorubicin is reduced cardiotoxicity. In essence, by enhancing drug accumulation in diseased tissue, nanoparticles can cause fewer side effects, thus reducing the dosage needed to achieve therapeutic efficacy. As an example, less than 0.01% of the injected dose of agents usually accumulates in the target region in the angstrom size range (e.g. antibodies) (Li et al. 2004), while the same value is approximately 1–5% for nanoparticles (Bae and Park 2011). The enhanced permeability and retention (EPR) effect is the main mechanism for increased deposition of nanoparticles in tumour tissue. Although small molecules can freely pass through any tissue's vasculature, nanoparticles' movement is more restrictive. The EPR effect occurs primarily because of variations between tumour vasculature and normal tissue (Matsumura and Maeda 1986; Maeda et al. 2000).

Another mechanism by which nanoparticles can reduce drug toxicity is associated with the administration of hydrophobic therapeutics. Namely, nanoparticles can serve as an alternative to toxic solubilizing agents (Celia et al. 2013; Norvaisas and Ziemys 2014; Wolfram et al. 2015). Indeed, to enable the delivery of non-water-soluble drugs, harmful

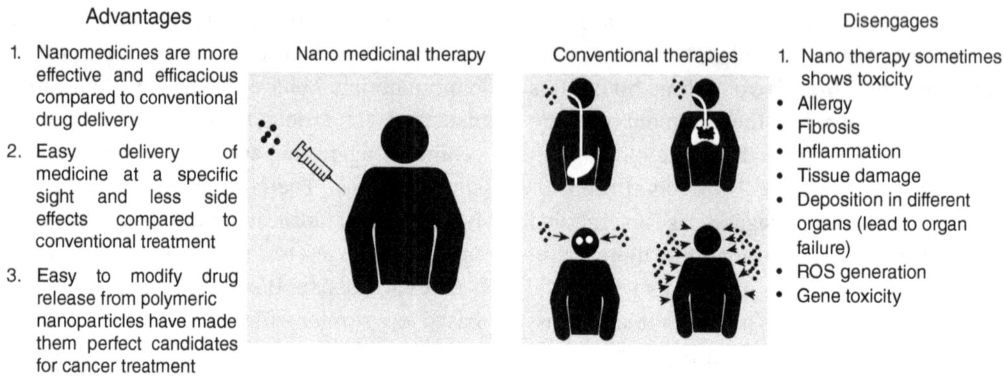

Figure 7.5 Safety concern of polymeric nanoparticles.

solvents have been used extensively *in vitro* and *in vivo*. Dimethyl sulfoxide (DMSO), for example, is typically the first cell culture agent of choice (Yang et al. 2012, 2013). However, DMSO has displayed toxicity in multiple studies (Ruiz-Delgado et al. 2009). In addition, toxic agents are used to formulate some clinically approved hydrophobic medicines. Examples include paclitaxel and docetaxel, which are treated with a dehydrated ethanol mixture and polyethoxylated castor oil (Cremophor EL) (Gelderblom et al. 2001). To relieve side effects resulting from these solubilizing agents, patients may need treatment with antihistamines and corticosteroids (Francis et al. 1994). Consequently, nanoparticles are a safer choice for the administration of poorly water-soluble drugs. For example, the nanotherapeutic drug Abraxane, consisting of protein-bound paclitaxel, was approved for pretreated metastatic breast cancer by the FDA in 2005 and has since also been approved for the treatment of other cancers (Miele et al. 2009). Accordingly, higher doses of Abraxane in comparison to Chremophor El-based paclitaxel can be tolerated by patients (Ibrahim et al. 2002). Sonaje et al. performed toxicity studies for nanoparticles made of poly-g-glutamic acid and chitosan in the presence of $MgSO_4$ and sodium tripolyphosphate. The nanoparticles were well tolerated after oral administrations for 14 days, lacking toxicity even when the administrated dose was 18 times higher than the dose used for pharmacodynamic/pharmacokinetic studies. Neither significant differences in clinical signs nor body weight were observed between the experimental and control groups. No relevant pathological changes were observed in the kidney, liver, and intestine and no inflammatory reactions were also observed. In addition, the toxicity of decanoic acid-grafted oligo chitosan nanoparticles has also been assessed in rats (Sonaje et al. 2009). Overall, we can summarize that nanoparticles have some safety concerns as compared to conventional therapy (Figure 7.5).

7.5.1 Nanoparticles Safety Study

Polymeric nanoparticle characterization is very important, not only in terms of its applicability but also to ascertain issues concerning nanotoxicology (Zielinska et al. 2020a). Risks associated with the use of nanoparticles are based on their possible toxicity and living cell interaction (Silva et al. 2019). The potential changes (chemical, physical) that the medium can induce in nanoparticles and the degradation that affects their bioavailability and *in vivo* behavior should also be taken into account (Bundschuh et al. 2018; Gupta and Xie 2018). The methods used to track the toxicity of nanoparticles need to be able to detect very low concentrations of toxicological biomarkers in order to ensure the accuracy of the results obtained from the toxicity assessment and should avoid the possible intervention of other compounds in the sample. For the elaboration of the appropriate analytical procedure, several elements should be considered, namely:

Sample treatment: A sampling of nanoparticle formulation and the laboratory procedures may change the state of dispersion. Due to the unavailability of sufficiently sensitive portable equipment, it is not possible to identify fluctuations *in situ* (Mourdikoudis et al. 2018).

Separation of nanoparticles: It is often required to submit the samples to pre-fractionation by centrifugation or filtration in order to remove unwanted particles (Robertson et al. 2016). For denser particles, centrifugation is a more effective process, whereas microfiltration is commonly used because of its simplicity. By collision or by electrostatic attraction, nanoparticles are deposited on a membrane. In relation to their diffusion coefficient, field flow fractionation may also be used to separate particles according to their size. Other efficient methods for separating and purifying nanoparticles by their size are size exclusion chromatography and capillary electrophoresis (Hanauer et al. 2007).

7.5.2 Methodological Considerations for Safety Assessment of Nanomedicines

It is important to select assays that are suitable for such applications when evaluating the cytotoxicity of nanoparticles *in vitro*. For conventional therapeutic agents, the majority of available safety tests have been established. Owing to their distinct size and surface properties, nanoparticles, however, behave differently than small molecules. Nano-sized objects, in particular, may accumulate, sediment, and exhibit distinct dynamics of diffusion (Zielinska et al. 2020b).

Nanoparticles may either be unsuitable for test conditions, or they may interfere with the results of the experiments. The Ames test, for instance, which uses bacteria to assess compound mutagenicity, is not especially suitable for nanoparticles. One research found that in the Ames test, nanofibers exhibiting genotoxicity in a mammalian assay failed to do so (Clift et al. 2013). However, nanofibers were found within the bacteria in the aforementioned case, implying that other factors are responsible for the decreased sensitivity of the Ames test to nanoparticles. In addition, in the MTT (3-(4,5-dimethylthiazol-2-yl)-2,5-diphenyltetra zolium bromide) assay, carbon nanotubes were found to interfere with the tetrazolium salt, interfering with precise viability measurements (Worle-Knirsch et al. 2006). Given that nanoparticles can interact with *in vitro* assays in some situations, it is good practice to perform several toxicity tests in parallel to validate the results of the safety assessment. Another significant point to remember is the use of appropriate doses, in addition to choosing appropriate assays for nanoparticles (Oberdorster 2010). Similarly, several *in vivo* measurements of nanoparticle toxicity have also been employed at excessively high doses. In addition, it is also necessary to take into account that different cell types exhibit varying rates of nanoparticle uptake concerning nanoparticle dosing. Nanoparticles should not be believed to be distributed uniformly *in vivo* in cells covering a given surface area. Therefore, the correlation of *in vitro* nanoparticle exposure doses with *in vivo* conditions can be difficult. But the purity of the sample is another point of concern for nanoparticle safety evaluation. Some carbon nanotube preparations, for example, contain high concentrations of dangerous metals that can influence measures of toxicity (Lam et al. 2004; Ge et al. 2012b). In summary, caution must be taken to identify the appropriate safety assessment assays, nanoparticle doses, and sample purity, to gain valuable insight into the potential toxicity of nanoparticles (Traeger et al. 2020).

7.6 Conclusion and Future Perspectives

There is continuing development in nanotechnology using polymers in nanomedicine. In addition, the relation between the properties of nanoparticles and their *in vivo* behaviors, long-term stability, and the way in which residual materials used for nanoparticle alteration will impact the human body, is still enormously uncertain. Since several nanoparticles have specific structures and compositions that contribute to their unique properties, there is also an unmet need to establish standardized test protocols and validation reference particles. The screening and creation of models may also lead to discoveries of potential nanomaterials and designs of nanoparticles. Therefore, machine learning and artificial intelligence would be a great asset to the realization of efficient translation from bench to bedside as well as customized nanomedicine with regard to the future of the production of nanomaterials in nanomedicine. Nevertheless, the inherent toxic properties of such nanoparticles could be exploited to ablate diseased tissue, as long as healthy organs are protected through selective targeting. Moreover, several strategies can be utilized to eliminate toxicity.

References

Abbad, S., Wang, C., Waddad, A.Y. et al. (2015). Preparation, in vitro and in vivo evaluation of polymeric nanoparticles based on hyaluronic acid-poly (butyl cyanoacrylate) and D-alpha-tocopheryl polyethylene glycol 1000 succinate for tumor-targeted delivery of morin hydrate. *International Journal of Nanomedicine* 10: 305–320.

Ai, J., Biazar, E., Jafarpour, M. et al. (2011). Nanotoxicology and nanoparticle safety in biomedical designs. *International Journal of Nanomedicine* 6: 1117–1127.

Akbarzadeh, A., Rezaei-Sadabady, R., Davaran, S. et al. (2013). Liposome: classification, preparation, and applications. *Nanoscale Research Letters* 1 (8): 102–108.

Bae, Y.H. and Park, K. (2011). Targeted drug delivery to tumors: myths, reality and possibility. *Journal of Controlled Release* 153: 198–205.

Barenholz, Y. (2012). Doxil(R)–the first FDA-approved nano-drug: lessons learned. *Journal of Controlled Release* 160: 117–134.

Bogart, L.K., Pourroy, G., Murphy, C.J. et al. (2014). Nanoparticles for imaging, sensing, and therapeutic intervention. *ACS Nano* 8: 3107–3122.

Bundschuh, M., Filser, J., Lüderwald, S. et al. (2018). Nanoparticles in the environment: where do we come from, where do we go to? *Environmental Sciences Europe* 30: 1–17.

Caster, J.M., Patel, A.N., Zhang, T., and Wang, A. (2017). Investigational nanomedicines in 2016: a review of nanotherapeutics currently undergoing clinical trials. *Wiley Interdisciplinary Reviews: Nanomedicine and Nanobiotechnology* 9 (1): 14–16.

Celia, C., Trapasso, E., Locatelli, M. et al. (2013). Anticancer activity of liposomal bergamot essential oil (BEO) on human neuroblastoma cells. *Colloids and Surfaces B: Biointerfaces* 112: 548–553.

Cho, K., Wang, X.U., Nie, S., and Shin, D.M. (2008). Therapeutic nanoparticles for drug delivery in cancer. *Clinical Cancer Research* 14 (5): 1310–1316.

Clancy, J.P., Dupont, L., Konstan, M.W. et al. (2013). Phase II studies of nebulised Arikace in CF patients with *Pseudomonas aeruginosa* infection. *Thorax* 68 (9): 818–825.

Clift, M.J., Raemy, D.O., Endes, C. et al. (2013). Can the Ames test provide an insight into nano-object mutagenicity? Investigating the interaction between nano-objects and bacteria. *Nanotoxicology* 7: 1373–1385.

De Salamanca, A., Diebold, Y., Calonge, M. et al. (2006). Chitosan nanoparticles as a potential drug delivery system for the ocular surface: toxicity, uptake mechanism and in vivo tolerance. *Investigative Ophthalmology and Visual Science* 47 (4): 1416–1425.

Ferrari, M. (2010). Frontiers in cancer nanomedicine: directing mass transport through biological barriers. *Trends in Biotechnology* 28: 181–188.

Francis, P., Schneider, J., Hann, L. et al. (1994). Phase II trial of docetaxel in patients with platinum-refractory advanced ovarian cancer. *Journal of Clinical Oncology* 12: 2301–2308.

Gaur, U., Sahoo, S.K., De, T.K. et al. (2000). Biodistribution of fluoresceinated dextran using novel nanoparticles evading reticuloendothelial system. *International Journal of Pharmaceutics* 202 (1): 1–10.

Ge, C., Li, Y., Yin, J.-J. et al. (2012a). The contributions of metal impurities and tube structure to the toxicity of carbon nanotube materials. *NPG Asia Materials* 4 (32): 1–10.

Ge, C., Meng, L., Xu, L. et al. (2012b). Acute pulmonary and moderate cardiovascular responses of spontaneously hypertensive rats after exposure to single-wall carbon nanotubes. *Nanotoxicology* 6: 526–542.

Gelderblom, H., Verweij, J., Nooter, K., and Sparreboom, A. (2001). Cremophor EL. the drawbacks and advantages of vehicle selection for drug formulation. *European Journal of Cancer* 37: 1590–1598.

Godin, B., Tasciotti, E., Liu, X. et al. (2011). Multistage nanovectors: from concept to novel imaging contrast agents and therapeutics. *Accounts of Chemical Research* 44: 979–989.

Gupta, R. and Xie, H. (2018). Nanoparticles in daily life: applications, toxicity and regulations. *Journal of Environmental Pathology, Toxicology, and Oncology* 37: 209–230.

Hajipour, M.J., Fromm, K.M., Ashkarran, A.A. et al. (2012). Antibacterial properties of nanoparticles. *Trends in Biotechnology* 30 (10): 499–511.

Hanauer, M., Pierrat, S., Zins, I. et al. (2007). Separation of nanoparticles by gel electrophoresis according to size and shape. *Nano Letters* 7: 2881–2885.

Ibrahim, N.K., Desai, N., Legha, S. et al. (2002). Phase I and pharmacokinetic study of ABI-007, a Cremophor-free, protein-stabilized, nanoparticle formulation of paclitaxel. *Clinical Cancer Research* 8: 1038–1044.

Jafar, A., Esmaeil, B., Mostafa, J. et al. (2011). S16603Nanotoxicology and nanoparticle safety in biomedical designs. *International Journal of Nanomedicine* 6: 1117–1127.

Jain, A.K., Thanki, K., and Jain, S. (2013). Co-encapsulation of tamoxifen and quercetin in polymeric nanoparticles: implications on oral bioavailability, antitumor efficacy, and drug-induced toxicity. *Molecular Pharmaceutics* 10 (9): 3459–3474.

Khan, I., Saeed, K., and Khan, I. (2017). Nanoparticles: properties, applications and toxicities. *Arabian Journal of Chemistry* 18: 1–8.

Kumari, A., Yadav, S.K., and Yadav, S.C. (2010). Biodegradable polymeric nanoparticles based drug delivery systems. *Colloids and Surfaces B: Biointerfaces* 75 (1): 1–8.

Lam, C.W., James, J.T., McCluskey, R., and Hunter, R.L. (2004). Pulmonary toxicity of single-wall carbon nanotubes in mice 7 and 90 days after intratracheal instillation. *Toxicological Sciences* 77: 126–134.

Leng, K.M., Vijayarathna, S., Jothy, S.L. et al. (2018). *In vitro* and *in vivo* anticandidal activities of alginate-enclosed chitosan-calcium phosphate-loaded Fe-bovine lactoferrin nanocapsules. *Future Science OA* 4 (2): 1–20.

Li, K.C., Pandit, S.D., Guccione, S., and Bednarski, M.D. (2004). Molecular imaging applications in nanomedicine. *Biomed Microdevices* 6: 113–116.

Lovestam, G., Rauscher, H., Roebben, G. et al. (2010). Considerations on a definition of nanomaterial for regulatory purposes. *Joint Research Centre (JRC) Reference Reports* 8: 1–4.

Maeda, H., Wu, J., Sawa, T. et al. (2000). Tumor vascular permeability and the EPR effect in macromolecular therapeutics: a review. *Journal of Controlled Release* 65: 271–284.

Matsumura, Y. and Maeda, H. (1986). A new concept for macromolecular therapeutics in cancer chemotherapy: mechanism of tumoritropic accumulation of proteins and the antitumor agent smancs. *Cancer Research* 46: 6387–6392.

Miele, E., Spinelli, G.P., Miele, E. et al. (2009). Albumin-bound formulation of paclitaxel (Abraxane ABI-007) in the treatment of breast cancer. *International Journal of Nanomedicine* 4: 99–105.

Moreno Vega, A.I., Gomez Quintero, T., Nunez Anita, R.E. et al. (2012). Polymeric and ceramic nanoparticles in biomedical applications. *Journal of Nanotechnology* 2012: Article ID 936041.

Mourdikoudis, S., Pallares, R.M., and Thanh, N.T. (2018). Characterization techniques for nanoparticles: comparison and complementarity upon studying nanoparticle properties. *Nanoscale* 10: 12871–12934.

Nagavarma, B.V., Yadav, H.K., Ayaz, A.V. et al. (2012). Different techniques for preparation of polymeric nanoparticles-a review. *Asian Journal of Pharmaceutical and Clinical Research* 5 (3): 16–23.

Norvaisas, P. and Ziemys, A. (2014). The role of payload hydrophobicity in nanotherapeutic pharmacokinetics. *Journal of Pharmaceutical Sciences* 103: 2147–2156.

Oberdorster, G. (2010). Safety assessment for nanotechnology and nanomedicine: concepts of nanotoxicology. *Journal of Internal Medicine* 267: 89–105.

Park, Y.H., Kim, J.N., Jeong, S.H. et al. (2010). Assessment of dermal toxicity of nanosilica using cultured keratinocytes, a human skin equivalent model and an in vivo model. *Toxicology* 267: 178–181.

Peer, D., Karp, J.M., Hong, S. et al. (2007). Nanocarriers as an emerging platform for cancer therapy. *Nature Nanotechnology* 2 (12): 751–760.

Richards, D.A., Richards, P., Bodkin, D. et al. (2005). Efficacy and safety of paclitaxel poliglumex as first-line chemotherapy in patients at high risk with advanced-stage non–small-cell lung cancer: results of a phase II study. *Clinical Lung Cancer* 7 (3): 215–220.

Robertson, J.D., Rizzello, L., Avila-Olias, M. et al. (2016). Purification of nanoparticles by size and shape. *Scientific Reports* 6: 1–9.

Rubin, L.F. (1975). Toxicity of dimethyl sulfoxide, alone and in combination. *Annals of the New York Academy of Sciences* 243: 98–103.

Ruiz-Delgado, G.J., Mancias-Guerra, C., Tamez-Gomez, E.L. et al. (2009). Dimethyl sulfoxide-induced toxicity in cord blood stem cell transplantation: report of three cases and review of the literature. *Acta Haematolgica* 122: 1–5.

Santiago, L., Hillaireau, H., Grabowski, N. et al. (2015). Compared in vivo toxicity in 13 mice of lung delivered biodegradable and non-biodegradable nanoparticles. *Nanotoxicology* 10 (3): 292–302.

Silva, A.M., Alvarado, H.L., Abrego, G. et al. (2019). In vitro cytotoxicity of oleanolic/ursolic acids-loaded in PLGA nanoparticles in different cell lines. *Pharmaceutics* 11 (8): 1–11.

Sonaje, K., Lin, Y.H., and Juang, J.H. (2009). In vivo evaluation of safety and efficacy of self-assembled nanoparticles for oral insulin delivery. *Biomaterials* 30: 2329–2339.

Tahara, K., Sakai, T., Yamamoto, H. et al. (2009). Improved cellular uptake of chitosan-modified PLGA nanospheres by A549 cells. *International Journal of Pharmaceutics* 382: 198–204.

Traeger, A., Voelker, S., Shkodra-Pula, B. et al. (2020). Improved bioactivity of the natural product 5-lipoxygenase inhibitor hyperforin by encapsulation into polymeric nanoparticles. *Molecular Pharmaceutics* 17: 810–816.

Wolfram, J., Motao, Z., Yong, Y. et al. (2005). Safety of nanoparticles in medicine. *Current Cancer Drug Targets* 16: 1671–1681.

Wolfram, J., Suri, K., Huang, Y. et al. (2015). Evaluation of anticancer activity of celastrol liposomes in prostate cancer cell. *Journal of Microencapsulation* 31 (5): 501–507.

Worle-Knirsch, J.M., Pulskamp, K., and Krug, H.F. (2006). Oops they did it again! Carbon nanotubes hoax scientists in viability assays. *Nano Letters* 6: 1261–1268.

Yadav, H.K., Almokdad, A.A., Sumia, I.M., and Debe, M.S. (2019). Polymer-based nanomaterial for drug-delivery carriers. *Nanocarriers for Drug Delivery* 1: 531–556.

Yang, Y., Wolfram, J., Shen, H. et al. (2012). Hesperetin: an inhibitor of the transforming growth factor-beta (TGF-beta) signaling pathway. *European Journal of Medicinal Chemistry* 58: 390–395.

Yang, Y., Wolfram, J., Boom, K. et al. (2013). Hesperetin impairs glucose uptake and inhibits proliferation of breast cancer cells. *Cell Biochemistry and Function* 31: 374–379.

Zielinska, A., Carreiro, F., Oliveira, A.M. et al. (2020a). Polymeric nanoparticles: production, characterization, toxicology and ecotoxicology. *Molecules* 25 (3731): 1–20.

Zielinska, A., Costa, B., Ferreira, M.V. et al. (2020b). Nanotoxicology and nanosafety: safety-by-design and testing at a glance. *International Journal of Environmental Research and Public Health* 17 (4657): 1–23.

8

Green Synthesis of Copper and Copper-Based Nanoparticles for Their Use in Medicine

Toxicity and Safety

Sadhucharan Mallick[1] *and Piyali Sabui*[2]

[1] *Department of Chemistry, Indira Gandhi National Tribal University (Central University), Amarkantak, MP, India*
[2] *Department of Zoology, The University of Burdwan, Golapbag, Bardhaman, WB, India*

8.1 Introduction

Nanotechnology, nanomedicine, and nanotoxicology are complementary disciplines engaged in the advancement of human life (Oberdörster et al. 2007). Nanotechnology is an innovation that deals with the design, manufacture, and application of nanostructured materials, which also involves a fundamental understanding of the relationship between the observed properties and the materials' dimensions, which typically range from a few nanometers to several hundred nanometers (Cao 2011). Nowadays, nanoscale science and technology research is finding remarkable improvements on many diverse sides.

Nanoparticles (NPs) of coinage metals (Cu, Ag, and Au) have attractive optical, physical, chemical, and biological properties (Link and El-Sayed 2000; Barnes et al. 2003; Liz-Marzan 2004; Murphy et al. 2005; Schwartzberg and Zhang 2008; Banerjee et al. 2010; Mallick et al. 2012). Coinage metals (Au, Ag and Cu) NPs furnish unique opportunity owing to their catalytic and biological applications. Among various metal nanoparticles, Cu and Cu-based NPs (Cu, Cu_2O, CuO, and CuI) are attractive not just in light of their optical, catalytic, magnetic, electrical, and conducting properties but also because of their significant anticancer, antibacterial, and antifungal properties (Vukojevic et al. 2005; Barreca et al. 2009; Mallick et al. 2012, 2015; Pariona et al. 2019; Mallick et al. 2019). Cu_2ONPs and CuONPs are ideal contestants for the improvement of super capacitors, solar cells, near IR filters, magnetic storage devices, sensors material, catalytic agent, and semiconductor nanomaterials. Importantly, as we know Cu does not accumulate in the human body due to a sequestering mechanism that helps its elimination from the body; hence copper-based NPs are beneficial in biological applications (Cater et al. 2006; Gaggelli et al. 2006; Musselman et al. 2012). Specific sizes and morphologies of Cu-based NPs can be produced using chemical and physical methods (Huang et al. 1997; Dhas et al. 1998; Vukojevic et al. 2005; Mallick et al. 2012). However, these methods employ potentially unsafe chemicals as reducing agents, extreme heat, organic solvents, and nonbiodegradable stabilizing agents, which are possibly unsafe to the environment and biological systems. There is a serious requirement for the development of a reliable, green, single-step eco-friendly method for the synthesis of Cu-based NPs, where hazardous chemicals are not used in synthesis protocol.

Nanotechnology in Medicine: Toxicity and Safety, First Edition. Edited by Mahendra Rai, Mrunali Patel, and Rashmin Patel.

Biological systems like alga, fungi, bacteria, yeasts, actinomycetes, viruses, and plants are reported to synthesize metal NPs over a variety of chemical compositions, sizes, shapes, and high monodispersity (Ahmad et al. 2003; Sastry et al. 2003; Lengke et al. 2007). Different microorganisms can synthesize NPs via both intracellular and extracellular pathways from precursor metal salt by using biochemical and enzymes that are available within microbes. Among these, the synthesis of NPs using plant extracts appears to be a really effective method in emerging a clean, nontoxic, quick, and eco-friendly technology (Ahmad et al. 2003; Sastry et al. 2003; Lengke et al. 2007; Rai et al. 2008).

In the present scenario, nanomedicine is the medical application of nanotechnology and is mostly associated with the field of oncology. It is suspected that modern medicine will provide potential assistance from nanoscience and nanotechnology (Ganguly et al. 2018; Halevas and Pantazaki 2018; He et al. 2019). It is known that the size of nanoscale materials (NPs, nanowires, nanofibers, and nanotubes) is comparable to that of most biomolecules and structures (e.g. proteins, peptides, liposomes, lipids, oligonucleotides, enzyme and polysaccharides). Therefore, NPs might be beneficial for both *in vitro* and *in vivo* biomedical exploration and applications (Halevas and Pantazaki 2018). The integration of NPs with biology, biotechnology, and pharmaceutical science has prompted the improvement of diagnostic gadgets, contrast agents for image enhancement, drug delivery vehicles, and therapeutic applications. These Cu and Cu-based NPs concomitantly assist human beings through destroying the single-cell pathogens, like bacteria, fungi, and viruses; so, these are utilized as potential bactericidal and fungicidal agents (Pariona et al. 2019). Generally, Cu and Cu-based NPs are used in fast-moving consumer goods such as personal protective equipment (PPE) with antimicrobial and antiviral properties (gloves, mask); antiviral (HIV-1 and other viruses) and antibacterial filters; antibacterial garments; wound dressing as well as in medicinal devices and other health-related products (Borkow and Gabbay 2005).

Cu-based NPs and nanocomposites show very high efficacy in killing cancer cells and are already through several clinical trials. Cu and CuONPs have been comprehensively studied for their anticancer properties for human epithelial cells derived from a lung carcinoma (A549), human epithelial cells derived from an epidermoid carcinoma of the larynx (HEp-2), human liver cancer cell (HEPG2), human breast cancer cells (MCF7, MDA-MB 231), human colorectal carcinoma (HCT-116), human cervical cancer (HeLa), human embryonic kidney (293T HEK), and human lung fibroblast (MRC5), neuronal (PC12) cell (Wang et al. 2012; Laha et al. 2014, 2015; Wongrakpanich et al. 2016; Chavez Soria et al. 2019). These Cu and CuONPs can induce ROS-mediated DNA and mitochondria damage, resulting in apoptosis in tumor cells (Wang et al. 2012; Wongrakpanich et al. 2016; Chavez Soria et al. 2019). The results are so encouraging that Cu and Cu-based NPs and nanocomposite might become an alternative to traditional cancer therapy since they permit cancer cells to be targeted accurately and assist detailed imaging of tissues, making planning of further therapy much easier (Laha et al. 2014, 2015). The use of nanomedicine has become advantageous due to its effective transformation within the cell and its role in cell recovery from intracellular harm.

The smaller size of NPs leads to varying high chemical reactivity and subsequently grave toxicity. Nanotoxicology or harmfulness evaluation of these Cu-based NPs is mandatory to know possible antagonistic impacts and their destiny in the interior of human body. Toxicity induced by Cu oxide NPs has been studied extensively and documented as biodegradation, bioaccumulation, ecotoxicity, genotoxicity, and cytotoxicity (Sharifi et al. 2012; Ganguly et al. 2018). There exist a few studies indicating the influence of Cu-based NPs in creating diseases, for example, Wilson disease, Alzheimer's diseases, pulmonary diseases, and exacerbations of prior respiratory illnesses, lung disease, liver disease hepatic and renal diseases (Oberdörster 2010; Studer et al. 2010; Sharifi et al. 2012).

We need to consider risks posed by NPs and nanotoxicology for finding possible hazards and potential risks associated with nanomedicine before widespread therapeutic and diagnostic applications (Oberdörster et al. 2007). There are basic requirements to protect workers' skin with personal protective equipment to escape the direct interaction with CuONPs, and extraordinary care should be applied for employees with an impaired pore and skin barrier (Zanoni et al. 2019). It is essential that Cu-based NPs must be biologically characterized for their health hazards to ensure risk-free, safe, and sustainable development of nanomaterials and nanotechnology applications.

8.2 Green Synthesis Methods of Copper and Copper-Based Nanoparticles

Copper NPs, Cu-based nano-size particles, and other metal NPs are synthesized through various strategies, usually categorized as top-down (physical strategy) and bottom-up (chemical strategies) approach. In top-down approaches, a massive solid is reduced to nano-sizes with the use of cutting, crushing, and carving methods (lithography techniques). In the bottom-up approach, the structure of NPs is generated employing atoms, molecules, or clusters (Wang et al. 2005; Merkel et al. 2010; Cao 2011). Over the past few decade's research had described on the green synthesis of Cu and Cu-based NPs and their biotechnological, biomedical, and therapeutic applications, are being carried out in various laboratories all over the world. Numerous physical and chemical approaches have been employed to produce nanomaterials. Over the past few years, biological systems such as fungi, bacteria, yeasts, actinomycetes, viruses, and plants have served as potential nano-factories for the eco-friendly and inexpensive synthesis of CuNPs, Cu_2ONPs, CuONPs, and other metal NPs (Gahlawat and Choudhury 2019; Yadav et al. 2019).

8.2.1 Fungi-Assisted Synthesis of Nanoparticles

Fungi are eukaryotic organisms which include yeasts and molds, are mostly tubular, elongated, and branching filamentous structures. In the past few years, fungi have been extensively used to grow microspheres, NPs, and nanostructures of different chemical compositions and sizes owing to their quick growth; relatively inexpensive, easy separation; environment-friendly nature; metal-accumulating capacity; and high tolerance to natural conditions by forming shells. Fungi-mediated Cu and Cu-based NPs are nontoxic and have easy scale-up for industrial applications. On the other hand, studies have indicated that secreted enzymes, protein, polymers by fungi help in producing metallic NPs. Honary et al. (2012) had described extracellular synthesis and stabilization of copper NPs utilizing yeast extract i.e. *Penicillium aurantiogriseum, Penicillium citrinum,* and *Penicillium waksmanii* and $CuSO_4$ salt solution in the ambient condition. These fungi were capable of synthesized polydispersed Cu NPs at different pH values (range of 5–9) and variations of salt concentration (Honary et al. 2012). The white-rot fungi *Stereum hirsutum* was used for the biosynthesis of copper oxide NPs in diverse pH conditions and using different copper salts solution, i.e. $CuCl_2$, $CuSO_4$, and $Cu(NO_3)_2$ (Cuevas et al. 2015).

8.2.2 Actinomycetes-Assisted Synthesis of Nanoparticles

Actinomycetes are microorganisms that share vital common characteristics of fungi and bacteria. Actinomycetes produce a broad range of bioactive compounds in huge amounts, spread widely, and their growth speed is slow as contrasted with different microorganisms, i.e. bacteria and fungi. Recently, Rasool and

Hemalatha (2017) reported the extracellular synthesis of copper NPs by marine endophytic actinomycetes isolated from seaweeds. Hassan et al. (2019) synthesized spherically shaped CuONPs using two endophytic actinomycetes: *Streptomyces zaomyceticus Oc-5* and *Streptomyces pseudogriseolus* Acv-11. They had isolated Oc-5 and Acv-11 from medicinal plant *Oxalis corniculata L* leaves. According to their transmission electron microscopy (TEM) analysis reports, the average size of CuONPs was ≈78 nm and ≈80 nm for Oc-5 and Acv-11, respectively. CuONPs were characterized by UV–Vis spectroscopy, transmission electron microscopy (TEM), energy dispersive X-ray (EDX), X-ray diffraction (XRD), Fourier transform infrared (FT-IR) spectroscopy, and X-ray photoelectron spectroscopy (XPS). Endophytic actinomycetes are capable of producing different secondary metabolites such as enzymes, organic acids, and proteins used for Cu^{+2} ion reduction and capping at nanoscale.

8.2.3 Bacteria-Assisted Synthesis of Nanoparticles

Bacteria are single-celled microbes and exist in three fundamental shapes, cocci, bacillus, and spirillum, which are acceptable sources to form micro and nano-sized materials. Bacterial cell wall contains enzyme reductase that catalyzes the reduction of Cu^{2+} ions into CuNPs. Syed Saif Hasan and co-workers had revealed the extracellular synthesis of CuONPs (10–30 nm) using a Gram-negative bacterium belonging to the genus *Serratia* which was isolated from the midgut of common Indian mosquitoes *Stibara* sp., an insect of the Cerambycidae family of beetles found in the Northwestern Ghats of India (Hasan et al. 2008). When *Serratia* sp. cells are exposed to the $CuSO_4$ salt, the stress is increased to a greater extent. The salt is internalized into the cells, and then the Cu^{2+} ions are reduced by precise biomolecules to Cu, CuO, Cu_2O, and Cu_4O_3, which are less harmful than the ionic forms (Hasan et al. 2008). Their detailed characterizations revealed that the CuNPs and different Cu oxides were formed. At room temperature, Varshney et al. (2010) had synthesized smaller CuNPs (8–15 nm) using non-pathogenic bacterial strain *Pseudomonas stutzeri* and aqueous $CuSO_4$ solution. Singh et al. (2010) reported extracellular synthesis of copper oxide NPs using trichloroacetic acid precipitated protein fraction of *Escherichia coli* under aerobic atmosphere, neutral pH, and room temperature.

8.2.4 Algae-Assisted Synthesis of Nanoparticles

Algae are a type of photosynthetic sea microorganisms that have been reported to have good capability to absorb heavy metal ions from the environment and to synthesize metallic NPs. Abboud et al. (2014) confirmed biosynthesis of highly stable CuONPs of dimensions 5–45 nm using *Bifurcaria bifurcata*, a brown alga, which was collected on marine rocks. They had documented that alga extract was added dropwise into aqueous copper (II) sulfate solution with constant stirring at 100–120°C. They observed that deep blue solution slowly became colorless in a few hours and then turned to brick red coloration, which turned dark after constant stirring for 24 hours (Abboud et al. 2014). Arya et al. (2018) had used an extract of green alga *Botryococcus braunii* for the synthesis of (10–70 nm) copper oxide and (40–100 nm) silver NPs.

8.2.5 Phytochemical-Assisted Synthesis of Nanoparticles

In recent years, the biosynthesis of NPs employing plant extract has been considered a favored strategy for the green fabrication of nanomaterials (Brumbaugh et al. 2014; Yadav et al. 2019). Plant-supported synthesis of NPs is more effective in terms of getting a greater yield than microbial synthesis (Rai et al. 2008; Singh et al. 2020). Several groups have reported the fabrication of metal NPs by use of extracts from medicinal plant leaf, latex, root, stem flower, and seed. Plant extracts have phytochemicals, mainly polyphenols, carbohydrates, flavones, etc., that act as stabilizing agents as well as reducing agents (Rai et al. 2008). Recently, Chen and co-workers (2019) had synthesized CuONPs utilizing papaya leaf broth and examined the bactericidal action of these CuONPs versus soil-borne pathogenic bacteria *Ralstonia solanacearum*. When papaya (*Carica papaya L.*) extract was mixed with cupric sulfate solution continuously reacted at room temperature until a black color was observed. Recently, Sukumar et al. (2020) synthesized rice grain-shaped CuONPs employing *Caesalpinia bonducella* seed extract for the study of antibacterial activities (Sukumar et al. 2020). They demonstrated that cupric nitrate trihydrate solution was mixed with freshly prepared seed extract and ammonia solution under constant stirring and observed dark brown precipitate formation around seven hours. Vijay Kumar and co-workers (2015) established the antibacterial activity of *Aloe vera* leaf-mediated CuONPs against bacterial fish pathogens at lower concentrations. Sutradhar and collaborators (2014) had reported the synthesis of CuO the usage of coffee powder and tea leaf in microwave irradiations. These spherical-shaped CuONPs (50–100 nm) are dispersible in water and stable in aqueous media for elongated periods and showed significant antibacterial activity against two pathogenic bacteria: *Shigella dysenteriae* as well as *Vibrio cholerae*. Recently, numerous attempts have been made for the green synthesis and stabilization of Cu NPs by using diverse leaf extracts such as *Aegle marmelos Correa* (Angajala et al. 2014), *Lawsonia inermis* (Cheirmadurai et al. 2014), *Ocimum sanctum* (Usha et al. 2017), and *Aerva lanata* leaves (Hariprasad et al. 2016). Cheirmadurai et al. (2014) had prepared reddish-brown color CuNPs using aqueous *L. inermis* leaf extract, which showed maximum absorption at 570 nm and also mentioned the larger size of Cu NPs (83 nm) in comparison to the calcined (43 nm) particles (Cheirmadurai et al. 2014). Only one report has been found in the literature for the synthesis of spherical-shaped CuNPs biofabricated from *Aegle marmelos* with larger particle size distribution of 50–100 and 100–200 nm (Angajala et al. 2014). Nagar and Devra (2018) synthesized crystalline, cubical CuNPs using the leaf broth of *Azadirachta indica* and cupric chloride dihydrate as a precursor salt (Nagar and Devra 2018). The biomolecules existing in the leaf broth of *A. indica* induce the reduction of Cu^{2+} ions to Cu NPs and also act as a capping agent and reductant. CuNP synthesis was reported by Nasrollahzadeh and co-workers using *Euphorbia esula L.* (Nasrollahzadeh et al. 2014 and *Ginkgo biloba* leaf extract (Nasrollahzadeh and Sajadi 2015). Further, the authors reported CuONP synthesis using aqueous *Tamarix gallica* leaf extract (Nasrollahzadeh et al. 2015a) and *Anthemis nobilis* flower extract (Nasrollahzadeh et al. 2015b).

8.2.6 Biomolecule and Biopolymer-Mediated Synthesis of Nanoparticles

Xiong et al. (2011) had synthesized stable CuNP dispersions with an average particle size of less than 2 nm the usage of nonhazardous L-ascorbic acid (vitamin C), which acts as a reducing agent and capping agent. Valodkar et al. (2011) synthesized microwave-assisted environmental benign starch-stabilized CuNPs, AgNPs, and Cu–Ag alloy NPs in the presence of ascorbic acid and cupric nitrate trihydrate solution. They tested the antibacterial activity of these NPs with both G-positive *Staphylococcus aureus* and Gram-negative *E. coli* bacteria (Valodkar et al. 2011). In recent years, Pariona et al. (2019) synthesized sodium citrate-stabilized Cu NPs using copper (II) sulfate pentahydrate, ascorbic acid, and sodium hydroxide for fungicides application. Ray and co-workers (2015) synthesized CuNPs of various sizes using the sugars dextrose (6–8 nm), dextrin (5–6 nm) and β-cyclodextrin (50–80 nm), ascorbic acid, and copper acetate as precursor salt.

8.3 Purification of Copper and Copper-Based Nanoparticles

CuNPs, CuONPs, and other NPs are separated from the reaction mixture through centrifugation using high-speed centrifugation machines due to the comfort of operation and less time consumption. The separation and purification of biologically produced NPs required repetitive washing and high-speed centrifugation to eliminate the unreacted bioactive molecules, anion, and excess metal ions from precursor salt (Mallick et al. 2012; Singh et al. 2020). The pellet was then redispersed in water for the study of biological activities.

8.4 Characterization of Green Synthesized Copper and Copper-Based Nanoparticles

The characterization of the NPs is most essential to realizing the control of synthesis and their catalytic, biomedical applications. The absorbance spectra of Cu-based (Cu, Cu_2O, CuO, and CuI) NPs can be recorded on a UV–Visible spectrophotometer within the wavelength range of 200–1100 nm which gives useful information on the size, shape, and stability of NPs. Cu NPs solution usually shows an absorbance band around 570–590 nm (Mallick et al. 2012, 2015). For particle size and surface morphology characterization, microscopy techniques such as SEM, FESEM, and TEM and AFM are extensively used. EDX of the selected area of nanoparticles provides an idea of the composition. The oxidation state of the metal present in the material can be concluded by using XPS (Mallick et al. 2015; Prajapati et al. 2021). The phase purity can be determined by using XRD with a comparison of those reported earlier. FTIR is utilized for the recognizable proof of obscure materials. The atomic absorption spectroscopy (AAS) technique is used to measure the concentration of released ions from metallic NPs. The above techniques are usually used for the characterization of synthesized nanostructured materials.

8.5 Copper and Copper-Based Nanoparticles as Nanomedicines

NPs are crucial components of nanomedicine. Cu-based NPs have grown fame in technological improvements, i.e. therapeutics, diagnostics, and nanomedicine fields, for their tunable physical, chemical, biological, and optoelectronic properties with greater performance over their bulk counterparts (Rosi and Mirkin 2005).

8.5.1 Application as Antibacterial Agents

In recent years, we had reported the synthesis of chitosan (CS)-supported AgNPs, CuNPs, core-shell Cu@AgNPs, and mixed metal (Cu and Ag) nanoparticles composite and studied their bactericidal activity on Gram-negative recombinant *E. coli* bacterium expressing green fluorescent protein (GFP) and Gram-positive *Bacillus cereus* (Banerjee et al. 2010; Mallick et al. 2012, 2015; Mallick 2018). Rasool and Hemalatha (2017) evaluated the antibacterial activity of Cu NPs against human infective bacteria *Klebsiella pneumoniae, E. coli, Salmonella typhimurium, Proteus mirabilis,* and Methicillin-resistant *Staphylococcus aureus* (MRSA) through inhibition zone measurements by well-plate methods. The results recommended that human pathogenic bacteria can be controlled using actinomycetes-assisted CuNPs. Recently, Recently, Hassan et al. (2019) evaluated the antimicrobial activity of actinomycetes-synthesized CuO-NPs and the results showed that the minimum inhibitory concentration (MIC) was 10 mM for *Bacillus diminuta, Bacillus subtilis, Staphylococcus aureus,* and

Pseudomonas aeruginosa while MIC was 15 mM for *E. coli*. Further, Abboud et al. (2014) demonstrated that brown alga extract-synthesized CuONPs were found to exhibit high bactericidal activity against Gram-negative (*Enterobacter aerogenes*) and Gram-positive (*S. aureus*) bacteria strains. Ingle et al. (2014) discussed in their review articles that CuNPs possess excellent antialgal, antimycotic, antiviral, antiparasitic, and anticancerous activities. Many authors reported that Cu, CuO, Cu_2O Ag, MgO, TiO_2 NPs and their nanocomposites have amazing antimicrobial potential, and hence can be utilized in alternative therapy (Banerjee et al. 2010; Rai et al. 2018; Alavi and Rai 2019). Mallick and his co-workers (2012) recently described the synthesis of a novel composite containing chitosan and CuNPs which was found to be more stable in the presence of molecular iodine and also have high bactericidal activities against Gram-negative green fluorescent expressing *E. coli* bacteria and Gram-positive *B. cereus* bacteria, shown in Figure 8.1 (Mallick et al. 2012). Moreover, Mallick et al. (2012) had discussed the mechanism of bactericidal action of the iodine-stabilized Cu NPs chitosan composite is in light of their experimental findings. Polycationic chitosan assisted to keep the bacterium attached, the Cu NPs present there possibly led to perforation of the cell wall causing irreparable injury to the membrane, finally leading to cell demise. Zeta potential measurements, electron microscopic images, and flow cytometric studies showed that the iodine-stabilized Cu nanoparticles chitosan composite was attached to the bacterial cell wall. These three components (chitosan, CuNPs, and iodine) working in tandem led the bacteria to become unviable and finally triggered cell death, which is shown in Figure 8.2 (Mallick et al. 2012). Iodinated mixed metal (Cu NPs ~11.5 nm and Ag NPs ~3.5 nm) composite at lower concentration was effective in killing *E. coli* (Table 8.1) and was found to be more effective in contrast to the individual components of the composite (Mallick 2018). Dr. Mallick and his co-workers had also reported a synthesis of spherical chitosan-supported core-shell Cu@Ag NPs (14.0 ± 3.4 nm) composite and their superior antibacterial activity at lower concentrations of Cu and Ag given in Table 8.1 (Mallick et al. 2015). Finally, it was ensured through the hard work of the last few years of ours (Banerjee et al. 2010; Mallick et al. 2012, 2015) a small contribution has been made for the development of nanoscience and nanotechnology,

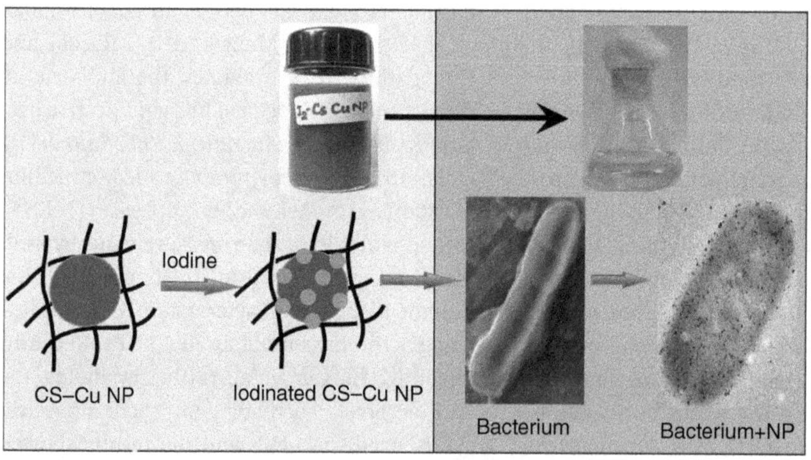

Figure 8.1 A schematic illustration of iodinated CS-Cu NP composite synthesis and bactericidal application. *Source:* Mallick et al. (2012) with permission of American Chemical Society.

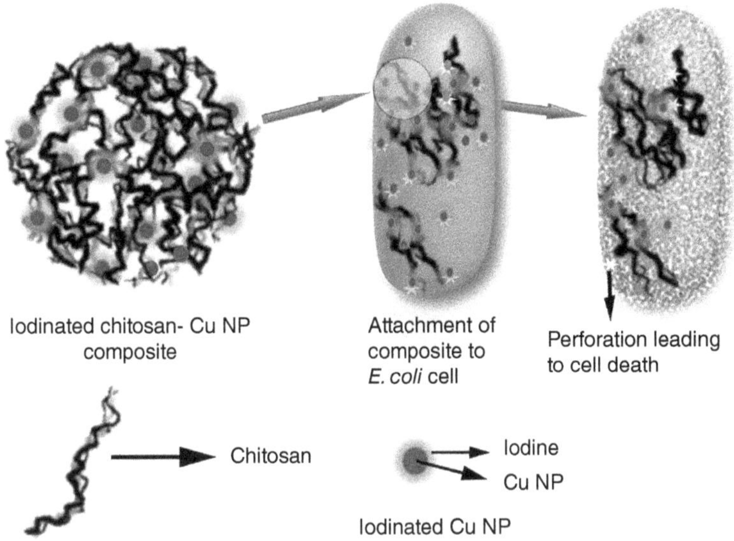

Figure 8.2 A schematic illustration of the proposed **mechanism** of **antibacterial action of the iodinated CS-Cu NP composite.** *Source:* Mallick et al. (2012) with permission of American Chemical Society.

Table 8.1 Antibacterial activity studies of chitosan–metal nanoparticle composites against green fluorescent expressing *Escherichia coli* bacteria.

Composite details		**Composite (μg/ml)**	**CS–Ag NPs (μg/ml)**	**Ag (μg/ml)**	**CS (μg/ml)**	**I2 (μg/ml)**
Iodinated CS–Ag NP composite Banerjee et al. (2010)	MIC	87.20	38.40	0.82	37.58	48.8
	MBC	119.00	51.70	1.1	50.60	67.30
		Composite (μg/ml)	**CS–Cu NPs (μg/ml)**	**Cu NPs (μg/ml)**	**CS (μg/ml)**	**I2 (μg/ml)**
Iodinated CS–Cu NP composite Mallick et al. (2012)	MIC	130.84	127.62	21.55	106.07	3.22
	MBC	239.4	233.50	39.43	194.07	5.9
		Composite (μg/ml)	**Cu (μg/ml)**	**Ag (μg/ml)**	**CS (μg/ml)**	**I2 (μg/ml)**
CS Cu@Ag NP composite Mallick et al. (2015)	MIC	63.40	3.03	1.47	58.90	0.0
	MBC	92.99	4.44	2.13	86.42	0.0
		Composite (μg/ml)	**Cu (μg/ml)**	**Ag (μg/ml)**	**CS (μg/ml)**	**I2 (μg/ml)**
I_2-CS-Ag NP-Cu NP composite (Cu: Ag 20:1) Mallick (2018)	MIC	63.50	9.98	0.74	52.64	0.16
	MBC	93.02	14.61	1.08	73.43	0.22

Sources: Based on Mallick (2018), Mallick et al. (2015), Banerjee et al. (2010) and Mallick et al. (2012).

which are summarized in Table 8.1, which would be applicable for the advancement of humanity.

8.5.2 Application as Antifungal Agents

Recently, Pariona et al. (2019) evaluated the fungicidal efficacy of CuNPs against plant pathogens *Fusarium solani, Neofusicoccum* sp., and *Fusarium oxysporum* and reported inhibition of radial growth (IRG) for each case. The IRG values had reported by treatment of different concentrations of Cu-NPs such as at 0.1 mg ml^{-1} IRG 55% for *F. solani*, at 0.5 mg ml^{-1} IRG 95% *Neofusicoccum* sp., at 0.5 mg ml^{-1} IRG 97% *F. oxysporum* (Pariona et al. 2019). Ponnusamy Ponmurugan evaluated the fungicidal efficacy of Cu NPs against *P. hypolateritia* causing red root-rot disease in tea plants (Ponmurugan et al. 2016). They had reported a 52.7% reduction in red root-rot disease in tea plants at 2.5 ppm of Cu NPs. Anju Arya had reported that biogenic alga (*B. braunii)*-mediated copper oxide NPs inhibit the growth of *F. oxysporum* at 250 μg/ml (Arya et al. 2018). The fungicidal activity of *Eichhornia*-mediated copper oxide NPs (28 ± 4 nm) was performed by Vanathi et al. (2016) on various plant pathogens and reported as the highest zone of inhibition for *Fusarium culmorum* (21.26 ± 1 mm), *Aspergillus niger* (18.33 ± 1 mm) at a concentration of 100 μg ml^{-1} and *Aspergillus fumigatus* (9.10 ± 1 mm) at a concentration of 25 μg ml^{-1}.

8.5.3 Application as Antiviral Agents

A highly infectious pathogen (influenza virus) that naturally spreads through people, birds, and pigs. The swine-origin influenza virus responsible for outbreaks around the world in the 2009 pandemic (H1N1) was a hereditarily hybridized mutant of the pig, avian, and human influenza virus strains. Thus, extra measures to suppress the spread of viral contamination were critically required. Horie et al. (2008) had detailed the antiviral influence of copper ions on avian flu infection (H9N2)-associated structural disorders of virus. According to their reports, electron microscopic and immunoelectron microscopic analysis confirmed that the morphological anomalies of the H9N2 virus deactivated by Cu^{2+} ions at 2.5–250 microM (μM) concentrations (Horie et al. 2008). Fujimori et al. (2012) had examined the antiviral activity of copper iodide (CuI) NPs (160 nm) against an influenza A virus of swine-origin (pandemic [H1N1] 2009) by plaque titration assay. In mechanistic approaches, they had established the inactivation of the virus with CuINPs owing to degradation of viral proteins hemagglutinin and neuraminidase by generated reactive oxygen species, i.e. hydroxyl radicals $^{\cdot}OH$ and O^{2-} (Fujimori et al. 2012). In another experiment, Hang et al. (2015) estimated the antiviral efficacy of CuONPs (92.4 ± 8.8 nm) against hepatitis C virus *in vitro* at a noncytotoxic concentration ranging from 0.5 μg/ml to 8 μg/ml. According to their information, CuONPs can work as an anti-hepatitis C agent by targeting multiple steps in the HCV viral life cycle and blocking hepatitis C virus infection both at the attachment and entrance periods (Hang et al. 2015).

8.5.4 Application as a Targeted Drug Delivery System

NPs and other nanoscale tools are also being developed for the pharmacological treatment of several human diseases by acting as delivery vehicles for pharmaceutical drugs (Chow and Ho 2013). Recently, many forms of NPs including liposomes, inorganic NPs, and polymeric micelles have been used as pharmaceutical drug delivery vehicles by transporting hydrophobic anticancer drugs into cancer cells (Zhang et al. 2019). It is well known that NP-based carrier systems can easily cross biological barriers. Thus, cellular uptake of NPs in tumors takes place by means of their localized leaky vasculature and are held due to destitute lymphatic seepage in the tumor microenvironment (Kagan et al. 2016; Pelaz et al. 2017). Dong et al. (2013) had shown that hydrophobic camptothecin (CPT) drug loading in hollow CuSNPs (CuS-CPT NPs) enhanced inhibition of tumor growth with irradiation through NIR (NIR, λ = 700–1100 nm) light (Dong et al. 2013). Hollow CuS NPs are considered an intelligent drug delivery system due to high stability, excellent dispersibility, large specific surface volume, numerous mesoporous pores, typical hydrophilicity, beneficial biocompatibility, and tunable drug-release property. They had demonstrated that hollow CuSNPs enhanced the NIR photothermal ablation efficiency of cancer cells *in vivo* against tumor-bearing mice. As importantly, due to the synergistic effects of Camptothecin (CPT), the and photoactivated Semiconductor CuS nanoparticles enabled tumoricidal efficiency can be achieved with a lesser drug quantity and mild irradiation conditions, which is the importance of moving this system closer to the clinical background CuS (130 nm in diameter and of ≈ 20 nm thickness) can quickly transform the 980 nm laser energy into environmental heat, which is due to the d–d transition of Cu^{2+} ions instead of surface plasmon resonance (SPR) band (Dong et al. 2013). Ramadan et al. (2012) had also described a hollow copper sulfide nanoparticles (HCuSNPs) photothermal ablation-enhanced transdermal drug delivery methodology of isothiocyanate-labeled dextran (Dextran-FITC) and human growth hormone (hGH) with intense photothermal coupling effects.

8.5.5 Application as Anticancer Agents

Cancer is one of the world's deadliest diseases that require high dedication and resilience to conventional treatments. Nanomedicine has been related to the use of NPs in oncology. In particular, nanotechnology and biotechnology, when challenged with cancer biology, have triggered new opportunities for improving targeted cancer treatment. In recent years, numerous researchers had evaluated anticancer activities of CuONPs in diverse human cell lines (like A549, MCF7, MRC5, HCT-116, HEp-2, HeLa, 293T HEK, PC12) showing a dose-dependent anticancer activity. Wang et al. (2012) had observed that CuONPs (20–40 nm) were more effective in reducing cell viabilities of human lung epithelial cells (A549 cell line) IC_{50} (24 hours 15 mg/l). In contrast, CuO bulk particles showed much less effectiveness (24 hours IC_{50}, 58 mg/l). They had experimentally explored the pathway of CuONP uptake and suggested that endocytosis was the prime route of CuONP uptake to the lysosome in the cytoplasm. The distribution and site of CuONPs in A549 cells were detected through TEM and TEM-EDX analysis (Wang et al. 2012). Experimentally, they had also shown that CuONPs inhibit lysosomal activity through lysosomal membrane damage, which may lead to subsequent translocation to other organelles such as mitochondria and nucleus. The CuO NPs inside the nucleus may be able to injury DNA directly. The cells CuONPs induced mitochondrial depolarization, possibly mediated by ROS generation which subsequently triggers the expression of p38 and p53, can cause cell cycle arrest and cell apoptosis. Laha et al. (2014) had demonstrated that a combination of CuONPs (~30 nm, 12 μg/ml) with the autophagy inhibitor 3-methyladenine (3-MA) is essential to trigger oxidative stress-induced apoptosis in human breast cancer MCF7 cells but CuO NPs alone only induce autophagy, which is a survival strategy for MCF7 cells. The same research group also measured that ROS production (1.6 fold) increased when MCF7 cells were treated with CuO–FA NPs compared to CuONPs-treated MCF7cells and reported that lethal dose (50%) i.e. LD_{50} is 0.6 mmol/l for CuONPs and 0.1 mmol/l for CuO–FA NPs (Laha et al. 2015). Fahmy and Cormier (2009) observed the cytotoxic effect on human laryngeal epithelial cells (HEp-2) when exposed to CuONPs. They found CuONPs significantly decreased cell viability by 60% at the exposure dose of 1.9 log μg/cm^2 (80 μg/cm^2). Chavez Soria et al. performed cell viability measurements against different cancerous and noncancerous mammalian cell lines: human colorectal carcinoma (HCT-116), human cervical cancer (HeLa), human embryonic kidney (293T HEK), and human lung fibroblast (MRC5) that are exposed to CuONPs for 24 hour. They found that HCT-116 revealed the least sensitivity (LC_{50} ~5 mg ml^{-1}) where MRC-5 and HeLa cell lines exhibited the highest sensitivity to CuONPs (LC_{50} ~1.25 mg ml^{-1}) among the 293T HEK (LC50 ~2.5 mg ml^{-1}) and other cell lines. They observed cell shrinkage and bigger cellular blebbing upon CuONP treatment on HCT-116 cells (Chavez Soria et al. 2019).

8.5.6 Applications in Molecular Imaging

Zhou and his coworkers (2015) had synthesized multifunctional ultrasmall copper sulfide nanodots (CuSNDs) for cancer imaging and therapy. These polyvinylpyrrolidone-coated CuSNDs (6 nm) can efficiently absorb near-infrared light (~990 nm) for photothermal ablation therapy using tightly incorporate the copper—64 radioisotope (^{64}Cu) for noninvasive positron emission tomography (PET). They had reported that these CuS NDs initially accumulated in the tumor, enabling tumor visualization at 2 hour after intravenous injection and that most of these (95%) particles were removed via the renal-urinary system within 24 hour after intravenous injection (Zhou et al. 2015). Zhou et al. and Ku et al. investigated copper-based chalcogenide semiconductor copper selenide, copper telluride, especially copper Sulfide NPs exhibit strong absorption in the near-infrared region demonstrate highly skilled in molecular imaging and photothermal ablation therapy of cancer cells (Ku et al. 2012; Zhou et al. 2015; 2016).

Perlman et al. (2015) demonstrated that CuONPs (7 nm diameter) are outstanding candidates for MRI-ultrasound dual imaging contrast agents. *In vitro* experiment and phantom specimens containing CuONP concentrations extending from 2.4 to 320 μg ml^{-1} were scanned using both 9.4 T MRI and through-transmission ultrasonic imaging. They reported CuONPs to induce shortening of the magnetic T1 relaxation time in magnetic resonance imaging (MRI) scans as well as increase the speed of sound and ultrasonic attenuation coefficient in case of ultrasound imaging (Perlman et al. 2015).

8.6 Copper and Copper-Based Nanoparticles and Their Toxicity

The good or bad impact of CuNPs and Cu-based NPs on living organisms depends on their size, shape, surface properties, and charge over the character of the actual metal ion utilized. Every day dietary intake of Cu in healthy individuals are varied between 0.07 and 0.1 mg kg^{-1} body weight (WHO 1998). US Environmental Protection Agency has reported that Cu is cytotoxic at doses greater than 1.3 mg l^{-1} in drinking water (U.S. EPA 1991).

8.6.1 Nanotoxicology

While the manufacture and use of NPs are rising dramatically, investigation into the toxicological influence and probable risk of NPs to human well-being and the environment is still in its earliest stages (ElSaesser and Howard 2012; Wang et al. 2015; Kamlesh et al. 2019). The public, in general, are more likely to be exposed to NPs occupationally or via consumer goods such as sunscreens and cosmetics (TiO_2, ZnO and ZnO_2); industries, especially the electrical sector (Cu); and the environment. Nanotoxicology is a reasonably new branch of toxicology that reports the gap in knowledge of toxicity prompted by NPs and nanostructured materials. In recent years, Ganguly et al. (2018) documented different types of toxicity induced by NPs, which are biodegradation, bioaccumulation, genotoxicity, cytotoxicity, and ecotoxicity.

8.6.2 Different Types of Toxicity

Biodegradation: Biomineralization and biotransformation of complex NPs occur by the biologically catalyzed procedure (Ganguly et al. 2018). Biodegradation is one of the plausible paths to result in toxicity within the environment, which affects humans, animals, and whole ecosystems. It is also depicted that biodegraded metal ions often end up accumulating within the globe and losing the usual enzymatic capability of numerous organisms.

Bioaccumulation: Bioaccumulation of metal ions and metal NPs is an additional mode of inducing toxicity in our surroundings and living organisms. It is the result of a dynamic equilibrium between uptake and elimination. The intake of xenobiotic ingredients often raises toxicity within the living organisms which consequently get accumulated within the soil and water sources. Fish, microorganisms, and other aquatic animals intake these NPs and, in consequence, get relocated into all living bodies through the food chain (Ganguly et al. 2018).

Genotoxicity: It is one of the prime ways for toxicity initiation in living beings. Toxicity causing damage to genetic information in cells is known as genotoxicity. It can lead to damage to the cell's DNA and chromosomes. Genotoxicity brings about various problems within the nucleus of the cell which might lead to transmissible sicknesses like diabetes, cystic fibrosis, sickle cell anemia, and hemophilia (Ganguly et al. 2018). Karlsson et al. (2008) had examined the toxicity effect of metal oxide particles CuONPs, $CuZnFe_2O_4$NPs, and nanotubes on human lung epithelial cell line (A549) and established that these NPs caused ominously increased levels of DNA damage and oxidative lesions (Karlsson et al. 2008)

Cytotoxicity: Cell cytotoxicity denotes the toxicity destroying the cell by certain chemicals or mediators. It causes the uncontrolled breakdown of cell constituents, injury of the cell membrane, and damage of cytosolic components. Cell cytotoxicity might furthermore start a programmed cell death phenomenon. Karlsson et al. (2008) had investigated the cytotoxicity of different metal oxide NPs (CuO, TiO_2, ZnO, $CuZnFe_2O_4$, Fe_3O_4, Fe_2O_3) on A549 cell line and found that CuONPs were most potent regarding cytotoxicity and DNA damage and mitochondrial dysfunction. Oxidative stress has been considered to be the primary reason for the cytotoxicity of CuONPs in mammalian cells. It has been manifested that NPs gain entry into our body often by way of airways, absorption through the skin, or ingestion. NPs were able to translocate by the blood reaching various organs such as the lung into the liver, spleen, heart, and possibly other organs (Sharifi et al. 2012). The lung is a genuine obstruction toward the uptake and spread of NPs. Inside the human respiratory tract, breathing in particles of diverse dimensions reveals distinctive fragmentary depositions, as ultrafine particles diameter $\leq 100\,nm$ are stored in all regions; however, particles dimensions $\leq 20\,nm$ deposit effectively within the nasopharyngeal region. Smaller dimension NPs $\leq 10\,nm$ stick within the tracheobronchial vicinity and NPs between 10 and 20 nm settle in the alveolar area (Sharifi et al. 2012).

Ecotoxicity: Ecotoxicity is the branch of toxicology concerned with the awareness of toxicity and its impact on the whole biotic system of living organisms including the human life cycle. It looks at various biological organism levels and proposes to protect populations and ecosystems.

8.6.3 Toxicity Effect of Copper and Copper-Based Nanoparticles

CuNPs and CuONPs can be exposed to our environment from many industries, i.e. power stations, smelters, metal foundries as well as asphalt, inkjet printers, and rubber tires (Chen et al. 2006; Karlsson et al. 2008, Ganguly et al. 2018). The Cu-based NPs when discharged in the environment suffer oxidation under ambient atmospheric conditions forming CuONPs and toxic cupric ions which might cause risk to nonpathogenic natural microbes and wildlife, aquatic organisms, and human health (Karlsson et al. 2008, Ganguly et al. 2018). Karlsson et al. (2008) investigated *in vitro* toxicity of CuO, $CuZnFe_2O_4$, Fe_3O_4, Fe_2O_3, ZnO, TiO_2, carbon NPs and multi-walled carbon nanotubes on human lung epithelial cell line (A549), and they found that CuONPs and $CuZnFe_2O_4$ NPs were most powerful concerning cytotoxicity, DNA destruction, and oxidative lesions (Karlsson et al. 2008). In a follow-up study, Meng et al. (2007) experimentally measured the biological toxicity of nano copper particles (23.5 nm) and micro copper particles (17.0 μm) in mice through *in vivo* and *in vitro* studies. The authors found CuNPs induced toxicological effects and substantial injuries on the kidney, liver, and spleen of treated mice, but did not cause any adverse effects of treated mice using the same mass of Cu microparticles. When mice were exposed to NPs, they had observed the changes of blood pH 1.75 to 6.04 as nano copper rapidly interacted with H^+ in artificial gastric juice (Meng et al. 2007). Recently, some *in vitro* studies indicated that CuONPs are particularly powerful in comparison to other metal oxides in terms of triggering cytotoxicity and genotoxicity (Perreault et al. 2012; Zanoni et al. 2019). Ahamed et al. (2010) established genotoxicity of CuONPs through the p53 pathway in human pulmonary epithelial cells (A549) and monitored the expression of DNA damage repair proteins Rad51 and MSH2 which lead to genome aberrations and mutations. Lei et al. (2008) had made a rapid *in vivo* screening of nano copper in rat kidneys through integrated metabolomic analysis. They found CuNPs prompted broad renal proximal tubule necrosis in rat kidneys with increasing levels of carbamide, nitrogen, and creatinine in serum or blood (Lei et al. 2008).

Sharma and Sharma (2007) had reported that when CuNPs (50–60 nm) enter into the central nervous system, they initiated neurotoxicity. According to their review, CuNPs are capable of inducing disruption of normal brain function in animals and aggravating the brain pathology caused by whole-body hyperthermia. The small size of NPs enables their uptake into cells and translocation across epithelial and endothelial cells. In a recent study, Zanoni et al. (2019) documented the toxic action of CuONPs on humans, with a special emphasis on the skin penetration route of access to the epidermis and dermis layer. They experimentally measured the capability of CuONPs to pass through epidermis and dermis layers in the skin and discharge Cu^{2+} ions toward the stratum corneum, arriving at the receptor fluid within the *in vitro* dispersion cell framework that causes a detrimental effect on the skin.

8.7 Safety Implications of Copper and Copper-Based Nanoparticles

The constantly increasing fabrication and use of NPs for diverse industrial and biomedical applications, i.e. therapeutics and diagnostics in addition to consumer goods, have increased serious worries about their protection for human fitness and the surroundings. In addition to using NPs and nanomedicine to identify and to treat infections, it is furthermore necessary to establish NPs' efficiency and safety in biotic systems. After the use of these human-made nanomaterials in the relevant field might release into our environment and interact with air, water, soil, and living organisms leading to nanotoxicity. There is a requirement to overcome nanotoxicity by developing a regulatory framework based on unbiased unprecedented research which will limit human exposure to undesirable Cu and Cu-based NPs in the environment to safe levels (Wang et al. 2015). However, the therapeutic usage of NPs in medicine needs an improved context that balances the therapeutic advantage against the possible hazard of detriment. In the meantime, it is of the greatest importance that companies working with NPs and nanometer-sized materials take proper precautions to protect their staff, environment, and stakeholders. Suitable ecological, health, and safety management ought to comprise of the following points (Friedrichsa and Schulteb 2007; Kagan et al. 2016; and Pelaz et al. 2017).

i) A case-by-case analysis is essential as soon as newly introduced nanomaterials and nanotechnology-enabled products appear. It is also important to consider systematic hazard and risk profiling of intrinsic nanomaterial to implement grouping or read-across.
ii) Suitable attention should be given to nanomaterial compositions that are already measured for safety by the U.S. Food and Drug Administration and contain degradable ingredients that do not lead to biopersistence or bioaccumulation.
iii) It is important to realize complete information about the pristine NPs or nano-enabled ingredients that have moved in the market, their total lifecycle analysis, and the exposure surroundings that could lead to adverse effects on employees, users, and the environment.
iv) It is important to develop novel methodological approaches as well as advanced instruments to empower the automation of barcode-based diagnostic assay and to position the therapeutic stimulus accurately.
v) It is essential to develop a regulatory environment in which NPs and nanostructured materials are considered to be different chemical substances despite their complicated one-dimensional (1D), two-dimensional (2D), three-dimensional (3D) structure, and complex physicochemical properties.

8.8 Future Perspectives

The synthesis of Cu and Cu-based NPs is undoubtedly attractive and challenging because of its catalytic, optical, and electrically conducting properties as well as significant antibacterial, antiviral, antifungal, and anticancer properties. In this chapter, an effort has been made to highlight the significant results in the synthesis and characterization of Cu and Cu-based (Cu, Cu_2O, CuO, and CuI) NPs and their theranostic applications. Among the metal-based materials, environment-friendly and reasonably low-priced Cu, Cu-based nanostructured materials are pretty catalysts for chemical reactions inclusive of the reduction of CO_2 to methane or methanol, carbon–carbon bond formation reactions and carbon–heteroatom (C–N, C–O, C–S) bond forming coupling reactions (Shen et al. 2011; Mallick et al. 2019). Because of their potential antimicrobial properties, copper and copper oxides, NPs have drawn attention to many biomedical and agricultural applications. In recent times, COVID-19 has caused common human morbidity and mortality through an infectious respiratory disease caused by a novel coronavirus SARS-CoV-2. Recent reports recommended that Cu_2O particles bound with polyurethane coating on glass and stainless steel cause a swift and dramatic decrease of SARS-CoV-2 virus within one hour (Behzadinasab et al. 2020). Hence, there is a huge opportunity for further advancement of the antimicrobial and antiviral surface coating of Cu and Cu-based NPs on consumer goods such as doorknobs, credit card buttons, respiratory masks, medical devices in addition to cell phone covers. The incorporation of Cu-based NP in chitosan polymer or combination of chitosan–CuNP composite with iodine can be used to prepare antibacterial food packaging materials and food storage equipment owing to its bactericidal and fungicidal activities. The combination of Cu-based NP along with bio-based oil (soybean oil) will be a promising antibacterial material for the growth inhibition of foodborne pathogens (Bhalerao et al. 2016). The abovementioned studies established that Cu-based NPs or their nanocomposite can be favorably utilized in food packaging to escape the rise in food spoilage microbes. This study can likewise be anticipated to inspire the usage of Cu-based bio-nanocomposites in biomedical applications, disinfectant formulation, food and agriculture industries. The novel formulation of nanoherbicides, nanopesticides, and nanofertilizers with Cu-based bio-nano composites will be a great boon to agriculture. Furthermore, Cu-based nanocomposites can also be used as an ingredients for the preparation of antibacterial paint coating for household materials and colloid ink for writing on substrates such as conducting lines. Many researchers, reports have demanded that CuNPs, CuONPs and Cu_2ONPs also demonstrated harmful effects on plants, the ecosystem, and consequently on human beings when used beyond a particular concentration. Therefore, several researchers have recommended that safer concentration of CuNPs be used in biomedical, agronomical fields, drug delivery systems, and biosensing applications.

8.9 Conclusion

Numerous studies have revealed that Cu and Cu-based NPs (Cu, Cu_2O, CuO, and CuI) of definite sizes and morphologies are produced through physical, chemical, and biological methods. The physical and chemical strategies are time-consuming, tedious, and expensive. However, chemical methods employ harsh and expensive chemicals and are non-biocompatible, therefore possibly dangerous to our environment and biotic systems. Nowadays, there is a serious requirement to develop a reliable, green, one-step eco-friendly procedure for the synthesis of Cu-based NPs for biomedical applications. Therefore, an eco-friendly, easy, reliable, green chemistry process is desirable. To overcome these problems, researchers have recently established new methods of green synthesis which are eco-friendly and cost-effective. In the green synthesis process, the cellular extract of living beings like algae, fungi, bacteria, actinomycetes, plants, and biopolymers have been reported to synthesize several CuNPs and CuONPs. These green methods can be essentially scaled up for the bulk manufacture of NPs rapidly without the generation of toxic byproducts. These Cu and Cu-based NPs are very promising antimicrobial and anticancer agents. When the NPs' size decreases, the proportion of the surface area increases, which promises superior biological activity, i.e. antifungal, antibacterial, anticancer, as compared to larger particles. Innovative research efforts resolve to assist the development of Cu and Cu-based NPs with better-quality and selective anticancer activity, greater biocompatibility and biodistribution, and low toxicity for normal tissues. Notably, it is proved that Cu-based NPs can be used as a conventional anticancer drug with reduced systemic toxicity at the nano-bio interfaces targeted to deliver anticancer drugs, contrast agents in molecular imaging techniques, and photothermal therapy of cancer cells. The awareness of the effects of Cu-based nanomaterials on human health should be generated. Nanotoxicology or toxicity evaluation of these Cu and Cu-based NPs is mandatory to understand their fate inside the human body. Several studies have shown that the high surface-to-volume ratio and higher reactivity of NPs are advantageous but sometimes cause significant toxicity. Nanotoxicology research has identified the numerous complex interactions at bio-nano interfaces between NPs and the living systems. In particular, complex interactions between nano-sized and nanostructured particles and the human immune system, with its diversity, make preclinical safety measurements extremely challenging. Furthermore, conclusions from different toxicological studies of NPs are not mass dependent but their surface area is the most suitable dose-metric for comparing the effects of different sizes and different types of NPs.

References

Abboud, Y., Saffaj, T., and Chagraoui, A. (2014). 'Biosynthesis, characterization and antimicrobial activity of copper oxide nanoparticles produced using brown alga extract (*Bifurcaria bifurcata*). *Applied Nanoscience* 4: 571–576.

Ahamed, M., Siddiqui, M.A., Akhtar, M.J. et al. (2010). Genotoxic potential of copper oxide nanoparticles in human lung epithelial cells. *Biochemical and Biophysical Research Communications* 396 (2): 578–583.

Ahmad, A., Senapati, S., Islam Khan, M. et al. (2003). Extracellular biosynthesis of monodisperse gold nanoparticles by a novel extremophilic actinomycete, *Thermomonospora* sp. *Langmuir* 19 (8): 3550–3553.

Alavi, M. and Rai, M. (2019). Recent advances in antibacterial applications of metal nanoparticles (MNPs) and metal nanocomposites (MNCs) against multidrug-resistant (MDR) bacteria. *Expert Review of Anti-Infective Therapy* 17 (6): 419–428.

Angajala, G., Pasupala, P., and Subashini, R. (2014). One-step biofabrication of copper nanoparticles from *Aegle marmelos* Correa aqueous leaf extract and evaluation of its anti-inflammatory and mosquito larvicidal efficacy. *RSC Advances* 4 (93): 51459–51470.

Arya, A., Gupta, K., Chundawat, T.S., and Vaya, D. (2018). Biogenic synthesis of copper and silver nanoparticles using green alga *Botryococcus braunii* and its antimicrobial activity. *Bioinorganic Chemistry and Applications* 2018: 7879403. 9 pages.

Banerjee, M., Mallick, S., Paul, A. et al. (2010). Heightened reactive oxygen species generation in the antimicrobial activity of a three component iodinated chitosan-silver nanoparticle composite. *Langmuir* 26 (8): 5901–5908.

Barnes, W., Dereux, A., and Ebbesen, T. (2003). Surface plasmon subwavelength optics. *Nature* 424 (6950): 824–830.

Barreca, D., Fornasiero, P., Gasparotto, A. et al. (2009). The potential of supported Cu_2O and CuO nanosystems in photocatalytic H_2 production. *ChemSusChem* 2 (3): 230–233.

Behzadinasab, S., Chin, A., Hosseini, M. et al. (2020). 'A surface coating that rapidly inactivates SARS-CoV-2. *ACS Applied Materials and Interfaces* 12 (31): 34723–34727.

Bhalerao, M.S., Patwardhan, A.V., Bhosale, M.A. et al. (2016). Epoxidised soybean oil–Cu/Cu_2O bio-nanocomposite material: synthesis and characterization with antibacterial activity. *RSC Advances* **6** (45): 38906–38912.

Borkow, G. and Gabbay, J. (2005). Copper as a biocidal tool. *Current Medicinal Chemistry* 12 (18): 2163–2275.

Brumbaugh, A.D., Cohen, K.A., and Angelo St, S.K. (2014). Ultrasmall copper nanoparticles synthesized with a plant tea reducing agent. *ACS Sustainable Chemistry Engineering* 2 (8): 1933–1939.

Cao, G. (2011). *Nanostructures and nanomaterials: synthesis, properties andapplications*, 2e. World Scientific Publishing Company.

Cater, M.A., Fontaine, S.L., Shield, K. et al. (2006). ATP7B mediates vesicular sequestration of copper: insight into biliary copper excretion. *Gastroenterology* 130 (2): 493–506.

Chavez Soria, N.G., Aga, D.S., and Atilla-Gokcumen, G.E. (2019). Lipidomics reveals insights on the biological effects of copper oxide nanoparticles in a human colon carcinoma cell line. *Molecular Omics* 15 (1): 30–38.

Cheirmadurai, K., Biswas, S., Ragothaman, M., and Palanisamy, T. (2014). Green synthesis of copper nanoparticles and conducting nanobiocomposites using plant and animal sources. *RSC Advances* 4 (37): 19507–19511.

Chen, Z., Meng, H., Xing, G. et al. (2006). Acute toxicological effects of copper nanoparticles in vivo. *Toxicology Letters* 163 (2): 109–120.

Chen, J., Mao, S., Xu, Z., and Ding, W. (2019). Various antibacterial mechanisms of biosynthesized copper oxide nanoparticles against soilborne *Ralstonia solanacearum*. *RSC Advances* 9 (7): 3788–3799.

Chow, E.K. and Ho, D. (2013). Cancer nanomedicine: from drug delivery to imaging. *Science Translational Medicine* 5 (216): 216rv4.

Cuevas, R., Durán, N., Diez, M.,.C. et al. (2015). 'Extracellular biosynthesis of copper and copper oxide nanoparticles by *Stereum hirsutum*, a native white-rot fungus from chilean forests. *Journal of Nanomaterials* 2015: 789089. 7 pages.

Dhas, N.A., Raj, C.P., and Gedanken, A. (1998). Synthesis, characterization, and properties of metallic copper nanoparticles. *Chemistry of Material* 10 (2): 1446–1452.

Dong, K., Liu, Z., Li, Z. et al. (2013). Hydrophobic anticancer drug delivery by a 980 nm laser-driven photothermal vehicle for efficient synergistic therapy of cancer cells in vivo. *Advanced Materials* 25 (32): 4452–4458.

Elsaesser, A. and Howard, C.V. (2012). Toxicology of nanoparticles. *Advanced Drug Delivery Reviews* 64 (2): 129–137.

Fahmy, B. and Cormier, S.A. (2009). Copper oxide nanoparticles induce oxidative stress and cytotoxicity in airway epithelial cells. *Toxicology in vitro* 23 (23): 1365–1371.

Friedrichsa, S. and Schulteb, J. (2007). Environmental, health and safety aspects of nanotechnology—implications for the R&D in (small) companies. *Science and Technology of Advanced Materials* 8 (1–2): 12–18.

Fujimori, Y., Sato, T., Hayata, T. et al. (2012). Novel antiviral characteristics of nanosized copper (I) iodide particles showing inactivation activity against 2009 pandemic H1N1 influenza virus. *Applied and Environmental Microbiology* 78 (4): 951–955.

Gaggelli, E., Kozlowski, H., Valensin, D., and Valensin, G. (2006). Copper homeostasis and neurodegenerative disorders (Alzheimer's, prion, and Parkinson's diseases and amyotrophic lateral sclerosis). *Chemical Reviews* 106 (6): 1995–2044.

Gahlawat, G. and Choudhury, A.R. (2019). A review on the biosynthesis of metal and metal salt nanoparticles by microbes. *RSC Advances* 9 (23): 12944–12967.

Ganguly, P., Breen, A., and Pillai, S.C. (2018). Toxicity of nanomaterials: exposure, pathways, assessment, and recent advances. *ACS Biomaterials Science and Engineering* 4 (7): 2237–2275.

Halevas, E.G. and Pantazaki, A.A. (2018). Copper nanoparticles as therapeutic anticancer agents. *Nanomedicine and Nanotechnology Journal* 2 (1): 119.

Hang, X., Peng, H., Song, H. et al. (2015). Antiviral activity of cuprous oxide nanoparticles against hepatitis C virus in vitro. *Journal of Virological Methods* 222: 150–157.

Hariprasad, S., Bai, G.S., Santhosh Kumar, J. et al. (2016). Green synthesis of copper nanoparticles by *Aerva lanata* leaves extract and their anti-microbial activities. *International Journal of ChemTech Research* 9 (2): 98–105.

Hasan, S.S., Singh, S., and Parikh, R.Y. (2008). Bacterial synthesis of copper/copper oxide nanoparticles. *Journal of Nanoscience Nanotechnology* 8 (6): 3191–3196.

Hassan, S.E., Fouda, A., Radwan, A.A. et al. (2019). Endophytic actinomycetes Streptomyces spp mediated biosynthesis of copper oxide nanoparticles as a promising tool for biotechnological applications. *Journal of Biological Inorganic Chemistry* 24 (3): 377–393.

He, H., Liu, L., Morin, E.E. et al. (2019). Survey of clinical translation of cancer nanomedicines—lessons learned from successes and failures. *Accounts of Chemical Research* 52 (9): 2445–2461.

Honary, S., Barabadi, H., Gharaeifathabad, E., and Naghibi, F. (2012). Green synthesis of copper oxide nanoparticles using *Penicillium aurantiogriseum, Penicillium citrinum and Penicillium waksmanii. Digest Journal of Nanomaterials and Biostructures* 7: 999–1005.

Horie, M., Ogawa, H., Yoshida, Y. et al. (2008). Inactivation and morphological changes of avian influenza virus by copper ions. *Archives of Virology* 153 (8): 1467–1472.

Huang, H.H., Yan, F.Q., Kek, Y.M. et al. (1997). Synthesis, characterization, and nonlinear optical properties of copper nanoparticles. *Langmuir* 13 (2): 172–175.

Ingle, A.P., Duran, N., and Rai, M. (2014). Bioactivity, mechanism of action, and cytotoxicity of copper-based nanoparticles: a review. *Applied Microbiology and Biotechnology* 98 (3): 1001–1009.

Kagan, C.R., Fernandez, L.E., Gogotsi, Y. et al. (2016). Nano day: celebrating the next decade of nanoscience and nanotechnology. *ACS Nano* 10 (10): 9093–9103.

Kamlesh, K., Prashant, S., Kuldeep, B. et al. (2019). Implications of metal nanoparticles on aquatic fauna: a review. *Nanoscience and Nanotechnology-Asia* 9 (1): 30–43.

Karlsson, H.L., Cronholm, P., Gustafsson, J., and Moller, L. (2008). Copper oxide nanoparticles are highly toxic: a comparison between metal oxide nanoparticles and carbon nanotubes. *Chemical Research in Toxicology* 21 (9): 1726–1732.

Ku, G., Zhou, M., Song, S. et al. (2012). Copper sulfide nanoparticles as a new class of photoacoustic contrast agent for deep tissue imaging at 1064 nm. *ACS Nano* 6 (8): 7489–7496.

Laha, D., Pramanik, A., and Maity, J. (2014). Interplay between autophagy and apoptosis mediated by copper oxide nanoparticles in human breast cancer cells MCF7. *Biochim Biophys Acta* 1840 (1): 1–9.

Laha, D., Pramanik, A., Chattopadhyay, S. et al. (2015). Folic acid-modified copper oxide nanoparticles for targeted delivery in vitro and in vivo systems. *RSC Advances* 5 (83): 68169–68178.

Lei, R., Wu, C., Yang, B. et al. (2008). Integrated metabolomic analysis of the nano-sized copper particle-induced hepatotoxicity and nephrotoxicity in rats: a rapid in vivo screening method for nanotoxicity. *Toxicology and Applied Pharmacology* 232 (2): 292–301.

Lengke, M.F., Fleet, M.E., and Southam, G. (2007). Biosynthesis of silver nanoparticles by filamentous cyanobacteria from a silver (I) nitrate complex. *Langmuir* 23 (5): 2694–2699.

Link, S. and El-Sayed, M.A. (2000). Shape and size dependence of radiative, non-radiative and photothermal properties of gold nanocrystals. *International Reviews in Physical Chemistry* 19 (3): 409–453.

Liz-Marzan, L.M. (2004). Nanometals: Formation and color. *Materialstoday* 7 (2): 26–31.

Mallick, S. (2018). Synthesis and characterization of Cu and Ag based mixed metal nanoparticle composite for antibacterial applications. *Indian Journal of Chemistry* 57A (10): 1249–1256.

Mallick, S., Sharma, S., Ghosh, S.S. et al. (2012). Iodine-stabilized Cu nanoparticle chitosan composite for antibacterial applications. *ACS Applied Materials and Interfaces* 4 (3): 1313–1323.

Mallick, S., Sanpui, P., Ghosh, S.S. et al. (2015). Synthesis, characterization and enhanced bactericidal action of chitosan supported core-shell copper-silver nanoparticle composite. *RSC Advances* 5 (16): 12268–12276.

Mallick, S., Mukhi, P., Kumari, P. et al. (2019). Synthesis, characterization and catalytic application of starch supported cuprous iodide nanoparticles. *Catalysis Letters* 149: 3501–3507.

Meng, H., Chen, Z., Xing, G. et al. (2007). Ultrahigh reactivity provokes nanotoxicity: explanation of oral toxicity of nano-copper particles. *Toxicology Letters* 175 (1–3): 102–110.

Merkel, T.J., Herlihy, K.P., Nunes, J. et al. (2010). Scalable, shape-specific, top-down fabrication methods for the synthesis of engineered colloidal particles. *Langmuir* 26 (16): 13086–13096.

Murphy, C.J., Sau, T.K., Gole, A.M. et al. (2005). Anisotropic metal nanoparticles: synthesis, assembly, and optical applications. *The Journal of Physical Chemistry B* 109 (29): 13857–13870.

Musselman, K.P., Marin, A., Schmidt-Mende, L., and MacManus-Driscoll, J.L. (2012). Incompatible length scales in nanostructured

Cu_2O solar cells. *Advanced Functional Materials* 22 (10): 2202–2208.

Nagar, N. and Devra, V. (2018). Green synthesis and characterization of copper nanoparticles using *Azadirachta indica* leaves. *Materials Chemistry and Physics* 213: 44–51.

Nasrollahzadeh, M. and Sajadi, S.M. (2015). Green synthesis of copper nanoparticles using *Ginkgo biloba L.* leaf extract and their catalytic activity for the Huisgen [3+2]cycloaddition oazides. *Journal of Colloid and Interface Science* 457 (1): 141–147.

Nasrollahzadeh, M., Sajadi, S.M., and Mehdi, K. (2014). Green synthesis of copper nanoparticles using aqueous extract of the leaves of *Euphorbia esula L* and their catalytic activity for ligand-free Ullmann-coupling reaction and reduction of 4-nitrophenol. *RSC Advance* 4 (88): 47313–47318.

Nasrollahzadeh, M., Sajadi, S.M., and Maham, M. (2015a). *Tamarix gallica* leaf extract mediated novel route for green synthesis of CuO nanoparticles and their application for N-arylation of nitrogen-containing heterocycles under ligand-free conditions. *RSC Advance* 5 (51): 40628–40635.

Nasrollahzadeh, M., Sajadi, S.M., and Rostami-vartooni, A. (2015b). Green synthesis of CuO nanoparticles by aqueous extract of *Anthemis nobilis* flowers and their catalytic activity for the A 3 coupling reaction. *Journal of Colloid and Interface Science* 459 (1): 183–188.

Oberdörster, G. (2010). Safety assessment for nanotechnology and nanomedicine: concepts of nanotoxicology. *Journal of Internal Medicine* 267 (1): 89–105.

Oberdörster, G., Stone, V., and Donaldson, K. (2007). Toxicology of nanoparticles: A historical perspective. *Nanotoxicology* 1 (1): 2–25.

Pariona, N., Mtz-Enriquez, A.I., Sanchez-Rangel, D. et al. (2019). Green-synthesized copper nanoparticles as a potential antifungal against plant pathogens. *RSC Advances* 9 (33): 18835–18843.

Pelaz, B., Alexiou, C., Alvarez-Puebla, R.A. et al. (2017). Diverse Applications of Nanomedicine. *ACS Nano* 11 (3): 2313–2381.

Perlman, O., Weitz, I.S., and Azhari, H. (2015). Copper oxide nanoparticles as contrast agents for MRI and ultrasound dual-modality imaging. *Physics in Medicine and Biology* 60 (15): 5767–5783.

Perreault, F., Pedroso Melegari, S., Henning da Costa, C. et al. (2012). Genotoxic effects of copper oxide nanoparticles in Neuro 2A cell cultures. *Science of The Total Environment* 441 (15): 117–124.

Ponmurugan, P., Manjukarunambika, K., Elango, V., and Gnanamangai, B.M. (2016). Antifungal activity of biosynthesized copper nanoparticles evaluated against red root-rot disease in tea plants. *Journal of Experimental Nanoscience* 11 (13): 1019–1031.

Prajapati, J.P., Das, D., Katlakunta, S. et al. (2021). Synthesis and characterization of ultrasmall Cu_2O nanoparticles on silica nanoparticles surface. *Inorganica Chimica Acta* 515: 120069. https://doi.org/10.1016/j.ica.2020.120069.

Rai, M., Yadav, A., and Gade, A. (2008). CRC675—current trends in photosynthesis of metal nanoparticles. *Critical Reviews in Biotechnology* 28 (4): 277–284.

Rai, M., Ingle, A.P., Pandit, R. et al. (2018). 'Copper and copper nanoparticles: role in management of insect-pests and pathogenic microbes. *Nanotechnology Reviews* 7 (4): 303–315.

Ramadan, S., Guo, L., Li, Y. et al. (2012). Hollow copper sulfide nanoparticle-mediated transdermal drug delivery. *Small* 8 (20): 3143–3150.

Rasool, U. and Hemalatha, S. (2017). Marine endophytic actinomycetes assisted synthesis of copper nanoparticles (CuNPs): characterization and antibacterial efficacy against human pathogens. *Materials Letters* 194: 176–180.

Ray, D., Pramanik, S., Prasad Mandal, R. et al. (2015). Sugar-mediated 'green' synthesis of copper nanoparticles with high antifungal activity. *Materials Research Express* 2 (10): 105002.

Rosi, N.L. and Mirkin, C.A. (2005). Nanostructures in biodiagnostic. *Chemical Reviews* 105 (4): 1547–1562.

Sastry, M., Absar, A., Islam Khan, M., and Kumar, R. (2003). Biosynthesis of metal nanoparticles using fungi and actinomycete. *Current Science* 85 (2): 162–170.

Schwartzberg, A.M. and Zhang, J.Z. (2008). Novel optical properties and emerging applications of metal nanostructures. *The Journal of Physical Chemistry C* 112 (28): 10323–10337.

Sharifi, S., Behzadi, S., Laurent, S. et al. (2012). Toxicity of nanomaterials. *Chemical Society Reviews* 41 (6): 2323–2343.

Sharma, H.S. and Sharma, A. (2007). 'Nanoparticles aggravate heat stress-induced cognitive deficits, blood-brain barrier disruption, edema formation, and brain pathology. *Progress in Brain Research* 162: 245–273.

Shen, S., Zhuang, J., Yang, Y., and Wang, X. (2011). Highly monodisperse Cu- and Ag-based bimetallic nanocrystals for the efficient utilization of noble metals in catalysis. *Nanoscale* 3, 2 (1): 72–279.

Singh, A.V., Patil, R., Anand, A. et al. (2010). Biological synthesis of copper oxide nano particles using Escherichia coli. *Current Nanoscience* 6 (4): 365–369.

Singh, A., Gautam, P.K., Verma, A. et al. (2020). Green synthesis of metallic nanoparticles as effective alternatives to treat antibiotics resistant bacterial infections: a review. *Biotechnology Reports* 25: e00427.

Studer, A.M., Limbach, L.K., and Van, D.L. (2010). Nanoparticle cytotoxicity depends on intracellular solubility: comparison of stabilized copper metal and degradable copper oxide nanoparticles. *Toxicology Letters* 197 (3): 169–174.

Sukumar, S., Rudrasenan, A., and Nambiar, D.P. (2020). Green-synthesized rice-shaped copper oxide nanoparticles using caesalpinia bonducella seed extract and their applications. *ACS Omega* 5 (2): 1040–1051.

Sutradhar, P., Saha, M., and Maiti, D. (2014). 'Microwave synthesis of copper oxide nanoparticles using tea leaf and coffee powder extracts and its antibacterial activity. *Journal of Nanostructure in Chemistry* 4: 86. 6 pages.

U.S. EPA (1991). National primary drinking water regulations: final rule. *Federal Register* 56 (20): 3526–3597.

Usha, S., Ramappa, K.T., Hiregouda, S. et al. (2017). Biosynthesis and characterization of copper nanoparticles from tulasi (*Ocimum sanctum L.*) leaves. *International Journal of Current Microbiology and Applied Sciences* 6 (11): 2219–2228.

Valodkar, M., Modi, S., Pal, A., and Thakore, S. (2011). Synthesis and anti-bacterial activity of Cu, Ag and Cu–Ag alloy nanoparticles. *A green approach Materials Research Bulletin* 46 (3): 384–389.

Vanathi, P., Rajiv, P., and Sivaraj, R. (2016). Synthesis and characterization of *Eichhornia*-mediated copper oxide nanoparticles and assessing their antifungal activity against plant pathogens. *Bulletin of Materials Science* 39 (5): 1165–1170.

Varshney, R., Bhadauria, S., Gaur, M.S., and Pasricha, R. (2010). Characterization of copper nanoparticles synthesized by a novel microbiological method. *The Journal of The Minerals, Metals and Materials Society* 62 (12): 102–104.

Vijay Kumar, P.P.N., Shameem, U., Kollu, P. et al. (2015). Green synthesis of copper oxide nanoparticles using Aloe vera leaf extract and its antibacterial activity against fish bacterial pathogens. *BioNanoScience* 5 (3): 135–139.

Vukojevic, S., Trapp, O., Grunwaldt, J.-D. et al. (2005). Quasi-homogeneous methanol synthesis over highly active copper nanoparticles. *Angewandte Chemie International Edition* 44 (48): 7978–7981.

Wang, X., Zhuang, J., Peng, Q., and Li, Y. (2005). General strategy for nanocrystal synthesis. *Nature* 437: 121–124.

Wang, Z., Li, N., Zhao, J. et al. (2012). CuO nanoparticle interaction with human epithelial cells: cellular uptake, location, export, and genotoxicity. *Chemical Research Toxicology* 25 (7): 1512–1521.

Wang, Y., Santos, A., Evdokiou, A., and Losic, D. (2015). An overview of nanotoxicity and nanomedicine research: principles, progress, and implications for cancer therapy. *Journal of Materials Chemistry B.* 3 (36): 7153–7172.

WHO (1998). *IPCS Environmental Health Criteria: Copper*. Geneva: World Health Organization.

Wongrakpanich, A., Mudunkotuwa, I.A., Geary, S.M. et al. (2016). Size-dependent cytotoxicity of copper oxide nanoparticles in lung epithelial cells. *Environmental Science: Nano* 3 (2): 365–374.

Xiong, J., Wang, Y., Xue, Q., and Wu, X. (2011). Synthesis of highly stable dispersions of nanosized copper particles using L-ascorbic acid. *Green Chemistry* 13 (4): 900–904.

Yadav, S., Jain, A., and Malhotra, P. (2019). A review on the sustainable routes for the synthesis and applications of cuprous oxide nanoparticles and their nanocomposites. *Green Chemistry* 21 (5): 937–955.

Zanoni, I., Crosera, M., Ortelli, S. et al. (2019). CuO nanoparticle penetration through intact and damaged human skin. *New Journal of Chemistry* 43 (43): 17033–17039.

Zhang, Y.-R., Lin, R., Li, H.-J. et al. (2019). Strategies to improve tumor penetration of nanomedicines through nanoparticle design. *WIREs Nanomedicine and Nanobiotechnology* 11 (1): e1519.

Zhou, M., Li, J., Liang, S. et al. (2015). CuS nanodots with ultrahigh efficient renal clearance for positron emission tomography imaging and image-guided photothermal therapy. *ACS Nano* 9 (7): 7085–7096.

Zhou, M., Tian, M., and Li, C. (2016). Copper-based nanomaterials for cancer Imaging and therapy. *Bioconjugate Chemistry* 27 (5): 1188–1199.

Part III

Nanotoxicology and Drug Delivery

9

Gene Delivery Using Nanocarriers

Toxicity and Safety Aspects

Imran Vhora[1] *and Nirav Khatri*[2]

[1] *Formulation R&D Consultant, Burnaby, British Columbia, Canada*
[2] *Dr. Reddy's Laboratories Ltd., Hyderabad, Telangana, India*

9.1 Introduction

Gene therapy has evolved recently to be a more effective alternative to treat life-threatening diseases. The delivery strategies have been designed with the aim of improving their effectiveness to carry the cargo to the target tissues and inside the target cells which are the ultimate regulators of the therapeutic effectiveness. Viral vectors which have natural machinery that can effectively deliver the genetic cargo inside the cell have been explored extensively preclinically and clinically. However, already identified issues of immunogenicity, mutagenicity, and carcinogenicity occur due to either viral infection or the presence of viral proteins in viral vectors that had led to current inclination toward nonviral vectors (Glover et al. 2005). Nanocarrier systems have already advanced in the field of small molecule drug delivery with successful results. This has induced a great deal of research in nanoparticles generating enough know-how for the development path to regulatory approval. A few examples are Abraxane, Doxil, and Myocet. This also induced the detailed evaluation of nanocarriers in gene delivery field as well. The advantages start from the availability to choose from various nanocarrier systems with different physicochemical properties made from a wide range of polymers, lipids, and combinations thereof. Moreover, better biocompatibility compared with viral systems provides a strong rationale for their development. Also, knowledge generated from approved small molecule nanocarriers allows for easily scalable and reproducible manufacturing with desired gene loading and particle characteristics. Success is seen in the form of a large number of clinical trials concluded/being carried out for lipidic nanocarriers (Zhao and Huang 2014; Vhora et al. 2015; Zatsepin et al. 2016) and polymeric nanocarriers (Davis et al. 2010; Zuckerman et al. 2014; Buscail et al. 2015). The recent approval of Onpattro™ (Lipid nanoparticulate formulation of siRNA for treatment of hereditary ATTR [hATTR] amyloidosis) by Alnylam Pharmaceuticals in the United States and EU in August 2018 and Japan in June 2019 has again boosted the research in nanocarrier gene delivery systems.

Nanocarrier systems provide a multitude of advantages such as protection of genetic cargo from nucleases, suitability for surface modification for tissue/cell targeting, composition modification for reducing immunogenic response, and impart discrete gene cargo release characteristics (Balazs and Godbey 2011; Patil et al. 2016, 2018; Vhora et al. 2019). Moreover, the tunable physicochemical characteristics allow

Nanotechnology in Medicine: Toxicity and Safety, First Edition. Edited by Mahendra Rai, Mrunali Patel, and Rashmin Patel.

for delivery of large genetic payloads (Mintzer and Simanek 2009), administration route-specific modifications of nanoparticle characteristics (Khatri et al. 2014a; Patil et al. 2018; Vhora et al. 2018), possibility to modify particle size for effective distribution in remote organs after intravenous administration (Vhora et al. 2014; Bhatt et al. 2018; Vhora et al. 2018), and possibility for surface modification for cellular targeting (Khatri et al. 2014b; Vhora et al. 2019).

Gene delivery nanocarriers are generally designed with the help of cationic ingredients such as cationic polymers, cationic lipids, or cationic inorganic salts through the electrostatic interactions with phosphates of therapeutic gene. Though cationic polymers such as poly-L-lysine, polyhistidine, and polyethyleneimine (PEI) have attracted much attention for their use in gene therapy, lipid-based gene delivery still remains the most widely explored field and has been successful in reaching clinical trials in the highest number as gene delivery nanocarriers. The recent approval of Onpattro™ by Alnylam Pharma marks the first instance of a lipid nanoparticle (LNP)-based siRNA delivery system being approved. Recently, the global pandemic due to COVID-19 has fueled the rapid development of nucleic acid-based vaccines as well as prompt authorizations by regulatory agencies of several clinical trials. These include >17 vaccine candidates in clinical trials (of which five are mRNA vaccines and two are DNA vaccines) and >15 RNA/DNA vaccine candidates in preclinical evaluation worldwide within the last 6 months (WHO 2020). In this chapter, we focus on the clinical investigations of nanocarrier-based gene therapy with a special focus on their safety and toxicity evaluations.

9.2 Nanocarrier Types

Considerable efforts have been put into the development of safe and effective *in vivo* gene delivery carriers as an alternative to viral vectors in the last few decades. Research so far has put forth a wide variety of gene delivery nanocarriers that span from lipid-based nanoparticles to polymeric nanoparticles, hybrid systems, and several other types. The majority of these nanocarriers are manufactured through ionic interaction with cationic excipients by taking advantage of the negatively charged nature of genetic cargoes. While there are many nanocarriers developed and studied *in vitro* and preclinically, only a few types have made it to the human clinical trials. Schematic representation of a few nanocarrier systems which are most commonly seen in the clinical trials is shown in Figure 9.1.

9.2.1 Lipid-Based Nanocarriers

Lipid nanocarriers so far have been in the largest in number to reach clinical trials for various critical and rare diseases. These nanocarriers use a wide range of nucleic acid complexing lipids which exist as cationic lipids at physiologic pH due to quaternary amine groups or as ionizable lipids showing pH-dependent ionization due to primary or secondary amine groups enabling them to complex with nucleic acid at acidic pH while remaining neutrally charged at physiological pH. Lipid nanocarriers exist in a wide range of configurations which exhibit different gene encapsulation characteristics and surface characteristics that govern their therapeutic effectiveness and safety profiles. Examples include lipid nanoparticles (LNPs), lipoplexes, and targeted LNPs. To date, lipid-based nanocarriers have been evaluated for delivery of majority of nucleic acid types including DNAs and RNAs (small RNAs - siRNAs, miRNAs, tRNAs as well as longer RNAs - long noncoding RNAs, mRNAs and self-amplifying mRNAs).

One important example in this category is Onpattro™ by Alnylam Pharmaceuticals, which is an LNP formulation of siRNA and is currently approved for the treatment of hATTR amyloidosis - a rare and serious liver disease. These LNPs are specifically designed for liver delivery of siRNA (Zhang et al. 2019) and are made using ionizable cationic lipid (i.e. DLin-MC3-DMA lipid or MC3 lipid in Onpattro™) which is designed to reduce the

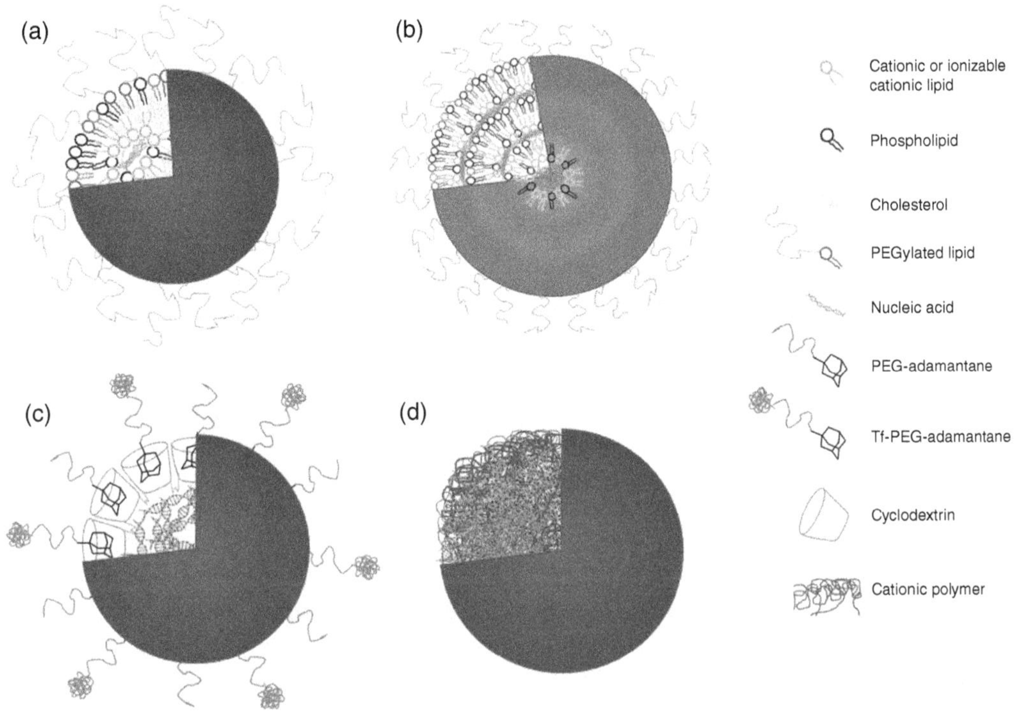

Figure 9.1 (a and b) represent different configurations of solid-core lipid nanoparticles used for delivery of nucleic acids. (a) Schematic representing the LNPs with nucleic acid loaded in water-filled cavities surrounded by lipid inverted micellar phases which, in turn, are surrounded by lipid monolayer surface with PEG chains of PEG lipids facing the exterior aqueous phase. (b) Schematic of lipid nanoparticles or lipoplex structure with nucleic acid loaded in an aqueous filled core and subsequent aqueous layers sandwiched between lipid bilayers made of cationic/ionizable cationic lipid, phospholipid, cholesterol, and PEG lipids. (c and d) represent different configurations of polyplex-based nanoparticles made using cationic polymers and nucleic acids. (c) Schematic of CALAA-1 nanoparticle. Adamantane-PEG and adamantane-PEG-transferrin modifies the nanoparticle surface through inclusion complex formation between adamantane group and cyclodextrin and cationic portion of the cationic cyclodextrin polymer (CDP) complexes with siRNA in core. (d) Schematic of polyplex-based nanoparticles made with any polycationic polymer such as polyhistidine, polylysine, HKP (histidine-lysine copolymer), PEI (polyethyleneimine), or other.

safety/toxicity issues with the first-generation cationic lipids. Also, Onpattro includes high levels of cholesterol and a specific PEG-lipid (DMG-PEG or PEG2000-C-DMG – which is methoxy-PEG2000-carbamoyl-dimyristyl glyceride) which can dissociate from the surface of LNPs after intravenous administration so that the LNPs behave similar to low-density lipoprotein (LDL) particles enabling them to recruit apolipoprotein-E (ApoE) and get internalized by hepatocytes through the LDL receptor pathway (Zhang et al. 2019). Other noteworthy examples in the LNP category are Moderna's and BioNTech/Pfizer's LNP formulations designed for intramuscular delivery of mRNA vaccines expressing corona virus antigen and currently are in the ongoing phase III clinical trials and are authorized for emergency clinical use (Jackson et al. 2020; Moderna Therapeutics 2020; Mulligan et al. 2020; Pfizer and BioNTech 2020).

There also have been several examples of liposomal/lipoplex-based gene delivery systems in clinical trials (Table 9.1). One such particular system has been targeted using an antibody fragment, for example, SGT-53, which is a liposomal formulation encapsulating pDNA expressing p53 targeted to metastatic lesions through transferrin receptors with the help of transferrin tagged DOPE (Senzer et al. 2013).

9.2.2 Polymeric Nanocarriers

Certain cationic polymer-based nanocarriers have also advanced to human gene therapy due to their variety and flexibility to impart certain characteristic features to nanocarriers. Examples in this category include cationic cyclodextrin polymer (CDP)-based transferrin receptor targeted nanoparticles for siRNA delivery in cancer (Zuckerman et al. 2014), PEI-based nanoparticles for pDNA as well as RNA oligonucleotide delivery for cancers (Buscail et al. 2015), and PEGylated poly-lysin nanoparticles for cystic fibrosis (CF) patients (Konstan et al. 2004). Among various polymer-based systems, CALAA-1 is the first polymer-based nanoparticle system for delivery of siRNA to ever reach clinical trial for use as an intravenous therapy. Another polymeric nanocarrier STP705 (also known as Cotsiranib) by Sirnaomics is a dry powder formulation of histidine-lysine copolymer (HKP)-based nanoparticles incorporating two siRNA oligonucleotides against TGF-β1 and Cox-2 mRNAs at 1 : 4 siRNA to peptide weight ratio and is being evaluated in phase I/II studies for hypertrophic scar treatment (Sirnaomics 2016) and in phase I/II studies for cutaneous squamous cell carcinoma *in situ* (isSCC, Bowen's disease) (ClinicalTrials.gov identifier NCT04293679). Cotsiranib is also being evaluated for its intravenous potential to deliver siRNAs to the liver. Protamine is another polymer that is used in development of mRNA vaccines by CurVac. CV7201 is CureVac's mRNA-based rabies vaccine that is a lyophilized formulation of rabies virus glycoprotein expressing mRNA in free and complex form with protamine (Alberer et al. 2017).

A study also evaluated PEGylated PEI-based lipopolymeric nanoparticles for delivery of pDNA expressing IL-12 for ovarian, fallopian tube, or peritoneal cancers taking advantage of polymeric and lipidic excipients to achieve safety and efficacy endpoints (Alvarez et al. 2014). There are a wide range of cationic polymer or polymer conjugates evaluated *in vitro* and *in vivo* with the aim of improving the therapeutic efficacy and to reduce the side effects. For example, chitosan nanoparticles, chitosan-protamine conjugate-based nanoparticles, etc.

A few nanocarrier systems which are developed based on viral vectors have also reached clinical trials for their immunostimulatory applications. For example, CYT003-QbG10 is virus like particles (VLPs) which are self-assembled coat proteins of bacteriophage Qb during recombinant expression (Kozlovska et al. 1993). It is evaluated in Phase II trials for the treatment of allergic conjunctivitis (Klimek et al. 2011). Another nanocarrier LSFV-IL12 is a cationic liposome encapsulating genetically modified non-replicative Semliki forest virus (SFV) carrying IL-12 gene for intratumoral administration in glioblastoma multiforme patients (Konstan et al. 2004).

9.3 Target Diseases and Routes of Administration

Table 9.1 gives a detailed description of nanocarrier systems evaluated/being evaluated in different phases of clinical trials for different therapeutic indications. Therapeutic indications for the gene delivery nanocarriers are usually genetically linked diseases and especially life-threatening or critical diseases where they can make the most impact. Among all the therapeutic indications, cancer remains the most explored disease area with a maximum number of clinical trials enrolled for gene delivery through viral vectors as well as nonviral nanocarriers. Clinical trials related to gene therapy with nanocarriers target melanomas, head and neck carcinoma, renal cancer, breast cancer, ovarian and other gynecological cancers, pancreatic cancer, lymphoma, hepatic cancer, and solid tumors. Most widely explored cancer types lie in advanced/metastatic solid tumors. Few examples include LNP-based delivery of siRNA against VEGF and KSP in

Table 9.1 Clinically evaluated gene delivery nanocarriers and their toxicity and safety evaluations.

Nanocarrier type and subtype, administration route	Carrier composition	Genetic cargo	Therapeutic indication (phase) – clilincaltrials.gov identifier	Clinical toxicity or adverse events data	References
Intravenous administration					
LNP-amino lipid, IV infusion	DLinDMA	siRNA targeting VEGF and KSP	Cancers (I)	• All patients were premedicated before each dose with dexamethasone, the night before and dexamethasone, acetaminophen, diphenhydramine, or hydroxyzine, and either ranitidine or famotidine, 30 min before infusion to minimize the risk of IRRs observed with liposomal products • Treatment was well tolerated with mostly low-grade fatigue/asthenia, nausea-vomiting, and fever occurring in 15–24% of patients without any clear dose dependence • IRRs (mainly facial flushing, chest tightness, back/abdominal pain, elevated heart rate, or sweating) occurred after first dose in 15% patients and were easily managed by temporary interruption of infusion and giving the remaining dose over 0.5–1 hr • Dose dependent increase in complement was linked with immediately occurring IRRs with predominant involvement of Bb complement • Other grade 1–2 chills/rigors were delayed AEs linked with transient cytokine induction such as IP-10, IL-1RA, G-CSF, and TNF-α. • dose limiting toxicities: grade 4 hepatic failure leading to death (1 out of 7) at 0.7 mg/kg dose, grade 3 thrombocytopenia (2 out of 11) at 1.25 mg/kg dose, grade 3 hypokalemia (1 out of 2) at 1.5 mg/kg • 1 death was found to be due to existing pancreatic neuroendocrine tumor (PNET) with >70% liver involvement with metastatic disease and prior splenectomy and prior hepatectomy, further trial excluded patients with >50% liver involvement or prior splenectomy • other 40 out of 41 patients showed no clinically significant liver function changes • recommended dose for phase II was 1.0 mg/kg	Tabernero et al. (2013)

(*Continued*)

Table 9.1 (Continued)

Nanocarrier type and subtype, administration route	Carrier composition	Genetic cargo	Therapeutic indication (phase) – clilincaltrials.gov identifier	Clinical toxicity or adverse events data	References
LNP-amino lipid, IV	DLinDMA	siRNA against PLK1	Solid tumors (I) – NCT01262235	• TKM-PLK1 was well tolerated with most common drug-related AEs of mild-to-moderate fever, chills, nausea, vomiting, and fatigue • MTD was determined to be 0.75 mg/kg body weight/week • No cumulative toxicity was observed in two patients who received therapy for six months	Ramanathan et al. (2013)
PNP-hTfR targeted cyclodextrin-based PEGylated nanoparticle, IV	Cationic cyclodextrin polymer/ adamantane-PEG/ adamantane PEG-hTf	siRNA against RRM2	Melanoma (I) – NCT00689065 cancers (Ia/Ib)	• CALAA-1 has been first polymer-based system to be translated to clinic for intravenous gene therapy • Surface modification through PEGylation and TfR targeting enabled C • Animal toxicology similarly extrapolated to humans in clinical trials except for animal kidney toxicities which were not observed in humans due to the use of predosing hydration protocol • Most common AEs related to treatment affecting >15% of patients were fatigue, chills, and fever • Grade 3/4 AEs occurring in multiple patients were lymphopenia ($n = 3$) and fatigue ($n = 2$) • Other grade 3/4 AEs were hypersensitivity, ischemic colitis, diarrhea, fever, and hyponatremia ($n = 1$ each) • conclusive association of plasma siRNA level with AEs and of elevated cytokine levels with severity was limited in this study • Dose-limiting toxicity were not causally related to drug as plasma levels were same in patients who experienced DLTs vs those who didn't	Davis et al. (2010) and Zuckerman et al. (2014)
LNP-anti-TfR scFV targeted cationic liposomes, IV	DOTAP/ DOPE/ DOPE-scFV	pDNA encoding p53	Advanced solid tumors (I-Ib) – NCT00470613	• Grade-3 fatigue observed in patients was causally related to massive tumor necrosis/tumor cell death due to therapy • Two serious adverse events: chest pain and tachycardia – only possibly related to drug • Fever – prophylactic regimen involving hydration, tylenol, histamine blockade, and indocin were able to manage fever	Senzer et al. (2013) and Pirollo et al. (2016)

Cationic liposomes, IV bolus	DODAB/EggPC/Chol	c-raf-1 antisense oligonucleotide	Advanced solid tumors (I)	• Dose-independent hypersensitivity reactions: all 22 patients across doses of 1, 2, 4, and 6 mg/kg/week experienced IRRs (hypersensitivity reactions including flushing, dyspnea, hypoxia, rigors, back pain, and hypotension) • Increasing the duration of infusion and premedication with acetaminophen, H1- and H2-antagonists, and corticosteroids reduced the rate and severity of IRRs. • DLT was thrombocytopenia at 6 mg/kg/week dose. • MTD was 4 mg/kg/week	Rudin et al. (2004)
Cationic liposomes, IV infusion	DODAB/EggPC/Chol	c-raf-1 antisense oligonucleotide	Advanced solid tumors (I)	• Dose-dependent IRRs (grade ≥2) experienced by most patients were back pain, chills, dyspnea, fatigue, fever, flushing, and hypertension and were managed by premedication with corticosteroids and antihistamines • Five patients experienced SAEs (acute infusion-related symptoms, abnormal liver function tests, hypoxia, dehydration, diarrhea, esophagitis, fever, hypokalemia, pharyngitis, and tachypnea) • A dose of 2.0 mg/kg of LErafAON administered twice weekly is tolerated with premedication and does not enhance radiation toxicity in patients • It became apparent early in this study (drug dose, 1.4 mg/kg daily) that infusion-related reactions attributable to the liposomal formulation (back pain, flushing, dyspnea, and hypotension) would require premedication to allow repeated infusions to be done. • IRRs were causally correlated to the induction of cytokine release due to cationic lipids of liposome and led to modification of liposomal composition for further clinical trials	Dritschilo et al. (2006)

(*Continued*)

Table 9.1 (Continued)

Nanocarrier type and subtype, administration route	Carrier composition	Genetic cargo	Therapeutic indication (phase) – clilincaltrials.gov identifier	Clinical toxicity or adverse events data	References
Cationic liposomes, IV	DOTAP/Chol	TUSC2/FUS1 pDNA	Lung cancer (I) – NCT00059605	• First of 22 patients observed grade 2 fever after 0.02 mg/kg dose infusion, hence all other patients received premedication with dexamethasone and diphenhydramine and no patient observed grade 1 or higher toxicities • FDA recommended lower dose 0.01 mg/kg without premedication; however, all three patients observed grade 2 or 3 fever and one patient showed grade 3 hypotension and hence FDA allowed trial with premedication • DLT observed in two patients were hypophosphatemia at 0.06 and 0.09 mg/kg dose • MTD was determined to be 0.06 mg/kg • Grade 2 toxicities were myalgias, hypophosphatemia, fever, nausea, and transaminase elevation	Lu et al. (2012)
Lipid nanoparticle, intravenous	DLin-MC3-DMA/DSPC/Chol/DMG-mPEG2000	siRNA against transthyretin	hATTR amyloidosis (I–IV complete) – NCT01148953 and NCT01559077	• Approved product in the United States, EU, and Japan	Coelho et al. (2013)
Lipid nanoparticles, IV	DLin-MC3-DMA/DSPC/Chol/DMG-mPEG2000	siRNA against proprotein convertase subtilisin/kexin type 9 (PCSK9)	Hypercholesterolaemia (I–II) NCT01437059	• Treatment was generally safe and well tolerated with no drug-related SAEs • Treatment-related AEs were mild to moderate in severity, with similar proportions affected in the treatment vs the placebo groups • No clinically significant dose-dependent changes were found in liver function tests, creatine prosphokinase C-reactive protein, hematological measures as well as concentrations of nine cytokines monitored • No clinically significant AEs were noted during the period of six months after treatment	Jayaraman et al. (2012) and Fitzgerald et al. (2014)

Lipoplex, intravenous	AtuFect-1(cationic lipid)/DPhyPE/DSPE-PEG2000	siRNA against PKN3	Advanced solid tumors (I)	• No cytokine activation (TNF alpha, ILb1, IFNg, IL6 was observed). Well tolerated. Adverse events, possibly tolerated with Atu027 were fatigue across all dose levels, hair loss grade 1, sweating grade 1, and abdominal pain. G3 AEs not dose limiting were elevated lipase activity and diarrhea. No dose-limiting toxicities were seen. No premedication required.	Strumberg et al. (2012)
LNP, intravenous	Arcturus Therapeutics's lipid (ATX)/DSPC/Chol/pegylated lipid	mRNA encoding ornithine transcarbamylase (OTC)	OTC deficiency (I, planned in 2020, orphan designation)	• Phase I studies planned in 2020	Arcturus Therapeutics Holdings Inc. (2019)
Lipid nanoparticles, LNP, intravenous	DLin-MC3-DMA/DSPC/Chol/DMG-mPEG2000	mRNAs encoding hemagglutinin (HA) proteins of H10N8 or H7N9 influenza viruses	Influenza (I) – NCT03076385	• AEs were mild to moderate with a few severe adverse events but no SAEs. • Majority of AEs were mild to moderate and comparable in frequency, nature, and severity with patients receiving unadjuvanted and adjuvanted H1N1 influenza vaccines, and included injection site pain, myalgia, headache, fatigue, and chills/common-cold-like symptoms. • Four severe AEs (injection site erythema, induration and chills, and common cold) were reported in 13% of patients; however, no SAEs were noted. • All AEs were expected and reversible.	Bahl et al. (2017)
Targeted lipid nanoparticles-liposomes, intravenous	S104 (ionizable lipid), vitamin A-conjugated targeting agent lipid (DiVA)	siRNA against HSP47	Hepatic fibrosis	• Infusion-related reactions occurred in 21% patients in phase Ia trial and hence, antihistaminics were administered as a preventive measure for phase II studies. • No deaths, serious adverse events or discontinuations due to adverse events were observed. Phlebitis, burning sensation, and headache were most common AEs. Infusion-related effects were seen in 1 (5%) of patients.	Soule et al. (2018)
Neutral liposomes, intravenous	DOPC/Tween 20	siRNA against EphA2	Advanced solid tumors (I-ongoing) NCT01591356	• Phase I – ongoing	Naing et al. (2017)

(Continued)

Table 9.1 (Continued)

Nanocarrier type and subtype, administration route	Carrier composition	Genetic cargo	Therapeutic indication (phase) – clilincaltrials. gov identifier	Clinical toxicity or adverse events data	References
Local delivery					
Cationic liposomes, intralesional	DMRIE/DOPE	A pDNA encoding HLA-B7 and β2-microglobulin	metastatic melanoma (II)	• Treatment was well tolerated with mild/moderate treatment-related events • Most common AEs were ecchymosis, pruritus (injection site reactions), and pneumothoraces	Stopeck et al. (2001) and Bedikian et al. (2010)
Cationic liposome, intralesional	DOTMA/Chol	pDNA encoding IL-2	Head and neck cancer (I)	• Immunological reactions were related to the bacterial origin of plasmid in this study (both control plasmid and plasmid encoding for IL-2 led to similar immunostimulatory response) • Immunostimulatory activity of bacterial origin plasmid (control plasmid as well as IL-2 plasmid)	O'Malley et al. (2005)
Cationic liposomes, intratumoral	DMRIE/DOPE	HLA-B7 pDNA	Renal cancer (I)	• Allovectin-7 is safe with minimal toxicity up on intratumoral administration • There was lack of tumor regression in this study, yet lower toxicity was compelling result for Phase II with renal and other cancers as well as for use with other genetic cargo	Rini et al. (1999)
Cationic liposomes, intratumoral	DMRIE/DOPE	pDNA encoding IL-2	Metastatic renal cancer	• Treatment was well tolerated with no grade 3/4 toxicities associated with drug • AEs included grade 1/2 reactions such as pain at injection site (20% of patients), constitutional symptoms (malaise/myalgia, low-grade fever, and chills in 74% of patients); fatigue (19%), nausea (10%), and allergy (1 patient)	Galanis et al. (2004)
Cationic liposomes, cutaneous and subcutaneous tumor nodules	TMAG/DLPC/ DOPE	DNA encoding IFN-β	Advanced melanoma (pilot)	• No adverse events were observed.	Kageshita et al. (2001) and Matsumoto et al. (2008)
Cationic liposomes, intratumoral	DC-Chol/ DOPE	DNA encoding E1A	Recurrent head and neck cancer (II)	• Intratumorally administered tgDCC-E1A was well tolerated with no significant toxicities related to drug	Villaret et al. (2002)

Cationic liposomes, intratumoral	DC-Chol/DOPE	DNA encoding E1A	Recurrent breast, head, and neck cancer (I)	• No DLT was observed in the dose groups (15, 30, 60, and 120 μg DNA/cm of tumor). All patients were able to tolerate the injections • Most common AEs were pain and bleeding at the injection site. • MTD was not reached in this study.	Yoo et al. (2001)
Liposomally encapsulated virus, intratumoral	Cationic liposome/Semliki forest virus	IL-12 gene	Recurrent glioblastoma multiforme (I/II)		Ren et al. (2003)
Polymeric nanoparticles, intratumoral	jetPEI	pDNA encoding somatostatin receptor subtype 2 (sst2) and deoxycitidine kinase::uridylmonophosphate kinase (dck::umk)	Advanced pancreatic carcinoma (I)	• Similar toxicity profiles were noted between combination therapy vs the gemcitabine alone. • 9% (2) patients reported grade 3–5 fever with 250 and 500 μg doses; however, this was not found to be dose limiting. • Intratumoral delivery was found to be well tolerated and safe in patients with pancreatic cancers.	Buscail et al. (2015)
Polymeric nanoparticles, intratumoral/intralesional	jetPEI	RNA oligonucleotide activator of the DDX58 (retinoic acid-inducible gene I/RIG-I) receptor	Advanced/metastatic solid tumors (I) – NCT03739138	• This trial is currently discontinued. • Favorable safety was observed at all dose levels without any dose-limiting toxicities. • Most common AEs were pyrexia, fatigue, headache, and nausea (all grade 1/2 except for one grade 3 pyrexia and one grade 3 pain and anxiety) • 47.7% of patients experienced grade 3–5 AEs and 53.3% experienced immune-mediated AEs • No mortalities were observed.	Middleton et al. (2018)
Lipid nanoparticle, intratumoral	Ionizable lipid/helper lipid/structural lipid/PEG-lipid (actual lipids not disclosed)	mRNAs encoding OX40L, IL-23, and IL-36γ	Relapsed/Refractory solid tumor malignancies or lymphoma (I-ongoing) NCT03739931	• mRNA-2752 is currently being evaluated in an ongoing clinical trial	Bauer et al. (2019)
Cationic liposomes, intracavitary	DC-Chol/DOPE	DNA encoding E1A	Breast and ovarian cancer (I)	• The most common treatment-related toxicities were fever, nausea, vomiting, and/or discomfort at the injection sites.	Hortobagyi et al. (2001)

(*Continued*)

Table 9.1 (Continued)

Nanocarrier type and subtype, administration route	Carrier composition	Genetic cargo	Therapeutic indication (phase) – clilincaltrials. gov identifier	Clinical toxicity or adverse events data	References
Cationic liposomes, nasal	EDMPC/Chol	pDNA encoding CFTR	Cystic fibrosis-CF (unknown)	• No AEs or toxicity or inflammation were observed in the nose or elsewhere in any of the patients	Knowles et al. (1998)
Cationic liposomes, nasal	DC-Chol/ DOPE	CFTR gene	Cystic fibrosis (unknown)	• No inflammation, toxicity, or an immune response was observed against DNA/liposomes or the expressed CFTR	Hyde et al. (2000)
PEG-substituted lysine peptides, nasal	PEG-substituted 30-mer lysine peptides	CFTR gene	Cystic fibrosis (safety and tolerability)	• Routine clinical assessment and lab tests were done and no SAEs were noted with no AEs were attributed to the DNA. • Moreover, toxicological evaluation of serum (levels of IL6, complement and C-reactive protein), nasal washings (cell-counts, protein, IL6, and IL8), pulmonary functions and hearing tests revealed no association to administration of DNA.	Konstan et al. (2004)
Cationic liposomes, nasal	GL67/DOPE/ DMPE-PEG5000	pDNA encoding CFTR	Cystic fibrosis (IIB) – NCT01621867	• Plasmid (pGM169) was designed to achieve prolonged duration of expression while eliminating CpG dinucleotides to reduce the inflammatory response (which is associated with flu-like illness) observed with plasmids rich in unmethylated CpG. • This single-dose study with aerosolized delivery of the plasmid using lipids demonstrated that an effective dose can be achieved with negligible inflammatory response.	Alton et al. (2013)
Cationic liposomes, pulmonary nebulized	GL67/DOPE/ DMPE-PEG5000	pDNA encoding CFTR	Cystic fibrosis (IIB) – NCT01621867	• No significant adverse events were recorded among different groups of the study	(Alton et al. 2015a)
Cationic liposomes, pulmonary+nasal	GL67/DOPE/ DMPE-PEG5000	pDNA encoding CFTR	Cystic fibrosis (unknown)	• Seven of the eight patients receiving the active complex reported mild influenza-like symptoms that resolved within 36 h. Six of eight patients in both the active and placebo groups reported mild airway symptoms over a period of 12 h following pulmonary administration. No specific treatment was required for either event.	Alton et al. (1999)

Cationic liposomes, airway	GL67/DOPE/DMPE-PEG5000	CFTR gene	Cystic fibrosis (safety trial)	• 50% (4) patients showed pronounced fever, myalgia, and arthralgia within six hours of administration. • Serum IL6 (but not IL8, IL1, TNF-α, or IFN-γ) were elevated in one to three hours post administration. • There were no antibodies detected against cationic liposome or pDNA. • pDNA (both eukaryotic pDNA and pDNA with unmethylated *E. coli*-derived CpG sequences) complexed with cationic lipid (but not the pDNA alone) led to brisk increase in proliferation of immune cells hence, it was concluded that pDNA and lipid are synergistic in this immune response and at least part of the immune response is independent of unmethylated CpG sequences.	Ruiz et al. (2001)
Cationic liposomes, intracranial postsurgical injections	TMAG/DLPC/DOPE	pDNA encoding IFN-β	High-grade gliomas (pilot safety and efficacy trial and phase I)	• Overall adverse events among patients were limited to grade I headache and brain edema • Various perioperative issues were scattered among patients (one patient with mild anemia, one with controllable intracerebral hemorrhage at a site distant to tumor and two patients with leakage cerebrospinal fluid). • No toxicity issues were found to be directly attributable to liposomes. • Acute toxicity with doses of 30 μg DNA were minimal hence, planned dosing schedule (30 μg injected after diagnosis and 15 μg doses injected on day 14, 17, 21, 24, and 28) was changed to 30 μg injected on weekly basis. This was to help reduce the pain associated with treatment as well.	Yoshida et al. (2004) and Wakabayashi et al. (2008)
Cationic liposomes, intracoronary	DC-Chol/DOPE	Plasmid encoding iNOS	Prevention of restenosis (I, extension study)	• There were no complications related to local application of CAR-MP583. In one patient, PCI procedure-related transient vessel occlusion occurred with consecutive troponin elevation. There were no signs of inflammatory responses or hepatic or renal toxicity. No dose relationship was seen with regard to adverse events across the dose groups.	von der Leyen et al. (2011)

(Continued)

Table 9.1 (Continued)

Nanocarrier type and subtype, administration route	Carrier composition	Genetic cargo	Therapeutic indication (phase) – clilincaltrials. gov identifier	Clinical toxicity or adverse events data	References
Cationic liposomes, oral	DOTIM/Chol	Plasmid encoding MnSOD	Damage induced by chemotherapy (paclitaxel/carboplatin) in cancer (I)	• Grade 2 dyspepsia and a grade 1 rash in one patient and grade 1 constipation and hyponatremia in another patient were considered to have a possible relationship to MnSOD PL. None of the other reported toxicities were considered possibly, probably, or definitely related to MnSOD PL. There were no DLTs reported in all three dosing tiers. The 30-mg dose was defined as the starting dose for the efficacy phase. • None of the treated patients has experienced a grade III or IV toxicity that was considered related to MnSOD PL. Therefore, an MTD was not defined, and the highest dose tested (based on preclinical data) (30 mg) was defined as the phase II recommended dose. • Most common adverse events were fatigue, anemia, neutropenia, thrombocytopenia, anorexia, and constipation.	Tarhini et al. (2011)
Lipopolymeric nanoparticle, intraperitoneal	PEG-PEI-Cholesterol conjugate	pDNA encoding IL-12	Persistent or recurrent epithelial ovarian, fallopian tube or primary peritoneal cancer (II)	• The most frequent drug-related AEs were grade 1/2 fatigue, fever, chills, abdominal pain, nausea, vomiting, anemia, thrombocytopenia, and leukopenia. • Most clinical and laboratory AEs were grade 1/2 in nature, transient, and medically manageable. • Grade 3 AEs possibly related to drug were nausea and vomiting (2), anemia (1), and decreased lymphocyte count (1).	Alvarez et al. (2014)
Virus-like particles, subcutaneous	Virus-like particles	G10 oligonucleotide with unmethylated CpG motifs as TLR9 agonist	Allergic rhinoconjunctivitis (IIb) – NCT00800332	• No adverse systemic effect on any specific organ system could be identified. • Local reactions were frequent, mostly mild or moderate and spontaneously resolving within days. Antinuclear or antiDNA antibodies did not change under therapy. • Major AEs were urticaria (three patients), local reactions with pain and flu-like symptoms (two patients), worsening of preexisting asthma (one patient), vitiligo (one patient), drug related influenza (one patient)	Klimek et al. (2011)

				• No drug-related SAE was observed. • Furthermore, erythema, lymphadenopathy, influenza-like illness, pyrexia, fatigue, and headache reached incidences of 45% of patients in any treatment group and were more frequent in the active treatment groups.	
LNP, intramuscular	ALC-0315/ DSPC/Chol/ ALC-0159	mRNA expressing RBD of corona virus	Prophylactic corona virus vaccine (I/II/ III) – NCT04368728	• BioNTech/Pfizer's two dose (prime and boost) vaccine BNT162b2 is the first mRNA vaccine authorized for emergency clinical use • Phase I/II study evaluated different formulations and different doses (low, mid, and high) and showed followings: • Local and systemic AEs were found to be dose dependent and generally mild to moderate and transient. • No serious AE were reported. • Mid dose was taken further for Phase III • Interim analysis of ~38 000 participants randomized 1 : 1 to vaccine dosed at 30 μg and placebo in Phase III study showed followings: – Common solicited AEs to be injection site reactions (84.1%), fatigue (62.9%), headache (55.1%), muscle pain (38.3%), chills (31.9%), joint pain (23.6%), fever (14.2%) – SAEs occurred in 0–4.6% participants and were more frequent after Dose 2 than after Dose 1, and were generally less frequent in participants ≥55 years of age as compared with younger participants – There were no anaphylactic or severe hypersensitivity reactions with close temporal relation to the vaccine. – No causal relationship was established for some reactions (Bell's palsy, myocardial infarction, cholecystitis, and nephrolithiasis) – Safety profile was generally similar across age groups, genders, ethnic, and racial groups – Studies currently ongoing and will evaluate long-term safety	Mulligan et al. (2020), and Pfizer and BioNTech (2020)

(Continued)

Table 9.1 (Continued)

Nanocarrier type and subtype, administration route	Carrier composition	Genetic cargo	Therapeutic indication (phase) – clilincaltrials.gov identifier	Clinical toxicity or adverse events data	References
LNP, intramuscular	SM-102/DSPC/Chol/DMG-PEG2000)	mRNA expressing perfusion stabilized spike protein of corona virus	Prophylactic corona virus vaccine (I/II/III) – NCT04283461	• Moderna's two dose (prime and boost) vaccine (mRNA-1273), received emergency use authorization for clinical use in December 2020 • Phase I/II studies evaluated different doses of vaccine 25, 100, and 250 μg and results showed followings: – Vaccine was found to be generally safe and well tolerated – Only one Grade 3 AE-erythema noted only in a single participant at 100 μg – Higher AEs were seen at 250 μg dose level, with 3 participants showing Grade 3 reactions and that was after booster dose. – All AEs were transient and self-resolving with no Grade 4 AE reported in this study. – 100 ug dose was taken further for Phase III • Interim analysis of the 30,350 participants randomized 1:1 to vaccine 100 μg dose vs placebo in Phase III study showed followings – Common solicited adverse reactions to be injection site pain (91.6%), fatigue (68.5%), headache (63.0%), muscle pain (59.6%), joint pain (44.8%), and chills (43.4%) – SAEs occurred in 0.2–9.7% participants and were more frequent after dose 2 than after dose 1, and were generally less frequent in participants ≥65 years of age as compared with younger participants. – There were no anaphylactic or severe hypersensitivity reactions with close temporal relation to the vaccine. – No causal relationship was established for some reactions (Bell's palsy, myocardial infarction, cholecystitis, and nephrolithiasis) – Safety profile was generally similar across age groups, genders, ethnic, and racial groups – Studies currently ongoing and will evaluate long-term safety	Jackson et al. (2020) and Moderna Therapeutics (2020)

LNPs, intramuscular	–	mRNA encoding rabies viral glycoprotein RABV-G	Rabies vaccination (I-ongoing)	• CureVav is currently carrying out the Phase I dose-escalation trial of its prophylactic mRNA-based rabies vaccine, CV7202 • Early interim data showed that low-dose vaccination (1 μg) was able to induce immune response and was well tolerated	CureVac AG (2020)
Polymeric nanoparticles, intravesical	PEI/PEG-PEI	pDNA expressing diphtheria toxin A	Superficial bladder cancer (II)	• No DLT was observed. • Most common AEs were mild-to-moderate bladder discomfort, dysuria, micturition urgency, urinary tract infection, diarrhea, hypertension, and asthenia.	Sidi et al. (2008)
LNP, pulmonary inhalation	–	mRNA encoding CFTR	Cystic fibrosis (orphan designation, I/II) – NCT03375047	• Product was well tolerated at low and mid dose levels (8 and 16 mg doses compared with 24 mg dose) • Five subjects (including 3 in the 24 mg dose cohort) experienced mild-to-moderate febrile reactions manifesting 4–10 h after dosing and resolving by 24 h.	Translate Bio, Inc. (2019)
Polymeric nanoparticles, polypeptide, intradermal	Histidine-lysine copolymer (HKP) peptide	siRNA against TGF-β1 and Cox-2	Hypertrophic scar (I–II ongoing) NCT02956317	• The product was well tolerated in patients	Sirnaomics (2016)
poloxamer-based gel, local	C32-447, Poloxamer	pDNA expressing TALEN or CRISPR/Cas9	HPV-related Cervical Intraepithelial Neoplasia I (I)	• First human trial of direct application of TALEN and CRISPR/Cas9	First Affiliated Hospital Sun Yat-Sen University (2017)

patients with gastrointestinal, gynecological, genitourinary tumors, and sarcoma with hepatic and extrahepatic tumors (Tabernero et al. 2013), and liposomal delivery of pDNA expressing p53 (SGT-53) for treatment of solid tumors of thyroid, rectum, colon, vagina and cervix (Senzer et al. 2013; Pirollo et al. 2016). Cationic liposomes made with TMAG were also evaluated for intracranial administration for the treatment of glioma (Yoshida et al. 2004; Wakabayashi et al. 2008).

Apart from cancer, other rare and critical diseases having major genetic predisposition have been treated with nanocarrier-mediated gene delivery in clinical trials. Two such diseases are CF and transthyretin amyloidosis. CF patients have a mutation in the cystic fibrosis transmembrane conductance regulator (CFTR) gene which causes irreversible lung damage. Several nasally administered and nebulized nanocarriers have been evaluated to correct the CFTR gene mutation in clinical trials. Most trials conducted involve the correct CFTR gene which restores the CFTR function in the airway cells. Some safety and/or efficacy trials involved cationic lipid–DNA complexes or ionizable lipid–DNA complexes (Knowles et al. 1998; Alton et al. 1999; Hyde et al. 2000; Ruiz et al. 2001; Alton et al. 2013; Alton et al. 2015a), while one study evaluated PEGylated lysine polymer-based nanoparticles (Konstan et al. 2004). One study also evaluated mRNA delivery through pulmonary inhalation and received orphan designation from the USFDA (Translate Bio. Inc. 2019). Transthyretin amyloidosis is another lethal disorder caused by amyloid deposits from hepatocytes in peripheral nerves and heart. RNAi-based therapeutic approaches have been investigated in its treatment using the LNP system. This became the first LNP-based therapeutic to be approved for the treatment of hATTR amyloidosis to go through vigorous phase I–IV clinical trials (Coelho et al. 2013). Ornithine transcarbamylase (OTC) deficiency is another inherited genetic disorder that leads to the accumulation of ammonia in the blood causing ammonia related neurotoxicity due to deficiency of key enzyme of the urea cycle (OTC) (Arcturus Therapeutics Holdings Inc. 2019). Arcturus therapeutics' ARCT-810 is a LNP-based mRNA delivery system which uses Arcturus' proprietary LUNAR lipid delivery platform for safe and effective delivery to hepatocytes (Arcturus Therapeutics Holdings Inc. 2019).

Nanocarrier-based gene delivery is also evaluated in other diseases for treatment as well as prevention. Examples include delivery of mRNA LNPs to express hemagglutinin proteins of influenza for vaccine application (Bahl et al. 2017), HSP-siRNA delivery using targeted liposomes in hepatic fibrosis (Soule et al. 2018), VLPs for SC delivery of immunostimulatory G10 oligonucleotide for the treatment of allergic rhinoconjunctivitis (Klimek et al. 2011), cationic liposome-mediated delivery of pDNA-expressing iNOS for prevention of restenosis (von der Leyen et al. 2011), treatment of chemotherapy-induced damage by cationic lipid-based delivery of MnSOD plasmid via oral route as antioxidant therapy (Tarhini et al. 2011), and siRNA delivery using LNPs for treatment of hypercholesterolemia (Jayaraman et al. 2012; Fitzgerald et al. 2014).

9.4 Learnings from Clinical Trials

Safety evaluation of the nonviral gene delivery nanocarriers in animal and human studies is of critical importance. The route of administration is directly related to the disease conditions, i.e. nanocarrier delivery in cancer is accomplished via intravenous, intratumoral, or intravesical administrations. Also, the route of administration for different nucleic acid-based nanocarriers is important as it can directly affect not only the efficacy of the gene therapy but the safety profile as well. Potential adverse events of gene delivery nanocarriers are attributed to the route of administration and can be described as acute and chronic effects.

Clinical toxicities of various gene delivery nanocarriers are divided herein into two categories, i.e. toxicity with systemic delivery (intravenous delivery) and toxicity with local delivery, as systemic administration often poses a higher risk of different adverse events compared with the local administration. In general, the administration site reactions and local inflammatory responses are more pronounced with local administration while systemic acute and delayed hypersensitivity reactions are common with intravenous administration. Table 9.1 details comprehensive analyses of clinical safety/toxicity evaluations of various gene therapy nanoformulations.

9.4.1 Toxicity with Systemic Delivery – Intravenous Administration

Majority of nanocarrier systems which have been evaluated for systemic delivery are lipid-based systems and hence, this section will mainly focus on their toxicity issues. Clinical evaluation of liposomal drugs such as Doxil and other lipid-based drug products (Taxol) has shown that systemic administration of these formulations leads to a variety of adverse events commonly referred to as infusion-related reactions (IRRs) related to immunogenicity of the excipients used. One of the underlying components of this is activation of complement (C) pathways or sometimes a specific hypersensitivity syndrome called C activation-related pseudoallergy (CARPA). CARPA belongs to an acute (Type I) hypersensitivity reaction which is mostly mild and transient in nature and can be prevented with precautionary measures such as premedications and proper hydration. However, occasionally, lethal reactions can also develop. C activation is mainly manifested as cardiopulmonary stress, hence becoming a concern for heart patients. A detailed review carried out previously has noted that there are considerable individual differences observed in *in vivo* findings showing involvement of almost all organ systems in manifested adverse events and underlying a complex immune phenomenon involving multiple pathways such as C activation, signal transduction in allergy-mediating cells and vasoactive mediators engaged at the effector cell level (Szebeni et al. 2011).

IRRs are the most commonly seen adverse events with systemically administered gene therapy nanocarriers as well and are easily manageable with the premedication or by management of infusion protocol. Clinical trials with ALN-VSP showed that IRRs are mainly fever and chills which were managed by premedication with anti-inflammatory drugs and antipyretic drugs. IRRs still not managed by premedication were managed with interruption of the infusion and continuing with the remaining dose over 0.5–1 hours (Tabernero et al. 2013). These early observations with LNPs also influenced the optimized clinical trial protocols used for development of patisiran (Onpattro). The incidence and severity of adverse events were similar in patients receiving patisiran and placebo. The most common adverse events that occurred more frequently with patisiran than with placebo were peripheral edema and IRRs. To reduce the risk of IRRs, patients received premedications prior to infusion (Adams et al. 2018). Similarly, IRRs are observed with other lipid-based as well as polymeric nanocarriers which correlate with the activation of toll-like receptors. Table 9.1 provides detailed account of specific IRRs linked with these nanocarriers as well as other adverse events linked to a particular product.

9.4.2 Toxicity with Local Delivery

Local delivery of gene-based nanocarriers is accomplished through intratumoral, intravesical/intralesional administration, nasal administration, ocular administration, pulmonary administration, intraperitoneal administration, etc. Generalization can be made that systemic adverse events that are seen with intravenous administration are usually minimized or eliminated through local administration. For example, lipid–DNA nanoparticles administered through the pulmonary route did not show any changes in systemic levels of cytokines (Chadwick et al. 1997).

9.4.2.1 Pulmonary Delivery

A dose escalation safety trial in eight CF patients of aerosolized pulmonary delivery of CFTR pDNA (cystic fibrosis transmembrane conductance regulator) gene complexed with lipid revealed that it rose serum IL-6 levels but not IL-8, IL-1, or IFNγ within one to three hours of administration (Ruiz et al. 2001). Moreover, administration of cationic lipid alone (GL-67/DOPE/DMPE-PEG5000, (Chadwick et al. 1997)) or administration of prokaryotic/eukaryotic DNA alone (Ruiz et al. 2001) produces much less clinical inflammatory response; however, lipids and pDNA in the lipid–DNA complex act synergistically in inducing an immune response (Ruiz et al. 2001). Chadwick et al. also found that there were no changes in systemic cytokines after pulmonary administration (1997).

A recent clinical trial of mRNA-LNPs developed by Translate Bio which was administered as a nebulized formulation for treatment of CF showed that the product was well tolerated at low and mid dose levels (8 and 16 mg doses compared with 24 mg dose) (Translate Bio. Inc. 2019). Five subjects (including three in the 24 mg dose cohort) experienced mild-to-moderate febrile reactions manifesting 4–10 hours after dosing and resolving by 24 hours. Similar adverse events profile has been noted for other lipid-based gene formulations in prior small-scale trials of CF (Alton et al. 1999; Alton et al. 2015b).

A dose escalation toxicity study in mice after pulmonary administration of cationic lipid (GL-67) showed that the administration led to neutrophil infiltration, multifocal pulmonary lesions with elevated levels of inflammatory cytokines such as IL-6, TNF-α, IFN-γ within one to two days (Scheule et al. 1997). The higher doses led to more toxicity; however, the toxicity was reversible. The study concluded that increased potency of the lipid : DNA complex (increased DNA loading) is desirable to reduce the cationic lipid-based toxicity.

9.4.2.2 Intratumoral and Other Local Delivery in Cancer

Another most widely explored delivery approach for gene delivery nanocarriers in cancer is intratumoral delivery. The disease areas usually include cancers of those organs which are easily reached by injection and include melanomas, bladder cancer, cervical/ovarian cancer, pancreatic cancer, renal cancer, and other solid tumors. Additionally, in some cases, local administration nearby affected site rather than intratumoral administration is employed. This includes intravesical delivery for treatment of bladder cancer, intracranial administration for gliomas, and intraperitoneal delivery for ovarian cancers. Intratumoral or local delivery in cancer has been explored for liposomal systems and LNPs (Rini et al. 1999; Hortobagyi et al. 2001; Kageshita et al. 2001; Stopeck et al. 2001; Yoo et al. 2001; Villaret et al. 2002; Matsumoto et al. 2008; Bedikian et al. 2010; Bauer et al. 2019) as well as polymeric nanocarriers (Buscail et al. 2015). This delivery approach has been shown to be more convenient for patients as it avoids IRRs commonly associated with systemic therapy with cationic lipid/polymeric nanocarriers while creating a profile of manageable local adverse events including injection site reactions.

9.4.2.3 Subcutaneous, Intradermal, and Intramuscular Delivery

Subcutaneous (SC) delivery opts for the delivery of nucleic acid cargos for systemic absorption. However, a suitable design of the delivery system is required. A preclinical study in mice using siRNA LNPs demonstrated that smaller particles (~35 nm) were able to accumulate in the liver after SC injection while ~80 nm particles failed to accumulate in larger amounts (Chen et al. 2014). This was further confirmed in other preclinical studies as well (Pardi et al. 2015). Regardless of the accumulation, ~35 nm particles were still not able to provide effective silencing activity in the liver and it was attributed to reduced uptake of these particles in liver cells. Intermediate size LNPs (~45 nm), however, are able to accumulate in the liver and have silencing activity as well (Chen et al. 2014). Hence, ligand targeting with trivalent GalNAc was further employed to induce cellular uptake and hence silencing activity of ~35 nm LNPs. Regardless of this, no clinical trial has so far been done for SC-administered LNPs. However, the systemic toxicity issues similar to those observed in IV-administered LNPs can be expected for SC delivery of LNPs, if the LNPs are designed to get absorbed in the systemic circulation. However, further studies should be designed to thoroughly evaluate the local toxicity issues with such applications. A study by Klimek et al. in phase IIb clinical trial evaluated SC delivery of VLPs for delivery of immunostimulatory G10 oligonucleotide for the treatment of allergic rhinoconjunctivitis (2011). The results showed no systemic adverse effects related to any specific organ systems.

Intradermal or intramuscular delivery is the route of choice for the delivery of mRNA-based vaccines. One study evaluated the LNPs which were originally developed for IV administration via IM route (Hassett et al. 2019). The study was rationalized after the possibility of use of mRNA-based vaccines in healthy individuals which might range from day-old newborn babies to elderly people who can receive the vaccines and hence, minimum injection site reaction and high tolerability were the primary endpoints of the research (Hassett et al. 2019). Recent COVID pandemic has fueled research in mRNA-based vaccines delivered using novel nanoformulations. Two important examples of such mRNA vaccine candidates for which interim phase III clinical safety and efficacy data are already available are Moderna's mRNA-1273 which expresses prefusion stabilized spike protein of nCoV-2019 (novel coronavirus 2019) (Jackson et al. 2020; Moderna Therapeutics 2020) and BioNTech/Pfizer's BNT162a which uses nucleoside-modified mRNA (modRNA) expressing the viral spike glycoprotein (S) of SARS-CoV-2 (Mulligan et al. 2020; Pfizer and BioNTech 2020).

These formulations are delivered as intramuscular LNP formulations which have shown local site reactions such as pain, swelling, erythema, redness, etc., and systemic adverse events such as fever, fatigue, headache, muscle pain (myalgia), chills, diarrhea, joint pain (arthralgia), etc. (Jackson et al. 2020; Mulligan et al. 2020). These studies have also noted that some systemic and local adverse events were more pronounced at the second booster dose compared with the priming dose. Also, there was a dose-dependent increase in adverse events profile which was the prime factor for the selection of doses for the next phases of studies.

9.5 Mechanisms of Toxicity of Gene Delivery Nanocarriers

Figure 9.2 represents the components of the LNP and polyplex-based nanoparticle systems which are implicated in various types of immunostimulatory effects that can affect the toxicological profile of the gene delivery nanocarriers. The most important component of the LNPs can be divided into cationic lipids, genetic cargo, and PEG lipids. Similar considerations also apply to polymeric nanocarrier systems as well in which cationic polymers and PEGylated polymer components play a role in the nanocarrier-related toxicities.

9.5.1 Cellular Damage and Inflammatory Mediators

Cytotoxicity of the cationic lipid or polymer-based nanocarriers is exerted by their interaction with the cellular membrane which leads to acute cell damage by pore formation in the cellular membrane and interaction with the cellular membrane proteins. Cell death by cell membrane damage is usually manifested in the form of hemolysis upon intravenous-administered nanocarriers. Moreover, once accumulated in the target tissue, the damage can occur to the other non-targeted normal cells as well. Such cellular damage is of more importance with local delivery systems where the localized concentration of cationic lipids and delivery systems can have more propensity of interacting with cells.

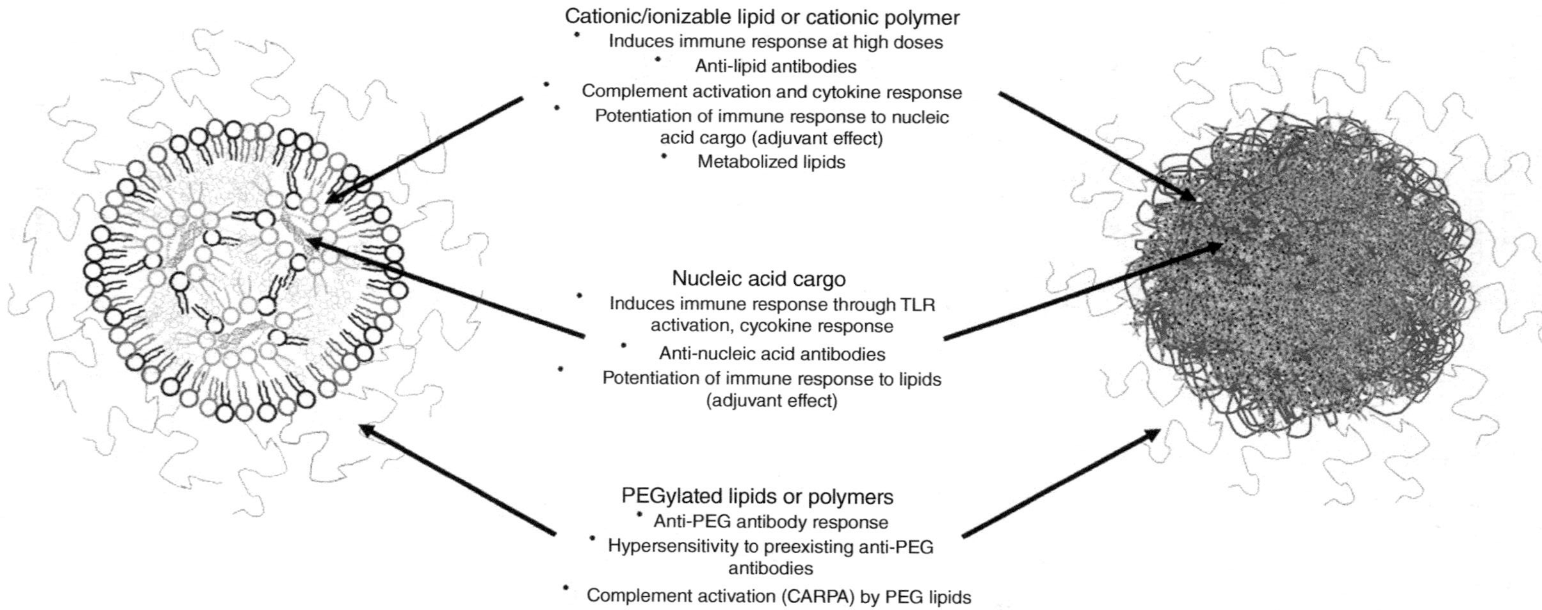

Figure 9.2 Toxicity mechanisms of components of gene delivery nanocarriers. For lipid-based nanocarriers (exemplified by lipid nanoparticles), cationic/ionizable cationic lipid, nucleic acid cargo, and PEGylate lipids are important. For polymeric nanoparticles, similar corresponding components, i.e. the cationic polymer, nucleic acid cargo, and PEGylated polymer, play a role in their toxicity.

9.5.2 ROS Induction

While cationic lipids in LNPs and cationic polymers enhance the transfection efficiency, they also can cause cellular toxicities through their cell membrane-based interactions (pore formation in membrane and disruption of normal membrane function by interaction with negatively charged phospholipids, proteins and proteoglycans such as toll like receptors [TLR4]) causing ROS generation and subsequent inflammation and apoptosis (Lonez et al. 2012). Hence, the rational design of lipids and selection of composition is of prime importance during the development of LNPs for gene delivery (Tousignant et al. 2000; Hafez et al. 2001; Semple et al. 2010).

9.5.2.1 Off-target Effects of the Genetic Cargo

Toll-like receptors are the key receptors that sense pathogenic stimuli recognizing pathogen-associated specific molecular patterns such as CpG motifs which are DNA sequences consisting of unmethylated CG sequences. Adenoviral vector genome and bacterial plasmid DNA bear the same characteristic of having unmethylated CpG motifs which are recognized by the TLR-9 (Hensley and Amalfitano 2007). This is the reason why the delivery of plasmid DNA induces a strong immune response independent of the expressed protein (Dow et al. 1999; Bramson et al. 2000; Lanuti et al. 2000). Also, the aim of moving from viral vectors to nonviral vectors was to reduce such toxicity typically associated with the virus. Clinically relevant acute dose-limiting toxic response has been noted upon systemic delivery of plasmid DNA: cationic lipid vector which includes an increase in proinflammatory cytokines such as IFN-γ, IL-12, and TNF-α (Tousignant et al. 2003). Other acute effects such as leukopenia, thrombocytopenia, and increased serum transaminase are linked with vector and not yet defined.

Single- as well as double-stranded RNA have also been found to be the potent inducers of the innate cytokine response. Some siRNA sequences yet induce production of IFN-α, independent of their GU content, in a specific subset of immune cells such as plasmacytoid dendritic cells (Hornung et al. 2005). This is similar to the immune response through CpG sequence-specific TLR9 activation. However, such cytokine response is mostly mediated by immune cells through the involvement of TLR7. Injection in mice of such siRNA complexed with cationic liposomes induced IFN-α in serum and activation of T cells and dendritic cells in spleen (Hornung et al. 2005). Sequence specificity was further validated by similar studies *in vivo* in mice and *in vitro* in human blood (Judge et al. 2005). Moreover, nonimmune cell activation may also occur due to cytoplasmic RNA sensors such as RIG1 (Robbins et al. 2009). These studies have also shown that delivery vehicles that facilitate cellular uptake of siRNA has a high tendency to induce inflammatory cytokine response and appropriate modification of the siRNA sequences can mediate effective RNAi with reduction in such cytokine response (Judge et al. 2005; Robbins et al. 2009).

Newer nucleic acid-based therapeutic agents such as Crisper systems are delivered as either plasmids or mRNAs, hence they have the toxicity potential as described above related to the CpG sequences of plasmid DNA or due to the innate immunogenicity of RNAs. However, apart from that, Crisper/Cas9 can cause additional toxicity issues. Crisper/Cas9 knockdowns specific gene sequences in the host genome completely eliminating its expression. This might cause a fatal loss of cellular protein. Hence, currently, Crisper/Cas9 is only used for external modification of cells that can be used for intravenous therapy in humans. However, with increasing understanding of underlying molecular pathways in order to minimize the toxicity issues while simultaneously improving nanocarrier properties to limit activity of these payloads to specific tissues, such therapeutic nucleic acids might offer even curative treatment for critical and fatal diseases.

9.6 Overcoming Toxicity Issues with Nanocarrier-Mediated Gene Therapy

9.6.1 Modification of Genetic Cargoes

As described earlier, some adverse/toxicity events are associated with the specific features of the genetic cargo the nanocarrier is carrying. These include TLR9 activation by unmethylated CpG sequences of plasmid DNA or TLR4/9 activation by siRNA or fatal removal of the target gene by CRISPER/cas9. Research has been targeted to modification of such features of therapeutic genes and some are described below. This leads to a reduction in the consequent toxicity profile of gene delivery nanocarriers. However, the toxicity issues are not always eliminated by such modifications and sometimes, such effects can be desirable.

Methylation of CpG sequences in therapeutic plasmid DNA leads to reduced acute cytokine response (Tousignant et al. 2003). In case of requirement of long-term therapeutic expression of proteins, such modification is necessary to avoid toxicity. However, if used for cancer treatment, such modification abrogates antitumor immunogenic response of pDNA (Rudginsky et al. 2001). Also, DNA vaccines elicit more immunogenic responses when they have such motifs (Klinman et al. 1997) and methylation of CpG motifs would not be a rational idea. Examples of non-CpG DNA sequences have also been identified which induce distinct toxicity profiles (Tousignant et al. 2003).

As described earlier, RNA can cause sequence-specific activation of TLRs as well as sequence-independent activation of cellular sensors in immune and nonimmune cells leading to induction of inflammatory cytokines (Judge et al. 2005; Robbins et al. 2009). In view of these, ALN-VSP was designed with VEGF siRNA and KSP siRNA which were chemically modified with 2′-*O*-methyl groups to minimize these reactions through TLR and other cellular sensors. In a clinical trial, ALN-VSP showed no induction of IFN-α which is most commonly associated with TLR as well as no induction of IL-1β and IFN-γ (Tabernero et al. 2013). This approach was also employed by cyclodextrin-based cationic polymer nanoparticles used for siRNA which is modified through 2′-*O*-methylation. Readers are encouraged to refer to other such modifications of siRNAs directed toward either improvement of their stability or to reduce their toxicity *in vivo* (Kanasty et al. 2013).

9.6.2 Modification of Nanocarrier System

The most widely explored avenue to overcome toxicity is the modification of the nanocarrier properties of nanocarrier such as optimizing the lipid or polymer chemistry, modification of the genetic cargo, modification of the nanocarrier morphology such as particle properties and surface properties.

9.6.2.1 Optimizing Excipient Chemistry – Lipid and Polymer Chemistry

Cationic lipid and polymer chemistry have long been found to be associated with various cellular toxicities. Hence, the majority of the approaches are directed toward rationalized modification of lipid or polymer structures which can provide better efficacy while minimizing the toxicity.

Close evaluation of structural properties of the lipids and their clinical toxicological data provide some generalization about their safety profiles. Figure 9.3 shows the structural diversity of first-generation cationic lipids with quaternary cationic head-groups (referred to as cationic lipids herein) and newer generation ionizable lipids with primary/secondary/tertiary amino head-groups (referred to as ionizable cationic lipids/ionizable lipids herein). Table 9.1 provides the clinical safety evaluation of the nanocarrier systems made with these lipids. Initially, the research efforts were more focused to increase the transfection potential of lipids and hence a large number of cationic lipids were developed and evaluated clinically.

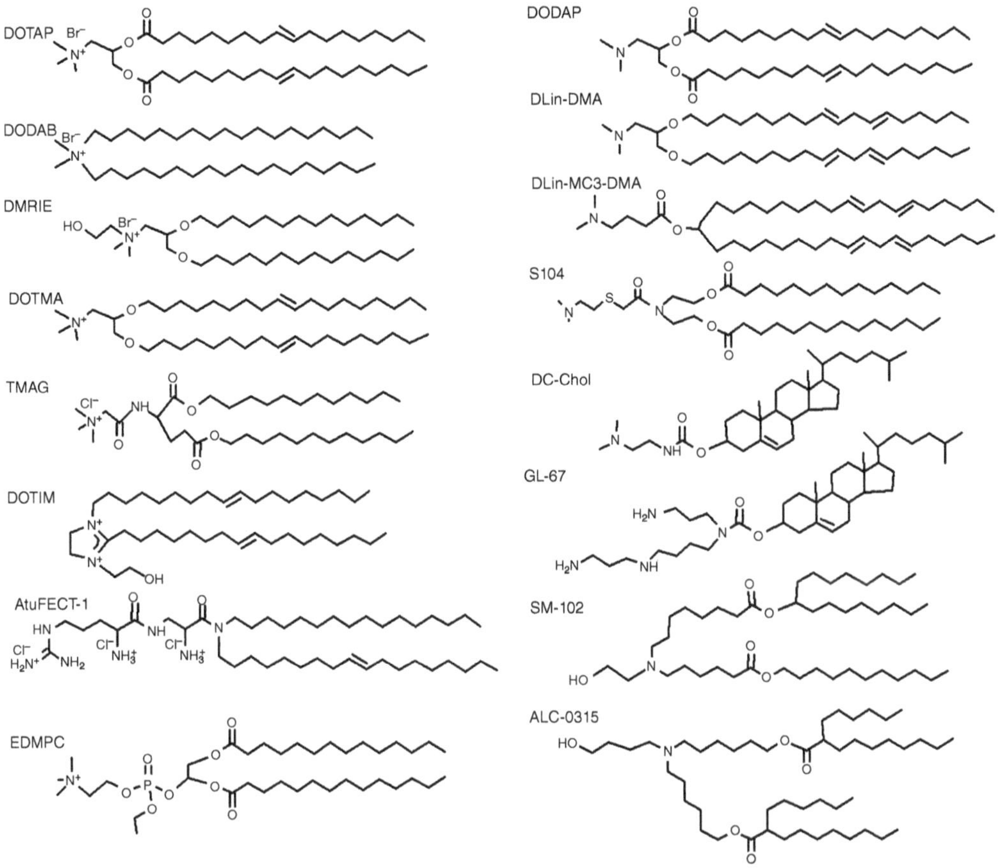

Figure 9.3 Structures of cationic lipids and ionizable cationic lipids evaluated or being evaluated in clinical trials. Lipids with a quaternary cationic nitrogen(s) ($-NR_3$) are usually considered as the first-generation cationic lipids while those with primary, secondary, or tertiary amine groups ($-NH_2$, $-NHR$, and $-NR_2$) and/or showing pK_a in the range of 6–7 are usually considered second-generation amino lipids or ionizable cationic lipids.

These studies provided a comprehensive correlation between the toxicological profile and the types of cationic lipids and their morphological properties and created the basis for the development of ionizable lipids. Now, the majority of pharmaceutical companies' preclinical and clinical pipeline includes a vast IP of such ionizable lipids and nanocarriers thereof. Development of optimized DLin-MC3-DMA lipid used in Onpattro® was achieved after a series of *in vitro* and *in vivo* evaluations of custom-synthesized lipids such as DDAB, DLinDMA, and DLin-KC2-DMA. Apart from the headgroup chemistry, the tail groups of the lipids were also considered for improving the biodegradability and hence the biocompatibility of the lipids. Several of such lipids can be noted in the Figure 9.3, which have multiple ester groups and shorter acyl chains which are easily metabolized and eliminated from the body.

A similar account of the polymers that have been translated to the clinic for gene delivery gives an idea about their toxicological profile. There is only a limited number of polymeric nanocarriers made of different classes of polymers (Figure 9.4) that have reached clinical trials. These include jetPEI, CDP, cationic polypeptides, and lipopolymers. A low number of clinical evaluations with polymer-based gene delivery nanocarriers and only a few polymers translating to late-stage clinical trials

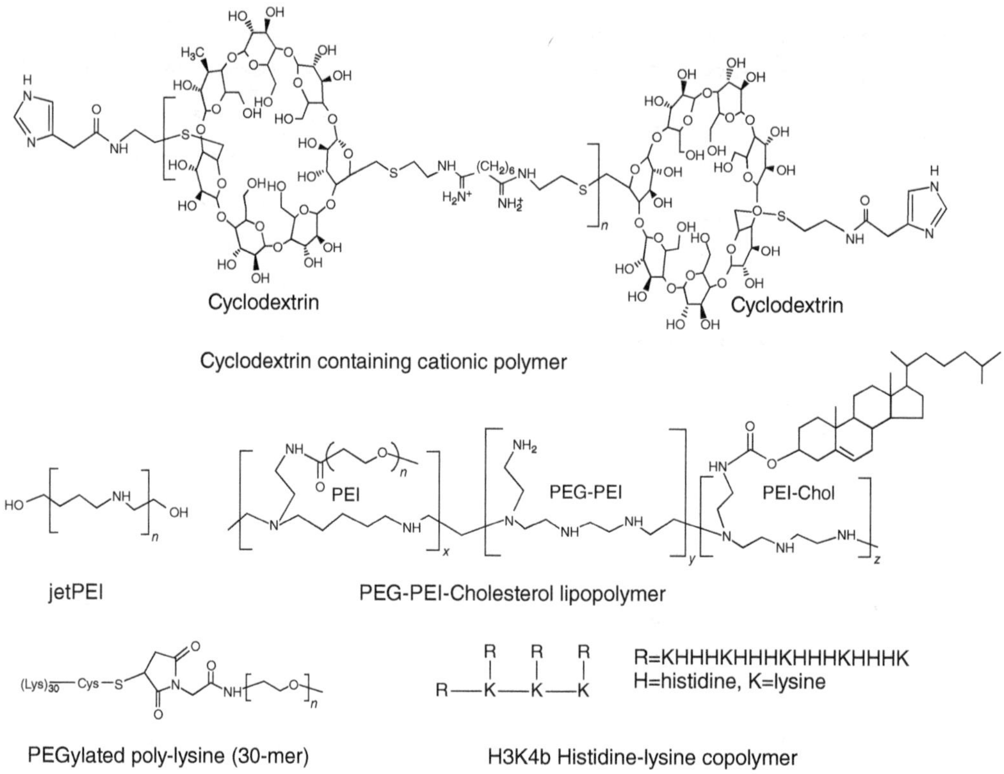

Figure 9.4 Structures of a few cationic polymers evaluated/being evaluated in clinical trials for gene delivery. Structures are redrawn from relevant publications. Cyclodextrin containing polymer (CDP) from Davis (2009), PEG-PEI-Cholesterol lipopolymer from Mater et al. (2015), lysine 30 mer-10 kDa PEG from Konstan et al. (2004), and H3K4b histidine-lysine copolymer (HKP) from Leng and Mixson (2005) and Zhou et al. (2017).

indicate a wide gap in the field and also points out to possible challenges. Cyclodextrin-based cationic polymeric nanoparticles were designed to take advantage of cyclodextrin's inclusion complex capability for surface modification of the nanoparticle (Zuckerman et al. 2014). From a series of HKP (see Figure 9.4), several copolymers were tested for their cytotoxicity and transfection potential (Leng and Mixson 2005; Zhou et al. 2017). Different polymers selected based on their transfection potential *in vitro/ in vivo* were HK4b, H2K4b, H3K4b, and H3K8b with different branches and same core (R = KHKHKHKHKGKHKHKHKHK, KHKHHKHHKHHKHHKHHKHHKHK, KHHHKHHHKHHHKHHHK and [HHHKH HHKHHHKHHH]2KH4NH4, respectively in Figure 9.4). A similar study done with peripheral blood mononuclear cells treated with different nanoplexes of siRNA made using H2K4b, H3K4b, H3K(+H)4b (with additional histidine) and H3K(+N)4b (with additional asparagine) showed that HK polymers with a higher number of histidines (H3K(+H)4b exhibited lower cytokine levels compared with other HK polymers. A similar pattern of cytokine induction was observed with these polymers when siRNA nanoplexes thereof were injected intravenously in the mouse model (Leng et al. 2012).

9.6.2.2 Modification of Nanocarrier Morphology

While lipid/polymer type governs the toxicity, the morphology of nanocarrier is also important. As described above, high cationic surface charge increases the interaction of the nanoparticles with RES and other cellular systems via cellular surface receptors such as TLR4, TLR9, and other proteoglycans or proteins leading to inflammatory pathways. Hence, modification of the surface properties can reduce such toxicities. Several approaches have been evaluated including preparation of neutral liposomes with neutral lipids such as DOPC without using any cationic lipids, use of ionizable lipids which have amino groups with $pK_a < 7$ so that nanoparticles flaunt neutral surface charge after intravenous administration at pH 7.4, or entrapping siRNA inside neutral liposomes through precipitation with calcium phosphate, etc.

All these approaches require suitable manufacturing techniques. For example, preparation of ionizable lipid-based nanoparticles (LNPs) with ethanol injection or microfluidic systems is accomplished at lower pH (≤4) with ionizable lipids entrapping nucleic acid cargo inside the solid core. The morphology of such kinds of LNPs is shown in Figure 9.1a. Similarly, entrapment of mRNA inside neutral liposomes made of DOPC has been accomplished using a lyophilization and reconstitution technique (Naing et al. 2017).

Several nanocarriers belong to LNP category, i.e. nucleic acid cargo trapped inside solid core of nanoparticle with neutral surface charge. Such LNPs can be made in two morphologically distinct neutral surface charge particles with (i) solid core either made of inverted micellar phases entrapping small water pockets containing nucleic acid (Figure 9.1a) or (ii) concentric multilamellar phases with nucleic acid sandwiched between the lamellas (Figure 9.1b). Many such clinically evaluated LNPs are developed by several pharmaceutical companies using structurally diverse ionizable lipids (Figure 9.3). For example, Arcturus Therapeutic's LUNAR LNP system uses its proprietary ionizable lipid (ATX) with DSPC, PEG-lipid, and cholesterol (Ramaswamy et al. 2017). The ionizable lipids are designed to contain a biodegradable lipid structure with ionizable amino head group with pK_a below 7. This makes the lipid become protonated at acidic pH and interact ionically with negatively charged RNA forming nanoparticles with RNA encapsulated inside the core. Similarly, Alnylam Pharma's DLin-MC3-DMA is an ionizable lipid which encapsulates siRNA inside the LNP core through a manufacturing process at lower pH. This approach creates nanoparticles which bear neutral surface charge and hence, reducing the toxicity issues commonly seen with cationic transfection vectors.

It is worth noting here the polymer-based siRNA delivery system CALAA-1 that uses a CDP with cationic moieties to complex with siRNA and cyclodextrin to modify the surface of nanoparticles via interaction with adamantane ring conjugated PEG or PEG-transferrin to allow for surface modification for active targeting of tumors (Figure 9.1c). Schematic of another peptide (HKP peptide) siRNA nanoparticle is shown in Figure 9.1d; however, it is a charge-based complex which does not have any surface functionalization.

Apart from that, physical and active targeting of lipidic as well as polymeric nanocarriers has served as a potential strategy to localize the effects of the genetic cargo in clinical use. Physical targeting such as intratumoral administration, intracranial administration, and intravesical administration may still pose different safety issues, but it definitely helps overcome toxicity associated with intravenous administration. Similarly, active targeting with surface ligands would provide preferred systemic distribution to target organs.

9.6.3 Preclinical Testing for Expected Toxicities

Despite the efforts made to reduce the toxicity with different nanocarriers, the development path toward a clinically approved product still includes evaluating toxicity on a case-by-case basis for each gene delivery nanocarrier.

Preclinical toxicological evaluation, hence, serves the best starting point to identify possible adverse events in future clinical trials. For local delivery, parameters such as blinded histopathological evaluation of the possibly affected organs or tissues can be done. Sectioned tissues such as the site of administration, peripheral tissues, and other systemic organs can be used to evaluate and find the possible root cause of adverse events. Moreover, systemic analysis of cytokine profile after any local administration can help correlate the adverse events with the possible components of the delivery system.

Such an important case study that can be cited here is the Moderna Therapeutics' mRNA LNP formulation development directed toward optimizing the LNPs for IM administration. Moderna carried out a thorough preclinical and clinical evaluation of their mRNA-LNP vaccine for the influenza vaccine to evaluate the safety and efficacy profile. They observed mild-to-moderate adverse events in clinical trials with MC3 lipid in this study (Bahl et al. 2017). This study also provided suitable evidence that intramuscular delivery of mRNA LNPs was more tolerable than intradermal delivery. However, follow-on studies were planned in order to design more suitable ionizable cationic lipids to develop a more efficacious and even safer LNP vaccine. Based on the previous nonhuman primate (NHP) studies that confirmed a slow clearance of MC3 lipid even after intravenous administration (Sabnis et al. 2018), it was hypothesized extended presence of ionizable cationic lipid at the IM injection site might be related to the adverse events. This was further corroborated by the mass spectrometric confirmation of reduced clearance of MC3 lipid from muscle tissue in mice after IM injection. Hence, a new series of ionizable lipids were developed and LNPs thereof were further evaluated in rats and nonhuman primates, and the selection of the appropriate ionizable cationic lipid was rationalized based on the adverse events profiling in comparison to MC3 lipid as a reference control (Hassett et al. 2019). In the NHPs, the injection sites were monitored for local adverse events (edema and erythema) one and three days after IM injection and scored based on severity for different ionizable lipids vs MC3 lipid used as a reference control. Also, results were supported by the analysis of systemic interleukin-6 (IL6) levels in all NHPs. To further evaluate the local tolerability of different ionizable lipid-based LNPs, different doses of mRNA LNPs for Zika virus vaccine were administered in rats and serum cytokines including eotaxin, GRO-alpha, IP-10, RANTES, and MCP1 were measured six hours post injection and a lipid (termed lipid H) was selected based on the lowest systemic cytokine induction. Further analysis was done on the injection site pathology to compare lipids. After 48 hours of administration, muscle and skin were collected from sacrificed rats and stained with H&E for pathological examination to rank the injection site reactions, i.e. mixed-cell inflammation, myofiber necrosis, and relative number of degenerated neutrophils. Hence, intramuscular administration of mRNA LNPs could lead to extended local and systemic lipid exposure which can be correlated to local and systemic adverse events. Moreover, this case study provides a very basic groundwork of the approach that needs to be taken on.

A similar example of a polymeric drug delivery system includes the preclinical development of CALAA-1 which successfully was seen to predict human toxicities. A similar toxicity profile was observed in human trials as seen in animal studies. Kidney toxicities were seen with animal studies and this led to the adoption of pre-dosing hydration protocol during human clinical trials. Hence, no kidney toxicities were observed in human patients. Current data on toxicity and safety of various gene delivery nanocarriers in clinical trials and their developmental studies have provided a detailed understanding of the pathway for identifying the toxicity concerns. A streamlined pathway with a detailed evaluation of toxicity at each stage of development sounds like a suitable development model. *In vitro* studies with human cell-lines (human dendritic cells, whole blood cultures, etc.) and *in vivo* developmental studies in rodent models followed by nonhuman primates can create comprehensive prediction of expected human toxicities and allow for the adoption of suitable study design and preventive measures.

9.7 Future Perspectives and Conclusion

While a lot of clinical evidence of safety and toxicity of various nanocarriers is available, there are a number of clinical trials currently undergoing or planned (Table 9.1) which provide further data on different fronts (nanocarrier excipients, nanocarrier morphology, nucleic acid cargo, delivery route, etc.) TranslateBio is advancing the development of MRT5005 for pulmonary delivery of mRNA expressing CFTR protein to lung epithelial cells to the next clinical phase, as it has shown positive results in phase I trial with suitable toxicological profile. This has been so far the most successful delivery approach in clinical trial to use mRNA therapeutic delivered to lung cells via nebulization. Advances in such applications are still needed to generate more understanding for pulmonary (any other local organ) delivery to overcome systemic toxicity issues and to localize the effects of nucleic acid therapeutics.

Additionally, improving the therapeutic index is the major objective that a lot of industries are trying to do through the development of effective lipid/polymeric excipients. Arcturus therapeutics has developed a proprietary lipid pool (termed ATX lipids) to develop a novel multilamellar LNPs into their so-called LUNAR (lipid-enabled and unlocked nucleic acid modified RNA) delivery platform. ATX lipids are ionizable lipids which have a wide therapeutic index and are biodegradable with low accumulation in cells after intracellular delivery. This platform uses Arcturus Therapeutics' lipid (ATX), cholesterol, a phospholipid 1,2-distearoyl-*sn*-glycero-3-phosphocholine (DSPC), and a pegylated lipid for delivery of nucleic acids (Ramaswamy et al. 2017). Arcturus has patented the LUNAR lipids of various structures over a wide range of patents (Arcturus Therapeutics Holdings Inc. 2019). The lipids are being evaluated in preclinical and clinical trials for delivery of siRNA, mRNA, CRISPR/Cas9 system, and pDNA for a wide range of therapeutic indications. Similar lipid pools and nanoparticles are being developed and evaluated in animals and humans by a lot of companies for a wide variety of applications and indications. A few examples include the works being done by Arbutus/Genevant, Moderna, BioNTech, Evonik, Acuitas, Vertex, Bayer, TranslateBio, CureVac, Precision NanoSystems/CanSino, etc. In view of this, however, novel approaches are still needed to advance the polymer-based nanocarriers toward the clinic.

While there is a wide range of LNP formulations made with different lipid compositions advancing to clinical trials for multiple applications, the current COVID pandemic has pushed and influenced mRNA-based drug delivery systems to the next level. This has made a lot of preclinical and clinical data available in the literature to advance these therapeutic systems in different disease applications. A further advancement in the mRNA field is the use of self-amplifying mRNA (replicon RNA or SAM), which uses certain viral machinery (noninfective) to allow the target mRNA sequences to be amplified intracellularly. This allows for a drastic reduction in the dose of SAM compared with conventional mRNA ultimately making it possible to dose lower levels of drug substance to achieve similar/higher therapeutic/prophylactic effectiveness and also to reduce the manufacturing burden as doses become 10–100 times lower which also reduced the toxicity issues related to the excipients required for delivery of such nucleic acids. An important example of such SAM vaccine is Imperial College London's LNP-based coronavirus vaccine (WHO 2020).

A wide range of clinical trials has already evaluated siRNA and pDNA in humans while a huge number of newer trials will be seen in this decade on delivery of mRNAs and other gene editing tools such as TALEN and CRISPR/Cas9. A new phase I trial is taking place in First Affiliated Hospital, Sun Yat-Sen University in China using direct application of CRISPR/Cas9 plasmid and TALEN plasmid (First Affiliated Hospital Sun Yat-Sen University 2017). So far, CRISPR/Cas9-based gene editing has been indirectly translated to clinical trial via gene editing outside the body by modifying the harvested human cells with CRISPR/Cas9 and injecting modified cells back in (Zhan et al. 2019). E6 and E7 are implicated in human papillomavirus (HPV) driven carcinogenesis leading to Cervical Intraepithelial Neoplasia (CIN) and cervical cancer and TALEN and CRISPR/Cas9 based genetic medicine has been shown to ameliorate HPV infection (Hu et al. 2014, 2015). Hence, the trial is designed to evaluate the safety and efficacy of TALEN-HPV E6/E7 and CRISPR/Cas9-HPV E6/E7 in treating HPV persistency and CIN. The trial will study the effects of TALEN plasmid and CRISPR/Cas9 plasmid dosed for three to six months (First Affiliated Hospital Sun Yat-Sen University 2017). Both gene editing tools are to be administered in the form of a gel with plasmid, C32-447, poloxamer 407 in distilled water. This study will provide the first safety data on the direct application of TALEN and CRISPR/Cas9 in humans. Apart from that, several other companies are actively working on optimization of LNP formulations of CRISPR/Cas9 tools, i.e. Intellia Therapeutics' LNP for Crispr sgRNA and Cas9 mRNA delivery (Finn et al. 2018), Silence therapeutics' lipoplex formulations for pDNA (Silence Therapeutics 2016) as well as CRISPR therapeutics' other LNP-based delivery pipeline (CRISPR Therapeutics 2017). Intellia Therapeutics' lead LNP candidate NTLA-2001 is currently being evaluated in Phase I clinical trial as a single dose therapy for in vivo CRISPR/Cas9 based editing of the TTR Gene for treatment of hereditary transthyretin amyloidosis with polyneuropathy. Data from these detailed safety evaluations in preclinical models and clinical trials will allow for better designing of future clinical trials and provide an assessment of the safety of these therapeutics.

In conclusion, the current industrial pipeline forecasts the widespread therapeutic application of gene delivery nanocarriers; however, such nanocarrier systems may pose challenging development path with high risks involved with increasing costs of development and high chances of failure. Hence, detailed *in vitro* and preclinical screening of the gene delivery nanocarrier remains an indispensable tool for assessment of safety and toxicity issues for clinical trial success. With current tools of rationalized designing of lipids or polymer chemistry, nucleic acid cargos, and optimizing the morphologic characteristic of nanocarriers, a future with more generalized and successful application of gene therapy can be expected.

References

Adams, D., Gonzalez-Duarte, A., O'Riordan, W.D. et al. (2018). Patisiran, an RNAi therapeutic, for hereditary transthyretin amyloidosis. *New England Journal of Medicine* 379 (1): 11–21.

Alberer, M., Gnad-Vogt, U., Hong, H.S. et al. (2017). Safety and immunogenicity of a mRNA rabies vaccine in healthy adults: an open-label, non-randomised, prospective, first-in-human phase 1 clinical trial. *Lancet* 390 (10101): 1511–1520.

Alton, E.W., Stern, M., Farley, R. et al. (1999). Cationic lipid-mediated CFTR gene transfer to the lungs and nose of patients with cystic fibrosis: a double-blind placebo-controlled trial. *Lancet* 353 (9157): 947–954.

Alton, E.W., Boyd, A.C., Cheng, S.H. et al. (2013). A randomised, double-blind, placebo-controlled phase IIB clinical trial of repeated application of gene therapy in patients with cystic fibrosis. *Thorax* 68 (11): 1075–1077.

Alton, E.W., Armstrong, D.K., Ashby, D. et al. (2015a). Repeated nebulisation of non-viral CFTR gene therapy in patients with cystic fibrosis: a randomised, double-blind, placebo-controlled, phase 2b trial. *Lancet Respir Med* 3 (9): 684–691.

Alton, E.W., Boyd, A.C., Porteous, D.J. et al. (2015b). A phase I/IIa safety and efficacy study of nebulized liposome-mediated gene therapy for cystic fibrosis supports a multidose trial. *Am J Respir Crit Care Med* 192 (11): 1389–1392.

Alvarez, R.D., Sill, M.W., Davidson, S.A. et al. (2014). A phase II trial of intraperitoneal EGEN-001, an IL-12 plasmid formulated with PEG-PEI-cholesterol lipopolymer in the treatment of persistent or recurrent epithelial ovarian, fallopian tube or primary peritoneal cancer: a gynecologic oncology group study. *Gynecol Oncol* 133 (3): 433–438.

Arcturus Therapeutics Holdings Inc. (2019). Arcturus therapeutics receives orphan drug designation. https://drug-dev.com/arcturus-therapeutics-receives-orphan-drug-designation/ (accessed 28 June 2020).

Bahl, K., Senn, J.J., Yuzhakov, O. et al. (2017). Preclinical and clinical demonstration of immunogenicity by mRNA vaccines against H10N8 and H7N9 influenza viruses. *Molecular Therapy* 25 (6): 1316–1327.

Balazs, D.A. and Godbey, W. (2011). Liposomes for use in gene delivery. *J Drug Deliv* 2011: 326497.

Bauer, T., Patel, M., Jimeno, A. et al. (2019). Abstract CT210: a Phase I, open-label, multicenter, dose escalation study of mRNA-2752, a lipid nanoparticle encapsulating mRNAs encoding human OX40L, IL-23, and IL-36γ, for intratumoral injection alone and in combination with immune checkpoint blockade. *Cancer Research* 79 (Suppl 13): CT210–CT210.

Bedikian, A.Y., Richards, J., Kharkevitch, D. et al. (2010). A phase 2 study of high-dose Allovectin-7 in patients with advanced metastatic melanoma. *Melanoma Res* 20 (3): 218–226.

Bhatt, P., Lalani, R., Vhora, I. et al. (2018). Liposomes encapsulating native and cyclodextrin enclosed paclitaxel: enhanced loading efficiency and its pharmacokinetic evaluation. *International Journal of Pharmaceutics* 536 (1): 95–107.

Bramson, J.L., Bodner, C.A., and Graham, R.W. (2000). Activation of host antitumoral responses by cationic lipid/DNA complexes. *Cancer Gene Therapy* 7 (3): 353.

Buscail, L., Bournet, B., Vernejoul, F. et al. (2015). First-in-man phase 1 clinical trial of gene therapy for advanced pancreatic cancer: safety, biodistribution, and preliminary clinical findings. *Molecular Therapy: The Journal of the American Society of Gene Therapy* 23 (4): 779–789.

Chadwick, S.L., Kingston, H.D., Stern, M. et al. (1997). Safety of a single aerosol administration of escalating doses of the cationic lipid GL-67/DOPE/DMPE-PEG5000 formulation to the lungs of normal volunteers. *Gene Ther* 4 (9): 937–942.

Chen, S., Tam, Y.Y., Lin, P.J. et al. (2014). Development of lipid nanoparticle formulations of siRNA for hepatocyte gene silencing following subcutaneous administration. *Journal of Controlled Release: Official Journal of the Controlled Release Society* 196: 106–112.

Coelho, T., Adams, D., Silva, A. et al. (2013). Safety and efficacy of RNAi therapy for transthyretin amyloidosis. *N Engl J Med* 369 (9): 819–829.

CRISPR Therapeutics (2017). News and press release. http://www.crisprtx.com/about-us/press-releases-and-presentations (accessed 28 June 2020).

CureVac AG (2020). CureVac announces positive results in low dose – 1 μg – rabies vaccine clinical phase 1 study. https://www.curevac.com/en/2020/01/07/curevac-announces-positive-results-in-low-dose-1-%c2%b5g-rabies-vaccine-clinical-phase-1-study/ (accessed 12 October 2020).

Finn, J.D., Rhoden Smith, A., Patel, M.C. et al. (2018). A single administration of CRISPR/Cas9 lipid nanoparticles achieves robust and persistent in vivo genome editing. *Cell Reports* 22: 2227–2235.

Davis, M.E. (2009). The first targeted delivery of siRNA in humans via a self-assembling, cyclodextrin polymer-based nanoparticle: from concept to clinic. *Molecular Pharmaceutics* 6 (3): 659–668.

Davis, M.E., Zuckerman, J.E., Choi, C.H.J. et al. (2010). Evidence of RNAi in humans from systemically administered siRNA via targeted nanoparticles. *Nature* 464 (7291): 1067–1070.

Dow, S.W., Fradkin, L.G., Liggitt, D.H. et al. (1999). Lipid-DNA complexes induce potent activation of innate immune responses and antitumor activity when administered intravenously. *The Journal of Immunology* 163 (3): 1552–1561.

Dritschilo, A., Huang, C.H., Rudin, C.M. et al. (2006). Phase I study of liposome-encapsulated c-raf antisense oligodeoxyribonucleotide infusion in combination with radiation therapy in patients with advanced malignancies. *Clin Cancer Res* 12 (4): 1251–1259.

First Affiliated Hospital Sun Yat-Sen University (2017). A safety and efficacy study of TALEN and CRISPR/Cas9 in the treatment of HPV-related Cervical Intraepithelial Neoplasia I. https://clinicaltrials.gov/ct2/show/NCT03057912 (accessed 28 June 2020).

Fitzgerald, K., Frank-Kamenetsky, M., Shulga-Morskaya, S. et al. (2014). Effect of an RNA interference drug on the synthesis of proprotein convertase subtilisin/kexin type 9 (PCSK9) and the concentration of serum LDL cholesterol in healthy volunteers: a randomised, single-blind, placebo-controlled, phase 1 trial. *Lancet* 383 (9911): 60–68.

Galanis, E., Burch, P.A., Richardson, R.L. et al. (2004). Intratumoral administration of a 1,2-dimyristyloxypropyl-3-dimethylhydroxyethyl ammonium bromide/dioleoylphosphatidylethanolamine formulation of the human interleukin-2 gene in the treatment of metastatic renal cell carcinoma. *Cancer* 101 (11): 2557–2566.

Glover, D.J., Lipps, H.J., and Jans, D.A. (2005). Towards safe, non-viral therapeutic gene expression in humans. *Nat Rev Genet* 6 (4): 299–310.

Hafez, I.M., Maurer, N., and Cullis, P.R. (2001). On the mechanism whereby cationic lipids promote intracellular delivery of polynucleic acids. *Gene Ther* 8 (15): 1188–1196.

Hassett, K.J., Benenato, K.E., Jacquinet, E. et al. (2019). Optimization of lipid nanoparticles for intramuscular administration of mRNA vaccines. *Molecular Therapy – Nucleic Acids* 15: 1–11.

Hensley, S.E. and Amalfitano, A. (2007). Toll-like receptors impact on safety and efficacy of gene transfer vectors. *Molecular Therapy* 15 (8): 1417–1422.

Hornung, V., Guenthner-Biller, M., Bourquin, C. et al. (2005). Sequence-specific potent induction of IFN-alpha by short interfering RNA in plasmacytoid dendritic cells through TLR7. *Nat Med* 11 (3): 263–270.

Hortobagyi, G.N., Ueno, N.T., Xia, W. et al. (2001). Cationic liposome-mediated E1A gene transfer to human breast and ovarian cancer cells and its biologic effects: a phase I clinical trial. *J Clin Oncol* 19 (14): 3422–3433.

Hu, Z., Yu, L., Zhu, D. et al. (2014). Disruption of HPV16-E7 by CRISPR/Cas system induces apoptosis and growth inhibition in HPV16 positive human cervical cancer cells. *BioMed Research International* 2014: 9.

Hu, Z., Ding, W., Zhu, D. et al. (2015). TALEN-mediated targeting of HPV oncogenes ameliorates HPV-related cervical malignancy. *The Journal of Clinical Investigation* 125 (1): 425–436.

Hyde, S.C., Southern, K.W., Gileadi, U. et al. (2000). Repeat administration of DNA/liposomes to the nasal epithelium of patients with cystic fibrosis. *Gene Ther* 7 (13): 1156–1165.

Jackson, L.A., Anderson, E.J., Rouphael, N.G. et al. (2020). An mRNA vaccine against SARS-CoV-2 – preliminary report. *New England Journal of Medicine* 383: 1920–1931.

Jayaraman, M., Ansell, S.M., Mui, B.L. et al. (2012). Maximizing the potency of siRNA lipid nanoparticles for hepatic gene silencing in vivo. *Angewandte Chemie* (International ed. in English) 51 (34): 8529–8533.

Judge, A.D., Sood, V., Shaw, J.R. et al. (2005). Sequence-dependent stimulation of the mammalian innate immune response by synthetic siRNA. *Nat Biotechnol* 23 (4): 457–462.

Kageshita, T., Mizuno, M., Ono, T. et al. (2001). Growth inhibition of human malignant melanoma transfected with the human interferon-β gene by means of cationic liposomes. *Melanoma Research* 11 (4): 337–342.

Kanasty, R., Dorkin, J.R., Vegas, A., and Anderson, D. (2013). Delivery materials for siRNA therapeutics. *Nature Materials* 12: 967.

Khatri, N., Baradia, D., Vhora, I. et al. (2014a). Development and characterization of siRNA lipoplexes: effect of different lipids, in vitro evaluation in cancerous cell lines and in vivo toxicity study. *AAPS PharmSciTech* 15 (6): 1630–1643.

Khatri, N., Baradia, D., Vhora, I. et al. (2014b). cRGD grafted liposomes containing inorganic nano-precipitate complexed siRNA for intracellular delivery in cancer cells. *Journal of Controlled Release* 182: 45–57.

Klimek, L., Willers, J., Hammann-Haenni, A. et al. (2011). Assessment of clinical efficacy of CYT003-QbG10 in patients with allergic rhinoconjunctivitis: a phase IIb study. *Clin Exp Allergy* 41 (9): 1305–1312.

Klinman, D.M., Yamshchikov, G., and Ishigatsubo, Y. (1997). Contribution of CpG motifs to the immunogenicity of DNA vaccines. *The Journal of Immunology* 158 (8): 3635–3639.

Knowles, M.R., Noone, P.G., Hohneker, K. et al. (1998). A double-blind, placebo controlled, dose ranging study to evaluate the safety and biological efficacy of the lipid-DNA complex GR213487B in the nasal epithelium of adult patients with cystic fibrosis. *Hum Gene Ther* 9 (2): 249–269.

Konstan, M.W., Davis, P.B., Wagener, J.S. et al. (2004). Compacted DNA nanoparticles administered to the nasal mucosa of cystic fibrosis subjects are safe and demonstrate partial to complete cystic fibrosis transmembrane regulator reconstitution. *Hum Gene Ther* 15 (12): 1255–1269.

Kozlovska, T.M., Cielens, I., Dreilinna, D. et al. (1993). Recombinant RNA phage Q beta capsid particles synthesized and self-assembled in *Escherichia coli*. *Gene* 137 (1): 133–137.

Lanuti, M., Rudginsky, S., Force, S.D. et al. (2000). Cationic lipid: bacterial DNA complexes elicit adaptive cellular immunity in murine intraperitoneal tumor models. *Cancer Research* 60 (11): 2955–2963.

Leng, Q. and Mixson, A.J. (2005). Small interfering RNA targeting Raf-1 inhibits tumor growth in vitro and in vivo. *Cancer Gene Ther* 12 (8): 682–690.

Leng, Q., Chou, S.-T., Scaria, P.V. et al. (2012). Buffering capacity and size of siRNA polyplexes influence cytokine levels. *Molecular Therapy* 20 (12): 2282–2290.

Lonez, C., Vandenbranden, M., and Ruysschaert, J.-M. (2012). Cationic lipids activate intracellular signaling pathways. *Adv Drug Deliv Rev* 64 (15): 1749–1758.

Lu, C., Stewart, D.J., Lee, J.J. et al. (2012). Phase I clinical trial of systemically administered TUSC2(FUS1)-nanoparticles mediating functional gene transfer in humans. *PloS One* 7 (4): e34833.

Matar, M., Fewell, J., Lewis, D.H., and Anwer, K. (2015). Nucleic acid-lipopolymer compositions. EP2178509B1.

Matsumoto, K., Kubo, H., Murata, H. et al. (2008). A pilot study of human interferon beta gene therapy for patients with advanced melanoma by in vivo transduction using cationic liposomes. *Jpn J Clin Oncol* 38 (12): 849–856.

Middleton, M.R., Wermke, M., Calvo, E. et al. (2018). Phase I/II, multicenter, open-label study of intratumoral/intralesional administration of the retinoic acid–inducible gene I (RIG-I) activator MK-4621 in patients with advanced or recurrent tumors. *Annals of Oncology* 28: vii712. https://www.annalsofoncology.org/article/S0923-7534(19)50407-0/fulltext.

Mintzer, M.A. and Simanek, E.E. (2009). Nonviral vectors for gene delivery. *Chemical Reviews* 109 (2): 259–302.

Moderna Therapeutics (2020). Vaccines and Related Biological Products Advisory Committee FDA Briefing Document, Moderna COVID-19 Vaccine.

Mulligan, M.J., Lyke, K.E., Kitchin, N. et al. (2020) Phase 1/2 Study to Describe the Safety and Immunogenicity of a COVID-19 RNA Vaccine Candidate (BNT162b1) in Adults 18 to 55 Years of Age: Interim Report. medRxiv, pp. 2020.2006.2030.20142570.

Naing, A., Lopez-Berestein, G., Fu, S. et al. (2017). EphA2 gene targeting using neutral liposomal small interfering RNA (EPHARNA) delivery: a phase I clinical trial. *Journal of Clinical Oncology* 35 (Suppl 15): TPS2604.

O'Malley, B.W. Jr., Li, D., McQuone, S.J., and Ralston, R. (2005). Combination nonviral interleukin-2 gene immunotherapy for head and neck cancer: from bench top to bedside. *Laryngoscope* 115 (3): 391–404.

Pardi, N., Tuyishime, S., Muramatsu, H. et al. (2015). Expression kinetics of nucleoside-modified mRNA delivered in lipid nanoparticles to mice by various routes. *Journal of Controlled Release: Official Journal of the Controlled Release Society* 217: 345–351.

Patil, S., Bhatt, P., Lalani, R. et al. (2016). Low molecular weight chitosan–protamine conjugate for siRNA delivery with enhanced stability and transfection efficiency. *RSC Advances* 6 (112): 110951–110963.

Patil, S., Lalani, R., Bhatt, P. et al. (2018). Hydroxyethyl substituted linear polyethylenimine for safe and efficient delivery of siRNA therapeutics. *RSC Advances* 8 (62): 35461–35473.

Pfizer and BioNTech (2020). Vaccines and Related Biological Products Advisory Committee FDA Briefing Document, Pfizer-BioNTech COVID-19 Vaccine.

Pirollo, K.F., Nemunaitis, J., Leung, P.K. et al. (2016). Safety and efficacy in advanced solid tumors of a targeted nanocomplex carrying the p53 gene used in combination with docetaxel: a phase 1b study. *Molecular Therapy: The Journal of the American Society of Gene Therapy* 24 (9): 1697–1706.

Ramanathan, R.K., Hamburg, S.I., Borad, M.J. et al. (2013). Abstract LB-289: a phase I dose escalation study of TKM-080301, a RNAi therapeutic directed against PLK1, in patients with advanced solid tumors. *Cancer Research* 73 (Suppl 8): LB-289.

Ramaswamy, S., Tonnu, N., Tachikawa, K. et al. (2017). Systemic delivery of factor IX messenger RNA for protein replacement therapy. *Proc Natl Acad Sci U S A* 114 (10): E1941–E1950.

Ren, H., Boulikas, T., Lundstrom, K. et al. (2003). Immunogene therapy of recurrent glioblastoma multiforme with a liposomally encapsulated replication-incompetent Semliki forest virus vector carrying the human interleukin-12 gene – a phase I/II clinical protocol. *J Neurooncol* 64 (1-2): 147–154.

Rini, B.I., Selk, L.M., and Vogelzang, N.J. (1999). Phase I study of direct intralesional gene transfer of HLA-B7 into metastatic renal carcinoma lesions. *Clinical Cancer Research* 5 (10): 2766.

Robbins, M., Judge, A., and MacLachlan, I. (2009). siRNA and innate immunity. *Oligonucleotides* 19 (2): 89–102.

Rudginsky, S., Siders, W., Ingram, L. et al. (2001). Antitumor activity of cationic lipid complexed with immunostimulatory DNA. *Molecular Therapy* 4 (4): 347–355.

Rudin, C.M., Marshall, J.L., Huang, C.H. et al. (2004). Delivery of a liposomal c-raf-1 antisense oligonucleotide by weekly bolus dosing in patients with advanced solid tumors: a phase I study. *Clin Cancer Res* 10 (21): 7244–7251.

Ruiz, F.E., Clancy, J.P., Perricone, M.A. et al. (2001). A clinical inflammatory syndrome attributable to aerosolized lipid-DNA administration in cystic fibrosis. *Hum Gene Ther* 12 (7): 751–761.

Sabnis, S., Kumarasinghe, E.S., Salerno, T. et al. (2018). A novel amino lipid series for mRNA delivery: improved endosomal escape and sustained pharmacology and safety in non-human primates. *Molecular Therapy: The Journal of the American Society of Gene Therapy* 26 (6): 1509–1519.

Scheule, R.K., St George, J.A., Bagley, R.G. et al. (1997). Basis of pulmonary toxicity associated with cationic lipid-mediated gene transfer to the mammalian lung. *Hum Gene Ther* 8 (6): 689–707.

Semple, S.C., Akinc, A., Chen, J. et al. (2010). Rational design of cationic lipids for siRNA delivery. *Nat Biotechnol* 28 (2): 172–176.

Senzer, N., Nemunaitis, J., Nemunaitis, D. et al. (2013). Phase I study of a systemically delivered p53 nanoparticle in advanced solid tumors. *Molecular Therapy: The Journal of the American Society of Gene Therapy* 21 (5): 1096–1103.

Sidi, A.A., Ohana, P., Benjamin, S. et al. (2008). Phase I/II marker lesion study of intravesical BC-819 DNA plasmid in H19 over expressing superficial bladder cancer refractory to bacillus Calmette-Guerin. *J Urol* 180 (6): 2379–2383.

Silence Therapeutics (2016). Silence is golden. https://www.edisoninvestmentresearch.com/?ACT=18&ID=16753&LANG= (accessed 28 June 2020).

Sirnaomics (2016). A randomized study to evaluate the safety and efficacy of various doses of STP705 in subjects with hypertrophic scar. https://clinicaltrials.gov/ct2/show/NCT02956317 (accessed 21 August 2020).

Soule, B., Tirucherai, G., Kavita, U. et al. (2018). Safety, tolerability, and pharmacokinetics of BMS-986263/ND-L02-s0201, a novel targeted lipid nanoparticle delivering HSP47 siRNA, in healthy participants: a randomised, placebo-controlled, double-blind, phase 1 study. *Journal of Hepatology* 68: S112.

Stopeck, A.T., Jones, A., Hersh, E.M. et al. (2001). Phase II study of direct intralesional gene transfer of allovectin-7, an HLA-B7/beta2-microglobulin DNA-liposome complex, in patients with metastatic melanoma. *Clin Cancer Res* 7 (8): 2285–2291.

Strumberg, D., Schultheis, B., Traugott, U. et al. (2012). Phase I clinical development of Atu027, a siRNA formulation targeting PKN3 in patients with advanced solid tumors. *Int J Clin Pharmacol Ther* 50 (1): 76–78.

Szebeni, J., Muggia, F., Gabizon, A., and Barenholz, Y. (2011). Activation of complement by therapeutic liposomes and other lipid excipient-based therapeutic products: prediction and prevention. *Adv Drug Deliv Rev* 63 (12): 1020–1030.

Tabernero, J., Shapiro, G.I., LoRusso, P.M. et al. (2013). First-in-humans trial of an RNA interference therapeutic targeting VEGF and KSP in cancer patients with liver involvement. *Cancer Discov* 3 (4): 406–417.

Tarhini, A.A., Belani, C.P., Luketich, J.D. et al. (2011). A phase I study of concurrent chemotherapy (paclitaxel and carboplatin) and thoracic radiotherapy with swallowed manganese superoxide dismutase plasmid liposome protection in patients with locally advanced stage III non-small-cell lung cancer. *Hum Gene Ther* 22 (3): 336–342.

Tousignant, J.D., Gates, A.L., Ingram, L.A. et al. (2000). Comprehensive analysis of the acute toxicities induced by systemic administration of cationic lipid:plasmid DNA complexes in mice. *Hum Gene Ther* 11 (18): 2493–2513.

Tousignant, J.D., Zhao, H., Yew, N.S. et al. (2003). DNA sequences in cationic lipid:pDNA-mediated systemic toxicities. *Hum Gene Ther* 14 (3): 203–214.

Translate Bio Inc. (2019). Study to evaluate the safety & tolerability of MRT5005 administered by nebulization in adults with cystic fibrosis (RESTORE-CF). https://clinicaltrials.gov/ct2/show/NCT03375047 (accessed 21 August 2020).

Vhora, I., Khatri, N., Desai, J., and Thakkar, H.P. (2014). Caprylate-conjugated cisplatin for the development of novel liposomal formulation. *AAPS PharmSciTech* 15 (4): 845–857.

Vhora, I., Patil, S., Amrutiya, J., and Misra, A. (2015). Liposomes and lipid envelope-type systems for systemic siRNA delivery. *Current Pharmaceutical Design* 21 (31): 4541–4555.

Vhora, I., Lalani, R., Bhatt, P. et al. (2018). Colloidally stable small unilamellar stearyl amine lipoplexes for effective BMP-9 gene delivery to stem cells for osteogenic differentiation. *AAPS PharmSciTech* 19: 3550–3560.

Vhora, I., Lalani, R., Bhatt, P. et al. (2019). Lipid-nucleic acid nanoparticles of novel ionizable lipids for systemic BMP-9 gene delivery to bone-marrow mesenchymal stem cells for osteoinduction. *International Journal of Pharmaceutics* 563: 324–336.

Villaret, D., Glisson, B., Kenady, D. et al. (2002). A multicenter phase II study of tgDCC-E1A for the intratumoral treatment of patients with recurrent head and neck squamous cell carcinoma. *Head and Neck* 24 (7): 661–669.

von der Leyen, H.E., Mugge, A., Hanefeld, C. et al. (2011). A prospective, single-blind, multicenter, dose escalation study of intracoronary iNOS lipoplex (CAR-MP583) gene therapy for the prevention of restenosis in patients with de novo or restenotic coronary artery lesion (REGENT I extension). *Hum Gene Ther* 22 (8): 951–958.

Wakabayashi, T., Natsume, A., Hashizume, Y. et al. (2008). A phase I clinical trial of interferon-beta gene therapy for high-grade glioma: novel findings from gene expression profiling and autopsy. *J Gene Med* 10 (4): 329–339.

WHO (2020). Draft landscape of COVID-19 candidate vaccines. https://www.who.int/publications/m/item/draft-landscape-of-covid-19-candidate-vaccines (accessed 28 June 2020).

Yoo, G.H., Hung, M.-C., Lopez-Berestein, G. et al. (2001). Phase I trial of intratumoral liposome E1A gene therapy in patients with recurrent breast and head and neck cancer. *Clinical Cancer Research* 7 (5): 1237.

Yoshida, J., Mizuno, M., Fujii, M. et al. (2004). Human gene therapy for malignant gliomas (glioblastoma multiforme and anaplastic astrocytoma) by in vivo transduction with human interferon β gene using cationic liposomes. *Hum Gene Ther* 15 (1): 77–86.

Zatsepin, T.S., Kotelevtsev, Y.V., and Koteliansky, V. (2016). Lipid nanoparticles for targeted siRNA delivery – going from bench to bedside. *International Journal of Nanomedicine* 11: 3077–3086.

Zhan, T., Rindtorff, N., Betge, J. et al. (2019). CRISPR/Cas9 for cancer research and therapy. *Seminars in Cancer Biology* 55: 106–119.

Zhang, X., Goel, V., and Robbie, G.J. (2019). Pharmacokinetics of patisiran, the first approved RNA interference therapy in patients with hereditary transthyretin-mediated amyloidosis. *J Clin Pharmacol* 60 (5): 573–585.

Zhao, Y. and Huang, L. (2014). Lipid nanoparticles for gene delivery. *Advances in Genetics* 88: 13–36.

Zhou, J., Zhao, Y., Simonenko, V. et al. (2017). Simultaneous silencing of TGF-β1 and COX-2 reduces human skin hypertrophic scar through activation of fibroblast apoptosis. *Oncotarget* 8 (46): 80651–80665.

Zuckerman, J., Gritli, I., Tolcher, A. et al. (2014). Correlating animal and human phase Ia/Ib clinical data with CALAA-01, a targeted, polymer-based nanoparticle containing siRNA. *Proc Natl Acad Sci U S A* 111 (31): 11449–11454.

10

Toxicity and Safety Evaluation of Lipid-Based Nanoparticles for Brain Delivery

Mitali Patel and Priyanshi Patel

Maliba Pharmacy College, Uka Tarsadia University, Bardoli, GJ, India

10.1 Introduction

Drug transport to the central nervous system (CNS) is imperative for treating various neurological diseases like encephalitis, parkinson's disease, meningitis, alzheimer's disease and/or tumors like glioblastoma, schizophrenia, etc. (Pathan et al. 2009). There are several drugs, peptides and monoclonal antibodies that have proven their efficacy in treating such diseases, but most of the diseases remain undertreated owing to the presence of the barriers, i.e. blood–brain barrier (BBB) and the blood cerebrospinal fluid barrier (BCSF) (Patel et al. 2016). The therapeutic efficacy of drug molecules is limited by their incapability to reach and maintain the desired concentration at the affected site. Hence, drug molecules are unable to reach the brain via the circulatory system (Gomes et al. 2016). The BBB is highly lipophilic in nature and because of this, only a small lipid-soluble drug can intervene the brain cells. It is strenuous to deliver the drug in the brain due to insufficient transport across the BBB and the efflux of drugs to blood from the brain.

By considering these obstacles, it is necessary for a researcher to explore an innovative approach that addresses the challenges associated with brain delivery. Currently, innovations in nanomedicine have fostered several opportunities which can effectively ameliorate the obstacles associated with brain delivery, namely nanoparticles (NPs). NPs offer potential for addressing all these challenges as they possess distinctive characteristics *viz* lipophilicity, size, surface charge, shape and coating (Kreuter 2001). It is one of the versatile drug delivery systems that protects the active ingredient, delivers drug to the target site, escapes reticuloendothelial system (RES) and enhances permeation across the BBB (Gomes et al. 2016; Saraiva et al. 2016; Zheng et al. 2018). Nanoparticulate carriers are effective in brain delivery as they facilitate the retention of molecules in blood–brain capillaries, which in turn produces a higher concentration gradient which results in the more uptake of NPs into the brain (Khosa et al. 2018). Moreover, modification of nanoparticle surfaces using polymer coating, ligand, polyethylene glycol (PEG), monoclonal antibodies, etc., prolongs the circulation time, increases plasma concentration and improves brain delivery (Dong 2018; Sonvico et al. 2018). Different approaches such as polymeric NPs (Saucier-Sawyer et al. 2015; Elzoghby et al. 2016), lipid NPs (Fang et al. 2008), dendrimers (Leyuan et al. 2014; Somani

Nanotechnology in Medicine: Toxicity and Safety, First Edition. Edited by Mahendra Rai, Mrunali Patel, and Rashmin Patel.

and Dufes 2014), carbon nanotubes (Guo et al. 2017), liposomes (Wang et al. 2019), etc., are used to deliver therapeutic ingredients to treat the CNS disorders. Distribution of small therapeutic molecules can be improved by incorporating into NPs using chemical methods that include adsorption, encapsulation and covalent linkage whereas that of macromolecules can be done by surface modifications (Zhang et al. 2015).

10.2 Barriers Across Brain Delivery

The human brain consists of capillaries having length and area of 650 km and 12 m^2, respectively. This acts as an effective barrier and does not allow lipid-insoluble compounds like polar molecules but some water-soluble drugs like levodopa are permeable across the BBB (Vyas and Khar 2002). The BBB and BCSF barriers impede drug diffusion in the middle of brain fluids and blood (Redzic 2011).

10.2.1 The Blood–Brain Barrier (BBB)

The BBB is dynamic, interacts with its microenvironment and responds to the needs of the CNS. The BBB is made up of extremely polarized endothelial cells, surrounded by pericytes, neurons, astrocytic foot processes, microglia, mast and immune cells. The BBB consists of tight junction proteins, transporters, adherence proteins, basal lamina and extracellular matrix at the molecular level (Warren 2018). This makes the tight junctions (TJs) in such a way that maintain the homeostasis in the middle of the CNS and blood circulation that prohibits entry of small and large molecules across this barrier. Various transporters are present at the BBB endothelium that control the entry and/or efflux of drug moieties into the CNS and brain. Examples of transporters that expedite the drug entrance into the brain are organic anion transporting polypeptides (OATPs), organic anion transporters (OATs), organic cation transporters (OCTs), nucleoside transporters, monocarboxylate transporters (MCTs) and putative transport systems for peptide transport. Efflux transporters are breast cancer resistance protein (BCRP), multidrug resistance proteins (MRPs) and P-glycoprotein (P-gp) (Engelhardt and Sorokin 2009; Warren 2018). The BBB hampers paracellular diffusion of hydrophilic drug moieties by TJs that interlink endothelial cells (Engelhardt and Sorokin 2009). The passive diffusion mechanism is responsible for the transportation of lipophilic drugs whereas active transport is linked to the transport of polar drug molecules across the BBB and hence they are considered as poor CNS agents. Furthermore, some drugs have inherent ability to cross the BBB by active transfer are transported again to vasculature via efflux transporters, for instance, P-glycoprotein 1, which belongs to ATP binding cassette family (van Tellingen et al. 2015).

10.2.2 The Blood Cerebrospinal Fluid (BCSF) Barrier

The BCSF barrier constitutes the choroid plexus, a linkage between systemic circulation and CSF. It has fenestrated capillaries, connected through TJs that provide linkage between adjoining choroid plexus epithelial cells and limits the diffusion of hydrophilic moieties (Warren 2018). The value of transepithelial electrical resistance (TEER) for the BCSF barrier is minute in contrast to that of the BBB, indicating that a high number of solutes can pass through the BCSF barrier (Sanchez-Covarrubias et al. 2014; van Tellingen et al. 2015).

The choroid plexus generates CSF, which gets discharged in the lateral, third and fourth ventricles. Thus, choroid plexus has an imperative part to play in fluid pressure regulation in the CSF (Abbott et al. 2006). CSF discharge is controlled by the swapping of ions through the epithelium and the driving force is the Na^+-K^+

ATPase pump and carbonic anhydrase enzyme (Strazielle and Ghersi-Egea 2000). CSF is continuously released and gets absorbed again in circulation so the total amount of CSF is restored four or five times each day. This causes a "sink effect" that decreases steady-state accumulation of molecules gaining entry in the CSF and brain (Brzica et al. 2017). The "sink effect" is augmented for molecules having higher molecular weight and hydrophilic nature.

10.3 Role of Lipid Nanoparticles in Brain Delivery

Lipid-based drug delivery systems have achieved more attention nowadays. The reason behind this is, 40–70% of new molecules possess limited aqueous solubility and poor absorption, which reduces the therapeutic concentration. Higher hydrophobicity of drug moiety leads to higher protein binding, causing a slowdown in drug expulsion and metabolism (Wu et al. 2005; Mundada et al. 2016). Hence, a suitable carrier is needed to overcome these pitfalls for effective treatment of any disease.

Lipid-based NPs are generally of 10–500 nm and consist of biodegradable lipids. They offer advantages such as controlled drug release, increase bioavailability, specific targeting, protect degradation of drugs, ease of scale up, low toxicity, depletion in dose and hence reduce adverse effects (Bondi et al. 2012; Shankar et al. 2018). Encapsulation of drugs in the lipid material also increases the solubility and permeation across biological membranes. Properties of lipids such as biodegradability and biocompatibility make them as ideal system to enhance solubility, bioavailability and effective therapy (Muller et al. 2002). Because the carrier used for its fabrication is of natural origin, it is associated with less toxicity in distinction to the polymeric particles. Lipid-based NPs are liposomes (Martins et al. 2007), solid lipid nanoparticles (SLNs) (Muller et al. 2000), nanostructured lipid carriers (NLCs) (Muller et al. 2002) and lipid–drug conjugates (LDCs) (Irby et al. 2017). All these lipid NPs are solid at body and room temperature. These lipid NPs are shown in Figure 10.1 and further discussed below.

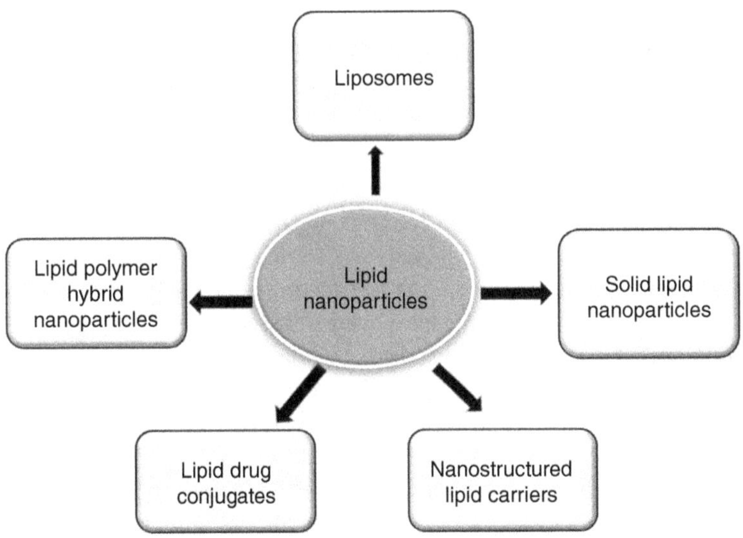

Figure 10.1 Types of lipid nanoparticles.

10.3.1 Liposomes

Liposomes are extensively used as carriers for the delivery of drugs since they were discovered by english hematologist named Alec Bangham in year 1961 (Lai et al. 2013). Liposomes are basically vesicles comprising phospholipid bilayers adjacent to an aqueous compartment. These vesicles are generally produced using cholesterols, sphingolipids, glycolipids, nontoxic surfactants, long chain fatty acids and also membrane proteins (Samad et al. 2007). Generally, lipophilic or hydrophobic moieties are incorporated into the hydrophobic core (lipid bilayers) of liposomes. Additionally, the adsorption of polyanions like DNA and RNA is facilitated by the use of cationic lipids (Dow 2008). They have a propensity to entrap and deliver more drugs and their surface can be modified with different ligands (Lestini et al. 2002; Nguyen et al. 2016). Different ways have been used by altering the bilayer of liposomes with amphiphiles to elongate the circulation time (e.g. stealth liposomes), improve the elasticity (e.g. transferosomes), or production of covalent drug–lipid complexes for increased transfer of drugs (e.g. pharmacosomes) (Zylberberg and Matosevic 2016). Cationic liposomes take the advantage of negative charge of BBB and so, initiate cell internalization process via electrostatic interactions (Agrawal et al. 2017).

Liposome's pharmacokinetics and biodistribution in the brain can further be boosted by surface-functionalization of liposomes. For example, coating around liposome with PEG imparts a protective covering and hampers the binding of drug to plasma proteins, prevent opsonization and hence clearance of liposomes (Schnyder and Huwyler 2005).

Gurturk et al. (2017) formulated levodopa-loaded liposomes modified with maltodextrin to increase its transport across BBB. Glutathione, an antioxidant, was used as a supportive agent and the consequence of its addition on the stability of liposome was determined. The addition of glutathione showed positive results on cell viabilities of 3T3 and SH-SY5Y. Results showed that liposomes modified with maltodextrin had elevated *in vitro* levodopa transit in the parallel artificial membrane permeability assay (PAMPA) and also showed more attachment to MDCK cells.

Soni et al. (2005) prepared transferrin-attached liposomes for brain delivery of 5-florouracil using cast film method. The size of liposome coupled with transferrin was higher and the drug entrapment was decreased. Higher *in vitro* drug release was obtained (i.e. $74.8 \pm 3.21\%$) with non-combined liposomes as that of combined liposomes. *In vivo* study showed the intake of transferrin-combined liposomes in the brain. The level of radioactivity of combined liposomes was 17 and 13 times higher in contrast to free radioactive agent and the non-combined liposomes.

10.3.2 Solid Lipid Nanoparticles (SLNs)

SLNs are colloidal dispersions, comprising of a matrix of solid biodegradable lipids and have particle size in the range of 10–500 nm (Sawant and Dodiya 2008; Patel et al. 2019). SLNs possess less toxicity due to biodegradable nature and are safe to be taken up by the brain. They can pass the BBB, get out from the RES and bypass the liver due to a narrow size of 100–200 nm. The lipid matrix gives stability to the drug, provides controlled release of drug and shows excellent reproducibility (Blasi et al. 2007; Kaur et al. 2008).

Neves et al. (2016) prepared functionalized SLNs of resveratrol using apolipoprotein E as they are overexpressed at the BBB and easily recognized by low-density lipoprotein (LDL) receptors. The particle size and zeta potential was less than 200 nm and −13 mV, respectively. No toxicity up to 50 μM was observed across the hCMEC/D3 cell line by MTT assay at the end of four hours. Functionalized NPs showed increase (1.8-fold higher) permeability across hCMEC/D3 in contrast non-functionalized NPs.

Zhou et al. (2015) developed venlafaxine (VLX)-loaded solid lipid nanoparticles (VLX-SLNs) and characterized VLX in distinct formulations by determining their pharmacokinetics and distribution in the brain following oral or intravenous administration. The particle size, entrapment efficiency (EE) and zeta potential of developed VLX-SLNs was 186.3 ± 69.26 nm, 74.9 ± 3.0% and −22.8 ± 7.78 mv, respectively. VLX-SLNs showed higher brain uptake after IV administration as compared with VLX solution, VLX solution with empty SLNs (VLX + empty SLNs) and VLX solution with verapamil (VLX + Ver), respectively. P-gp protein mass was found to be lowest in the group treated with VLX-SLNs. All the results conclusively proved that SLNs could effectively overcome P-gp efflux and target the brain cells successfully by intravenous administration.

10.3.3 Nanostructured Lipid Carriers (NLCs)

NLCs, the second generation of lipid NPs overcome the disadvantages of SLNs, namely low drug loading and removal of drugs throughout storage (Mukherjee et al. 2009; Ganesan and Narayanasamy 2017). The NLCs consist of liquid and solid lipids, creating more imperfections in the matrix so that higher drug can be incorporated. NLCs stay as solid matrix just like SLNs at the body as well as room temperature (Gartziandia et al. 2015; Chella and Shastri 2017). NLCs with small particle size and high drug loading are better than SLNs.

Nasiri et al. (2019) formulated and evaluated Indinavir-loaded NLCs for targeting the brain to reduce P-gp efflux. The NPs had particle size, PDI, zeta potential, EE and drug loading of 161.02 ± 4.8 nm, 0.293 ± 0.07, −40.62 ± 2.21 mV, 93 ± 1.58% and 9.15 ± 0.15%, respectively. On the surface of NLCs, human Transferrin (TR) was attached to enhance the drug transport. This TR-modified NLCs showed particle size, zeta potential and PDI of 185.29 ± 6.7 nm, −28.68 ± 3.37 mV and 0.247 ± 0.06, respectively. The brain uptake clearance of NLCs and TR-NLCs was found to be 6.5 and 32.75 times higher than free drug.

Madane and Mahajan (2016) investigated potential of curcumin (CRM)-loaded NLCs for brain cancer treatment. The CRM-loaded NLCs had particle size, PDI, zeta potential and EE of 146.8 nm, 0.18, −21.4 mV and 90.86%, respectively. CRM-NLCs showed higher cytotoxicity to astrocytoma-glioblastoma cell line (U373MG) in contrast to that of CRM. Higher drug concentration was determined with CRM-NLCs in the brain with NLCs after intranasal administration in biodistribution studies.

10.3.4 Lipid–Drug Conjugates (LDCs)

LDCs consist of drug molecules that are covalently linked with lipids. LDCs exhibit numerous advantages like improve bioavailability, high tumor selectivity, diminish toxicity and high drug loading (Banerjee and Kundu 2018). LDCs have high therapeutic power especially in the case of hydrophilic drug targeting to the site of action (Irby et al. 2017). In LDCs, through covalent or noncovalent bond, drug is affixed to lipid, namely diglyceride, phosphoglyceride, or fatty acid (Muchow et al. 2008; Adhikari et al. 2017).

Olbrich et al. (2002) developed and evaluated LDCs of antitrypanosomal, hydrophilic drug diminazenediaceturate. It was confirmed by the FTIR and DSC study that diminazene and fatty acids reacted to form water-insoluble salts such as diminazenedistearate and dioleate. The particle size and PDI were 364 nm and 0.233 for diminazenedistearate and 442 nm and 0.268 for diminazenedioleate, respectively. The drug loading of 33% (w/w) proved that the loading capacity of SLNs for hydrophilic drugs was increased.

Formulation of nanosized LDC of water-soluble drug facilitates prolong drug release and target the particular site after i.v. administration.

10.3.5 Lipid Polymer Hybrid Nanoparticles (LPHNPs)

LPHNPs incorporate all attributes of liposomes and polymeric nanoparticles, namely physical stability and biocompatibility (Mukherjee et al. 2019). LPHNPs are core–shell nanoparticle comprise of lipid or lipid–PEG shells and polymer cores. LPHNPs have higher *in vivo* cell distribution potential in contrast to polymeric NPs and liposomes (Hadinoto et al. 2013; Dehaini et al. 2016).

The inner core is a polymer matrix that contains API with high drug loading capacity and the lipidic portion that is enclosed by polymer core which gives biocompatibility and facilitates the holding of the drug in a polymer-containing core. The exterior coating consists of lipid and PEG coating the lipoidal part which further increases stability, enhances circulation time and provides steric stabilization (Dave et al. 2019).

Ishak et al. (2017) prepared LPHNPs of rutin (RU) containing polymeric core and a phospholipid exterior using PEG based surfactants (SAA), i.e. TPGS or Solutol HS 15 in contrast to Tween 80. The target was to increase the brain delivery, avoid opsonization and prove its anti-alzheimer activity. The particle size of Tween 80 LPH NPs, TPGS LPH NPs and Solutol HS 15 LPH NPs was found to be 272.50 ± 3.39, 203.00 ± 2.20 and 232.4 ± 4.01 nm, respectively. The *in vitro* drug release study demonstrated that all NPs had sustained release of drug upto six hours. The hemolysis test showed the biocompatibility of the developed NPs at SAA levels <2 mg/ml. The pharmacokinetic study showed a significant improvement in RU bioavailability in NPs as compared with the drug solution. No significant difference in concentration of RU aggregation inside the brain was reported in biodistribution studies.

10.4 Transport Mechanisms Involved for Brain Delivery

It is well known that the brain does not allow random entry of any compounds inside it. It only allows selective nutrients and some regulatory factors to enter the brain. Endothelial cells allow the transfer of molecules either passively, reliant on concentration gradient, or actively in an energy-reliant manner (Georgieva et al. 2014). The various mechanisms involved in the transport of lipid NPs are discussed below.

10.4.1 Paracellular Transport

It is a pathway for tiny hydrophilic substances and TJs which are present between endothelial cells hamper the entry of these substances (Rooy et al. 2011).

10.4.2 Transcellular Transport

Lipophilic substances (barbiturates and benzodiazepines) with molecular weight <400 Da are transported by transcellular diffusion through the endothelial plasma membranes (Cecchelli et al. 2007; Rooy et al. 2011). The accumulation rate in the brain is determined by the physical properties like lipophilicity and molecular weight.

10.4.3 Carrier-mediated Transport

Carrier-mediated transfer is involved in the transportation of polar molecules like amino acids, glucose and nucleosides into the CNS that cannot easily diffuse through the cell membrane. This pathway is substrate peculiar, i.e. drugs those strictly imitate the endogenic substrates will only be transferred.

10.4.4 Receptor-mediated Endocytosis

Receptor-mediated endocytosis begins with binding to the receptor-specific ligand. It is extremely particular and energy-driven transport that enables specific intake of macromolecules. For targeting the brain, BBB receptor-specific ligands are helpful in carrying endogenous peptides comprising of transferrin, insulin, albumin and opioid peptides (Masserini 2013).

10.4.5 Adsorptive-mediated Endocytosis

When polycationic molecules attach to negatively charged biological membrane, an adsorption-driven endocytosis begins. The mechanism behind this is electrostatic relation comprising a positively charged molecule (e.g. albumin) and a negatively charged location on the brain endothelial cell (BEC) surface (e.g. glycoprotein) which leads to the formation of endosomes (Bhaskar et al. 2010; Malhotra and Prakash 2011).

10.5 Toxicity of Lipid Nanoparticles

Regulatory science associated with nanotechnology necessitates the utilization of particular tools to detect nanomaterials in FDA-controlled products and also in the development of new methods for analysis or for confirming properties and size of these NPs in toxicity studies. In this viewpoint, the National Center for Toxicological Research (NCTR) has developed analytical procedures and tools for the purpose of quantifying NPs present in matrices and for performing toxicity studies on NPs (FDA's Approach to Regulation of Nanotechnology Products 2018).

Injury to the brain and CNS is the primary trouble for nanocarrier applications. The BBB protects the brain from a vast number of molecules that circulate continuously in the blood by hindering their entry. Small-sized drugs can easily permeate the BBB, resulting in potential neurological effects; a few NPs can also retain in brain tissue. Several factors are responsible for lipid nanoparticle toxicity such as composition of NPs, excipients used, particle size, surface coating, particle shape, slow clearance of NPs from the body, etc. (Yildirimer et al. 2011).

Cationic NPs are attracted toward the negative charge of the biological membrane and have greater cellular internalization in comparison to anionic or neutral NPs. The extent of cellular uptake of lipidic NPs determines the extent of toxicity caused by NPs (Pacheco-Blandino et al. 2012). An important issue regarding nanoparticles is that they are degraded in the cellular vicinity and hence they can induce cellular reactions. They might assemble in cells and lead to the disordering of integrity of organelle or gene modifications that leads to acute toxicity (Hossain et al. 2011). Smaller particles are more dangerous than larger ones because smaller ones are not efficiently removed and can cross the cell membranes more easily (Zoroddu 2017).

NPs smaller than cells and organelles can easily permeate cell membrane, gain entry into cytoplasm and organelles and disturb cellular mechanisms ultimately causing the disease (Buzea et al. 2007). Because they are tiny, NPs can transmit from the entrance site into circulatory, lymphatic and other systems and eventually to body tissues and organs. Some of them, based on their composition, charge and size, can cause irreparable destruction to cells either by increasing oxidative stress and/or by causing injury to organs.

Cellular uptake of NPs and residence in organelles leads to cytotoxicity. Biological membrane can be physically destructed by NPs directly or indirectly by producing reactive oxygen species (ROS) which can cause cell death. More the nanoparticles accumulation, more will be the damage (Buzea and Pacheco 2019). NPs accumulate in organs, create ROS and induce inflammation. The mechanism of these has been postulated that oxidative stresses cause changes in intracellular calcium amount, trigger transcription factors and activate cytokine generation (Oberdorster et al. 2009; Zoroddu et al. 2014). The reaction of NPs

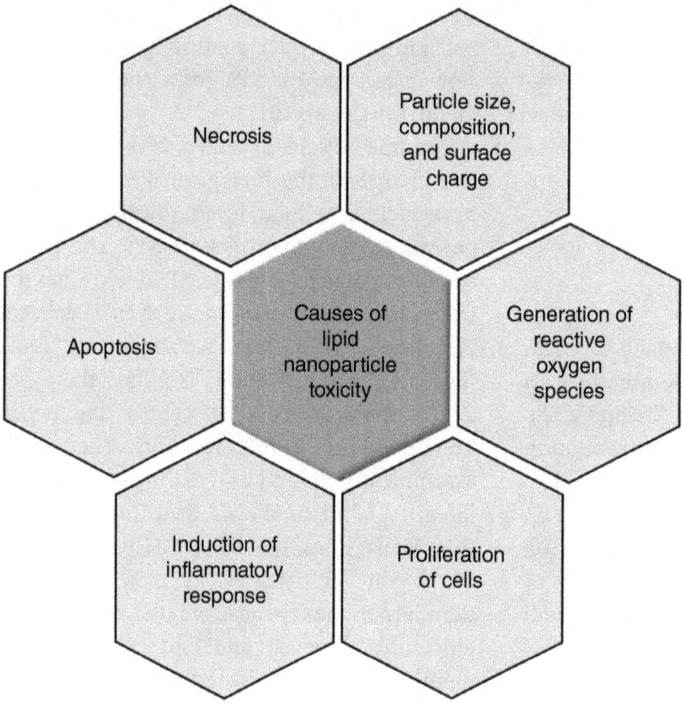

Figure 10.2 Toxicity due to nanoparticles.

with cells causes a prooxidant effect resulting in ROS generation, mitochondrial respiration and NADPH-regulated enzyme activation. When internalization of NPs occurs, phagocytosis causes production of ROS (Kumar et al. 2017). All the reasons of toxicity are depicted in Figure 10.2.

Different techniques used to determine the toxicity of lipid NPs *in vitro* and *in vivo* are outlined in Table 10.1.

10.6 Safety of Lipid Nanoparticles

In the current healthcare practice, the toxicity of drugs is an important aspect that leads to the majority of adverse drug reactions. In order to improve drug safety, one approach that can be used is the use of lipid-based nanocarriers in drug formulation (Kuntsche et al. 2008). Lipid-based nanocarriers are most commonly formulated using lipidic constituents such as cholesterol, phospholipids, fatty acids, triglycerides and waxes (Copland et al. 2005; Rawat et al. 2008). The materials are biocompatible and biodegradable when administered *in vivo* as these are obtained or derivatized from natural sources.

Numerous lipidic excipients have already been utilized in the formulation of pharmaceuticals which are approved by the FDA and also these excipients possess a proven safety along with familiar toxicity-related data (Koo et al. 2005). However, though considered safe, lipids portray toxicity relying on the dose, especially when they are taken at elevated doses. However, the amount of lipid that reaches the human body by formulating lipid-based delivery system is significantly lower. Also, as the formulation is administered according to the already prescribed therapeutic dosage of the active ingredient, the chances of overdosing with lipids are ruled out.

Table 10.1 Tests for accessing toxicity of nanoparticles.

Test	Description
NMR, FT-IR, XRD, NTA, AFM, TEM, DLS, MS	It determines the chemical composition, particle size and charge on surface of nanoparticles (Dhawan and Sharma 2010).
In vitro methods	
MTT cell proliferation assay	It calculates the depletion in cell proliferation rate and cell viability due to metabolic events owing to apoptosis or necrosis (Jones and Grainger 2009; Silva et al. 2011).
Cellular interaction assays	It determines the potential of nanoparticles to interact and cross the cellular barriers (Kumar et al. 2017).
Apoptosis assay	Production of free radical causes DNA damage and ultimately apoptosis and this is considered as a vital marker linked with toxicity of lipid nanoparticles (Mulik et al. 2010).
Oxidative stress assays	As nanoparticles possess high surface area to volume ratio, they form reactive oxygen species which increases intracellular damage.
Endotoxin assays	Nanoparticles have ability to produce endotoxins which can cause the inflammatory response in humans (Smulders et al. 2012).
Hemolytic and platelet aggregation tests	Nanoparticles when administered by i.v. route, can interfere with normal blood cell functioning and this test ascertains the thrombogenic or anticoagulant properties of nanoparticles (Hall et al. 2007).
In vivo methods	
Single/repeat dose acute toxicity measurement	It is used to measure the dose-dependent toxicity in mice or rats (Grainger 2009).
Organ distribution and clearance study	Absorption, distribution, metabolism and excretion determination of nanoparticles is a prerequisite for estimating toxic outcomes of nanoparticles (Hall et al. 2007; Grainger 2009).
Immunotoxicity study	FDA recommends immunotoxicity study for examining prospective of nanoparticles for immunotoxicity.

AFM, atomic force microscopy; DLS, dynamic light scattering; FTIR, Fourier transform infrared; MS, mass spectroscopy; NMR, nuclear magnetic resonance; NTA, nanoparticle tracking and analysis; TEM, transmission electron microscopy; XRD, X-ray diffraction.

In comparison to normal vasculature, the vasculature of tumors is comparatively porous and vasculature of inflamed tissue has a size around 200–780 nm (Torchilin 2007). So, the particles smaller than this cut off range increases accumulation of drugs at the targeted tumor or to a particular site and causes a decrease in the accumulation of drugs in surrounding healthy tissues, leading to an increase in potency along with a reduction in adverse reactions due to drugs. This is called the enhanced permeability and retention (EPR) effect. Generally, the lipid-based nanocarriers have a diameter of >10 nm and these are not removed by the kidneys but are removed through capillaries present at locations where microvasculature is leaky.

The pharmacokinetic profile of drug moieties is altered due to its entrapment into lipid-based nanocarrier. Depending upon the drug diffusion rate across the lipid and/or lipid degradation rate, the lipid-based nanocarriers show a controlled, sustained, or prolonged release profile (Onyuksel et al. 2006). Drug aggregation and drug concentration are two important factors that affect drug release kinetics. Drug delivery through lipid nanocarriers results in a sustained release of drug that eventually diminishes the drug dosing frequency, plasma fluctuation in drug concentration and ultimately drug toxicity (Gupta and Xie 2018). Furthermore, by surface modification of nanocarriers, a reduction in immunogenicity of drug–carrier complex is obtained. Also, stealth properties are imparted by reducing opsonization and the elimination of nanocarriers by phagocytes and additionally, the chances of interaction of nanocarrier with blood components that are constantly circulating is avoided. In view of increasing consciousness on drug safety in cure of patients, lipidic nanocarriers have great potential and is a promising tool in order to formulate pharmaceutical products with safety and efficacy.

10.7 Conclusion and Future Perspectives

The neurological diseases have become destructive to human being. The greatest challenge for successful delivery to brain disorders is the presence of the BBB and BCSF. So, it is necessary to discover a carrier that can effectively deliver the drug to the desired site with concentration necessary to elicit therapeutic response with minimal side effects. Generally, brain delivery of macromolecules is impeded owing to the presence of a defense mechanism, i.e. the BBB.

Nanotechnology-based formulation is a promising approach used in the treatment of CNS disorders. The versatile property of lipid-based NPs makes them an attractive carrier for the brain delivery to enhance the therapeutic efficacy. Lipid NPs manifest higher possibilities in curing neurological problems. The characteristics of NPs, i.e. biological, chemical and physical, make them an appropriate carrier for treatment of brain disorders. They are considered as a suitable carrier for the treatment of CNS disorders owing to their small size and innate potential to cross the BBB. The lipid NPs such as liposomes, SLNs, NLCs and LDCs have shown promising results in the treatment of diseases such as alzheimer's, parkinson's and brain tumors. Moreover, lipid NPs could also provide the opportunity to encapsulate therapeutic substances such as enzyme, gene, DNA/RNA, etc. The LNPs are easily decorated with ligands, antibody, or peptide to achieve site-specific action. All the recent literature have proved the potential in treating the neurological diseases as higher concentration can be easily achieved at the brain cells. But, at the same time, it is imperative to take into account the safety or probable risk of nanoparticle use and toxicity should be minimum. Thus, the lipid NPs open a new path for the neurological disorders. But, the future remains more shining if the issues of the toxicity arising due to NPs can be effectively regulated.

References

Abbott, N.J., Ronnback, L., and Hansson, E. (2006). Astrocyte-endothelial interactions at the blood-brain barrier. *Nature Reviews Neuroscience* 7: 41–53.

Adhikari, P., Pal, P., Das, A.K. et al. (2017). Nano lipid-drug conjugate: an integrated review. *International Journal of Pharmaceutics* 529: 629–641.

Agrawal, M., Ajazuddin, Tripathi, D.K. et al. (2017). Recent advancements in liposomes targeting strategies to cross blood-brain barrier (BBB) for the treatment of Alzheimer's disease. *Journal of Controlled Release* 260: 61–77.

Banerjee, S. and Kundu, A. (2018). Lipid-drug conjugates: a potential nanocarrier system for oral drug delivery applications. *DARU Journal of Pharmaceutical Sciences* 26: 65–75.

Bhaskar, S., Tian, F., Stoeger, T. et al. (2010). Multifunctional nanocarriers for diagnostics, drug delivery and targeted treatment across blood-brain barrier: perspectives on tracking and neuroimaging. *Particle and Fibre Toxicology* 7: 1–25.

Blasi, P., Giovagnoli, S., Schoubben, A. et al. (2007). Solid lipid nanoparticles for targeted brain drug delivery. *Advanced Drug Delivery Reviews* 59: 454–477.

Bondi, M.L., Di Gesu, R., and Craparo, E.F. (2012). Lipid nanoparticles for drug targeting to the brain. In: *Methods in Enzymology* (ed. N. Düzgüneş), 229–251. Academic Press.

Brzica, H., Abdullahi, W., Ibbotson, K., and Ronaldson, P.T. (2017). Role of transporters in central nervous system drug delivery and blood-brain barrier protection: relevance to treatment of stroke. *Journal of Central Nervous System Disease* 9: 1–12.

Buzea, C. and Pacheco, I. (2019). Toxicity of nanoparticles. In: *Nanotechnology in Eco-efficient Construction* (eds. F. Pacheco-Torgal, M.V. Diamanti, A. Nazari, et al.), 705–754. Woodhead Publishing.

Buzea, C., Pacheco, I.I., and Robbie, K. (2007). Nanomaterials and nanoparticles: sources and toxicity. *Biointerphases* 2: MR17–MR71.

Cecchelli, R., Berezowski, V., Lundquist, S. et al. (2007). Modelling of the blood-brain barrier in drug discovery and development. *Nature Reviews Drug Discovery* 6: 650–661.

Chella, N. and Shastri, N.R. (2017). Lipid carriers: role and applications in nano drug delivery. In: *Particulate Technology for Delivery of Therapeutics* (eds. J. Sougata and J. Subrata), 253–289. Singapore: Springer.

Copland, M.J., Rades, T., Davies, N.M., and Baird, M.A. (2005). Lipid based particulate formulations for the delivery of antigen. *Immunology and cell biology* 83: 97–105.

Dave, V., Tak, K., Sohgaura, A. et al. (2019). Lipid-polymer hybrid nanoparticles: synthesis strategies and biomedical applications. *Journal of Microbiological Methods* 160: 130–142.

Dehaini, D., Fang, R.H., Luk, B.T. et al. (2016). Ultra-small lipid-polymer hybrid nanoparticles for tumor-penetrating drug delivery. *Nanoscale* 8: 14411–14419.

Dhawan, A. and Sharma, V. (2010). Toxicity assessment of nanomaterials: methods and challenges. *Analytical and Bioanalytical Chemistry* 398: 589–605.

Dong, X. (2018). Current strategies for brain drug delivery. *Theranostics* 8: 1481–1493.

Dow, S.W. (2008). Liposome-nucleic acid immunotherapeutics. *Expert Opinion on Drug Delivery* 5: 11–24.

Elzoghby, A.O., Abd-Elwakil, M.M., Abd-Elsalam, K. et al. (2016). Natural polymeric nanoparticles for brain-targeting: implications on drug and gene delivery. *Current Pharmaceutical Design* 22: 3305–3323.

Engelhardt, B. and Sorokin, L. (2009). The blood–brain and the blood–cerebrospinal fluid barriers: function and dysfunction. *Seminars in Immunopathology* 31: 497–511.

Fang, J.Y., Fang, C.L., Liu, C.H., and Su, Y.H. (2008). Lipid nanoparticles as vehicles for topical psoralen delivery: solid lipid nanoparticles (SLN) versus nanostructured lipid carriers (NLC). *European Journal of Pharmaceutics and Biopharmaceutics* 70: 633–640.

FDA's Approach to Regulation of Nanotechnology Products (2018). https://www.fda.gov/science-research/nanotechnology-programs-fda/national-center-toxicological-research-nanotechnology-programs (accessed 10 July 2020).

Ganesan, P. and Narayanasamy, D. (2017). Lipid nanoparticles: different preparation techniques, characterization, hurdles, and strategies for the production of solid lipid nanoparticles and nanostructured lipid carriers for oral drug delivery. *Sustainable Chemistry and Pharmacy* 6: 37–56.

Gartziandia, O., Herran, E., Pedraz, J.L. et al. (2015). Chitosan coated nanostructured lipid carriers for brain delivery of proteins by intranasal administration. *Colloids and Surfaces: B Biointerfaces* 134: 304–313.

Georgieva, J.V., Hoekstra, D., and Zuhorn, I.S. (2014). Smuggling drugs into the brain: an overview of ligands targeting transcytosis for drug delivery across the blood–brain barrier. *Pharmaceutics* 6: 557–583.

Gomes, M.J., Mendes, B., Martins, S., and Sarmento, B. (2016). Nanoparticle functionalization for brain targeting drug delivery and diagnostic. In: *Handbook of Nanoparticles* (ed. M. Aliofkhazraei), 941–959. Cham: Springer.

Grainger, D.W. (2009). Nanotoxicity assessment: all small talk? *Advanced Drug Delivery Reviews* 61: 419–421.

Guo, Q., Shen, X.T., Li, Y.Y., and Xu, S.Q. (2017). Carbon nanotubes-based drug delivery to cancer and brain. *Journal of Huazhong University of Science and Technology Medical Sciences* 37: 635–641.

Gupta, R. and Xie, H. (2018). Nanoparticles in daily life: applications, toxicity and regulations. *Journal of Environmental Pathology, Toxicology and Oncology* 37: 209–230.

Gurturk, Z., Tezcaner, A., Dalgic, A.D. et al. (2017). Maltodextrin modified liposomes for drug delivery through the blood-brain barrier. *Medchemcomm* 8: 1337–1345.

Hadinoto, K., Sundaresan, A., and Cheow, W.S. (2013). Lipid–polymer hybrid nanoparticles as a new generation therapeutic delivery platform: a review. *European Journal of Pharmaceutics and Biopharmaceutics* 85: 427–443.

Hall, J.B., Dobrovolskaia, M.A., Patri, A.K., and McNeil, S.E. (2007). Characterization of nanoparticles for therapeutics. *Nanomedicine (London)* 2: 789–803.

Hossain, S., Chowdhury, E.H., and Akaike, T. (2011). Nanoparticles and toxicity in therapeutic delivery: the ongoing debate. *Therapeutic delivery* 2: 125–132.

Irby, D., Du, C., and Li, F. (2017). Lipid-drug conjugate for enhancing drug delivery. *Molecular Pharmaceutics* 14: 1325–1338.

Ishak, R.A.H., Mostafa, N.M., and Kamel, A.O. (2017). Stealth lipid polymer hybrid nanoparticles loaded with rutin for effective brain delivery – comparative study with the gold standard (Tween 80): optimization, characterization and biodistribution. *Drug Delivery* 24: 1874–1890.

Jones, C.F. and Grainger, D.W. (2009). In vitro assessments of nanomaterial toxicity. *Advance Drug Delivery Review* 61: 438–456.

Kaur, I.P., Bhandari, R., Bhandari, S., and Kakkar, V. (2008). Potential of solid lipid nanoparticles in brain targeting. *Journal of Controlled Release* 127: 97–109.

Khosa, A., Reddi, S., and Saha, R. (2018). Nanostructured lipid carriers for site-specific drug delivery. *Biomedicine and Pharmacotherapy* 103: 598–613.

Koo, O.M., Rubinstein, I., and Onyuksel, H. (2005). Role of nanotechnology in targeted drug delivery and imaging: a concise review. *Nanomedicine: Nanotechnology, Biology and Medicine* 1: 193–212.

Kreuter, J. (2001). Nanoparticulate systems for brain delivery of drugs. *Advanced Drug Delivery Reviews* 47: 65–81.

Kumar, V., Sharma, N., and Maitra, S.S. (2017). In vitro and in vivo toxicity assessment of nanoparticles. *International Nano Letters* 7: 243–256.

Kuntsche, J., Bunjes, H., Fahr, A. et al. (2008). Interaction of lipid nanoparticles with human epidermis and an organotypic cell culture model. *International Journal of Pharmaceutics* 354: 180–195.

Lai, F., Fadda, A.M., and Sinico, C. (2013). Liposomes for brain delivery. *Expert Opinion on Drug Delivery* 10: 1003–1022.

Lestini, B.J., Sagnella, S.M., Xu, Z. et al. (2002). Surface modification of liposomes for selective cell targeting in cardiovascular drug delivery. *Journal of Controlled Release* 78: 235–247.

Leyuan, X., Hao, Z., and Yue, W. (2014). Dendrimer advances for the central nervous system delivery of therapeutics. *ACS Chemical Neuroscience* 5: 2–13.

Madane, R.G. and Mahajan, H.S. (2016). Curcumin-loaded nanostructured lipid carriers (NLCs) for nasal administration:

design, characterization, and in vivo study. *Drug Delivery* 23: 1326–1334.

Malhotra, M. and Prakash, S. (2011). Targeted drug delivery across blood-brain-barrier using cell penetrating peptides tagged nanoparticles. *Current Nanoscience* 7: 81–93.

Martins, S., Sarmento, B., Ferreira, D.C., and Souto, E.B. (2007). Lipid-based colloidal carriers for peptide and protein delivery – liposomes versus lipid nanoparticles. *International Journal of Nanomedicine* 2: 595–607.

Masserini, M. (2013). Nanoparticles for brain drug delivery. *ISRN Biochemistry* 2013: 1–18.

Muchow, M., Maincent, P., and Muller, R.H. (2008). Lipid nanoparticles with a solid matrix (SLN, NLC, LDC) for oral drug delivery. *Drug Development and Industrial Pharmacy* 34: 1394–1405.

Mukherjee, A., Waters, A.K., Kalyan, P. et al. (2019). Lipid–polymer hybrid nanoparticles as a next-generation drug delivery platform: state of the art, emerging technologies, and perspectives. *International Journal of Nanomedicine* 14: 1937–1952.

Mukherjee, S., Ray, S., and Thakur, R.S. (2009). Solid lipid nanoparticles: a modern formulation approach in drug delivery system. *Indian Journal of Pharmaceutical Sciences* 71: 349–358.

Mulik, R.S., Monkkonen, J., Juvonen, R.O. et al. (2010). Transferrin mediated solid lipid nanoparticles containing curcumin: enhanced in vitro anticancer activity by induction of apoptosis. *International Journal of Pharmaceutics* 398: 190–203.

Muller, R.H., Mader, K., and Gohla, S. (2000). Solid lipid nanoparticles (SLN) for controlled drug delivery – a review of the state of the art. *European Journal of Pharmaceutics and Biopharmaceutics* 50: 161–177.

Muller, R.H., Radtke, M., and Wissing, S.A. (2002). Solid lipid nanoparticles (SLN) and nanostructured lipid carriers (NLC) in cosmetic and dermatological preparations. *Advanced Drug Delivery Reviews* 54: S131–S155.

Mundada, V., Patel, M., and Sawant, K. (2016). Submicron emulsions and their applications in oral delivery. *Critical Reviews in Therapeutic Drug Carrier Systems* 33: 265–308.

Nasiri, M., Azadi, A., Zanjani, M.R.S., and Hamidi, M. (2019). Indinavir-loaded nanostructured lipid carriers to brain drug delivery: optimization, characterization and neuropharmacokinetic evaluation. *Current Drug Delivery* 16: 341–354.

Neves, A.R., Queiroz, J.F., and Reis, S. (2016). Brain-targeted delivery of resveratrol using solid lipid nanoparticles functionalized with apolipoprotein E. *Journal of Nanobiotechnology* 14: 1–11.

Nguyen, T.X., Huang, L., Gauthier, M. et al. (2016). Recent advances in liposome surface modification for oral drug delivery. *Nanomedicine (London)* 11: 1169–1185.

Oberdorster, G., Elder, A., and Rinderknecht, A. (2009). Nanoparticles and the brain: cause for concern. *Journal of Nanoscience and Nanotechnology* 9: 4996–5007.

Olbrich, C., Gessner, A., Kayser, O., and Muller, R.H. (2002). Carrier system for the hydrophilic antitrypanosomal drug diminazenediaceturate. *Journal of Drug Targeting* 10: 387–396.

Onyuksel, H., Sejourne, F., Suzuki, H., and Rubinstein, I. (2006). Human VIP-α: a long-acting, biocompatible and biodegradable peptide nanomedicine for essential hypertension. *Peptides* 27: 2271–2275.

Pacheco-Blandino, I., Vanner, R., and Buzea, C. (2012). Toxicity of nanoparticles. In: *Toxicity of Building Materials* (eds. F. Pacheco-Torgal, S. Jalali and A. Fucic), 427–475. Woodhead Publishing.

Patel, M., Mundada, V., and Sawant, K. (2019). Enhanced intestinal absorption of asenapine maleate by fabricating solid lipid nanoparticles using TPGS: elucidation of transport mechanism, permeability across Caco-2 cell line and in vivo pharmacokinetic studies. *Artificial Cells Nanomedicine and Biotechnology* 47: 144–153.

Patel, M.R., Patel, M.H., and Patel, R.B. (2016). Preparation and in vitro/ex vivo evaluation of nanoemulsion for transnasal delivery of paliperidone. *Applied Nanoscience* 6: 1095–1104.

Pathan, S.A., Syed, M.A., Talegaonkar, S., and Vohra, D. (2009). CNS drug delivery systems: novel approaches. *Recent Patents on Drug Delivery and Formulation* 3: 71–89.

Rawat, M., Singh, D., Saraf, S., and Saraf, S. (2008). Lipid carriers: a versatile delivery vehicle for proteins and peptides. *Yakugaku Zasshi* 128: 269–280.

Redzic, Z. (2011). Molecular biology of the blood-brain and the blood-cerebrospinal fluid barriers: similarities and differences. *Fluids Barriers of the CNS* 8: 1–23.

Rooy, I.V., Tascioglu, S.C., Hennink, W.E. et al. (2011). In vivo methods to study uptake of nanoparticles into the brain. *Pharmaceutical Research* 28: 456–471.

Samad, A., Sultana, Y., and Aqil, M. (2007). Liposomal drug delivery systems: an update review. *Current Drug Delivery* 4: 297–305.

Sanchez-Covarrubias, L., Slosky, L.M., Thompson, B.J. et al. (2014). Transporters at CNS barrier sites: obstacles or opportunities for drug delivery. *Current Pharmaceutical Design* 20: 1422–1449.

Saraiva, C., Praca, C., Ferreira, R. et al. (2016). Nanoparticle-mediated brain drug delivery: overcoming blood-brain barrier to treat neurodegenerative diseases. *Journal of Controlled Release* 235: 34–47.

Saucier-Sawyer, J.K., Deng, Y., Seo, Y.E. et al. (2015). Systemic delivery of blood-brain barrier-targeted polymeric nanoparticles enhances delivery to brain tissue. *Journal of Drug Targeting* 23: 736–749.

Sawant, K.K. and Dodiya, S.S. (2008). Recent advances and patents on solid lipid nanoparticles. *Recent Patents on Drug Delivery and Formulation* 2: 120–135.

Schnyder, A. and Huwyler, J. (2005). Drug transport to brain with targeted liposomes. *NeuroRx* 2: 99–107.

Shankar, R., Joshi, M., and Pathak, K. (2018). Lipid nanoparticles: a novel approach for brain targeting. *Pharmaceutical Nanotechnology* 6: 81–93.

Silva, A.C., Gonzalez-Mira, E., García, M.L. et al. (2011). Preparation, characterization and biocompatibility studies on risperidone-loaded solid lipid nanoparticles (SLN): high pressure homogenization versus ultrasound. *Colloids and Surfaces B: Biointerfaces* 86: 158–165.

Smulders, S., Kaiser, J.P., Zuin, S. et al. (2012). Contamination of nanoparticles by endotoxin: evaluation of different test methods. *Particle and Fibre Toxicology* 9: 1–11.

Somani, S. and Dufes, C. (2014). Applications of dendrimers for brain delivery and cancer therapy. *Nanomedicine (London, England)* 9: 2403–2414.

Soni, V., Kohli, D.V., and Jain, S.K. (2005). Transferrin coupled liposomes as drug delivery carriers for brain targeting of 5-florouracil. *Journal of Drug Targeting* 4: 245–250.

Sonvico, F., Clementino, A., Buttini, F. et al. (2018). Surface-modified nanocarriers for nose-to-brain delivery: from bioadhesion to targeting. *Pharmaceutics* 10: 1–34.

Strazielle, N. and Ghersi-Egea, J.F. (2000). Choroid plexus in the central nervous system: biology and physiopathology. *Journal of Neuropathology and Experimental Neurology* 59: 561–574.

Torchilin, V.P. (2007). Targeted Pharmaceutical Nanocarriers for Cancer Therapy and Imaging. *The AAPS Journal* 9: E128–E147.

Van Tellingen, O., Yetkin-Arik, B., de Gooijer, M.C. et al. (2015). Overcoming the blood-brain tumor barrier for effective glioblastoma treatment. *Drug Resistance Update* 19: 1–12.

Vyas, S.P. and Khar, R.K. (2002). *Targeted and Controlled Drug Delivery, Novel Carrier Systems*. New Delhi: CBS Publishers and Distributors Pvt. Ltd.

Wang, Z.Y., Sreenivasmurthy, S.G., Song, J.X. et al. (2019). Strategies for brain-targeting liposomal delivery of small hydrophobic molecules in the treatment of neurodegenerative diseases. *Drug Discovery Today* 24: 595–605.

Warren, K.E. (2018). Beyond the blood brain barrier: the importance of central nervous system (CNS) pharmacokinetics for the treatment of CNS tumors, including diffuse intrinsic pontine glioma. *Frontiers in Oncology* 8: 1–11.

Wu, Y., Simonovsky, F.I., Ratner, B.D., and Horbett, T.A. (2005). The role of adsorbed fibrinogen in platelet adhesion to polyurethane surfaces: a comparison of surface hydrophobicity, protein adsorption, monoclonal antibody binding, and platelet adhesion. *Journal of Biomedical Material Research Part A* 74: 722–738.

Yildirimer, L., Thanh, N.T., Loizidou, M., and Seifalian, A.M. (2011). Toxicology and clinical potential of nanoparticles. *Nano Today* 6: 585–607.

Zhang, F., Xu, C.L., and Liu, C.M. (2015). Drug delivery strategies to enhance the permeability of the blood-brain barrier for treatment of glioma. *Drug Design Development and Therapy* 9: 2089–2100.

Zheng, M., Tao, W., Zou, Y. et al. (2018). Nanotechnology-based strategies for siRNA brain delivery for disease therapy. *Trends in Biotechnology* 36: 562–575.

Zhou, Y., Zhang, G., Rao, Z. et al. (2015). Increased brain uptake of venlafaxine loaded solid lipid nanoparticles by overcoming the efflux function and expression of P-gp. *Archives of Pharmacal Research* 38: 1325–1335.

Zoroddu, M.A. (2017). Toxicity of nanoparticles: etiology and mechanisms. In: *Antimicrobial Nanoarchitectonics* (ed. A.M. Grumezescu), 511–546. Elsevier.

Zoroddu, M.A., Medici, S., Ledda, A. et al. (2014). Toxicity of nanoparticles. *Current Medicinal Chemistry* 21: 3837–3853.

Zylberberg, C. and Matosevic, S. (2016). Pharmaceutical liposomal drug delivery: a review of new delivery systems and a look at the regulatory landscape. *Drug Delivery* 23: 3319–3329.

11

Risk Assessment of Injectable Nanoparticles Used as Nanomedicine

Marco Vinícius Chaud[1,2], *Fernando Batain*[1], *Eliana Barbosa Souto*[3], *Patrícia Severino*[4], *Aleksandra Zielińska*[5], *and Thais Francine Ribeiro Alves*[1,2]

[1] *Laboratory of Biomaterials and Nanotechnology (LaBNUS), University of Sorocaba, Sorocaba, Brazil*
[2] *Sorocaba Technology Park, Innovation Agency, Sorocaba, Brazil*
[3] *Faculty of Pharmacy, Department of Pharmaceutical Technology, University of Coimbra, Coimbra, Portugal*
[4] *Nanomedicine and Nanotechnology Laboratory, University of Tiradentes, Aracaju, Brazil*
[5] *Institute of Human Genetics, Polish Academy of Sciences, Poznań, Poland*

11.1 Introduction

Despite the risks of cellular and tissue toxicity, there is a permanent development of nanotechnology in the pharmaceutical sciences field, expanding the use of nanomedicines for all dosage forms and all administration routes. Scientists picked up the nanotechnology first for oral and transdermal drug delivery system, and the success came fast through the improvement of critical biopharmaceutical properties for bioavailability, such as dissolution rate, increased surface area, unique physicochemical and biological properties for new medicinal applications including, parenteral injections, implants, and targeting cell (Sheikh et al. 2016; Tyagi and Subramony 2018).

Injections are invasive procedures of excessive cost; nevertheless, they are widely used in health-care procedures, and at least 16 billion injections are administered worldwide each year. Therefore, it is imperative to protect against avoidable risks and meet minimum safety standards. The World Health Organization (WHO) through the Safe Injection Network, developed strategies for reducing the improper use and stimulating the administration of risk-free injections. The turn of alternatives for drug administration strategies that provide more or less frequent administration intervals is new and may reduce some shortcoming of current oral drug delivery rules (Elhoseeny and Mourad 2014; WHO 2016; Tatham et al. 2018a). These strategies have driven long-acting injectable (LAIs) nanoformulations which may solve some of these problems (Tatham et al. 2018a). Currently, the characterization of biological safety has been an arduous task considering the complexity of nanomaterials which makes them difficult to assess, universalize, and predict important aspects of toxicity.

Emerging nanostructures engineered like germicide nanoparticles, nano protein, target delivery, diagnostic, and the novel polymeric nano vaccines should be subjected to rigorous cytotoxicity and genotoxicity trial to set up harmless mechanisms of application and dosage levels. To mitigate or prevent the potential hazards of engineered nanostructures in consumer products, an action of independent

Nanotechnology in Medicine: Toxicity and Safety, First Edition. Edited by Mahendra Rai, Mrunali Patel, and Rashmin Patel.

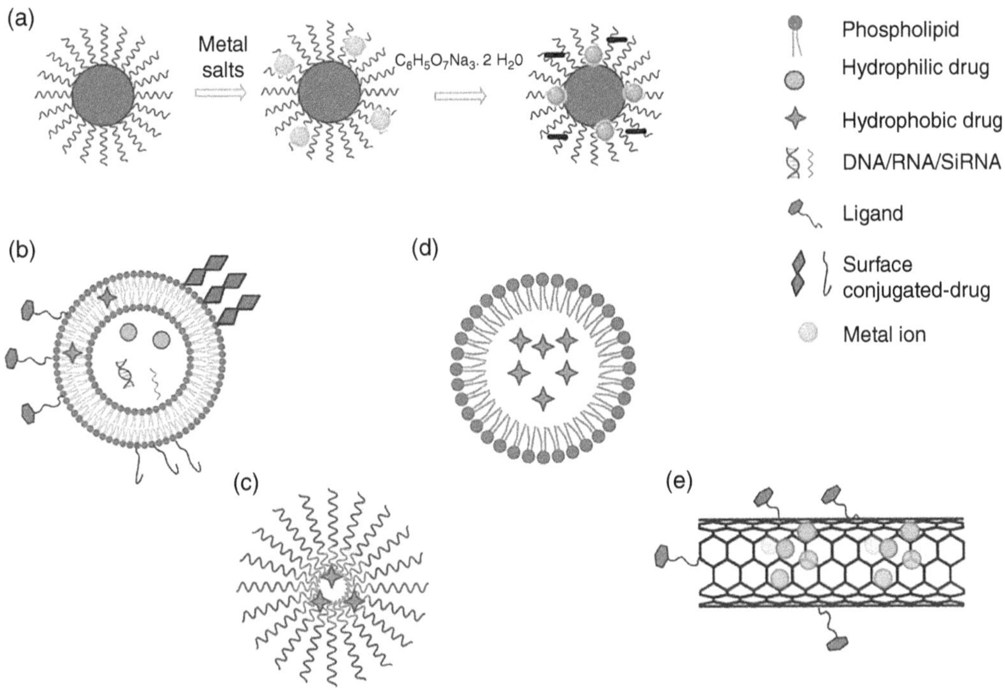

Figure 11.1 Illustrations of carriers nanoparticulated and surface's functionality. (a) Metallic nanoparticle. (b) Liposome. (c) Polymeric micelle. (d) Lipidic micelle. (e) Carbon nanotube.

agencies government must protect the people from unreasonable risks.

This chapter will discuss about the risks assessment of injectable nano-based drug delivery systems by parenteral routes weighing up the pros and cons associated with nanoscale materials, especially those that are often overlooked.

11.2 Nanomaterials, Nanoparticles, and Nanoformulation

Although there is a miscellany of nomenclatures to conceptualize different types of engineered nanomaterials, the definition is unequivocal "nanomaterials are materials with any internal or external structure on the nanoscale dimension, which have been characterized with multiples and specific physical and chemical properties and particular biological interactions when it comes to nanomedicines" (Figure 11.1a–e). The nanomaterials are widely used in health-care systems for drug delivery systems, including, theranostic, vaccinal agents and even vaccine adjuvants (Chaud et al. 2020a). Although the future for nanomedicine appears promising, the permeation of xenobiotic across biobarriers is a significant concern as it can cause irreversible damage for vital organs such as the brain, eye, and for the fetus (Zielińska et al. 2020).

In living systems, nanoparticles with the exact dimensions as biological molecules are a safety concern surrounding the apparent ability of some nanomaterials to cross biological barriers and spread for vital organs, increasing the potentiality for cytotoxicity and genotoxicity (Huang et al. 2017). Nanomaterials with ultrafine particle size (1–100 nm), especially in a range less than 60 nm, are dangerous since they can cross biological barriers such as blood–ocular, blood–placental, blood–brain, enter cells, and trigger oxidative stress due to ROS (reactive oxygen species) formation. The size is not the only toxicity factor of nanomaterials; other key factors related to fundamental properties of nanomaterial show a complex dependence such as morphology, defects, chemical composition, chemical stability, and a capping of proteins overlay the particle, which may give biomimetic functionality (Figure 11.1b–e). In the latter, nanomaterials would not be identified by humoral defence factors and would pass freely through biological barriers (Tedesco and David 2012). In this context, the free transit of nanomaterials by the human body can be considered as the major risk factor for the toxicity of different types of nanostructures. The physical and chemical characteristics of nanoparticles such as size, surface area, shape, charge, presence, or absence of specific active groups on the surface and composition are major factors in the toxicity and cytotoxicity of nanoparticles for mammals other animal species (Ema et al. 2017; Jeevanandam et al. 2018; Khan 2020).

Lipid nanoparticle-based formulations have become the best option for the development of many complex parenteral products such as genetic material including mRNA and other biologically active agents, recombinant protein, drug combinations, anticancer agents for modified release and controlled delivery to a target cell or organ site, and target-bioaccumulation (Figure 11.1b, d). These nanoparticles like lipid nanocapsule, liposome, liquid crystal, self-emulsifying drug delivery (SEDDS) are delivery systems pharmacologically inactive with minimal intrinsic toxicity and safety for topical, oral, and inhalation administration routes (Alves et al. 2020). The parenteral formulations of lipid nanoparticles have been used to overcome the challenges of current therapy and supply important added benefits. However, the cytotoxicity of the lipidic system for drug delivery has been reported for parenteral preparations; it is tightly related to the type of lipid, surface properties, exposure time, and dose (Sercombe et al. 2015; U.S. Food and Drug Administration 2015; Wang and Grainger 2019; Inglut et al. 2020).

Polymeric nanoparticles (Figure 11.1c) are currently used for intravenous, intramuscular, and intracutaneous injections for drug delivery, bioluminescence, fluorescence, and photoacoustic imaging and as conjugated carriers for nonviral gene delivery with potential benefited for therapeutic treatment and diagnostic (Joralemon et al. 2010; Li and Liu 2014). Amphiphilic copolymer in aqueous media generate a core-shell structure, like lipid nanoparticle-based formulations of the amphiphilic polymers tend to self-assemble into nanoclusters and micelles in an aqueous solution. Although these systems show biological tolerability, in some cases they do not keep their loading following parenteral administration, which has restrained their ability to target tumoral cells. Polyester-based nanoparticles have shown good stability and drug targeting; however, they display a burst effect whereby a significant part of the load is released from the nanoparticle upon injection. The stability and ability to retain the drug loaded upon parenteral injection is an important factor to avoid cytotoxicity risks (Gaucher et al. 2010; Bachhav et al. 2011; Liu et al. 2014). Furthermore, the use of organic solvents to dissolve preformed polymers is strongly associated with toxicity risk and environmental pollution.

Carbon nanotubes (Figure 11.1e) are a type of nanostructure easily functionalized and an efficient drug delivery system displaying considerable potential in cancer therapy. Systems of this type attract interest, thanks to a combination of multiple physicochemical properties unique for implant and a wide range of biomedical application because their stability and versatility such as target drug delivery system, biosensors, and theranostic with unique selectivity for tumor sites. The cytotoxicity of carbon nanotubes has been attributed to factors such as concentration, length, stiffness exposure duration, state of aggregation, and adjuvants used to disperse the nanotubes (Francis and Thiyagarajan 2018).

Overall material that plays an essential and distinctive role in protecting the structural integrity of nanoformulation during the synthesis process, storage, and subsequent shelf life, such as bulking agents, lyoprotectants, cryoprotectants, chemical stabilizers, osmotic balancers, bacteriostatic injection water also have potential-associated toxicity risks. The major drawback of protecting agent is that the total interaction between them is mostly unknown and requires a risk assessment.

11.3 Injectable Nanoparticles Toxicity

Advantages for injectable nanoparticles are sustained drug release, reduced frequency of administration, an increase of bioavailability, dose accuracy, and reduced dosage. The disadvantages include cellular and systemic toxicity, costs more elevated when compared with other less invasive routes of administration. Besides, a suitable place and people qualified to administer injectable drugs are needed. Another critical problem for injection of nanoparticles associated with an extended–release for days or months is the increase of toxicity risk factor, due to inflammatory response of the body at the site of application or bioaccumulation, especially for specific target delivery systems (Skalska and Lidia 2015). The inflammatory process can be generated distinctly by the release of material used to synthesize nanoparticles, molecular weight, shape, size, charge, and surface area.

The toxic effects and risk factor of the cumulative of nanoparticles toward mammalian cell and tissue are not yet completely understood; also, the injection speed and clearance speed of nanostructure is misunderstood. These underscore that the dosing of nanoparticles must be technically perfected to avoid untimely clearance and increase the injectable dosage form's therapeutic benefits (Chaud et al. 2020b). So, the assessment of risk for injectable nanoparticles makes up intrinsic factors associated with design and technique perfected of injection administration either by perfusion or *in bolus*. Even so, the injectable nanoparticles stand for an important active area of technological research and an economic sector in permanent expansion.

Gelatine, albumin, and polylactide polymers and copolymers are approved for human use as injectable nanoformulation for drug delivery changed system (Table 11.1). Once injected by the parenteral route, the nanoformulations release the encapsulated therapeutic agent over extended periods of time depending on hydrophobicity, molecular weight, drug loading, and particle size. Because these formulations release drug over prolonged periods, the nanoparticles require long-term studies about stability, toxicity, and cytotoxicity.

Currently, nanoparticle-based therapeutics are thoroughly investigated in clinical trial studies. The side effects of nanoparticle release systems are the primary concern in preclinical evaluation studies about the drug's toxic potential in prolonged-release systems and the material with which nanoparticles are synthesized. The same attention should be mandatory for diluent and stabilizer of injection; however, a few studies have been conducted with this approach (Wang and Grainger 2019).

The electrokinetic potential of nanoparticle surfaces influences both cellular uptake and cytotoxicity. In general, the cationic nanostructured carries condense nucleic acids by electrostatic interactions with anionic nucleic cargo that cannot be understood without a sophisticated appreciation of ionic interaction. On the other hand, it is known that cationic nanoparticles are more cytotoxic than neutral or negative nanoparticles.

Clinical studies based on cationic nanostructure delivery systems reported side events related to inflammatory toxicity, chills, fever, and weariness (Knudsen et al. 2015; Liu et al. 2018). In a particular way for gene delivery systems, a liposomal surface charge less than 30 mV should have viewed to reduce the cytotoxicity (Inglut et al. 2020).

Table 11.1 Current marketed formulations with polymeric and lipidic nanoparticles.

Product	Nanomaterial	API	Diluent/Stabilizer	Route	Company
Decapeptyl SR	PLA/PLDLA	Triptorelin	Mannitol, carmellose, polysorbate 80, water	IM	Ipsen
Risperdal	PLGA	Risperidone	Polysorbate 20, CMC, citric acid, disodium phosphate, sodium chloride, water	IM	Janssen
Sandostatin LAR	PLGA-glucose	Human growth hormone	Mannitol, carmellose, poloxamer-188, CMC, water	SC/IM	Genentech Novartis
Genexol-PM	mPEG-PDLA	Paclitaxel	Sodium chloride, dextrose, water	IV	Samyang
Abraxane	Human albumin	Paclitaxel	Sodium chloride, water	IV	AstraZeneca
Myocet	Liposome	Doxorubicin	Dextrose, water	IV	Elan
Aroplatin	Liposome	Cisplatin	*a*	IV	Antigenic
Atragen	Liposome	Trans-retinoic acid	*b*	IV	Aronex

[a/b] In clinical study phase II/III; API, active pharmaceutical ingredient; PM, polymeric micelle; LAR, long-acting repeatable.
Source: From Food and Drug Administration. Public Domain.

Cationic nanostructures like micelles, liposomes, SEDDS, lyotropic liquid crystal, and cationic materials like polyethene imines, polysaccharides, proteins, cellulose, and cyclodextrins are widely used as a carrier in gene delivery, nonviral vectors for nucleic acids, theranostics, target delivery, and nanomedicine. However, necrotic cell, tumor necrosis factor-α (TNF-α), interferon-γ, interleukin-1β, and interleukin-6 are growing in the serum after parenteral treatment with cationic nanoparticles (Tabernero et al. 2013; Wei et al. 2015; Zancheta et al. 2015; Souza et al. 2020; Zielińska et al. 2020). Following intravenous injection, the cationic liposomes are sequestered by the mononuclear phagocyte system leading to bioaccumulation in organs such as the liver and spleen, inducing cytotoxicity and causing cell depletion, which affects the proper functioning of the immune system. The liposomal nanoparticles are an attractive tool to reduce the dose-limiting adverse effects; however, the liposomes themselves can add new toxicities, scientifically misunderstood, raising concerns regarding their use by parenteral route (Inglut et al. 2020).

Efforts have been made to explain the immune responses and cytotoxicity that cationic nanoparticles may raise. The most rational approach found is that cationic nanomedicines induced acute cell necrosis via interacting with Na^+/K^+-ATPase and subsequently stimulated the release of organelles and mitochondrial contents as mitochondrial DNA from necrotic cells as damage-associated molecular patterns to stimulate the immune response further (Liu et al. 2018).

Anionic liposomes for injection have potential as drug, protein, and gene delivery carriers, due to their biocompatibility, biomimetic and modular properties, which can attenuate the risk of systemic toxicity from the encapsulated therapeutic agent. Paclitaxel formulation using block copolymers as solubilizer forming a polymeric micelle (Genexol-PM) offers clinical advantages combining increased doses, improvement efficacy, and low toxicity (Kim et al. 2001). However, sufficient integrity of polymeric nanostructures after intravenous injection is decisive for successful drug release to solid tumors, but the right measurement of the *in vivo* micellar integrity remains troublesome (Gaucher et al. 2010).

Nanocomplex (<10 nm) Amphotericin B-Albumin (AmB-BSA) in blood circulation induces the bioaccumulation in kidney and nephrotoxicity. To increase the diameter of the nanocomplex that far exceeds the size range of kidney excretion seems to be an up-and-coming technological alternative to avoid nephrotoxicity by bioaccumulation. AmB-BSA nanocomplex previously loaded regulates nanoparticles' size from 10 to 60 nm, increasing hepatic distribution and decreasing the bioaccumulation in the kidney. AmB-BSA notably increased the brain concentration of AmB in mice with *Cryptococcus neoformans*--caused central nervous system infection and enhanced 100-fold antifungal activity *in vivo* (Liu et al. 2020).

Carbon nanotubes are used in many fields of drug therapy such as implants and systems for modified drug release. However, several toxic effects associated with this nanostructure are reported, either due to biological properties or due to physical and chemical properties. The safety of clinical application of nanotubes as targeting drug delivery has been a subject of concern. The carbon nanotubes like metal nanoparticles having high cytotoxicity to mammalian mainly stem from similarities in structure to asbestos fibers (Ema et al. 2017; Jeevanandam et al. 2018; Tatham et al. 2018b; Yan et al. 2019).

11.4 Safety of Nanoparticles in Acute and Chronic Studies

The toxicity of injectable nanostructures for the treatment of chronic disease and the high dosage is not sufficiently known; despite all-knowing about lipid, polymeric, emulsion nanoparticle systems, few *in vivo* studies have been performed. Fonseca-Gomes et al. reported encouraging results for long-term therapies using liposome, lipid nanoparticles, and polymeric nanoparticles with a diameter less than 200 nm with PDI below 0.22 (2020). Nevertheless, it needs to be highlighted that all nanostructures evaluated were electrostatically negatives, and the bioaccumulation is more common for nanostructures smaller than 60 nm, and cytotoxicity is higher for positive than a negative surface charge.

With advances in the metallic nanostructure approaches, many biological and biomedical applications are recognized as being nontoxic. In particular, silver and gold nanoparticles have been widely used from the biomedical, agricultural, and environmental fields for their effect on plasmon field single optical properties arising from free electrons' oscillation (Huang et al. 2007; Cho et al. 2009; Huang et al. 2017). Despite this potential, little is known about the cytotoxicity, toxicity, bioavailability, and bioaccumulation of metallic nanoparticles. However, it has been reported that the safety of metal and bimetal nanoparticles depends on the size, shape, and surface properties, reductor agent, and like other nanoparticles the electrokinetic potential.

Metallic nanoparticles can be found under certain conditions as xenobiotics by the immune cells to arise an inflammatory response and attract more cells to kill the unknown substances. The persistency of metallic and bimetallic nanoparticles in biological systems increases the risk of carcinogenicity, and thus it is essential to investigate the tissue distribution, cytotoxicity, and genotoxicity (Li et al. 2013).

AuNPs (gold nanoparticles) synthesized with $HAuCl_4$ and citric acid (1 wt%) raised AuNPs from 11 to 15 nm. AuNPs were PEGylated using thiol and poly (ethylene) glycol of MW 5000. *in vivo* toxicity studies show acute and chronic toxicity regarding inflammation in the liver and spleen and histopathological lesions (Cho et al. 2009). AgNPs (silver nanoparticles) synthesized with $AgNO_3$ and sodium citrate (1 wt%) and raised AgNPs over 100 nm, zeta potential reported to be 100 mV. It was proved *in vivo* that AgNP induces cytotoxicity and genotoxicity in primary Syrian Hamster Embryo cells. In addition, studies suggest that AgNPs could induce apoptosis and the inhibition of cell proliferation (Li et al. 2013; Zhang et al. 2014).

Amphiphilic copolymer micelles are synthesized with a hydrophilic shell, and a hydrophobic core is suggested as promising carriers, water-soluble drugs and poorly water-soluble drugs, and gene transfection. However, the cytotoxicity is dependent on the polymer's properties, size, structure, and surface charge like lipid nanoparticles. The toxicity for intravenous injection for polymeric micelles has been attributed to lack of targeting specificity. However, there is a potential risk when polymeric nanomaterials are used, which are associated with inflammatory processes, cytotoxicity, and genotoxicity (Moller et al. 2013; Knudsen et al. 2015).

Carbon nanotubes increase the reactive oxygen species level during the process of entry into the cell for degradation, wreaking the cytotoxicity by oxidative stress affecting the cellular structure. The principal targets of carbon nanotubes' toxicity are cell membrane, lysosome, mitochondria, nucleus, and the cytoskeleton. Studies have registered that carbon nanotubes blocked by the reticuloendothelial system are primarily concentrated in the liver with macrophage damage, inflammation, blood coagulation, and swelling of the cells (Liu et al. 2006; Deng et al. 2008; Awasthi et al. 2013; Yan et al. 2019).

11.5 Future Perspectives and Conclusion

Synthetic nanomaterials engineered or not engineered are present in all the ecosystems, and they have mechanisms to cause toxic effects among living organisms. Research advancements have found severe and dangerous cytotoxic effects that require the implementation of rigid laws and regulations worldwide by government agencies. Although the general standard of toxicity does not reveal any levels of severe systemic changes related to nanoparticles, studies have shown that cytotoxicity due to oxidative stress, liver and kidney bioaccumulation causes severe health damages. The more severe side effects are genotoxicity and gene expression responses in target organs and the inflammatory cascade nanoparticle induced. Other occurrences reported in the scientific literature have led to different views about the overuse and safety of nanoparticles. Recent results about the toxicity of nanoparticles have forced scientists to review consolidated concepts on environmental damage and severe consequences on human and animal health, especially those that are on the top of the alimentary chain.

Future research with new methodological strategies designed to answer this kind of question must be done in order to clarify the relationship between impurity, reactivity, charge surface, size, surface area, shape, absence of specific active groups on the surface, chemical composition, and aging. In this context, the same nanoparticles are primarily nontoxic, while others having reduced toxicity may also have progressive side effects. In this context, the nanoparticles even those that are primarily nontoxic, others for as much as having reduced toxicity may have progressive side effects. So, the risk assessment for all nanostructured dosage forms, with particular attention to the parenteral route and inhalation airway, must be carefully evaluated.

The cationic nanocarrier-induced multicellular host responses to infection, injury or inflammatory stimuli lead to the formation of a comprehensive range of chemical mediators which are negatively regulated. Therefore, research investigating the management and resolution of cationic nanoparticle-induced inflammation is essential for distinguishing the probable mechanisms underlying this kind of toxicity and providing new approaches to prevent cytotoxicity, genotoxicity, and promoting the employment of nanotechnology for life-saving drugs and environmental responsibility.

Metallic nanoparticles and carbon nanotubes, when administrated by parenteral route, are, respectively, associated with genotoxicity and side effect. When intercepted by the reticuloendothelial system and concentrated in the liver cause a cascade of cytotoxic events due to oxidative stress with macrophage damage, inflammation, blood coagulation, cell swelling, and cellular apoptosis. This observation is in agreement with studies that indicate as the main cytotoxic targets of carbon nanotubes are cell membrane, lysosome, mitochondria, nucleus, and the cytoskeleton (Yan et al. 2019).

In cytotoxicity study, the oxidative stress of AgNPs was comparable to that of Ag^+ ions; the findings suggest that AgNP-induced toxicity is an intrinsic effect of AgNPs that is independent of free Ag^+ ions. However, the expression of oxidative stress-related mRNA species is regulated differentially by AgNPs and Ag^+ ions (Kim et al. 2009). Tissue Ag levels determined in the offspring of rat dams showed that intravenous-injected AgNPs accumulated in maternal tissues and increased in the endometrium and visceral yolk sac. Although they did not cross the placental barrier in large amounts, the AgNPs potentially affected embryonic growth, and accumulate in some organs to cause renal toxicity and hepatotoxicity when administered through subcutaneous injection.

Carbon nanotubes are present in the bronchoalveolar lavage solution from asthmatic people. Further studies reported that carbon nanotubes into cells could generate granulomatous disease, including sarcoidosis and sarcoid-like reaction, and bronchospasm leading to lungs' fibroplasia and neoplasia.

The toxicity identification of metallic nanoparticles and carbon nanotubes is a highly demanded research area. In recent times, AuNPs and multi-walled carbon nanotubes are reported to induce transplacental genotoxicity after maternal administration in mice. These findings are a proximal causal factor with the high rise of reactive oxygen species for the reproductive and developmental toxicity nanostructured systems.

Apoptosis and acute inflammation caused by PEG-functionalized AuNPs occurred due to nanoparticles' accumulative effect in the liver and spleen for up to seven days after injection (Ema et al. 2017). Incremental advancement of polymeric nanocarriers has led to widespread research of nanoparticles for parenteral release. However, most clinical studies based on liposomal or polymer systems for parenteral release have reported side effects, such as fatigue, chills, and fever related to inflammatory toxicity.

From this review study, the overall conclusion is that engineered nanomaterials, when administered parenterally, display unsafety, and it requires strict both *in vitro* and *in vivo* research into the toxicity, genotoxicity and bioaccumulation effects of the nanostructured systems in the organism and environment. Even for those materials that previously had limited or no toxicity at low doses, the intrinsic toxicity must be carefully investigated. Our understanding is that the nonexistence of toxicity needs to be proved for all materials and drugs loaded into all types of nanoparticulate systems.

References

Alves, T., Barros, C., and Chaud, M. (2020). Lipid nanoparticles: advances in research and applications. In: *Lipid Nanoparticles* (ed. S. Ziegler), 1–42. New York: Nova Medicine & Health.

Awasthi, K., John, P., Awasthi, A., and Awasthi, K. (2013). Multi walled carbon nano tubes induced hepatotoxicity in Swiss albino mice. *Micron* 44: 359–364.

Bachhav, G., Kalia, Y., Gurny, R. et al. (2011). Novel micelle formulations to increase cutaneous bioavailability of azole antifungals. *Journal of Controlled Release* 153 (2): 126–133.

Chaud, M., Amaral, V., Batain, F. et al. (2020a). Nanobitechnological strategies for treatment of tegumentry and visceral leishmaniasis including resistance strains. In: *Nanotechnology in Skin, Soft Tissue, and 'Bone Infections'* (ed. M. Rai), 183–204. Berlin: Springer, Cham.

Chaud, M.V., Santos, C., Torqueti, C. et al. (2020b). Nanostructure self-assembly for direct nose-to-brain drug delivery. In: *Nanomycotoxicology* (eds. R. Mahendra and K. Abd-Elsalam), 449–480. Elsevier Inc.

Cho, W., Cho, M., Jeong, J. et al. (2009). Acute toxicity and pharmacokinetics of 13 Nm-sized PEG-coated gold nanoparticles. *Toxicology and Applied Pharmacology* 236: 16–24.

Deng, X., Yang, S., Nie, H. et al. (2008). A generally adoptable radiotracing method for tracking carbon nanotubes in animals. *Nanotechnology* 19: 1–6.

Elhoseeny, T. and Mourad, J. (2014). Assessment of the safety of injection practices and injection-related procedures in family health units and centers in Alexandria. *Journal of the Egyptian Public Health Association* 89 (2): 66–73.

Ema, M., Hirokazu, O., Masashi, G., and Kazumasa, H. (2017). A review of reproductive and developmental toxicity of silver nanoparticles in laboratory animals. *Reproductive Toxicology* 67: 149–164.

Fonseca-Gomes, J., Loureiro, J., Tanqueiro, S. et al. (2020). *In vivo* bio-distribution and toxicity evaluation of polymeric and lipid-based nanoparticles: a potential approach for chronic diseases treatment. *International Journal of Nanomedicine* 15: 8609–8621.

Francis, A. and Thiyagarajan, D. (2018). Toxicity of carbon nanotubes: a review. *Toxicology and Industrial Health* 34 (3): 200–210.

Gaucher, G., Robert, M., and Jean-Christophe, L. (2010). Polyester-based micelles and nanoparticles for the parenteral delivery of taxanes. *Journal of Controlled Release* 143 (1): 2–12.

Huang, L., Hu, J., Huang, S. et al. (2017). Nanomaterial applications for neurological diseases and central nervous system injury. *Progress in Neurobiology* 157: 29–48.

Huang, W., Wei, Q., Prashant, J., and El-sayed, M. (2007). The effect of plasmon field on the coherent lattice phonon oscillation in electron-beam fabricated gold nanoparticle pairs. *Nano Letters* 7 (10): 3227–3234.

Inglut, C., Sorrin, A., Kuruppu, T. et al. (2020). Immunological and toxicological considerations for the design of liposomes. *Nanomaterials* 10 (190): 1–24.

Jeevanandam, J., Barhoum, A., Chan, Y. et al. (2018). Review on nanoparticles and nanostructured materials: history, sources, toxicity and regulations. *Beilstein Journal of Nanotechnology* 9 (1): 1050–1074.

Joralemon, M., Emrick, T., and Mcrae, S. (2010). PEGylated polymers for medicine: from conjugation to self-assembled systems. *Chemical Communications* 46: 1377–1393.

Khan, A. (2020). Metal nanoparticles toxicity: role of physicochemical aspects. In: *Metal Nanoparticles for Drug Delivery and Diagnostic Applications* (ed. M. Deans), 1–11. Amsterdam: Elsevier Inc.

Kim, S., Eun, J., Choi, J. et al. (2009). Toxicology *in vitro* oxidative stress-dependent toxicity of silver nanoparticles in human hepatoma cells. *Toxicology in vitro* 23: 1076–1084.

Kim, S., Kim, D., Shim, Y. et al. (2001). *In vivo* evaluation of polymeric micellar paclitaxel formulation: toxicity and efficacy. *Journal of Controlled Release* 72 (1–3): 191–202.

Knudsen, K., Northeved, H., Pramod Kumar, E. et al. (2015). *In vivo* toxicity of cationic micelles and liposomes. *Nanomedicine: Nanotechnology, Biology, and Medicine* 11 (2): 467–477.

Li, K. and Liu, B. (2014). Polymer-encapsulated organic nanoparticles for fluorescence and photoacoustic imaging. *Chemical Society* 43: 6570–6597.

Li, X., Xu, L., Shao, A. et al. (2013). Cytotoxic and genotoxic effects of silver nanoparticles on primary syrian hamster embryo (SHE) cells. *Journal of Nanoscience and Nanotechnology* 13 (1): 161–170.

Liu, L., Liu, Y., Xu, B. et al. (2018). Negative regulation of cationic nanoparticle-induced inflammatory toxicity through the increased production of prostaglandin E2 via mitochondrial DNA-activated Ly6C+ monocytes. *Theranostics* 8 (11): 3138–3152.

Liu, Y., Han, Y., Fang, T. et al. (2020). Turning weakness into strength: albumin nanoparticle-redirected Amphotericin B biodistribution for reducing nephrotoxicity and enhancing antifungal activity. *Journal of Controlled Release* 324: 657–668.

Liu, Y., Hsuan-ying, C., Jay-an, Y., and Li-fang, W. (2014). Novel micelle formulations to increase cutaneous bioavailability of azole antifungals. *RSC Advances* 4: 59548–59557.

Liu, Z., Cai, W., He, L. et al. (2006). *In vivo* biodistribution and highly efficient tumour targeting of carbon nanotubes in mice. *Letters* 2: 47–52.

Moller, P., Danielsen, P., Jantzen, K. et al. (2013). Oxidatively damaged DNA in animals exposed to particles. *Critical Reviews in Toxicology* 43 (2): 96–118.

Sercombe, L., Veerati, T., Moheimani, F. et al. (2015). Advances and challenges of liposome assisted drug delivery. *Frontiers in Pharmacology* 6: 1–13.

Sheikh, A., Sheikh, S., and Zaheer, Z. (2016). Injectable controlled release drug delivery systems. *Asian Journal of Pharmaceutics* 10 (4): 8–10.

Skalska, J. and Lidia, S. (2015). Toxic effects of silver nanoparticles in mammals – does a risk of neurotoxicity exist? *Folia Neuropathologica* 53 (4): 281–300.

Souza, J., Pontes, K., Alves, T. et al. (2020). Structural comparison, physicochemical properties, and *in vitro* release profile of curcumin-loaded lyotropic liquid crystalline nanoparticle: influence of hydrotrope as interface stabilizers. *Journal of Molecular Liquids* 306: 112861.

Tabernero, J., Shapiro, G., LoRusso, P. et al. (2013). First-in-humans trial of an RNA interference therapeutic targeting VEGF and KSP in cancer patients with liver involvement. *Cancer Discovery* 3 (4): 406–417.

Tatham, L., Savage, A., Dwyer, A. et al. (2018a). Towards a maraviroc long-acting injectable nanoformulation. *European Journal of Pharmaceutics and Biopharmaceutics* 138: 92–98.

Tatham, L., Savage, A., Dwyer, A. et al. (2018b). Impact of nanoparticles on brain health: an up to date overview. *Journal of Clinical Medicine* 7 (12): 1–14.

Tedesco, S. and David, S. (2012). Nanomaterials as emerging environmental threats. *Current Chemical Biology* 4 (2): 151–160.

Tyagi, P. and Subramony, A. (2018). Nanotherapeutics in oral and parenteral drug delivery: key learnings and future outlooks as we think small. *Journal of Controlled Release* 272: 159–168.

U.S. Food and Drug Administration (2015). Drug quality assurance. Compliance program guidance, 1–38. https://www.fda.gov/drugs/drug-safety-and-availability (accessed 10 April 2020).

Wang, Y. and Grainger, D. (2019). Lyophilized liposome-based parenteral drug development: reviewing complex product design strategies and current regulatory environments. *Advanced Drug Delivery Reviews* 151–152: 56–71.

Wei, X., Shao, B., He, Z. et al. (2015). Cationic nanocarriers induce cell necrosis through impairment of Na^+/K^+-ATPase and cause subsequent inflammatory response. *Cell Research* 25 (2): 237–253.

WHO, Safe Injections Global Network (SIGN) (2016). *Report of the SIGN 2016 Meeting* (ed. World Health Organisation). Geneva: World Health Organisation.

Yan, H., Xue, Z., Xie, J. et al. (2019). Toxicity of carbon nanotubes as anti-tumor drug carriers. *International Journal of Nanomedicine* 14: 10179–10194.

Zancheta, B., Chaud, M., and Santana, M. (2015). Self-emulsifying drug delivery systems (SEDDS) in pharmaceutical development.

Journal of Advanced Chemical Engineering 5 (3): 1–7.

Zhang, T., Liming, W., Qiang, C., and Chunying, C. (2014). Cytotoxic potential of silver nanoparticles. *Yonsei Medical Journal* 55 (2): 283–291.

Zielińska, A., Costa, B., Ferreira, M. et al. (2020). Nanotoxicology and nanosafety: safety-by-design and testing at a glance. *International Journal of Environmental Research and Public Health* 17 (13): 1–22.

12

Dermatological Delivery of Nanodrugs

Applications, Toxicity, and Safety

Saurabh Shivalkar, Arushi Verma, Vishal Singh, and Amaresh Kumar Sahoo

Department of Applied Sciences, Indian Institute of Information Technology, Allahabad, UP, India

12.1 Introduction

The development of nanodrugs in the last few decades has dramatically improved the extent of drug delivery mechanisms. Dermatological delivery routes for these nanodrugs have attracted much attention in the recent past for several topical and transdermal therapies. Since skin acts as the defense mechanism, the anatomy and physiology of the skin need to be properly understood for advanced formulation and efficient dermatological delivery of nanodrugs. The anatomy of skin from the outer to the inner layer constitutes the epidermis, dermis, and subcutaneous layer. The outer layer, the epidermis itself is divided into five layers. These include stratum corneum, stratum lucidum, stratum granulosum, stratum spinosum, and stratum basale from outside to inside (Zhao et al. 2016). The stratum corneum primarily acts as a barrier function of the skin and helps in the regulation of the transport of essential molecules and compounds through the skin. Therefore, the formulation of drugs that can easily penetrate through the skin is difficult and challenging. The epidermis is a non-vascularized structure and the majority of nutrition it takes is through diffusion. However, if formulated drugs can efficiently pass through the epidermis, it reaches the vascularized region (dermis) of skin from where these drugs are instantly taken up by the body. Dermatological delivery is broadly categorized into topical delivery and transdermal delivery. The nanodrugs that are delivered by overcoming the skin barrier are known as topical delivery, while nanodrugs when delivered directly into subcutaneous tissues and taken up by the body directly are termed as transdermal delivery. These nanodrugs are formulated from various potential nanoparticles of sizes mostly less than 500 nm. Therefore, the unique property of the nanoparticles such as a high surface area to volume ratio plays an important role in overcoming the biological barrier offered by the skin and the cell membranes during the transport of the nanodrugs through the skin. Various types of nanoparticles can be used for the development of nanodrugs for dermatological delivery. Nanoparticles such as nanospheres, solid lipid nanoparticles, micellar nanoparticles, liposomes, and nanoemulsions are most commonly used. However, natural polymer-based nanoparticles such as chitosan and synthetic polymer-based nanoparticles such as poly-(lactide-co-glycolide), poly-(ε-caprolactone), and polyacrylates were also reported (Zhang et al. 2013). Preliminary

Nanotechnology in Medicine: Toxicity and Safety, First Edition. Edited by Mahendra Rai, Mrunali Patel, and Rashmin Patel.

studies suggested that the nanodrugs formulated from these nanoparticles have shown faster repair of skin scar. Therefore, they are frequently used in creams, ointments, and unguents for medical imaging for the diagnosis of dermatological diseases. Moreover, various dermatological delivery systems have been explored until now to achieve better biodistribution of these nanodrugs. However, the biggest challenge in the nanoformulation of nanodrugs is to understand their pharmacodynamics and pharmacokinetics (Tey et al. 2008). Several researchers have been working on developing complex dermatological delivery mechanisms that have the potential to transport these nanodrugs through the cell membranes and into the cytoplasm of the cells. These properties are highly variable and differ frequently among patients. The use of nanoparticles for the nanoformulation of nanodrugs has significantly improved dermatological drug delivery. However, safety is of utmost priority and thus the danger of nano toxicity of these drugs needs to be analyzed and understood properly before medical applications (Zangabad et al. 2017). The aim of this chapter is to provide recent advancements in therapeutics using nanodrugs through various dermatological delivery routes.

12.2 An Overview of Dermatology and Nanodrugs

Transdermal drug delivery (TDD) is considered an effective drug delivery route due to its convenience, easy to use, noninvasiveness, and improved patients' compliance. Other advantages of using dermatological delivery are:

a) Reduction of drug concentration fluctuation in the blood.
b) Provide steady plasma levels.
c) Higher chances of preventing overdose.
d) Avoids conflicting gastrointestinal environment such as enzymatic activities and pH for the drugs.
e) Improves the efficacy of the drugs due to its noninterference with the food particles.
f) Evades "first-pass effect" by the liver.
g) Provides long-term stability to the drug molecules.

However, apart from these benefits of dermatological delivery, the challenge of efficiently transporting the drugs through skin is still under stringent research. The skin is the largest and most exposed organ of the human body, thus more prone to the harsh environment, radiations, and microorganisms. Therefore, as a preventive measure, it forms a protective layer against the harsh environment and foreign molecules such as drugs to maintain the integrity of its structure and curtails any damages to the functions of the human body. Conventional drug molecules are impermeable to the skin due to epidermal cell cohesion and lipids present in the stratum corneum layer (Bodeker et al. 2017). Thus, there is a requirement to introduce an advanced and efficient dermatological delivery route that has the potential to pass this barrier.

The introduction of nanotechnology to the field of dermatological delivery has removed several obstacles to conventional drug delivery mechanisms. Nanomaterials being small in size have the potential to interact at the subatomic level with dermal and transdermal tissues. Therefore, it promises to transform the dermatological delivery mechanisms as well as the diagnosis and therapeutics of dermatological ailments. The use of nanotechnology has also improved drug penetration by regulating the release of active substances and increasing the period of the drug permanency on the skin without any interaction with the stratum corneum. Consequently, it prevents the drug from being exposed to any chemical stress and physical instability. Various nanomaterials such as inorganic nanoparticles, polymer-based nanoparticles, nano-emulsions, etc., have been explored to formulate nanodrugs of dermatological delivery for multiple areas of therapeutics and diagnosis (Mordorski et al. 2016).

Transdermal delivery mechanisms of nanodrugs are gaining popularity over topical delivery due to its wide range of applications. Topical delivery mechanisms are suited best for the treatment of the main skin disease, while in transdermal delivery, nanodrugs are taken up by the body directly and thus can be used to treat diseases other than skin too. A TTD system was devised to formulate and deliver the nanodrugs in a controlled manner through skin permeation. The technologies that are used in the advanced permeation enhancement of the skin are electroporation, microneedles, iontophoresis, and ultrasound (Iqbal et al. 2018). Nanoformulation matrices for the nanodrugs used in the transdermal delivery system differ greatly from the topical nanoformulations. These are the following ways in which the nanoformulations of transdermal delivery systems differ from conventional topical ones:

a) The nanoformulations matrix of the patch used in the transdermal nanodrugs delivery mechanism maintains the proper drug concentration, providing sustained drug release.
b) An impermeable backing film prevents excessive moisture loss from the skin under the patch.
c) An adhesive layer of the patch maintains the contact of the nanodrugs along with the continued drug release.

The nanodrugs permeation into the skin requires diffusion through the tough epidermis layer and reaching the follicles and glands. The follicles and glands present underneath the skin develop a shunt pathway for the permeation of these nanodrugs. However, these shunted pathways are just 0.1% of the total human body, and the stratum corneum mostly resists drug permeation into the skin (Makhmalzade and Chavoshy 2018). Therefore, preferred drug permeation routes are intracellular and transcellular routes. Major dermatological delivery routes are (Figure 12.1):

The intercellular lipid route: Hydrophobic and amphiphilic nanodrugs can diffuse through the interlamellar regions along with the linker region in the stratum corneum. This is because these regions possess very less ordered lipids and have more versatile hydrophobic chains, thus creating a nonplanar space between crystalline lipid lamellae and their adjacent cells present in the outer membranes (Geinoz et al. 2004). However, the intracellular macromolecular matrix has a large amount of keratin present within the stratum corneum, thus mostly contributing to its mechanical stability and support instead of directly as a diffusion barrier of the skin.

The transcellular route: A region of poor cellular and intercellular lipid packaging is observed in narrow aqueous trans-epidermal pathways that coincide with the wrinkles on the skin. This forms a site at the lowest resistance point on the skin for the transport of hydrophilic nanodrugs (Cevc and Vierl 2010). The efficacy of the transcellular route can be enhanced by subjecting the stratum corneum to an electrical stimulus like electroporation or iontophoresis, a mechanical stimulus such as sonoporation, or some thermal stimulus suitable for increasing skin permeation.

Follicular/transfollicular route: As discussed earlier, follicles and glands underneath the skin form a shunted pathway for the delivery of nanodrugs. Since follicles occupy very less area compared with the total surface of the skin, therefore it was considered unimportant for dermatological delivery. In the recent past, researchers have shown interest in this type of delivery route and administered polystyrene nanoparticles. The study suggested that these nanoparticles were localized mostly in the follicular regions of the skin due to its small size (Escobar-Chavez et al. 2012). It was also found that the time-dependent distribution of the drugs was enhanced substantially. This confirms that nanodrugs can efficiently be delivered using the follicular route in dermatological delivery.

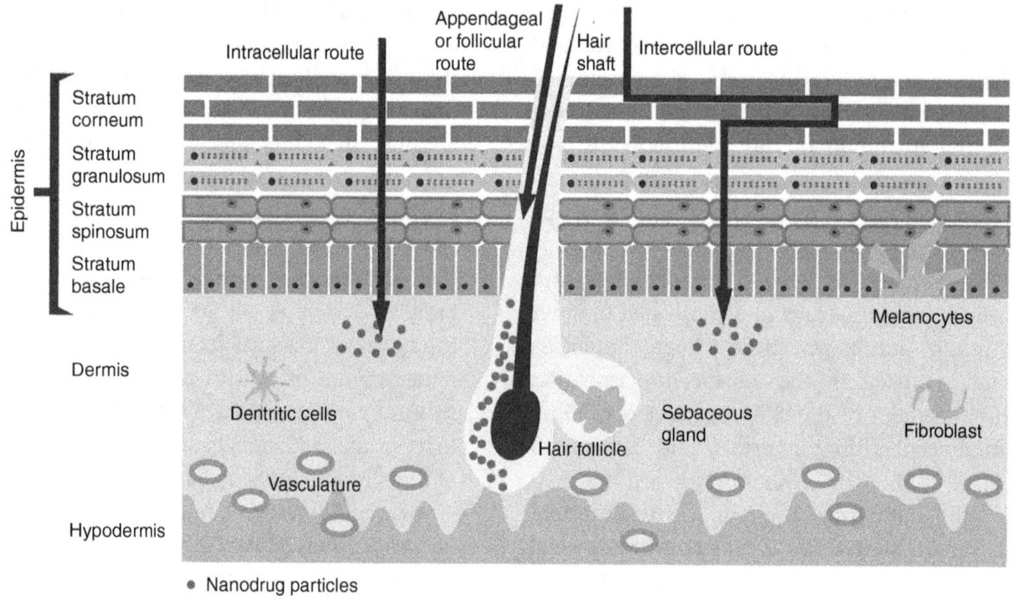

Figure 12.1 Illustration of different layers of the skin and various routes of dermatological delivery of nanodrugs.

Drug delivery systems are generally designed to remove the limitations of interfering pharmacokinetics and bio-distribution of the drugs. However, the problem of higher variance in pharmacokinetics and pharmacodynamics of nanodrugs when administered among various patients is a debatable issue. Researchers developing nanodrugs from several nanoparticles to evade skin barrier and defense mechanisms have found them beneficial in improving the drug permeation and delivery to the targeted sites. Complex drug delivery mechanisms were introduced to transport nanodrugs to the cytoplasm and to do so, more efficiently triggered responses were used. These triggered responses are derived from the specific signal generated against the nanodrugs administration. For example, to improve the solubility of poorly soluble nanodrugs, these drugs are administered to the place where a hydrophilic and hydrophobic environment is present (Sá et al. 2016). However, the challenge of targeting specific organs through the dermatological delivery of nanodrugs persists due to the complex host's reaction to nanomaterials. Nevertheless, researchers are putting a lot of effort into understanding the potential drawbacks of these nano-particulate systems.

12.3 Nanomaterials for Dermatologically Active Nanodrugs

Nanomaterials for the development of dermatologically active nanodrugs are gaining popularity due to their several benefits over the conventional formulations used in the drug delivery system. The advantages of using these nanoparticulate systems are:

a) The small size range of 10–1000 nm assists in providing targeted delivery.
b) Sustained release.
c) Enhanced drug adhesion to the skin.
d) More stable as it prevents the degradation of labile groups.
e) Low toxicity.

Apart from the abovementioned benefits, these nanomaterial-based dermatologically active nanodrugs also provide effective, specific, and dynamic treatment options. Incorporating nanodrugs allows producing a versatile, customized, and cost-effective dermatological delivery system. Most of the nanoparticles that are used in the formulation of nanodrugs for dermatological delivery are generally polymer-based nanoparticles, inorganic nanoparticles, and carbon-based nanomaterials (Table 12.1). Nanodrugs formulated as a nanostructured carrier were developed from solid lipid nanoparticles, liposomes, micelles, or different types of sub-micrometer range magnetic nanoparticles (Figure 12.2). They are found to be highly effective compared with conventional drugs. For instance, zinc oxide particles are generally opaque and greasy, but when its nanoparticles are formed and formulated as an emulsion, they become almost nongreasy with better texture. This zinc oxide nanoparticle (ZnO)-based emulsion has potentially higher and deeper permeation into the skin (DiNardo and Downs 2018). Therefore, it became an essential formulation for ointments, emollients, and conditioners. Another aspect that is important for the enhanced skin permeation is the physicochemical characteristics like charge, size, hydrophobicity, and rigidity of these nanocarriers. Therefore, it directly affects the cellular uptake of the nanodrugs. In the recent past, most investigated nanomaterials for the formulation of nanodrugs in dermatological delivery systems are:

Polymer-based nanoparticles: Nanosphere or nanocapsule from polymer-based nanoparticles has been used to encapsulate the active nanodrugs for dermatological delivery. These are vesicular structures with greasy/oil core with a thermal-sensitive polymeric wall that is stabilized by a stearic compound. Formulated nanodrugs are encapsulated within the oily core of nanospheres or nanocapsules. These structures have attracted researchers due to their ability to diffuse nanodrugs in a controlled manner through the polymeric matrix to penetrate the skin. These are also structurally very stable, thus maintaining longer retentivity of the nanodrugs. For instance, dexamethasone as an active nanodrug enclosed in a hydrogel has been used to cure psoriasis. Similarly, small inhibitor ribonucleic acid was nanoencapsulated in polymeric nanoparticles for targeted drug delivery and to selectively suppress the gene expression (inhibit the test gene expression) in melanoma and regulate the expression in pachyonychia congenita (Nasir 2010; Majumdar et al. 2020).

Inorganic nanoparticles: Recent advancement in the area of inorganic nanoparticle-based nanodrugs is gaining much attention due to their potential for specifically targeted drug delivery with enhanced cellular uptake. The major properties of using inorganic nanoparticle-based nanodrugs for dermatological drug delivery are:

a) Excellent cellular actions
b) Targeted drug delivery
c) Widely available
d) Good biocompatibility
e) Controlled release of the nanodrugs
f) Rich in functionality

Dermatological delivery of inorganic nanoparticle-based nanodrugs has a wide range of applications. It can be used for topical delivery to enhance cosmetics by treating acne, hydrating skin along with the development of a variety of skincare products. For example, zinc oxide and titanium dioxide nanoparticles of less than 100 nm are commonly used in sunscreens due to their ability to block UVA and UVB (De Matteis 2017; Singh et al. 2020). Furthermore, silver nanoparticle-based nanodrugs have been known for improved antiseptic properties and found to be effective against diseases caused by *Onychomycosis, Trichophyton, Staphylococcus aureus,* etc. On the other hand, they are also found to be effective in curing chronic skin diseases like vitiligo and skin cancer through transdermal delivery of these nanodrugs. In the recent past, gold nanoparticles (Au NPs) have shown promising results as an intracellular dermatological delivery. The Au NPs

Table 12.1 List of different nanomaterials and their respective drug nanoencapsulation used in dermatological delivery.

S.No.	Polymer-based nanoparticles	Origin	Size range (nm)	Drugs nanoencapsulation	Applications	References
1.	Chitosan	Natural	50–200	Retinol	Antiaging, acne	Kim et al. (2006)
2.	Chitosan-Lecithin	Natural	95–250	Clobetasol-17-propionate, quercetin	—	Şenyiğit et al. (2010) and Tan et al. (2011)
3.	Chitosan-TPP	Natural	50–287	Antisense oligonucleotides, aciclovir, Plasmid DNA	Overexpressed genes, antiviral drug to cure herpes infections, genetic immunization	Ozbas-Turan et al. (2010) and Hasanovic et al. (2011)
4.	Gelatin	Natural	300–500	Pilocarpine HCl	Ophthalmic application	Vandervoort and Ludwig (2004)
5.	Polystyrene	Synthetic	28–31	Nile red (hydrophobic dye)	To study drug release mechanism	Alvarez-Roman et al. (2004)
6.	Polyacrylate	Synthetic	40	*N*-thiolated β-lactam	Antibiotic for wound healing	Turos et al. (2007)
7.	PLGA	Synthetic (biodegradable)	100–210	Hinokitiol, 6-benzylaminopurine, indomethacin	Hair growth, analgesic for transdermal delivery	Tsujimoto et al. (2007)
8.	PLGA-Chitosan	Synthetic (biodegradable)	169	Spantide II and Ketoprofen	Anti-inflammatory	Tomoda et al. (2011) and Shah et al. (2012)
9.	Poly(ε-caprolactone)	Synthetic (biodegradable)	255–427	Octyl methoxycinnamate	Sunscreen	Alvarez-Roman et al. (2001)
10.	Poly(methyl methacrylate)	Synthetic	68–100	Nile red (hydrophobic dye)	To study drug release mechanism	Wu et al. (2009b)
11.	Poly(ε-caprolactone)-block-PEG	Synthetic (biodegradable)	40–130	Minoxidil	Anti-hair loss	Shim et al. (2004)

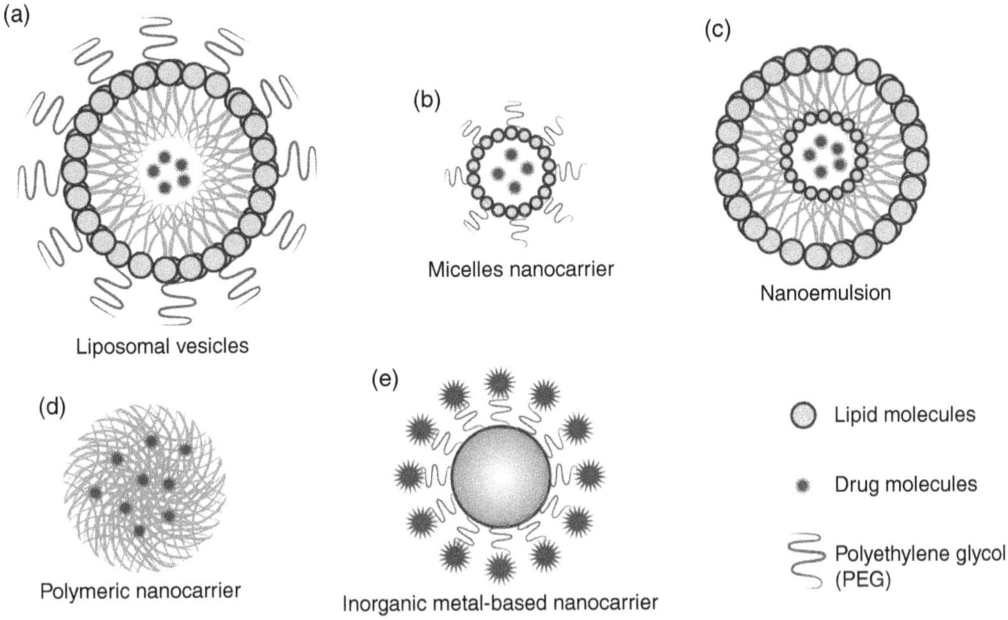

Figure 12.2 Schematics of different types of nanoformulations used in topical and transdermal delivery of nanodrugs (a) Liposomal vesicles (b) micelles nanocarrier (c) nanoemulsion (d) polymeric nanocarrier (e) inorganic metal based nanocarrier.

offer a wide size range, i.e. from 0.8 to 200 nm with excellent biocompatibility for the formulation of nanodrugs. It can be used to treat cancer using the tumor photothermolysis technique where the tumor present in the cutaneous region is targeted (Arafa et al. 2018).

Carbon-based nanomaterials: Carbon-based nanomaterials like nanotubes and fullerenes have been explored extensively in the dermatological delivery of nanodrugs. Carbon nanotubes are of less than 100 nm and the hollow structure provides a larger volume for drug loading while exteriors provide higher surface area that can be functionalized as per the applications. It is assumed that carbon nanotubes can be effective in the targeted delivery of genes and proteins through transdermal delivery mechanisms. Tagging or functionalization of these nanotubes can help in biomarker sensing for early diagnosis of malignancies of the skin (Mousavi et al. 2017). On the other hand, fullerenes are the arrangement of 60 carbon atoms into a hollow spherical structure of around 1 nm. These structures are generally hydrophobic but can be tagged with hydrophilic functional groups to transform into a water-soluble structure. Therefore, they can also be loaded with proteins, genes, or nanodrugs for targeted dermatological delivery. Moreover, carbon quantum dots were most frequently used in medical studies due to their photostability and photoemission properties. Therefore, transdermal delivery of these carbon quantum dots has innovative potential for advanced medical imaging (Abramovits et al. 2010).

12.4 Nanoformulations for Topical and Transdermal Delivery

The introduction of nanomaterials has instigated the nanoformulations for the preparation of nanodrugs. The dermatological delivery of these nano-sized materials as nanoformulations has shown enhanced specificity, retention, and on-site targeting. Being smaller than the pore size of the human skin, it neither alters the physiology of the stratum corneum nor affects physiochemical functions of the skin barrier. The nanoformulations can be broadly classified into two categories: (i) vesicles and (ii) nanoemulsions.

Vesicles: Vesicles are structures that are generally used for the nano-encapsulation of the active biomolecules. There are different types of vesicles that include liposomes, niosomes, invasomes, ethosomes, and transfersomes. Out of all these types, liposomes are considered the main type of vesicles and mostly it is used for the study of dermatological permeation. For example, liposomal nanoformulation of an anti-inflammatory agent, dipotassium glycyrrhizinate, has shown promising results in the treatment of dermatitis (acute as well as chronic) through their transdermal delivery. However, as compared with polymeric nanoparticles, these are relatively unstable when administered on the skin surface (Shabbir et al. 2020).

Nanoemulsions: In contrast to vesicles, nanoemulsions have shown enhanced skin permeation capability. This is due to their flexible nature as they can adjust according to the surroundings and evade interference from the stratum corneum due to the absence of polymers. Therefore, nanoemulsions can be used for topical as well as transdermal applications. Moreover, the lipophilic nature of the nanoemulsions' interior can efficiently deliver the hydrophobic nanodrugs in the aqueous environment (Figure 12.2). Nanoemulsion-based hydrophobic active biomolecules such as antioxidants or retinol can be delivered to the deepest region of the skin where it can be taken up by the body directly. Considering such merits of nanoemulsions, several antiaging products were introduced that can be delivered directly to the cutaneous or subcutaneous regions. For instance, gamma-amino-butyric-acid is an inhibitory neurotransmitter that assists in muscle relaxation. A nanoemulsion of gamma-amino-butyric-acid was formulated for topical applications to reduce wrinkles (Souza et al. 2018). The advantage of using nanoemulsions over conventional macroemulsions is not facing any challenges of sedimentation, flocculation, creaming, and coalescence. Thus, nanoemulsion-based nanodrugs are more stable and effective for dermatological delivery.

Nanoformulation of a nanodrug is a complex process. It requires an understanding of multiple factors that are biologically and chemically in sync to develop an effective nanodrug and prevent unprecedented side effects. The factors that needed to be explored thoroughly for the nanoformulation of the nanodrugs for dermatological delivery are:

a) Pharmacokinetics.
b) Pharmacodynamics.
c) Biodistribution of the nanodrugs.
d) Tolerance of skin against active nanomaterials.
e) The extent of side-effect impact on the skin or body.
f) Low toxicity.
g) Does not severely affect the lifestyle.

Nanoformulations for topical applications: The benefits of nanoformulations for topical delivery systems have taken over the conventionally FDA-approved drugs for various skin conditions like psoriasis, dermatitis, skin cancer, etc., as well as other physiological conditions. Lately, a few FDA-approved topical nanoformulations are micellar nanoparticle-based nanodrug: Estrasorb for menopause therapy and albumin-bound paclitaxel nanoparticle-based nanodrug; Abraxane for breast cancer. Recently, focus has been shifted toward nanoformulations of nanoparticles derived from natural sources. Among natural polymeric nanoparticles, chitosan-based nanoparticles are most frequently in the topical nanodrug delivery. Nanoformulations of chitosan nanoparticles and chitosan-tripolyphosphate nanoparticles were successfully used in topical gene therapy as they were shown to potentially deliver antisense oligonucleotides and plasmid DNA. In a study, it was reported that in a dose range of 15–90 mg of antisense oligonucleotide-loaded chitosan nanoparticles has shown the highest inhibition of Beta-gal expression at six-days post-transfection in rats. In another report, chitosan-tripolyphosphate nanoparticles have shown high loading efficiency for Plasmid DNA pSV-Beta-Gal and that the DNA with these nanoparticles exhibited transfection in baby rats. Although, these techniques have a profound impact on biomedical applications, still there is unavailability of sufficient data to validate their toxicity (Zhang et al. 2013). Moreover, nanoformulations of ultra-deformable liposomes with Vismodegib (UDL-V) drug were effectively employed in the treatment of skin cancer. The *in vitro* assays on human skin explants showed that these UDL-V formulations have better permeation through the stratum corneum. When incubated for about eight hours, it can be delivered to both the membranes and inner aqueous space content in the epidermis. The efficacy of UDL-V nanoformulations for topical delivery can be understood by comparing it with orally administered Vismodegib drug. When around 150 mg of Vismodegib was administered orally, the estimated concentration of this drug on the epidermis was around 3 mg/ml, whereas 8.4 mg/ml was present in the epidermis through topical delivery of UDL-V after eight hours incubation. Therefore, around three times higher dose of drug can be delivered to the epidermis using topical delivery of UDL-V for the treatment of basal cell carcinoma (Calienni et al. 2019).

Nanoformulations for transdermal delivery: Lipid-based nanoemulsions/nanocarriers are most commonly used in TTD. Some of the drugs delivered through transdermal delivery systems are Ketoprofen, Spantide II, Aceclofenac, Tretinoin, etc. Surface modifications of lipid nanocarrier with a peptide containing 11 arginine moieties have significantly improved the delivery of Ketoprofen and Spantide II into the deeper region of the skin. Consequently, they efficiently relieved the allergic conditions of dermatitis in the mouse model by reducing the inflammation (Zhang et al. 2016). In another example, antibacterial activity against the acne causing bacteria was significantly improved by combining chitosan with solid lipid particles by increasing the loading with tretinoin. This combination of chitosan–solid lipid particles–tretinoin has significantly increased the stability of nanoformulation and also nontoxic to human skin keratinocyte cell line (HaCaT cells). Recently, an iontophoretic TTD mechanism is devised that has the potential to deliver a charged drug nanocarrier using disposable reverse electrodialysis. An electroconductive polypyrrole-incorporated poly(vinyl alcohol) hydrogel (PV-H) facilitated electron transfer from the reverse electrodialysis battery that accelerates the release of the charged drug nanocarrier from PV-H. Based on this system, a Fluconazole (0.5% fluconazole)-loaded drug nanocarrier (F-DN) functionalized with a charge-inducing agent was used for the transdermal delivery. About 3 mg of F-DN was applied to the skin patch and the amount of drug loaded within the skin patch was calculated based on the swelling ratio and specified drug concentration of 0.5% (An et al. 2020). Later, the amount of drug permeated within the skin tissues was determined by the HPLC technique. In another example, a paintable oligonucleotide hydrogel incorporating Paclitaxel encapsulated cell-penetrating peptide-modified transferosomes was investigated for enhanced transdermal delivery of Paclitaxel to treat melanoma (Jiang et al. 2018).

12.5 Applications of Dermatological Nanodrugs and Its Delivery Mechanisms

Nanoparticles (NPs) diffusion through the skin depends on various physical parameters such as its size, surface charge, and the pathway it follows to penetrate inside the skin cells. The pathway it adapts is dependent upon the physicochemical properties of the NPs or formulation of the topically applied agent (Verma et al. 2020a). Generally, two pathways may be followed by a drug to penetrate inside the skin surface. It may be via the epidermal route, i.e., stratum corneum or through skin appendages and transappendageal route or transfollicular route. The stratum corneum is the primary barrier to any kind of absorption and is strongly dependent upon the polarity of the absorbent applied topically. The absorption here might take place either through corneocytes/bricks or through the intercellular spaces. Intracellular pathway generally allows the drug to pass through cell membranes. Endocytosis, transcytosis, and diffusion of polar molecules follow the mentioned route passively. However, the lipid layer in the stratum corneum restricts the movement and permeability of the drug (Hofmeier and Surber 2016; Verma et al. 2020b). Bhowmick et al. (2013) in their research established that the permeability of drug or NPs is inversely proportional to size and hydrophilicity. Thereby, usage of penetration enhancers is promoted to overcome this hindrance and enhance permeability. The channel-like structures present in the intercellular route along the lipid matrix is a more convenient method of drug/NP diffusion.

It is also to be noted that many studies have established the fact that though particulate formulations cannot pass the stratum corneum, they can penetrate preferentially into the hair follicle canals. This increases the concentration of formulation inside the follicular infundibulum, thus enabling the use of NPs as a carrier system by changing the penetration pathway and kinetics. As a result of particle accumulation within the follicular infundibulum, the desired drug loaded on the particle is thus released increasing its diffusion gradient, penetration depth, and drug permeation rate (Vogt et al. 2005; Papakostas et al. 2011).

Researchers have also defined several mathematical models for diffusion through the skin barrier. These models are based particularly on Fick's laws which states that diffusion of particles occurs from higher concentration to lower concentration. Laplace transformation, homogeneous membrane model, and scaled particle theory also find their role in these models (Verma et al. 2020b). Following these models, numerous researches have supported the fact that NPs are many efficient candidates for permeability and specific drug delivery (Couto et al. 2014; Verma et al. 2020b).

12.5.1 Prevention

12.5.1.1 Antisepsis

The topical application of bactericidal ointments is the first line of treatment for any superficial wound or cut in the skin. Antiseptics are commonly used to prevent infections. Nanoparticles containing silver, nitric oxide, chlorhexidine, etc., find a major role in the formulations of these antiseptics owing to their antimicrobial and wound-healing properties (Nhung et al. 2007; Hasanzadeh et al. 2016; Hofmeier and Surber 2016; Rong et al. 2019). Silver (Ag) from ancient ages has been known to possess efficient antimicrobial assets and are known to cause bacterial DNA damage by binding to it and inhibiting bacterial enzymes. Silver ions usually cause irreparable damage to the bacterial cell wall and cytoplasmic membrane leading to bacterial cell death. Recent studies also show combination therapy of Ag NPs with carbon dots (Cdots) causing bacterial cell death by a generation of enhanced reactive oxygen species (ROS) (Hasanzadeh et al. 2016). Martinez et al. (2009) demonstrated the synthesis and use of nitric oxide-releasing NPs that acted efficiently against methicillin-resistant *S. aureus* infection on the skin of a mouse (Nhung et al. 2007). Hasanzadeh Kafshgari et al. (2016) similarly synthesized Nitric oxide-releasing porous silicon nanoparticles that killed both gram-negative and gram-positive bacteria without causing any ill effect on mammalian cells. Nhung et al. (2007) tested chlorhexidine-loaded nanocapsule-based gel (Nanochlorex®) as a hand rub gel against the resident skin flora and found effective bactericidal property. The effect was prolonged due to the sustained release property of the synthesized nanosystem, whereas the immediate effect was due to the rapid desorption of chlorhexidine from the wall of the capsule (Nhung et al. 2007; Martinez et al. 2009; Verma et al. 2020b).

12.5.1.2 Cosmetics and Photoprotection

Exposure of skin to UV radiation from the sun is a major cause of skin cancer among the population (Linos et al. 2016). Therefore, cosmetics such as sunscreens have become the need of the hour for skin protection by providing it a temporary topical covering. NPs such as zinc oxide (ZnO) and titanium dioxide (TiO_2) are extensively being used in sunscreen formulations as physical sun blockers. ZnO NPs absorb the UVA spectrum and thus prevent photoaging and photoallergy of skin as well as prevent the formation of cancerous skin lesions (Hofmeier and Surber 2016). Similarly, TiO_2 is also known to combat against photoallergy due to its high photostability, besides being an effective antimicrobial agent as well. Thus, these NPs upon topical applications in the form of sunscreen might usually reflect, scatter, or absorb UV radiations (Smijs and Pavel 2011). Due to the nanometric size of these particles in the formulation that is less than that of visible light range, i.e. 400–700 nm, they are not perceptible to the naked eye and are visibly transparent leaving no white residue like in the case of bulk ZnO or TiO_2 particles. However, some studies showed that they also have certain drawbacks on human dermal fibroblasts which might be overcome by surface functionalization of the NPs with certain hydrophobic polymers like dimethicone or silicone (Pan et al. 2009). Studies are still in progress regarding the stability of these surface-modified NPs.

Various other cosmetic agents also contain nanoscale materials in their formulations. For instance, antiaging creams contain gold and Ag NPs for skin tissue regeneration purposes; facial cleansers may comprise of gold, silver, zinc, and titanium dioxide NPs reflecting antibacterial and anti-inflammatory properties. Even the deodorants have been reported to contain Zn and TiO_2 NPs owing to their antimicrobial properties.

12.5.2 Diagnosis

Nanotechnological interventions for diagnostic purposes are another popular field of work among researchers. The major advantage of NP-based diagnosis is that analysis can be performed on a smaller tissue sample due to the enhanced sensitivity of permissible detection methods. Molecular imaging is the most prevalent area where Au NPs, Cdots, QDs, and nanoparticles with magnetic properties are predominantly used as biosensors or labels for biosensing (Hia and Nasir 2011). Clothes made up of optical fiber containing fabric is also another way of nanodiagnosis as they may be employed for measuring the size of skin lesions, monitoring atopic dermatitis and psoriasis, etc. (Eden et al. 2005).

Au NPs have been extensively explored for nanodiagnosis as they can be traced due to inherent properties of fluorescence and optical absorption, thereby making them efficient labels for sensors (Zuo et al. 2007). QDs are also highly fluorescent nano-scaled semiconductor particles that inherit the property of absorbing light over a wide frequency range and then re-emitting as a single color as per the particle size. They exhibit very strong fluorescence that is stable for a longer time (Shiohara et al. 2004). They have been explored in detecting sentinel nodes showing fluorescence in the infrared region. This is extremely useful in managing tumors like melanoma (Kim et al. 2004; Nuzhatun Nisa 2015).

12.5.3 Therapeutic Applications

12.5.3.1 Phototherapy

Researchers working in this field are exploring several therapeutic applications of nanoparticles in dermatology. One of them is phototherapy, in which, for instance, melanosomes are targeted with short laser pulses for the treatment of hyperpigmentation disorders. Specific cell targeting is being explored using immune conjugates of Au NPs and iron oxide microparticles. The particles absorb light from the incident laser pulse and liberate the absorbed energy in the form of heat that damages the target cell and tissue (Pitsillides et al. 2003). Besides, it is also useful in photothermal therapy, which may use agitated Au NPs to prevent tumor growth with the advantage of damaging lesser surrounding cells. Cure for Murine melanoma is being tracked by studying the effect of Melanocyte-Stimulating Hormone (MSH) analogs in combination with gold nanospheres for selective ablation of affected cells (Dickerson et al. 2008). Similarly, photodynamic therapy using nanoparticles is also being explored. This technique works upon the principle of ROS or free radical generation via optical activation of photosensitizer causing local tissue damage in tumor cells (Henderson and Dougherty 1992). These techniques also have a promising scope in the treatment of skin cancer (Nuzhatun Nisa 2015).

12.5.3.2 Sebaceous Gland Disorders

Blockage in sebaceous glands is one of the most common causes of many dermatological diseases including acne in adolescence. Several NPs like biodegradable poly-lactic acid (PLA), solid lipid nanoparticles, and liposome loaded with drug particles can be employed as topical agents to target hair follicle-related diseases via the follicular route (Rolland et al. 1993; Patel et al. 2000; Nuzhatun Nisa 2015). They increase the follicular uptake, which fosters an increase in local drug concentration, thereby enhancing the therapeutic effect. Microspheres are being used lately for the treatment of acne. Benzoyl peroxide in microsphere formulation with different concentrations (5.5 and 7%) has been launched in the market already (Bikowski and Del Russo 2008). These formulations have shown improved skin tolerability leading to superior patient satisfaction. Besides microsphere-based creams, liposomal formulations containing antiandrogens are also used which target specifically the sebaceous glands by going deeper inside the hair follicle (Hofmeier and Surber 2016).

12.5.3.3 Treatment of Inflammatory Skin Diseases

Various steroid-based topical creams are used in the treatment of dermal diseases like eczema, atopic dermatitis, etc. However, their continuous application has several side effects like striae, skin atrophy, perioral dermatitis, hypertrichosis, pigment alteration, and several other skin conditions (Coondoo et al. 2014; Nuzhatun Nisa 2015). NP-based formulations are being explored here to minimize these adverse effects. Cutaneous atrophy is being addressed by liposomes containing topical steroid creams. Liposomal formulations of tacrolimus in the treatment of immune-mediated skin disorders have also shown promising effects (Erdogan et al. 2002). Encouraging results have been obtained on animal models while clinical trials of many such formulations are under review for safety and efficacy concerns (Hofmeier and Surber 2016).

12.5.3.4 Other Therapeutic Applications

There are several other therapeutic applications of nanotechnology in the dermatological aspect. Among these, its usage in hair treatment offers huge scope. Follicular delivery of nanoformulations is way more advantageous than conventional delivery methods. Researchers have reported enhanced bioavailability, increased penetration of drugs, and improved residence time (Hofmeier and Surber 2016). NP-based formulations have been reported to be better than water- or alcohol-based ones for treating alopecia areata and androgenic alopecia. Liposomal cyclosporine is being explored for treating alopecia areata (Verma et al. 2004). Polyethylene glycol (PEG)-encapsulated minoxidil has shown enhanced permeation inside the hair follicles (Shim et al. 2004). Nanoformulations are also employed in treating several autoimmune diseases of hair follicles. Nanoparticle-loaded immunosuppressants are efficient in reducing the side effects of these medications (Nuzhatun Nisa 2015).

Skin malignancies or cancer is also a potential target of nanotechnology. Several studies have claimed the effectiveness of polymeric nanoparticle-based formulations such as 5-fluorouracil, curcumin, 5-aminolevulinic acid, quercetin, etc., in skin cancer (Jain et al. 2009; Sabitha et al. 2013). Magnetic/paramagnetic/superparamagnetic nanoparticles are also employed in localized drug delivery to precise locations (Mangalathillam et al. 2012; Rao et al. 2015; Wu and Huang 2017).

12.6 Toxicity Evaluation of Dermatologically Active Nanodrugs

Toxicity of a drug is the adverse effect associated with its dose of that drug being applied. It implies that with any variation in drug dose concerning its number or concentration, we can change the effect of toxicity on the organism or target organ. Toxicity of any drug is species specific and also the degree of toxicity also depends on the individual health or age within the species. Thus, it is significant to mention that discovery of new/novel drugs and investigation or analysis of its absorption, distribution, metabolism, and excretion (ADME) properties cannot be completed without toxicity evaluation of the drug. The toxicity of drugs can generally be categorized as genotoxicity and cell toxicity (Bhadar et al. 2016). However, based on site, the toxicity can be classified as specific target organ toxicity and systemic toxicity. Specific target organ toxicity refers to the toxicity effect of drugs confined to only one site, whereas systemic toxicity is related to toxicity at multiple sites. Further, it can be divided into acute and chronic toxicity. Toxicity can be referred to as acute if the adverse effect occurs due to single or multiple doses taken orally or dermally within 24 hours (Colerangle 2017). In the case of chronic toxicity, adverse effect occurs due to repeated exposure of the drug for a longer time period (De Jong et al. 2012). In recent decades, the pharmaceutical and cosmetic industry has witnessed a revolution in the manufacture of medicine and cosmetics. Nanoparticles, the constituent particles of active nanodrugs, are being prepared from metal oxides, nonmetal oxides, and polymers. In the field of health or medicine, NPs are generally being used for their novelty and specificity in drug delivery systems (Lewinski et al. 2008; Nowrouzi et al. 2010; Bahadar et al. 2016). Several different types of nanoparticles like dendrimers, fullerenes, liposomes, micelles, and nanospheres have applications in nanodrug synthesis. Skin is the first-line defense barrier of our body. Nanodrug, because of its nanoscale size, easily penetrates through the biological barriers, i.e., layers of skin and can enter into various vital organs and circulate in body fluid (Pourmand and Abdollahi 2012; Bahadar et al. 2016). Many research groups have proposed that small-sized particles cause more toxicity to human organs than larger sized particles synthesized from the same substance; that is, toxicity is inversely proportional to size of the nanodrug (Scheuplein 1967; Mostafalou et al. 2013).

12.6.1 Nanodrugs and Toxicity Due to Penetration

Mostly, nanodrug enters the body through the skin or inhalation. There are mainly four ways to penetrate through the skin: (i) transappendageal,

(ii) intracellular and transcellular, (iii) through sweat glands, and (iv) through hair follicles. These nanodrugs penetrate passively through the layers of the healthy dermis and free radicals and it is worth mentioning that these free radicals result in oxidative stress in the cell/organ and also cause collagen depletion, skin wrinkling, keratinization, and atrophy in the skin layers (Baroli et al. 2007; Zhao and Ng 2014; Silva et al. 2017). Recent investigations do not clearly describe that nanodrugs pass through the SC or pass through the circulation to the dermal tissue. However, penetration of nanodrug is doubled if the layers of the dermis are damaged or pore size is altered. A group of researchers says that Fe nanoparticles penetrate through the skin layer and accumulate in the epidermis as well as in SC layers (Baroli et al. 2007). Whereas another research group of Wu et al. (2009a) proposed that titanium dioxide (TiO_2) nanoparticles (NPs) can easily penetrate through SC but not cross the dermis layer of skin. But after 40 days of continuous exposure of TiO_2 nanoparticles, it penetrates through the dermal layer and reaches the deeper part of the skin and different parts of the various organs and causes pathological changes (Wu et al. 2009a). Another research says that TiO_2 NPs being used in photoprotector products effectively safeguard skin from the UV exposure of the sunlight. Moreover, along with the TiO_2 and zinc oxide (ZnO), nanoparticles have been reported to promote lesser whitening of the skin and have the capability to disperse or reflect the UV radiation (Hudson et al. 2008).

Another similar research proposed that naked TiO_2 nanoparticles when exposed to UV light exhibit antibacterial activity due to their photocatalytic activity and promoting peroxidation of unsaturated phospholipid of lipid membrane in prokaryotic (Von Montfort et al. 2015). The most commercially used antibacterial nanoparticles are silver nanoparticles (Ag NPs). Its mechanism of being antibacterial lies in the fact that silver nanoparticle interacts with the thiol group of various cell wall proteins. The report suggested that Ag NPs cause bacterial cell wall perforation and elevation of the ROS level that leads to antibacterial activity. However, a higher dose of Ag NPs also exhibit cytotoxicity for humans including skin cells; they interact with the membrane proteins to cause oxidative stress and result in mitochondrial toxicity (Plytycz and Seljelid 2003). In another *in vivo* investigation, Ag NPs' toxicity to skin layers after 14 days of its exposure was detected superficially at the SC surface of the skin layer and was also confirmed by the transmission electron microscope (TEM) images (Samberg et al. 2010). A recent study also showed the gold nanoparticle (Au NP) penetration and its metabolic effect studied by Labouta et al. (2011). However, the Au NPs were found to be less cytotoxic as compared with Ag NPs. They demonstrated that 15 nm-sized Au NPs after 24 hours of exposure in aqueous medium accumulated on the SC surface, whereas smaller size Au NPs (~6 nm) in toluene deeply penetrated the SC as well as the epidermal layer of skin. Another research group did similar work to study the penetration and interaction of luminescent QDs having a size less than 10 nm, on the skin and keratinocytes cells of humans. The aim of this study was to regulate the cellular uptake, toxicity, and inflammation (Schek et al. 2004). QDs were applied on the skin for 24 hours and examined using confocal microscopy and transmission electron microscopy. Confocal microscopy illustrated that the fluorescence was mainly in the upper layer of SC near hair follicles and the TEM image also confirmed and detected its presence between the lipid bilayers of the SC surface. The above results were corroborated with another study that investigated the toxic effect of Ag NPs of size less than 100 nm. They examined clinically and histopathologically the acute as well as sub-chronic toxic effect of Ag NPs with variable concentration. The study showed a clear indication of inflammation. They also evaluated the amino acid-derived fullerene responses on the skin by evaluating the cellular viability and inflammation in keratinocytes of the epidermal layer of skin (concentrations of 0.4–0.00004 mg/ml). There was a decrease in the cell viability and an increase in the level of secretion of pro-inflammatory cytokines like IL-6, IL-8, and IL-1β. This increase in the cytokines indicates that amino acid-derived fullerenes can stimulate the secretion of cytokines in epidermal keratinocytes (Korani et al. 2011). The above studies revealed that the toxicity of NPs on the skin hugely relies on their size, shape, and surface chemistry. Thus, judicious modification of the NPs might reduce the toxicity of the NPs. Furthermore, toxicity due to transdermal delivery of carbon quantum dots is not fully explored until now. However, their toxicity is primarily dependent on the behavior in the body such as physiochemical properties and environmental factors affecting absorption, distribution, metabolism, and excretion of the body (Abramovits et al. 2010).

12.6.2 Genotoxicity

Genotoxicity is a subclass of systemic toxicity that results due to any kind of damage to DNA caused by mutagens (may be nanoparticles). This toxicity can be considered as the most prominent and potential cause of cancer. The most common cause can be due to the free radical generation by mutagens/nanoparticles. Several research groups have been working in these areas; their studies showed that uptake of NPs might affect the DNA directly or indirectly. In the case of the indirect pathway, NPs inhibit and/or hinder the function of several vital enzymes required for DNA replications and repair. This also causes DNA damage, which is one of the prominent drawbacks of direct use of NPs for skin care. The research claimed that Au NPs interact with DNA as well as interrupt DNA replication (Connor et al. 2005). Another study also revealed that genetic abnormalities occurred due to the interaction of Ag NPs (5–10 nm) with lymphocyte DNA (Huk et al. 2015). The cerium oxide (CeO_2) nanoparticles of size 7 nm interacted with DNA and it potentially damaged the DNA as well as broke the chromosome (Auffan et al. 2009). A study found that synthesized copper oxide (CuO) and cobalt oxide (CoO) NPs of size 10–40 and 6 nm, respectively, caused considerable DNA damage due to a higher level of ROS generation. Further, the study claimed that the CoO of size 6 nm was highly toxic to humans (Seabra and Durán 2015). Other investigations with metal oxide (ZnO NPs) claimed that ZnO NPs induced apoptosis by interfering with the function of mitochondria and stimulating the activation of p53 and p38 and also induced genotoxicity in keratinocyte cells. A toxicological examination of cosmetics reveals that the product contains TiO_2 NPs. Different ranges of TiO_2 NPs size were investigated and the study revealed that smaller sized TiO_2 NPs (~20 nm) caused complete damage to the DNA at a very lower dose and in the absence of UV light also (Hashempour et al. 2019).

12.7 Safety Considerations

Nanomedicine is a field of nanotechnology where nanomaterials like lipids, metal, silicon and silica, polymers, proteins, and carbon are used for the construction of nanoparticles that can be used for diagnostic and therapeutic purposes (Mohajerani et al. 2019). A drug has to overcome several biological barriers to reach its location; also the drug load has to be protected by nanocarriers, whose toxicity also needs to be considered (Ferrari 2010). Properties of nanomaterials and bulk materials are different, thus we have to perform suitable assays to evaluate nanoparticle toxicity. The approved science of nanotherapeutics is based on similar principles but has low toxicity levels, for example, a drug named Doxil has less cardiotoxicity in comparison to free doxorubicin (Barenholz 2012). Difference in vascular arrangements significantly impacts Enhanced Permeability and Retention (EPR) effect, therefore providing access to nanoparticles into the tumor tissues (Matsumura and Maeda 1986). Hydrophobic therapeutics like dimethyl sulfoxide can also be used (Wolfram et al. 2014a). However, most of the study showed that hydrophobic therapeutics cause less toxicity and thus are clinically approved (Busatto et al. 2019). The toxicity also depends upon the morphology and size of the nanoparticle. Research found that Au NPs of size 1.4 nm diameter showed significant toxicity, while a similar particle of diameter 1.5 nm did not show any prominent toxicity (Pan et al. 2007). The nanoparticles' properties can be changed due to their interaction with their biological ambiance, for example, in interacting with protein corona (Wolfram et al. 2014b). Some endogenous biomolecules' exposure to a nanoparticle can lead to functional and structural alterations (Zuo et al. 2013; Wolfram et al. 2014b). Nanoparticles can also cause folding and unfolding aggregation of some proteins which might trigger inflammatory response (Linse et al. 2007). The formation of a protein corona layer decreases the toxicity of a nanoparticle as it might cause some damage to circulating protein that leads to a reduction in the overall toxicity level (Tenzer et al. 2013). Nanoparticles' toxicity can also be seen as the disruption of membrane-bound organelles like Endoplasmic Reticulum (ER) and mitochondria by lysosomal damage which may cause oxidative stress due to the release of ions and protons (Stern et al. 2012).

Immunological recognition is one of the major challenges faced by nanodrugs as cells have toll-like receptors (TLRs) which help cells to recognize unwanted elements or organisms. However, the immunological activation can be partially prevented by using PEG on the surface of the nanoparticles, which results in great protection from protein binding (Bartucci et al. 2002). Some antifouling agents like high- and low-density lipoprotein can be used to inhibit complete activation of an immunological response that can be caused by nanostructure due to interactions with the biological environment (Hamad et al. 2013).

12.8 Limitations and Risk

Despite having several potential benefits associated with them, there are considerable limitations and risks in using nanomaterials in dermatology. One of the key aspects is the limitation of absorption through intact skin due to the relative impermeability of the stratum corneum. The major concern about toxicity is related to their dimensions, i.e. their size and surface to volume ratio as well as their functionalization (Hofmeier and Surber 2016). For instance, TiO_2 NPs if not properly functionalized may trigger ROS generation leading to nucleic acid or protein damage. Further studies claim that in the case of nanomaterials, risk assessment has to be done for each class and type of particles for evaluating the concerns regarding particle tissue interactions. This is mainly because the surface modifications and manufacturing processes may strongly influence the interaction of these nanomaterials with tissues in the biological environment (Couto et al. 2014).

12.9 Conclusion

The dermatological delivery of nanodrugs is a promising approach in diagnostics and therapeutics. Topical delivery of drugs has been practiced for a long time to treat skin ailments. However, the introduction of nanodrug formulations has taken dermatological delivery to another level that can potentially be as effective as oral drug delivery or intravenous routes. This level is the advanced transdermal delivery of nanodrugs, where drugs are delivered noninvasively to the dermis of the skin through various mechanisms like transcellular, intercellular, or follicular route and taken up by the body instantly. Further, modifications such as controlled mechanical skin permeation techniques were incorporated to improve the drug retention and controlled release of dermatological delivery systems. This helped in various conditions of dermatological treatment like contamination by microorganisms and chemical exposure, cosmetics like skin care and protection, and drug delivery through several transdermal treatments. Moreover, a variety of nanomaterials such as polymer-based, inorganic, naturally derived, and carbon-based nanoparticles were explored to formulate nanodrugs for different diseases and their diagnostic purposes. These nanoformulations are dermatologically active and are specific to diseases. The complexity and specificity of these nanoformulations have assisted in developing a dermatological delivery system for targeted drug delivery. However, the limiting factors such as pharmacodynamics and pharmacokinetics of the nanodrugs that relatively alter among various patients are still under research.

References

Abramovits, W., Granowski, P., and Arrazola, P. (2010). Applications of nanomedicine in dermatology: use of nanoparticles in various therapies and imaging. *Journal of Cosmetic Dermatology* 9 (2): 154–159.

Alvarez-Román, R., Barré, G., Guy, R.H., and Fessi, H. (2001). Biodegradable polymer nanocapsules containing a sunscreen agent: preparation and photoprotection. *European Journal of Pharmaceuties and Biopharmaceutics* 52 (2): 191–195.

Alvarez-Román, R., Naik, A., Kalia, Y.N. et al. (2004). Skin penetration and distribution of polymeric nanoparticles. *Journal of Controlled Release* 99 (1): 53–62.

An, Y.H., Lee, J., Son, D.U. et al. (2020). Facilitated transdermal drug delivery using nanocarriers-embedded electroconductive hydrogel coupled with reverse electrodialysis-driven iontophoresis. *ACS Nano* 14 (4): 4523–4535.

Arafa, M.G., El-Kased, R.F., and Elmazar, M.M. (2018). Thermoresponsive gels containing gold nanoparticles as smart antibacterial and wound healing agents. *Scientific Reports* 8 (1): 13674.

Auffan, M., Rose, J., Orsiere, T. et al. (2009). CeO_2 nanoparticles induce DNA damage towards human dermal fibroblasts in vitro. *Nanotoxicology* 3 (2): 161–171.

Bahadar, H., Maqbool, F., Niaz, K., and Abdollahi, M. (2016). Toxicity of nanoparticles and an overview of current experimental models. *Iranian Biomedical Journal* 20 (1): 1–11.

Barenholz, Y. (Chezy) (2012). Doxil® – the first FDA-approved nano-drug: lessons learned. *Journal of Controlled Release* 160 (2): 117–134.

Baroli, B., Ennas, M.G., Loffredo, F. et al. (2007). Penetration of metallic nanoparticles in human full-thickness skin. *Journal of Investigative Dermatology* 127 (7): 1701–1712.

Bartucci, R., Pantusa, M., Marsh, D., and Sportelli, L. (2002). Interaction of human serum albumin with membranes containing polymer-grafted lipids: spin-label ESR studies in the mushroom and brush regimes. *Biochimica et Biophysica Acta (BBA) – Biomembranes* 1564 (1): 237–242.

Bhowmick, P., Pancsa, R., Guharoy, M., and Tompa, P. (2013). Functional diversity and structural disorder in the human ubiquitination pathway. *PLoS One* 8 (5): e65443.

Bikowski, J. and Del Rosso, J.Q. (2008). Benzoyl peroxide microsphere cream as monotherapy and combination treatment of acne. *Journal of Drugs in Dermatology* 7 (6): 590–595.

Bodeker, G., Ryan, T.J., Volk, A. et al. (2017). Integrative skin care: dermatology and traditional and complementary medicine. *The Journal of Alternative and Complementary Medicine* 23 (6): 479–486.

Busatto, S., Pham, A., Suh, A. et al. (2019). Organotropic drug delivery: synthetic nanoparticles and extracellular vesicles. *Biomedical Microdevices* 21 (2): 46.

Calienni, M.N., Febres-Molina, C., Llovera, R.E. et al. (2019). Nanoformulation for potential topical delivery of Vismodegib in skin cancer treatment. *International Journal of Pharmaceutics* 565: 108–122.

Cevc, G. and Vierl, U. (2010). Nanotechnology and the transdermal route: a state of the art review and critical appraisal. *Journal of Controlled Release, vol.* 141 (3): 277–299.

Colerangle, J.B. (2017). Preclinical development of nononcogenic drugs (small and large molecules). In: *A Comprehensive Guide to Toxicology in Nonclinical Drug Development*, 2ee (ed. A.S. Faqi), 659–683. Academic Press.

Connor, E.E., Mwamuka, J., Gole, A. et al. (2005). Gold nanoparticles are taken up by human cells but do not cause acute cytotoxicity. *Small* 1 (3): 325–327.

Coondoo, A., Phiske, M., Verma, S., and Lahiri, K. (2014). Side-effects of topical steroids: a long overdue revisit. *Indian Dermatol Online Journal* 5 (4): 416.

Couto, A., Fernandes, R., Cordeiro, M.N.S. et al. (2014). Dermic diffusion and stratum corneum: a state of the art review of mathematical models. *Journal of Controlled Release* 177: 74–83.

De Jong, W.H., Carraway, J.W., and Geertsma, R.E. (2012). In vivo and in vitro testing for the biological safety evaluation of biomaterials and medical devices. In: *Biocompatibility and Performance of Medical Devices* (ed. J.-P. Boutrand), 120–158. Sawston, Cambridge: Woodhead Publishing.

De Matteis, V. (2017). Exposure to inorganic nanoparticles: routes of entry, immune response, biodistribution and in vitro/in vivo toxicity evaluation. *Toxics* 5 (4): 29.

Dickerson, E.B., Dreaden, E.C., Huang, X. et al. (2008). Gold nanorod assisted near-infrared plasmonic photothermal therapy (PPTT) of squamous cell carcinoma in mice. *Cancer Letters* 269 (1): 57–66.

DiNardo, J.C. and Downs, C.A. (2018). Dermatological and environmental toxicological impact of the sunscreen ingredient oxybenzone/benzophenone-3. *Journal of Cosmetic Dermatology* 17 (1): 15–19.

Eden, J.G., Park, S.J., Ostrom, N.P., and Chen, K.F. (2005). Recent advances in microcavity plasma devices and arrays: a versatile photonic platform. *Journal of Physics D: Applied Physics* 38 (11): 1644–1648.

Erdogan, M., Wright, J.R., and McAlister, V.C. (2002). Liposomal tacrolimus lotion as a novel topical agent for treatment of immune-mediated skin disorders: experimental studies in a murine model. *British Journal of Dermatology* 146 (6): 964–967.

Escobar-Chávez, J.J., Díaz-Torres, R., Rodríguez-Cruz, I.M. et al. (2012). Nanocarriers for transdermal drug delivery. *Research and Reports in Transdermal Drug Delivery* 1: 3–17.

Ferrari, M. (2010). Frontiers in cancer nanomedicine: directing mass transport through biological barriers. *Trends in Biotechnology* 28 (4): 181–188.

Geinoz, S., Guy, R.H., Testa, B., and Carrupt, P.A. (2004). Quantitative structure-permeation relationships (QSPeRs) to predict skin permeation: a critical evaluation. *Pharmaceutical Research* 21 (1): 83–92.

Hamad, I., Hunter, A.C., and Moghimi, S.M. (2013). Complement monitoring of Pluronic 127 gel and micelles: suppression of copolymer-mediated complement activation by elevated serum levels of HDL, LDL, and apolipoproteins AI and B-100. *Journal of Controlled Release* 170 (2): 167–174.

Hasanzadeh Kafshgari, M., Delalat, B., Harding, F.J. et al. (2016). Antibacterial properties of nitric oxide-releasing porous silicon nanoparticles. *Journal of Materials Chemistry B* 4 (11): 2051–2058.

Hashempour, S., Ghanbarzadeh, S., Maibach, H.I. et al. (2019). Skin toxicity of topically applied nanoparticles. *Therapeutic Delivery* 10 (6): 383–396.

Henderson, B.W. and Dougherty, T.J. (1992). How does photodynamic therapy work? *Photochem Photobiol* 55 (1): 145–157.

Hia, J. and Nasir, A. (2011). Photonanodermatology: the interface of photobiology, dermatology and nanotechnology: photonanodermatology. *Photodermatology, Photoimmunology and Photomedicine* 27 (1): 2–9.

Hofmeier, K.S. and Surber, C. (2016). *Nanomedicine in Dermatology: Nanotechnology in Prevention, Diagnosis, and Therapy, in Nanoscience and Nanotechnology for Human Health*, 329–356. Weinheim, Germany: Wiley-VCH Verlag GmbH and Co. KGaA.

Hudson, S.P., Padera, R.F., Langer, R., and Kohane, D.S. (2008). The biocompatibility of mesoporous silicates. *Biomaterials* 29 (30): 4045–4055.

Huk, A., Izak-Nau, E., el Yamani, N. et al. (2015). Impact of nanosilver on various DNA lesions and HPRT gene mutations – effects of charge and surface coating. *Particle Fibre Toxicology* 12 (1): 25.

Iqbal, B., Ali, J., and Baboota, S. (2018). Recent advances and development in epidermal and dermal drug deposition enhancement technology. *International Journal of Dermatology* 57 (6): 646–660.

Jain, B., Singh, B., Katare, O.P., and Vyas, S.P. (2009). Development and characterization of minoxidil-loaded liposomal system for delivery to pilosebaceous units. *Journal of Liposome Research* 20 (2): 105–114.

Jiang, T., Wang, T., Li, T. et al. (2018). Enhanced transdermal drug delivery by transfersome-embedded oligopeptide hydrogel for topical chemotherapy of melanoma. *ACS Nano* 12 (10): 9693–9701.

Kim, S., Lim, Y.T., Soltesz, E.G. et al. (2004). Near-infrared fluorescent type II quantum dots for sentinel lymph node mapping. *Nature Biotechnology* 22 (1): 93–97.

Kim, D.G., Jeong, Y.I., Choi, C. et al. (2006). Retinol-encapsulated low molecular water-soluble chitosan nanoparticles. *International Journal of Pharmaceutics* 319 (1–2): 130–138.

Korani, M., Rezayat, M., and Gilani, K. (2011). Acute and subchronic dermal toxicity of nanosilver in guinea pig. *International Journal of Nanomedicine* 6: 855–862.

Labouta, H.I., El-Khordagui, L.K., Kraus, T., and Schneider, M. (2011). Mechanism and determinants of nanoparticle penetration through human skin. *Nanoscale* **3** (12): 4989–4999.

Lewinski, N., Colvin, V., and Drezek, R. (2008). Cytotoxicity of nanoparticles. *Small* 4 (1): 26–49.

Linos, E., Katz, K.A., and Colditz, G.A. (2016). Skin cancer: the importance of prevention. *JAMA Internal Medicine* 176 (10): 1435.

Linse, S., Cabaleiro-Lago, C., Xue, W.F. et al. (2007). Nucleation of protein fibrillation by nanoparticles. *Proceedings of the National Academy of Sciences* 104 (21): 8691–8696.

Majumdar, M., Shivalkar, S., Pal, A. et al. (2020). Nanotechnology for enhanced bioactivity of bioactive compounds. In: *Biotechnological Production of Bioactive Compounds* (eds. M.L. Verma and A.K. Chandel), 433–466. Amsterdam, Netherlands: Elsevier.

Makhmalzade, B.S. and Chavoshy, F. (2018). Polymeric micelles as cutaneous drug delivery system in normal skin and dermatological disorders. *Journal of Advanced Pharmaceutical Technology and Research* 9 (1): 2–8.

Mangalathillam, S., Rejinold, N.S., Nair, A. et al. (2012). Curcumin loaded chitin nanogels for skin cancer treatment via the transdermal route. *Nanoscale* 4 (1): 239–250.

Martinez, L.R., Han, G., Chacko, M. et al. (2009). Antimicrobial and healing efficacy of sustained release nitric oxide nanoparticles against *Staphylococcus aureus* skin infection. *Journal of Investigative Dermatology* 129 (10): 2463–2469.

Matsumura, Y. and Maeda, H. (1986). A new concept for macromolecular therapeutics in cancer chemotherapy: mechanism of tumoritropic accumulation of proteins and the antitumor agent smancs. *Cancer Research* 46 (12 Pt 1): 6387–6392.

Mohajerani, B., Smith, K., Milas, A. et al. (2019). Nanoparticles in construction materials and other applications, and implications of nanoparticle use. *Materials* 12 (19): 3052.

Mordorski, B., Landriscina, A., and Friedman, A. (2016). An overview of nanomaterials in dermatology. In: *Nanoscience in Dermatology* (eds. M.R. Hamblin, P. Avci and T.W. Prow), 31–46. Boston: Academic Press.

Mostafalou, S., Mohammadi, H., Ramazani, A., and Abdollahi, M. (2013). Different biokinetics of nanomedicines linking to their toxicity: an overview. *Daru Journal of Pharmaceutical Sciences* 21 (1): 14, pp. 2008-2231-21-14.

Mousavi, S.Z., Nafisi, S., and Maibach, H.I. (2017). Fullerene nanoparticle in dermatological and cosmetic applications. *Nanomedicine: Nanotechnology, Biology and Medicine* 13 (3): 1071–1087.

Nasir, A. (2010). Nanodermatology: a bright glimpse just beyond the horizon – part I. *Skin Therapy Letter* 15 (8): 1–4.

Nhung, D., Freydiere, A., Constant, H. et al. (2007). Sustained antibacterial effect of a hand rub gel incorporating chlorhexdine-loaded nanocapsules (Nanochlorex®). *International Journal of Pharmaceutics* 334 (1–2): 166–172.

Nowrouzi, A., Meghrazi, K., Golmohammadi, T. et al. (2010). Cytotoxicity of subtoxic AgNP in human hepatoma cell line (HepG2) after long-term exposure. *Iranian Biomedical Journal* 14 (1–2): 23–32.

Nuzhatun Nisa, T.A. (2015). Therapeutic and diagnostic applications of nanotechnology in dermatology and cosmetics. *Journal Nanomedine Biotherapeutic Discovery* 5 (3): 1–24.

Ozbaş-Turan, S., Akbuğa, J., and Sezer, A.D. (2010). Topical application of antisense oligonucleotide-loaded chitosan nanoparticles to rats. *Oligonucleotides* 20 (3): 147–153.

Pan, Y., Neuss, S., Leifert, A. et al. (2007). Size-dependent cytotoxicity of gold nanoparticles. *Small* 3 (11): 1941–1949.

Pan, Z., Lee, W., Slutsky, L. et al. (2009). Adverse effects of titanium dioxide nanoparticles on human dermal fibroblasts and how to protect cells. *Small* 5 (4): 511–520.

Papakostas, D., Rancan, F., Sterry, W. et al. (2011). Nanoparticles in dermatology. *Archives of Dermatological Research* 303 (8): 533–550.

Patel, V.B., Misra, A., and Marfatia, Y.S. (2000). Topical liposomal gel of tretinoin for the treatment of acne: research and clinical implications. *Pharmaceutical Development and Technology* 5 (4): 455–464.

Pitsillides, C.M., Joe, E.K., Wei, X. et al. (2003). Selective cell targeting with light-absorbing microparticles and nanoparticles. *Biophysical Journal* 84 (6): 4023–4032.

Plytycz, B. and Seljelid, R. (2003). From inflammation to sickness: historical perspective. *Archiwum Immunologiae et Therapiae Experimentalis*, (Warsz.) 51 (2): 105–109.

Pourmand, A. and Abdollahi, M. (2012). Current opinion on nanotoxicology. *DARU, Journal of Pharmaceutical Sciences* 20 (1): 95, pp. 2008-2231-20-95.

Rao, Y., Chen, W., Liang, X. et al. (2015). Epirubicin-loaded superparamagnetic iron-oxide nanoparticles for transdermal delivery: cancer therapy by circumventing the skin barrier. *Small* 11 (2): 239–247.

Rolland, A., Wagner, N., Chatelus, A. et al. (1993). Site-specific drug delivery to pilosebaceous structures using polymeric microspheres. *Pharmaceutical Research* 10 (12): 1738–1744.

Rong, F., Tang, Y., Wang, T. et al. (2019). Nitric oxide-releasing polymeric materials for antimicrobial applications: a review. *Antioxidants* 8 (11): 556.

de Sá, D.C., Festa Neto, C., de Sá, D.C., and Festa Neto, C. (2016). Inflammasomes and dermatology. *Anais Brasileiros de Dermatologia* 91 (5): 566–578.

Sabitha, M., Sanoj Rejinold, N., Nair, A. et al. (2013). Development and evaluation of 5-fluorouracil loaded chitin nanogels for treatment of skin cancer. *Carbohydrate Polymers* 91 (1): 48–57.

Samberg, M.E., Oldenburg, S.J., and Monteiro-Riviere, N.A. (2010). Evaluation of silver nanoparticle toxicity in skin in vivo and keratinocytes in vitro. *Environmental Health Perspectives* 118 (3): 407–413.

Schek, R.M., Hollister, S.J., and Krebsbach, P.H. (2004). Delivery and protection of adenoviruses using biocompatible hydrogels for localized gene therapy. *Molecular Therapy* 9 (1): 130–138.

Scheuplein, R.J. (1967). Mechanism of percutaneous absorption. *Journal of Investigative Dermatology* 48 (1): 79–88.

Seabra, A. and Durán, N. (2015). Nanotoxicology of metal oxide nanoparticles. *Metals* 5 (2): 934–975.

Şenyiğit, T., Sonvico, F., Barbieri, S. et al. (2010). Lecithin/chitosan nanoparticles of clobetasol-17-propionate capable of accumulation in pig skin. *Journal of Controlled Release* 142 (3): 368–373.

Shabbir, M., Nagra, U., Zaman, M. et al. (2020). Lipid vesicles and nanoparticles for non-invasive topical and transdermal drug delivery. *Current Pharmaceutical Design* 26 (18): 2149–2166.

Shah, P.P., Desai, P.R., Patel, A.R., and Singh, M.S. (2012). Skin permeating nanogel for the cutaneous co-delivery of two anti-inflammatory drugs. *Biomaterials* 33 (5): 1607–1617.

Shim, J., Seokkang, H., Park, W. et al. (2004). Transdermal delivery of mixnoxidil with block copolymer nanoparticles. *Journal of Controlled Release* 97 (3): 477–484.

Shiohara, A., Hoshino, A., Hanaki, K. et al. (2004). On the cyto-toxicity caused by quantum dots. *Microbiology and Immunology* 48 (9): 669–675.

Silva, S.A.M., Michniak-Kohn, B., and Leonardi, G.R. (2017). An overview about oxidation in clinical practice of skin aging. *Anais Brasileiros de Dermatologia* 92 (3): 367–374.

Singh, A., Gautam, P.K., Verma, A. et al. (2020). Green synthesis of metallic nanoparticles as effective alternatives to treat antibiotics resistant bacterial infections: a review. *Biotechnology Reports* 25: e00427.

Smijs, T.G. and Pavel, S. (2011). Titanium dioxide and zinc oxide nanoparticles in sunscreens: focus on their safety and effectiveness. *Nanotechnology, Science and Applications* 4: 95–112.

de Souza, M.L., Oliveira, D.D., de Pereira, N.P., and Soares, D.M. (2018). Nanoemulsions and dermatological diseases: contributions and therapeutic advances. *International Journal of Dermatology* 57 (8): 894–900.

Stern, S.T., Adiseshaiah, P.P., and Crist, R.M. (2012). Autophagy and lysosomal dysfunction as emerging mechanisms of nanomaterial toxicity. *Particle and Fibre Toxicology* 9 (1): 20.

Tan, Q., Liu, W., Guo, C., and Zhai, G. (2011). Preparation and evaluation of quercetin-loaded lecithin-chitosan nanoparticles for topical delivery. *International Journal of Nanomedicine* 6: 1621–1630.

Tenzer, S., Docter, D., Kuharev, J. et al. (2013). Rapid formation of plasma protein corona critically affects nanoparticle pathophysiology. *Nature Nanotech* 8 (10): 772–781.

Tey, H.L., Tian, E.L., and Tan, A.W. (2008). Drug interactions in dermatological practice. *Clinical and Experimental Dermatology* 33 (5): 541–550.

Tomoda, K., Terashima, H., Suzuki, K. et al. (2011). Enhanced transdermal delivery of indomethacin-loaded PLGA nanoparticles by iontophoresis. *Colloids and Surfaces B: Biointerfaces* 88 (2): 706–710.

Tsujimoto, H., Hara, K., Tsukada, Y. et al. (2007). Evaluation of the permeability of hair growing ingredient encapsulated PLGA nanospheres to hair follicles and their hair growing effects. *Bioorganic and Medicinal Chemistry Letters* 17 (17): 4771–4777.

Turos, E., Shim, J.Y., Wang, Y. et al. (2007). Antibiotic-conjugated polyacrylate nanoparticles: new opportunities for development of anti-MRSA agents. *Bioorganic and Medicinal Chemistry Letters* 17 (1): 53–56.

Vandervoort, J. and Ludwig, A. (2004). Preparation and evaluation of drug-loaded gelatin nanoparticles for topical ophthalmic use. *European Journal of Pharmaceutics and Biopharmaceutics* 57 (2): 251–261.

Verma, D.D., Verma, S., McElwee, K.J. et al. (2004). Treatment of alopecia areata in the DEBR model using Cyclosporin A lipid vesicles. *European Journal of Dermatology* 14 (5): 332–338.

Verma, A., Shivalkar, S., Sk, M.P. et al. (2020a). Nanocomposite of Ag nanoparticles and catalytic fluorescent carbon dots for synergistic bactericidal activity through enhanced reactive oxygen species generation. *Nanotechnology* 31 (40): 405704.

Verma, A., Singh, V., and Sahoo, A.K. (2020b). Metal nanoparticle based antibacterial nanocomposites for skin infections. In: *Nanotechnology in Skin, Soft Tissue, and Bone Infections* (ed. M. Rai), 25–48. Cham: Springer International.

Vogt, A., Mandt, N., Lademann, J. et al. (2005). Follicular targeting–a promising tool in selective dermatotherapy. *Journal of Investigative Dermatology Symposium Proceedings* 10 (3): 252–255.

Von Montfort, C., Alili, L., Teuber-Hanselmann, S., and Brenneisen, P. (2015). Redox-active cerium oxide nanoparticles protect human dermal fibroblasts from PQ-induced damage. *Redox Biology* 4: 1–5.

Wolfram, J., Suri, K., Huang, Y. et al. (2014a). Evaluation of anticancer activity of celastrol liposomes in prostate cancer cells. *Journal of Microencapsulation* 31 (5): 501–507.

Wolfram, J., Yang, Y., Shen, J. et al. (2014b). The nano-plasma interface: implications of the protein corona. *Colloids and Surfaces B: Biointerfaces* 124: 17–24.

Wu, M. and Huang, S. (2017). Magnetic nanoparticles in cancer diagnosis, drug delivery and treatment. *Molecular and Clinical Oncology* **7** (5): 738–746.

Wu, J., Liu, W., Xue, C. et al. (2009a). Toxicity and penetration of TiO_2 nanoparticles in hairless mice and porcine skin after subchronic dermal exposure. *Toxicology Letters* 191 (1): 1–8.

Wu, X., Price, G.J., and Guy, R.H. (2009b). Disposition of nanoparticles and an associated lipophilic permeant following topical application to the skin. *Molecular Pharmaceutics* 6 (5): 1441–1448.

Zangabad, P.S., Karimi, M., Mehdizadeh, F. et al. (2017). Nanocaged platforms: modification, drug delivery and nanotoxicity. Opening synthetic cages to release the tiger. *Nanoscale* 9 (4): 1356–1392.

Zhang, Z., Tsai, P.C., Ramezanli, T., and Michniak-Kohn, B.B. (2013). Polymeric nanoparticles-based topical delivery systems for the treatment of dermatological diseases. *Wiley Interdisciplinary Reviews: Nanomedicine and Nanobiotechnology* 5 (3): 205–218.

Zhang, R., Li, Y., Hu, B. et al. (2016). Traceable nanoparticle delivery of small interfering RNA and retinoic acid with temporally release ability to control neural stem cell differentiation for Alzheimer's disease therapy. *Advance Materials* 28 (30): 6345–6352.

Zhao, Y. and Ng, K.W. (2014). Nanotoxicology in the skin: how deep is the issue? *Nano LIFE* 04 (01): 1440004.

Zhao, C., Quan, P., Liu, C. et al. (2016). Effect of isopropyl myristate on the viscoelasticity and drug release of a drug-in-adhesive transdermal patch containing blonanserin. *Acta Pharmaceutica Sinica B* 6 (6): 623–628.

Zuo, L., Wei, W., Morris, M. et al. (2007). New technology and clinical applications of nanomedicine. *Medical Clinics of North America* 91 (5): 845–862.

Zuo, G., Kang, S., Xiu, P. et al. (2013). Interactions between proteins and carbon-based nanoparticles: exploring the origin of nanotoxicity at the molecular level. *Small* 9 (9–10): 1546–1556.

13

Nanonutraceuticals

Considerations for Toxicity and Safety Assessment

Shah Esha Bhavin[1,2] *and Anuradha Ketan Gajjar*[3]

[1] *Babaria Institute of Pharmacy, Vadodara, GJ, India*
[2] *Ramanbhai Patel College of Pharmacy, Charotar University of Science and Technology, Changa, GJ, India*
[3] *L. M. College of Pharmacy, Ahmedabad, GJ, India*

13.1 Introduction

Nature, an abundant resource of drugs for the wellbeing and disease prevention, is too huge to be fully explored and is nothing less than an abundant reserve. The International Food Information Council (IFIC) has identified "functional foods" as foods that afford health benefits beyond basic nutrition. Newer diseases are emerging every day and majority of such conditions are stress induced, which makes the search for drugs – natural or synthetic even more intriguing. Be it antioxidants, vitamin-enriched products, enzymes and proteins, sports drinks, probiotics; nature has time and again, proven itself as a potential reserve! (Huang et al. 2010). A paradigm shift has occurred in healthcare from treatment of diseases to their prevention largely due to greater awareness and better affordability (Braithwaite et al. 2014). This shift has created a momentum for nutraceuticals with opportunities for pharma and food industries. Hence, nutraceuticals are definitely the future of intelligent foods! Nutraceuticals – a conjoint word made up of "Nutrition" and "Pharmaceuticals" defines a new category which plays a pivotal role as functional food between drugs and food to promote health and prevent diseases (Santini et al. 2017). They are defined as foodstuffs that provide health benefits in addition to their basic nutritional value such as fortified foods or dietary supplements (Chaudhari 2017).

The nutraceuticals, although having large health benefits, have issues pertaining to delivery due to low solubility profiles. The oral route, being the most favored route of administration, has been maximally explored for the delivery of a majority of nutraceuticals. However, the challenges in their oral delivery involve the poor oral absorption leading to low bioavailability and/or their site-specific delivery. There are abundant references in literature describing the attempts by researchers for solubility enhancement of nutraceuticals, each posing unique challenges (Ghosh and Smarta 2016). Nanosizing techniques are the latest techniques to be employed with promising results. The advantages offered include stability on shelf and *in-vivo*, enhancement in solubility and dissolution, increased intestinal permeability, intracellular and subcellular targeting, and extended circulation half-life resulting in greater bioavailability (Zaki 2014).

The word "Nano" upholds a "mega" position in today's era. Nanotechnology offers

Nanotechnology in Medicine: Toxicity and Safety, First Edition. Edited by Mahendra Rai, Mrunali Patel, and Rashmin Patel.

paramount applications related to therapeutics, diagnosis, and research in the healthcare sector. Nano pharmaceuticals, i.e. pharmaceuticals designed using nanotechnology cover a plethora of nano biomaterials and nanosized delivery systems, including nanoparticles, nanocrystals, nanoemulsions, nanoconjugates, liposomes, dendrimers, nanogels, nanocapsules, lipid nanospheres, etc. (Jaiswal et al. 2014). Such nano pharmaceuticals have shown promising benefits to increase the surface area per unit volume, enhanced aqueous solubility, target specificity, controlled release potential, biocompatibility, precise control of particle size, and ability for drug combination therapy(Tingetal.2014).Nanopharmaceuticals have permeated across all the domains of medical sciences including therapeutics and diagnostics and made notable impact in areas like cancer, CNS disorders, cardiovascular diseases, diabetes, orthopedic diseases, immunological disorders, and infection control (Acosta 2009). The increasing number of federally approved nanomedicines and/or nano pharmaceuticals and many more under clinical trials pipeline is a strong proof of the paramount applications of nano pharmaceuticals (Kalepu and Nekkanti 2015).

Unique advantages are offered by nano pharmaceuticals over conventionally available systems like improved bioaccesibility and efficacy, tailored release, faster onset of action, site-specific targeting, etc. (Kalepu and Nekkanti 2015). Broadly, there are two strategies to make any systems into nanometric scale, bottom-up technique or top-down technique, which involves controlled precipitation or size reduction, respectively. With the recent advances, nanocarriers have a tremendous potential to revolutionize drug delivery applications, thus overcoming obstacles and fostering clinical needs.

From physicochemical perspectives, these nano formulations have the ability to incorporate a wide variety of drug substances and offer advantages to meet the tailored needs like greater solubilization, superior encapsulation, altered absorption pathway, drug, and site-specific targeting (Patil 2011). Nanocarriers entrap the drug physically or by chemical conjugation and thereby protect the drug from biotransformation, improve intracellular permeation and accumulation with betterment of the pharmacokinetic profile of the drug. Market surveys have projected nanocarriers to take up 60% of the drug delivery market by 2020. Targeted therapy is the most promising therapy where the active agent is preferentially directed to its site of action circumventing the other organs, tissues or cells, for safe and effective pharmacological response (Punia et al. 2019). Development of such products is challenging to scientists as majority of new molecular entities (drugs/nutraceuticals) exhibit solubility and bioavailability issues (Chen and Hu 2020). In this regard, nanostructured systems (lipid based, polymer based, and lipid–polymer hybrid based) are available and offer potential increase in bioavailability (Bhatia 2016).

13.2 Novel Carrier-Based Drug Delivery Systems

The main highlighting feature of nanotechnology-aided drug delivery is that it offers drugs in the nanometer size range, offering advantages resulting from nano sizing like increased surface area leading to enhanced solubility and oral bioavailability, along with targeted drug release. Nano pharmaceuticals have dramatically altered drug delivery and the total medical and pharmaceutical service system (Bhatia 2016).

Nanocarriers comprise of an outer shell made of polymeric material and an inner pocket for drug carriage. The outer core determines the stability, particle circulation time, the interaction of nanoparticle with its surrounding environment and cell surfaces. Nanocarriers, generally sized between 1 and 500 nm and stacked with therapeutically active molecules, provide an array of opportunities for developing drug-delivery strategies. Various novel approaches such as solid lipid nanoparticles, nano micelles, hydrophilic nanogels, and nanosponges are explored recently in applications like taste masking, delivery of insoluble actives, prolonging the continuance, reducing side effects, and, thereby enhancing bioaccesibility and therapeutic efficacy of the drugs. Various nanostructured drug delivery systems are depicted in Figure 13.1.

13.2.1 Lipidic Nanosystems

The lipidic nanosystems can be categorized as vesicular and non-vesicular systems. The vesicular systems are highly ordered assemblies of mono- or multiple concentric bilayers like liposomes, niosomes, and nerbosomes. The non-vesicular systems lack the tiny cavity as in vesicular systems and include emulsifying and particulate systems.

13.2.1.1 Non-Vesicular Systems

Self-emulsifying drug delivery system (SEDDS) consists of isotropic mixtures of one or more hydrophilic solvents and cosolvents/surfactants forming fine oil-in-water (o/w) microemulsions. When SEDDS formulation is released into the lumen of the gastrointestinal tract after oral administration, they encounter the gastric fluid, forming a fine (micro or nano) emulsion. This *in situ* emulsification or self-emulsification further leads to solubilization of the drug and subsequent absorption by lymphatic pathways, bypassing the hepatic first-pass effect. These systems have the advantage of enhancing the rate and uniformity of drug absorption and hence, aid in reducing the total dose of drug.

Solid lipid nanoparticles and nanospheres represent the particulate systems (Mehnert and Mader 2012). These differ in the core matrix composition; solid lipid nanoparticles possess solid-lipid core matrix, while nanospheres contain polymeric matrix (Garud et al. 2012). The particulate systems are spherical in shape with diameters ranging from 10 to 1000 nm (nanoparticles) and between 10 and 200 nm (nanospheres) (Pachpute et al. 2017). Solid lipid nanoparticles have attracted wide attention of researchers. The nanospheres dissolve, entrap, encapsulate, or attach the drug to the polymeric matrix resulting in uniform dispersion of the drug. Nanospheres can be amorphous or crystalline in nature and protect the drug from enzymatic and chemical degradation (Singh et al. 2010). This helps in controlled drug release and enhancing local drug delivery.

13.2.1.2 Vesicular Systems

Vesicular drug delivery systems are highly organized assemblies containing the drug within the core covered by concentric bilayers. These are widely investigated for a variety of drugs by virtue of their ability to localize the drug at the site of action. These are further classified based on either the bilayer material or the drug loaded. A few such systems are discussed here.

Liposomes are lipid-based vesicles for the targeted delivery of drugs like doxorubicin, verapamil, and insulin. Hydration of dry phospholipids yields the vesicles that are either unilamellar or multilamellar depending on the conditions and the use of other excipients (Akbarzadeh et al. 2013). Entrapment of the drug molecules is in the aqueous core or within the lipid bilayers. Niosomes or nonionic surfactant vesicles are formed from the self-assembly of nonionic amphiphiles in aqueous media (Uchegbu and Vyas 1998). The resultant delivery system has the hydrophobic portions of the molecule shielded from the aqueous solvent within the closed bilayer structures (Uchegbu and Vyas 1998). Herbosome technology incorporates phospholipids into standardized herbal extracts to enhance the bioavailability of hydrophilic bioactives such as flavonoids, glycosides, etc. (Dewan et al. 2016).

13.2.2 Polymeric Nanosystems

These systems are composed of simple or modified polymers, block copolymers, and synthetic or natural polymers. Nanoparticles,

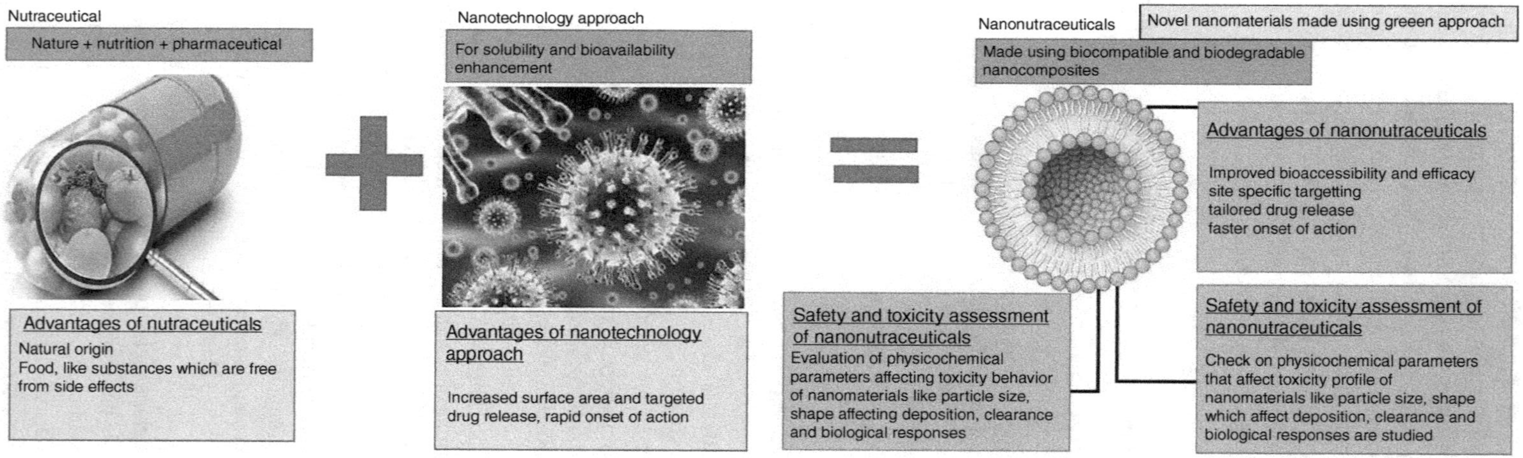

Figure 13.1 Nanostructured drug delivery systems.

nanospheres, and nanomicelles, to name a few, characterize the simple polymeric nanosystems. Nanoscopic supramolecular core-shell structures composed of amphiphilic block copolymers, arranged in spheroidal structure are polymeric micelles (Mandal et al. 2013). Nanomicelles are nanosized colloidal dispersions and solubilize hydrophobic drugs by entrapping the drugs within a mixed micellar hydrophobic core with an outer shell composed of hydrophilic chains radiating outward, resulting in a clear aqueous formulation (Vadlapudi and Mitra 2013). Polymeric nanoparticles are composed of biodegradable polymers and hence, are biocompatible and nontoxic. These features make polymeric nanoparticles suitable for applications in tissue engineering and gene and vaccine delivery in addition to the delivery of a variety of drugs. Nanocapsules are another variant of the polymeric nanosystems, which confine the drug in a cavity surrounded by polymeric membrane (Radhika et al. 2011). On the other hand, nanospheres have the drug dispersed in the polymer matrix they are composed of (Singh et al. 2010). Nanosponges represent the modified polymeric nanosystems and are discussed in complete detail in Section 13.5.

13.2.3 Lipid Polymer Nanosystems

Liposomes offer biomimicking advantages while polymeric nanoparticles offer the diversity in the use of biodegradable polymers to expand the scope of application. Hence, a combination of the two in the form of lipid polymer nanosystems is a promising delivery platform (Akbarzadeh et al. 2013). These have the core consisting of biodegradable polymer or polymer hydrids and the shell consists of phospholipids. This dual feature assembly offers advantages such as precise control of particle size and surface functionality, greater drug loading with multiple actives, desirable drug release profile, and better stability *in-vivo*. The applications of various nanocarriers in drug delivery are discussed with appropriate examples in Table 13.1.

13.3 Safety and Toxicity Assessment of Nanoparticles

The recent past has witnessed innovative and diverse use of nanoparticles. These are composed of a variety of materials, which has made it difficult to correlate the toxicity of nanoparticles and their physicochemical properties. The chemical composition of nanoparticles has a major influence on the safety as well as toxicity profile of nanoparticles (Hsiao and Huang 2011). Little evidence exists to establish the safety of these materials and their biological fate in living systems. Nanoparticles gain entry into the body relatively easily, by inhalation, ingestion or dermal penetration after administration, or due to unintended exposure. The increasing use of biodegradable and biocompatible materials for nanoparticle-based drug delivery is an area, which has garnered a lot of interest and publications (Garnett and Kallinteri 2006). It is important to develop novel nanomaterials made from biodegradable polymers using green approach after elaborate studies of their biological characterization and safety assessment on prolonged exposure. These safeguards will help to ensure risk-free and sustainable implementation of nanotechnology (Mazzarino et al. 2015). The schematic representation of safety and toxicity assessment of nanoparticles is depicted in Figure 13.2.

The physicochemical parameters that affect the toxicity profile of nanoparticles are the particle size and shape since they influence the deposition, clearance, and biological responses (Powers et al. 2006). Small particle size results in higher surface area and increased surface reactivity, which, in turn, produce pronounced toxic effects, in comparison to larger particles (Aillon et al. 2009; Luyts et al. 2013). The shape of the nanoparticle could govern cellular

Table 13.1 Applications of nanocarriers for drug delivery.

Sr. no.	Type of nanocarrier	Examples	Advantages of nanocarriers	References	
1	Liposomes	a	Actinomycin D	These carriers increase the efficacy and therapeutic index of drug.	Akbarzadeh et al. (2013)
		b	Amphotericin B and Taxol	Reduced toxicity of the encapsulated drug	
2	Niosomes	a	Methotrexate	For targeted delivery of many other anticancer drugs.	Uchegbu and Vyas (1998)
		b	Doxorubicin	For targeting and retention of drug in blood circulation	
3	Herbosomes	a	Large molecules like peptides and proteins.	Increased drug permeation via skin; hence, one of the most favorable carriers for dermal and transdermal drug delivery systems.	Dewan et al. (2016)
		b	Phytoconstituents like flavanoids	The phospholipid carrier forms a complex with phytoconstituents and stabilizes against gastric secretion, thereby increasing the life of drug.	
4	Self-emulsifying drug delivery system (SEDDS)		Cyclosporin A, ritonavir, and saquinavir	Oral delivery of hydrophobic drugs presents a challenge because of the low aqueous solubility. SEDDS improves the oral absorption of highly lipophilic drug compounds.	Neslihan Gursoy and Benita (2004)
5	Solid lipid nanoparticles (SLN)		Cyclosporine A, Camptothecin	Improved bioavailability of drugs and sustained release of lipophilic drugs.	Mehnert and Mader (2012)
6	Nanospheres		Valproic acid, Dipeptide kytorphin, Doxorubicin, Tacrine	Easily pass through the smallest capillary vessels to organs like liver, spleen, lungs, spinal cord, and provide targeted release. Also suitable for brain targeting.	Singh et al. (2010)
7	Nanomicelles		Voclosporin	Topical administration of water-insoluble drugs, noninvasive drug delivery to back-of-the-eye disorders such as age-related macular degeneration, diabetic retinopathy, and achieve high concentration of drug at targeted site, eliminating frequent dosing.	Vadlapudi and Mitra (2013)
8	Polymeric nanoparticles		Anticancer drugs, DNA delivery, Vaccine adjuvants.	Site-specific targeting achieved by attaching targeting ligands to the surface of the particles or through magnetic guidance for oral, nasal, parenteral, intraocular administration.	Jawahar and Meyyanathan (2012)
9	Nanocapsules		Diclofenac sodium, Insulin delivery, Camptothecin, and Artimisinin	Used widely as drug carriers for active targeting of drugs.	Radhika et al. (2011)

uptake and dissemination within the body; however, the relationship is not clearly established. The surface chemistry and surface charge of nanoparticles is a determinant of not only the toxicity but also other product attributes like shelf life and loading capability of the nutraceutical (Powers et al. 2006; Mura et al. 2011). The surface characteristics determine the extent and manner of interactions with the charged biological membranes leading to adverse effects or toxicity (Luyts et al. 2013).

A vast majority of the toxicity studies on nanomaterials involves cellular systems in order to avoid unnecessary exposure of animals to potentially toxic material (Barabadi et al. 2019; Chen et al. 2020). Cell viability assay estimates the cell death and survival rate as well as metabolic activity during toxicity assessment of nanomaterials (Castell and Gómez-Lechón 1997; Eisenbrand et al. 2002). Cytotoxicity and genotoxicity studies aim to study the effect of the nanoparticles on the intact cells as well as their influence on DNA once they enter the cells (AshaRani et al. 2009). The most frequently used *in vitro* and *in vivo* test for genotoxicity is the sensitive and reliable comet assay (Landsiedel et al. 2009; Karlsson 2010).

Hemocompatibility evaluation is essential for the materials used in nanoparticulate drug delivery systems intended for *in vivo* use (Goncharov et al. 2019). Erythrocytes act as osmometers, which can detect disturbances in osmotic and physical changes in blood and express the same by their lysis. Erythrocyte lysis results in the hemoglobin release into the blood and subsequent hemoglobinuria, an indication of blood poisoning by the nanomaterial. Another effect of the nanoparticles is heagglutination, caused by adhesion of erythrocytes. The consequences of heagglutination in humans include obstruction of blood cells, imbalance in osmolarity of blood, and unavailability of erythrocytes to carry out normal functions (Evani and Ramasubramanian 2011).

The enlarged scope of applications of nanoparticles as therapeutics and imaging tools has necessitated the study of local toxicity (as in lung toxicity for inhalation nanoparticles) as well as systemic toxicity (Bakand et al. 2012; Seiffert et al. 2015). The systemic toxicity studies could involve preliminary acute, subacute and intraperitoneal toxicity studies (Aillon et al. 2009). These studies would assess factors like daily behavioral observations, weight control, hematological, biochemical, and histopathological analysis. Daily behavioral observations like appearance, social behavior, irritability, and lack of interest in food and water may indicate signs of toxicity in study animals (Crosera et al. 2009; Sadrieh et al. 2010; Smijs and Bouwstra 2010; Hadrup and Lam 2014; Yun et al. 2015).

A study by the National Centre for the Replacement, Refinement and Reduction of Animals in Research (NC3Rs) established a body weight loss limit of 10% for rats and dogs in toxicity studies of 1-week duration and body weight loss limit of 15 and 20%, respectively, for studies with longer duration.

Physiological conditions of study animals during toxicity studies are well assessed by the hematological and biochemical parameters. Biochemical parameters like aspartate aminotransferase (AST) and alanine aminotransferase (ALT) are the biomarkers of hepatocellular injury and alterations in these are associated with hepatocyte necrosis and inflammation. Urea and creatinine values are indicators of nephrotoxicity and any alterations are indicative of alterations in the glomerular filtration process and glomerular and tubular injury (Diezi and Biollaz 1979).

Toxicity considerations are significant when selecting nanoparticles for a particular application, as toxicity of nanoparticle-based drugs is a major concern today (Elsaesser and Howard 2012). Lipid-based nanoparticles or liposomes find major applications in drug delivery but they have inherent toxicity-related problems. Toxicity can occur due to the liposome composition, particle size, or charge. Optimization of the liposome formulation in terms of the lipids used and charge of liposomes to reduce the adverse effects is a

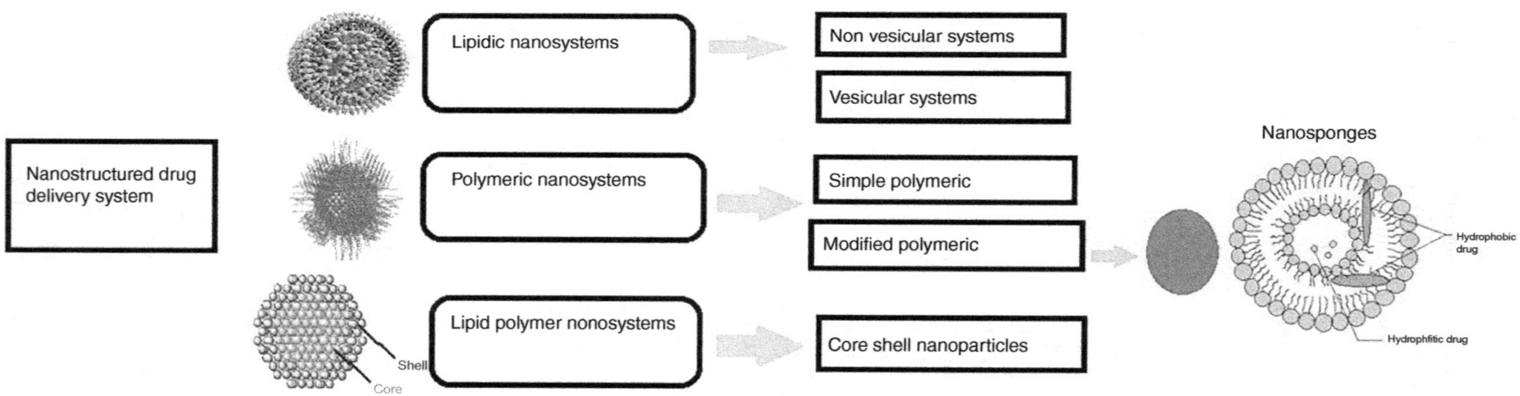

Figure 13.2 Nanonutraceuticals – considerations for toxicity and safety assessment.

crucial step in formulation development. The constituents of cationic liposomes control the toxicity and cause membrane solubilization, poration, and lysis due to their surfactant activity. The positive charge of cationic liposomes is responsible for hepatic impairment and macrophage-mediated toxicity on long-term exposure.

The glucose polymer, dextrans, finds widespread therapeutic applications in the form of nanoparticles. However, it brings along serious adverse effects like anaphylaxis, volume overload, platelet dysfunction, and pulmonary and cerebral edema. Dextran-based nanoparticles are contraindicated for patients with diabetes mellitus or renal insufficiency due to acute renal failure.

Dendrimers are polymeric, highly branched, spherical nanoscopic structures having tree-like, highly organized structure with a core and inner and outer shells. Cationic dendrimers as those synthesized using PAMAM (polyamidoamine) can interact with the anionic groups of biologic membranes and destabilize them causing increase in permeability and cell lysis. Various studies have investigated the cytotoxicity of PAMAM dendrimers using Caco-2 cells in comparison to other anionic dendrimers. These studies have shown the cationic dendrimers are much more cytotoxic than anionic or PEGylated dendrimers (Mazzarino et al. 2015).

Hence, it is crucially important that novel nanomaterials are made using biodegradable polymers and green approaches. These novel nanomaterials, once biologically characterized for their health hazards, will ensure risk-free and sustainable implementation of nanotechnology in disease prevention and mitigation.

13.4 Approaches for Biodegradable Nanoparticles

Numerous reports are available in literature describing the use of a variety of nonbiodegradable polymers and lipids for the delivery of drugs and genes (Patel et al. 2011). However, these pose inherent unintended cytotoxicity and/or genotoxicity challenges. Biodegradable polymers, as the name indicates, can be degraded or disintegrated by the normal physiologic processes. On degradation, these polymers produce water, inorganic salts, biomass, and gases such as carbon dioxide, nitrogen, etc. The physicochemical properties and the route of degradation evidently are dictated by the chemical composition of the polymer. Esters, amides, and ethers are the functional groups constituting the biodegradable polymers, facilitating their degradation by enzymatic or hydrolytic cleavage. The biodegradable polymers are either of natural, synthetic, or semi-synthetic origin.

The biopolymers (natural biodegradable polymers) are chemically either polysaccharides (e.g. starch, cellulose, chitin) or proteins such as collagen, laminin, and fibronectin (Brar and Kaur 2014). Numerous references are available in literature elaborating the diverse medical and pharmaceutical applications exploiting the biocompatibility and biodegradability of biopolymers (Hosseinkhani et al. 2013). Exploration into blends of natural polymers with synthetic polymers as advanced materials with enhanced mechanical properties and biocompatibility find applications in cell-based transplantation, tissue engineering, and gene therapy (Sionkowska 2011). Hyaluronic acid and gelatin, after chemical modification, are studied to repair the osteochondral defects in rabbit models for delivery of mesenchymal stem cells with satisfactory improvement subsequent to 12 weeks of observational study (Liu et al. 2006).

Poly(hydroxylbutyrate), poly(orthoester)s, poly(amidoester)s, and poly-lactide-co-glycolide (PLGA) are a few synthetic biodegradable polymers extensively studied. United State Food and Drug Administration (FDA) has approved PLGA for various pharmaceutical applications (Mooguee et al. 2010; Matthaiou et al. 2014; Kapoor et al. 2015). It offers enormous potential as drug carrier due to the feasibility for tunable release characteristics (sustained, controlled, or pulsatile-release) in addition to biodegradability and biocompatibility. The degradation rate of PLGA can be manipulated by varying the ratio of lactic acid to glycolic acid; the greater the content of hydrophilic glycolic acid, the faster is the degradation. The features rendering PLGA suitable for nanosystems include small size, high structural integrity, stability, ease of fabrication, and tunable release capability (Mooguee et al. 2010; Matthaiou et al. 2014). Nonetheless, the features that limit the use of PLGA include poor loading efficiency for hydrophobic drugs, burst release, uptake by reticuloendothelial system, shortened *in-vivo* circulation time, and scale-up problems (Danhier et al. 2012). PLGA coated with beta-tricalcium phosphate (β-TCP) scaffold, loaded with vascular endothelial growth factor (VEGF) has been investigated for release of VEGF for bone regeneration *in-vitro*. Polylactic acid (PLA) polymer is A biocompatible and biodegradable material that undergoes scission in the body to monomeric units of lactic acid as a natural intermediate in carbohydrate metabolism. Solvent evaporation, solvent displacement, salting out, and solvent diffusion techniques are widely used for the preparation of PLA nanoparticles (Hans and Lowman 2002; Panyam and Labhasetwar 2003; Danhier et al. 2012; Shargh et al. 2016).

Chemical and morphological alterations of the natural biopolymers give semisynthetic degradable polymers with better physicochemical profiles to suit specific applications. Chitosan is a modified natural carbohydrate polymer prepared by the partial N-deacetylation of crustacean-derived natural biopolymer chitin (Hanan and Ahmad 2020). Other variants of chitosan such as carboxylate, thiolate, and acetyl derivatives also find applications in drug and gene delivery. Chitosan is a positively charged linear polymer and holds potential for loading of negatively charged drugs or nucleic acids into nanoparticles prepared from it (Kumari et al. 2010). Polyelectrolyte complex of chitosan and gelatin hydrogels at pH 6.5 are investigated for 3D bioprinting at room temperature. Chitosan modified as diacetate and triacetate forms is used as matrix to make nanoparticles for sustained delivery of doxorubicin with obvious advantages. Nanoparticles obtained from chitosan offer advantages like loading of large genes for delivery in addition to their non-immunogenicity.

Biodegradable polymers are challenging to prepare since it is not sufficient to just have the polymers disintegrating into simple absorbable end products. There are hosts of other features that need careful consideration. These include ability to elicit little or no inflammatory response; degradation time to correspond their functional time *in-vivo*; appropriate mechanical and tensile strength for the intended use; production of degradation products that can be easily absorbed, metabolized, or excreted; appropriate permeability; and scale-up processing. Nature of the polymer and linkers, molecular weight, hydrophilicity/hydrophobicity, charge on the surface, adsorption to water, erosion mechanism, etc., mainly control the behavior of the biodegradable polymer (Kumari et al. 2010).

It is unlikely to find a single polymer exhibiting all the ideal characteristics. The currently available biodegradable polymers offer advantages in comparison to the nonbiodegradable nanosystems but suffer from a few drawbacks that need proper resolution. Enormous efforts are required for the design and synthesis of biodegradable polymers to meet the specific application demands. The challenges in the synthesis of biodegradable polymers include developing unique chemistries to enhance the polymer diversity, developing biosynthetic processes for biomimetic polymers, and implementing computational and combinatorial approaches to develop libraries of polymers that can be used as biodegradable polymers. The future ahead is suggestive of applying the natural polymers from the human body and engineering the polymeric scaffolds for nanosystem-based targeted delivery (Fathi and Barar 2017).

13.5 Modified Nanocarriers (Nanosponges)

Nanosponges are a type of nanoparticles that hold the drug within the core. These nanoparticles may be conjugating, encapsulating, or complexing in nature, based on their association with the active molecules (Trotta et al. 2014; Patil et al. 2017). They have gained limelight in drug delivery-related applications due to their tiny sponge size, which holds a capacity to accommodate a wide range of drugs within its core (Shringirishi et al. 2014). The success of these nanosponges as delivery vehicles for drugs has generated a lot of interest for the delivery of hydrophobic molecules like the nutraceuticals (Lembo et al. 2013).

13.5.1 Cyclodextrin-Based Nanosponges

Biological entities of cyclic oligosaccharides, cyclodextrins, contain α-1,4 linked glucopyranose monomers (Gharakhloo et al. 2020). Their molecular structure depicts a truncated cone or inverted bucket, consisting of a hydrophilic external surface and a nonpolar interior core (Allahyari et al. 2019). α, β, and γ are the main functional moieties of cyclodextrins consisting of six, seven, and eight glucopyranose blocks, respectively (Chilajwar et al. 2014; Trotta et al. 2014). These cyclodextrin monomers are ideal carriers for the synthesis of nanosponges (Sherje et al. 2017). Major applications of cyclodextrin-based nanosponges in drug and/or bioactive delivery include warding off drug–drug interactions (drug–bioactive interactions, food–bioactive interactions), offering programmable release, converting lipophilic drugs into microcrystalline powders, diminishing ocular and gastrointestinal irritation, and masking of unpleasant taste and smell (Emeje et al. 2012). This genre of drug delivery offers a propitious platform in the use of nutraceuticals for targeted delivery. Nanosponges adhere to the target site and stick to it to offer drug/nutraceutical release in a predictable manner as they are synthesized by treating cyclodextrins, in a specific ratio, with suitable crosslinking agents (Cavalli et al. 2010). Predictable release profile of the poorly water-soluble drugs and nutraceuticals and higher amounts of loading capacity are the major advantages of nanosponges. Nutraceutical-loaded nanosponges can be administered through a variety of routes like oral, transdermal, topical, parenteral, etc. The net effect is formation of spherical particles with hydrophobic cavities and hydrophilic channels where drug and nutraceutical molecules can be entrapped. A single nanosponge system consists of innumerable interconnecting voids within a non-collapsible structure capable of holding a wide variety of substances. Cyclodextrin-based nanosponges are spherical in shape, having a negative surface charge and sizes in the range of 200–500 nm (Tejashri et al. 2013). They are effective vehicles for targeted delivery, by overcoming the issues related to poor bioavailability and toxicity, due to their ability to amalgamate water-soluble and water-insoluble bioactives into their cavities.

An active carbonyl compound also termed as cross-linker, e.g. carbonyldimiidazole, triphosgene, diphenyl carbonate, or organic dianhydride, provides promising results as carriers as per literature survey. Cross-linkers link the cyclodextrins together to form the innumerable interconnections and voids resulting in the nanosponges. The commonly used cross-linkers for the synthesis of nanosponges include diphenyl carbonate, diaryl carbonates, diisocyanates, carbonyldiimidazoles, glutaraldehyde, carboxylic acid dianhydrides, 2,2-bis(acrylamido) acetic acid, etc. Hypercross-linked polystyrenes, cyclodextrins, and its derivatives like methyl β-cyclodextrin (β-CD), alkyloxy carbonyl cyclodextrins, 2-hydroxy propyl β-CDs, and copolymers like poly (Valero lactone-allyl Valero lactone oxypanedione) and ethylcellulose and polyvinyl alcohol are the commonly used polymers in the synthesis of nanosponges (Pushpalatha et al. 2018).

Table 13.2 Applications of nanosponges in targeted delivery.

Sr. No.	Advantage/Property	Application of the advantage	References
1	Taste masking	Gabapentin, an antiepileptic, has a bitter taste, short biological half-life (5–7 h), and low bioavailability. Nanosponge-based pediatric, controlled release dry suspension overcame these limitations.	Rao and Bhingole (2015)
2	Solubility enhancement	Itraconazole, a BCS Class II drug, when loaded into nanosponges showed solubility enhancement.	Swaminathan et al. (2007)
3	Protection from GI degradation	Enzymatic breakdown of sensitive drugs in gastrointestinal tract is overcome by nanosponge approach as the drug is sealed inside the pocket of the polymer.	Chilajwar et al. (2014)
4	Formation of nontoxic, non-mutagenic complexes	Cyclodextrins yield inclusion complexes with guest molecules, while nanosponges offer complexation potential with both hydrophobic and hydrophilic compounds and hence are suitable for the delivery of proteins, enzymes, and peptides.	Allahyari et al. (2019)
5	Sustained drug delivery	Antiviral agent, Acyclovir exhibits slow and incomplete absorption in the gastrointestinal tract giving suboptimal concentrations at the target site. Nanosponges were effective in sustained drug delivery.	Lembo et al. (2013)
6	Extended release up to 12–24 h	Modifying the polymer to cross-linker ratio, helps control the release characteristics for loaded drug. This was applied for the natural antioxidant, Quercetin.	Anandam and Selvamuthukumar (2014)
7	Extended stability over high temperatures and pH range	Literature reports nanosponges incorporating proteolytic enzymes, to treat cancer or type I mucopolysaccharidosis, while DNA- and oligonucleotide-loaded, nanosponges are used in gene therapy.	Ahmed et al. (2013)
8	As a carrier for delivery of gases	Nanosponge approach in the delivery of oxygen via topical route offers the ability to entrap and liberate oxygen periodically over a specified period.	Trotta et al. (2011) and Cavalli et al. (2010)
10	Immobilization of enzymes	Cyclodextin-based nanosponges adsorb certain enzymes like *Pseudomonas fluorescens* lipase.	Boscolo et al. (2010)
11	Delivery of Nutraceuticals/ Bioactives	Cyclodextrin-based nanosponges increase the solubility of nutraceutical Curcumin and antitumor agent, Camptothecin, with controlled release for the treatment of tumors.	Swaminathan et al. (2010) and Darandale and Vavia (2013)

The distinguishing feature of nanosponges is that the dimensions of their cavity can be customised within a range of dimensions from 1 µm or less. The process conditions are very important in the preparation of nanosponges. Appropriate polymer to cross-linker ratio can result in favorable-sized nanosponges (Trotta et al. 2014).

The crystal structure of nanosponges determines their loading efficiency. Paracrystalline forms show variable loading capacities while crystalline forms exhibit a higher degree of loading. The loading capacity of nanosponges is greatly enhanced due to the availability of multiple cavities in the 3D network-like structure. Nanosponges are shown to withstand high temperatures over a wide range of pH (1–11). In water, nanosponges form clear opalescent suspensions and show negligible solubility in organic solvents. They are nontoxic, and pervious in nature (Trotta et al. 2014).

Table 13.2 clearly illustrates the various applications of nanosponges in targeted drug delivery.

13.6 Conclusion and Future Perspectives

The past few decades are witness to increased awareness toward prevention of diseases rather than mitigation of diseases. This has resulted in enormous exploration of nature for hence unheard of nutraceuticals and health foods/super foods. The nutraceuticals have paved the way for a paradigm shift. The challenges posed by nutraceuticals propelled the quest for better delivery systems like the nanosystems explored for the delivery of drugs and more recently proteins and genes. Moreover, the increasing awareness about the sustainable development has caused the researchers over the world to think about natural and biodegradable polymers over synthetic polymers for various applications. Hence, adoption of Nanotechnology Approach in Drug Delivery keeping in mind the safety and toxicity aspects has emerged as the key highlight of research in today's era. Cyclodextrin-based nanosponges are relatively new members of the family of nanosystems offering superiority in terms of greater bioaccessibility, biodegradability, and tailored release profile. They offer complexation capacity with a plethora of hydrophilic and lipophilic drugs and nutraceuticals and have the potential to be formulated into a wide range of dosage forms such as topical, tablets and capsules, parenteral and aerosol. The advances in molecular biology and polymer engineering continue to provide further impetus to find biopolymers having obvious advantages. Nevertheless, either nanosystems, made of nonbiodegradable or biodegradable polymers pose special issues due to their size. Hence, the nanosystems used for the delivery of both the drugs and the nutraceuticals have to be biologically characterized for their long-term health hazards.

References

Acosta, E. (2009). Bioavailability of nanoparticles in nutrient and nutraceutical delivery. *Current Opinion in Colloid and Interface Science* 14 (1): 3–15. https://doi.org/10.1016/j.cocis.2008.01.002.

Ahmed, R.Z., Patil, G., and Zaheer, Z. (2013). Nanosponges-a completely new nano-horizon: pharmaceutical applications and recent advances. *Drug Development and Industrial Pharmacy* 39 (9): 1263–1272. https://doi.org/10.3109/03639045.2012.694610.

Aillon, K.L., Xie, Y., El-Gendy, N. et al. (2009). Effects of nanomaterial physicochemical properties on *in vivo* toxicity. *Advanced Drug Delivery Reviews* 61 (6): 457–466. https://doi.org/10.1016/j.addr.2009.03.010.

Akbarzadeh, A., Rezaei-Sadabady, R., Davaran, S. et al. (2013). Liposome: classification, preparation, and applications. *Nanoscale Research Letters* 8 (1): 102–111. https://doi.org/10.1186/1556-276X-8-102.

Allahyari, S., Trotta, F., Valizadeh, H. et al. (2019). Cyclodextrin-based nanosponges as promising carriers for active agents. *Expert Opinion on Drug Delivery* 16 (5): 467–479. https://doi.org/10.1080/17425247.2019.1591365.

Anandam, S. and Selvamuthukumar, S. (2014). Fabrication of cyclodextrin nanosponges for quercetin delivery: physicochemical characterization, photostability, and antioxidant effects. *Journal of Materials Science* 49 (23): 8140–8153. https://doi.org/10.1007/s10853-014-8523-6.

Asharani, P.V., Low, G., Mun, K. et al. (2009). Cytotoxicity and genotoxicity of silver. *ACS Nano* 3 (2): 279–290. https://doi.org/10.1021/nn800596w.

Bakand, S., Hayes, A., and Dechsakulthorn, F. (2012). Nanoparticles : a review of particle toxicology following inhalation exposure. *Inhalation Toxicology* 24 (November 2011): 125–135. https://doi.org/10.3109/08958378.2010.642021.

Barabadi, H., Najafi, M., Samadian, H. et al. (2019). A systematic review of the genotoxicity and antigenotoxicity of biologically synthesized metallic nanomaterials: are green nanoparticles safe enough for clinical marketing? *Medicina* 55 (8): 439–444. https://doi.org/10.3390/medicina55080439.

Bhatia, S. (2016). Nanoparticles types, classification, characterization, fabrication methods and drug delivery applications. In: *Natural Polymer Drug Delivery Systems*, 33–93. Cham: Springer https://doi.org/10.1007/978-3-319-41129-3_2.

Boscolo, B., Trotta, F., and Ghibaudi, E. (2010). High catalytic performances of pseudomonas fluorescens lipase adsorbed on a new type of cyclodextrin-based nanosponges. *Journal of Molecular Catalysis B: Enzymatic* 62 (2): 155–161. https://doi.org/10.1016/j.molcatb.2009.10.002.

Braithwaite, M., Tyagi, C., Tomar, L. et al. (2014). Nutraceutical-based therapeutics and formulation strategies augmenting their efficiency to complement modern medicine: an overview. *Journal of Functional Foods* 6: 82–99. https://doi.org/10.1016/j.jff.2013.09.022.

Brar, V. and Kaur, G. (2014). Biopolymers as carriers for nasal drug delivery. *Polymer-Plastics Technology and Engineering* 53 (14): 1518–1531. https://doi.org/10.1080/03602559.2014.912327.

Castell, J. and Gómez-Lechón, M. (1997). *In vitro Methods in Pharmaceutical Research*. San Diego: Academic Press.

Cavalli, R., Akhter, A.K., Bisazza, A. et al. (2010). Nanosponge formulations as oxygen delivery systems. *International Journal of Pharmaceutics* 402 (1–2): 254–257. https://doi.org/10.1016/j.ijpharm.2010.09.025.

Chaudhari, S. (2017). Nutraceuticals: a review. *World Journal of Pharmacy and Pharmaceutical Sciences*: 681–739. https://doi.org/10.20959/wjpps20178-9825.

Chen, J. and Hu, L. (2020). Nanoscale delivery system for nutraceuticals: preparation, application, characterization, safety, and future trends. *Food Engineering Reviews* 12 (1): 14–31. https://doi.org/10.1007/s12393-019-09208-w.

Chen, R., Chen, Y., Liao, M. et al. (2020). The current understanding of autophagy in nanomaterial toxicity and its implementation in safety assessment-related alternative testing strategies. *International Journal of Molecular Sciences* 21 (7): 2387. https://doi.org/10.3390/ijms21072387.

Chilajwar, S.V., Pednekar, P.P., Jadhav, K.R. et al. (2014). Cyclodextrin-based nanosponges: a propitious platform for enhancing drug delivery. *Expert Opinion on Drug Delivery* 11 (1): 111–120. https://doi.org/10.1517/17425247.2014.865013.

Crosera, M., Bovenzi, M., Maina, G. et al. (2009). Nanoparticle dermal absorption and toxicity: a review of the literature. *International Archives of Occupational and Environmental Health* 82 (9): 1043–1055. https://doi.org/10.1007/s00420-009-0458-x.

Danhier, F., Ansorena, E., Silva, J.M. et al. (2012). PLGA-based nanoparticles : an overview of biomedical applications. *Journal of Controlled Release* 161: 505–522. https://doi.org/10.1016/j.jconrel.2012.01.043.

Darandale, S.S. and Vavia, P.R. (2013). Cyclodextrin-based nanosponges of curcumin: formulation and physicochemical characterization. *Journal of Inclusion Phenomena and Macrocyclic Chemistry* 75 (3–4): 315–322. https://doi.org/10.1007/s10847-012-0186-9.

Dewan, N., Dasgupta, D., Pandit, S., and Ahmed, P. (2016). Review on-herbosomes, a new arena for drug delivery. *Journal of Pharmacognosy and Phytochemistry* 5 (4): 104–108.

Diezi, J. and Biollaz, J. (1979). Renal function tests in experimental toxicity studies. *Pharmacology and Therapeutics* 5 (1–3): 135–145.

Eisenbrand, G., Pool-zobel, B., Baker, V. et al. (2002). Methods of *in vitro*. *Toxicology* 40: 193–236.

Elsaesser, A. and Howard, C.V. (2012). Toxicology of nanoparticles. *Advanced Drug Delivery Reviews* 64 (2): 129–137. https://doi.org/10.1016/j.addr.2011.09.001.

Emeje, M.O., Obidike, I.C., Akpabio, E.I., and Ofoefule, S.I. (2012). Nanotechnology in drug delivery. *Recent Advances in Novel Drug Carrier Systems* 1 (4): 69–106. http://dx.doi.org/10.5772/51384.

Evani, S.J. and Ramasubramanian, A. (2011). Hemocompatibility of nanoparticles. In: *Nanobiomaterials Handbook*, (ed. B. Sitharaman), 1–17. Boca Raton: CRC Press https://doi.org/10.1201/b10970-32.

Fathi, M. and Barar, J. (2017). Perspective highlights on biodegradable polymeric nanosystems for targeted therapy of solid tumors. *Tabriz University of Medical Sciences* 7 (1): 49–57. https://doi.org/10.15171/bi.2017.07.

Garnett, M.C. and Kallinteri, P. (2006). Nanomedicines and nanotoxicology: some physiological principles. *Occupational Medicine* 56 (5): 307–311. https://doi.org/10.1093/occmed/kql052.

Garud, A., Singh, D., and Garud, N. (2012). Solid lipid nanoparticles (SLN): method, characterization and applications. *International Current Pharmaceutical Journal* 1 (11): 384–393. https://doi.org/10.3329/icpj.v1i11.12065.

Gharakhloo, M., Sadjadi, S., Rezaeetabar, M. et al. (2020). Cyclodextrin-based nanosponges for improving solubility and sustainable release of curcumin. *Chemistry Select* 5 (5): 1734–1738. https://doi.org/10.1002/slct.201904007.

Ghosh, D. and Smarta, R.B. (2016). *Pharmaceuticals to Nutraceuticals: A Shift in Disease Prevention*. CRC Press.

Goncharov, N., Sobolev, V., Terpilowski, M. et al. (2019). Safety and toxicity evaluation of nutraceuticals in animal models. In: *Nutraceuticals in Veterinary Medicine*, 675–684. Cham: Springer https://doi.org/10.1007/978-3-030-04624-8_48.

Hadrup, N. and Lam, H.R. (2014). Oral toxicity of silver ions, silver nanoparticles and colloidal silver–a review. *Regulatory Toxicology and Pharmacology* 68 (1): 1–7. https://doi.org/10.1016/j.yrtph.2013.11.002.

Hanan, E. and Ahmad, F.J. (2020). Nutraceutical-loaded chitosan nanoparticles for healthcare applications. In: *Nanomedicine for Bioactives*, 231–257. Singapore: Springer https://doi.org/10.1007/978-981-15-1664-1_8.

Hans, M.L. and Lowman, A.M. (2002). Biodegradable nanoparticles for drug delivery and targeting. *Current Opinion in Solid State and Materials Science* 6 (4): 319–327. https://doi.org/10.1016/S1359-0286(02)00117-1.

Hosseinkhani, H., He, W.J., Chiang, C.H. et al. (2013). Biodegradable nanoparticles for gene therapy technology. *Journal of Nanoparticle Research* 15 (7): 179–187. https://doi.org/10.1007/s11051-013-1794-z.

Hsiao, I.L. and Huang, Y.J. (2011). Effects of various physicochemical characteristics on the toxicities of ZnO and TiO2 nanoparticles toward human lung epithelial cells. *Science of the Total Environment* 409 (7): 1219–1228. https://doi.org/10.1016/j.scitotenv.2010.12.033.

Huang, Q., Yu, H., and Ru, Q. (2010). Bioavailability and delivery of nutraceuticals using nanotechnology. *Journal of Food Science* 75 (1): R50–R57. https://doi.org/10.1111/j.1750-3841.2009.01457.x.

Jaiswal, M., Dudhe, R., and Sharma, P. (2014). Nanoemulsion: an advanced mode of drug delivery system. *3 Biotech* 5 (2): 123–127. https://doi.org/10.1007/s13205-014-0214-0.

Jawahar, N. and Meyyanathan, S. (2012). Polymeric nanoparticles for drug delivery and targeting: a comprehensive review. *International Journal of Health and Allied Sciences* 1 (4): 217–223. https://doi.org/10.4103/2278-344x.107832.

Kalepu, S. and Nekkanti, V. (2015). Insoluble drug delivery strategies: review of recent advances and business prospects. *Acta Pharmaceutica Sinica B* 5 (5): 442–453. https://doi.org/10.1016/j.apsb.2015.07.003.

Kapoor, D.N., Kaur, R., Sharma, R., and Dhawan, S. (2015). PLGA: a unique polymer for drug delivery. *Therapeutic Delivery* 6: 41–58. https://doi.org/10.4155/tde.14.91.

Karlsson, H.L. (2010). The comet assay in nanotoxicology research. *Analytical and Bioanalytical Chemistry* 398 (2): 651–666. https://doi.org/10.1007/s00216-010-3977-0.

Kumari, A., Yadav, S.K., and Yadav, S.C. (2010). Biodegradable polymeric nanoparticles based drug delivery systems. *Colloids and Surfaces B: Biointerfaces* 75 (1): 1–18.

Landsiedel, R., Diana, M., Schulz, M. et al. (2009). Author's personal copy mutation research/reviews in mutation research genotoxicity investigations on nanomaterials : methods, preparation and characterization of test material, potential artifacts and limitations – many questions, some answers. *Mutation Research-Reviews in Mutation Research* 681 (2–3): 241–258. https://doi.org/10.1016/j.mrrev.2008.10.002.

Lembo, D., Swaminathan, S., Donalisio, M. et al. (2013). Encapsulation of acyclovir in new carboxylated cyclodextrin-based nanosponges improves the agent's antiviral efficacy. *International Journal of Pharmaceutics* 443 (1–2): 262–272. https://doi.org/10.1016/j.ijpharm.2012.12.031.

Liu, Y., Shu, X.Z., and Prestwich, G.D. (2006). Osteochondral defect repair with autologous bone marrow–derived mesenchymal stem cells in an injectable, *in situ*, cross-linked synthetic extracellular matrix. *Tissue Engineering* 12 (12): 3405–3416. https://doi.org/10.1089/ten.2006.12.3405.

Luyts, K., Napierska, D., Nemery, B., and Hoet, P.H. (2013). How physico-chemical characteristics of nanoparticles cause their toxicity: complex and unresolved interrelations. *Environmental Science: Processes and Impacts* 15 (1): 23–38. https://doi.org/10.1039/C2EM30237C.

Mandal, B., Bhattacharjee, H., Mittal, N. et al. (2013). Core–shell-type lipid–polymer hybrid nanoparticles as a drug delivery platform. *Nanomedicine: Nanotechnology, Biology and Medicine* 9 (4): 474–491. https://doi.org/10.1016/j.nano.2012.11.010.

Matthaiou, E.I., Barar, J., Sandaltzopoulos, R. et al. (2014). Shikonin-loaded antibody-armed nanoparticles for targeted therapy of ovarian cancer. *International Journal of Nanomedicine* 9: 18551–18557. https://doi.org/10.2147/IJN.S51880.

Mazzarino, L., Loch-Neckel, G., Bubniak, L. et al. (2015). Nanoparticles made from xyloglucan-block-polycaprolactone copolymers: safety assessment for drug delivery. *Toxicological Sciences* 147 (1): 104–115. https://doi.org/10.1093/toxsci/kfv114.

Mehnert, W. and Mader, K. (2012). Solid lipid nanoparticles: production, characterization and applications. *Advanced Drug Delivery Reviews* 64: 83–101.

Mooguee, M., Omidi, Y., and Davaran, S. (2010). Synthesis and *in vitro* release of adriamycin from star-shaped poly (lactide-co-glycolide) nano-and microparticles. *Journal of Pharmaceutical Sciences* 99 (8): 3389–3397. https://doi.org/10.1002/jps.22106.

Mura, S., Hillaireau, H., Nicolas, J. et al. (2011). Influence of surface charge on the potential toxicity of PLGA nanoparticles towards calu-3 cells. *International Journal of Nanomedicine* 6: 2591–2605. https://doi.org/10.2147/ijn.s24552.

Neslihan Gursoy, R. and Benita, S. (2004). Self-emulsifying drug delivery systems (SEDDS) for improved oral delivery of lipophilic drugs. *Biomedicine and Pharmacotherapy* 58 (3): 173–182. https://doi.org/10.1016/j.biopha.2004.02.001.

Pachpute, T., Dwivedi, J., Shelke, T., and Jeyabalan, G. (2017). A review on nanospehere formulation. *International Journal of Pharmaceutical and Biological Science Archive* 5 (03): 1–4. http://www.ijpba.in/index.php/ijpba/article/view/146.

Panyam, J. and Labhasetwar, V. (2003). Biodegradable nanoparticles for drug and gene delivery to cells and tissue. *Advanced Drug Delivery Reviews* 55: 329–347. https://doi.org/10.1016/S0169-409X(02)00228-4.

Patel, S.C., Patel, R.C., and Saiyed, M.A. (2011). Toxicology perspective of nanopharmaceuticals: a critical review. *International Journal of Pharmaceutical Sciences and Nanotechnology* 4 (1): 1287–1295. https://doi.org/10.37285/ijpsn.2011.4.1.1.

Patil, C.S. (2011). Current trends and future prospective of nutraceuticals in health promotion. *BIOINFO Pharmaceutical Biotechnology* 1 (1): 1–7.

Patil, T.S., Nalawade, N.A., Kakade, V.K., and Kale, S.N. (2017). Nanosponges : a novel targeted drug delivery for cancer treatment. *International Journal of Advance Research and Development* 2 (4): 55–62.

Powers, K.W., Brown, S.C., Krishna, V.B. et al. (2006). Research strategies for safety evaluation of nanomaterials. Part VI. Characterization of nanoscale particles for toxicological evaluation. *Toxicological Sciences* 90 (2): 296–303. https://doi.org/10.1093/toxsci/kfj099.

Punia, S., Sandhu, K.S., Kaur, M., and Siroha, A.K. (2019). Nanotechnology: a successful approach to improve nutraceutical bioavailability. In: *Nanobiotechnology in Bioformulations*, 119–133. Cham: Springer https://doi.org/10.1007/978-3-030-17061-5_5.

Pushpalatha, R., Selvamuthukumar, S., and Kilimozhi, D. (2018). Hierarchy analysis of different cross-linkers used for the preparation of cross-linked cyclodextrin as drug nanocarriers. *Chemical Engineering Communications* 205 (6): 759–771. https://doi.org/10.1080/00986445.2017.1416354.

Radhika, P.R., Sasikanth, and Sivakumar, T. (2011). Nanocapsules: a new approach in drug delivery. *International Journal of Pharmaceutical Sciences and Research* 2 (6): 1426–1429.

Rao, M.R.P. and Bhingole, R.C. (2015). Nanosponge-based pediatric-controlled release dry suspension of gabapentin for reconstitution. *Drug Development and Industrial Pharmacy* 41 (12): 2029–2036. https://doi.org/10.3109/03639045.2015.1044903.

Sadrieh, N., Wokovich, A.M., Gopee, N.V. et al. (2010). Lack of significant dermal penetration of titanium dioxide from sunscreen formulations containing nano- and submicron-size TiO_2 particles. *Toxicological Sciences* 115 (1): 156–166. https://doi.org/10.1093/toxsci/kfq041.

Santini, A., Tenore, G., and Novellino, E. (2017). Nutraceuticals: a paradigm of proactive medicine. *European Journal of Pharmaceutical Sciences* 96: 53–61. https://doi.org/10.1016/j.ejps.2016.09.003.

Seiffert, J., Hussain, F., Wiegman, C. et al. (2015). Pulmonary toxicity of instilled silver nanoparticles: influence of size, coating and rat strain. *PLoS One* 10 (3): 1–17. e0119726. https://doi.org/10.1371/journal.pone.0119726.

Shargh, V.H., Hondermarck, H., and Liang, M. (2016). Antibody-targeted biodegradable nanoparticles for cancer therapy. *Nanomedicine* 11 (1): 63–79. https://doi.org/10.2217/nnm.15.186.

Sherje, A.P., Dravyakar, B.R., Kadam, D., and Jadhav, M. (2017). Cyclodextrin-based nanosponges: a critical review. *Carbohydrate Polymers* 173 (1): 37–49. https://doi.org/10.1016/j.carbpol.2017.05.086.

Shringirishi, M., Prajapati, S.K., Mahor, A. et al. (2014). Nanosponges: a potential nanocarrier for novel drug delivery-a review. *Asian Pacific Journal of Tropical Disease* 4 (S2): S519–S526. https://doi.org/10.1016/S2222-1808(14)60667-8.

Singh, A., Garg, G., and Sharma, P.K. (2010). Nanospheres: a novel approach for targeted drug delivery system. *International Journal of Pharmaceutical Sciences Review and Research* 5 (3): 84–88.

Sionkowska, A. (2011). Progress in polymer science current research on the blends of natural and synthetic polymers as new biomaterials : review. *Progress in Polymer Science* 36 (9): 1254–1276. https://doi.org/10.1016/j.progpolymsci.2011.05.003.

Smijs, T.G. and Bouwstra, J.A. (2010). Focus on skin as a possible port of entry for solid nanoparticles and the toxicological impact. *Journal of Biomedical Nanotechnology* 6 (5): 469–484. https://doi.org/10.1166/jbn.2010.1146.

Swaminathan, S., Vavia, P.R., Trotta, F., and Torne, S. (2007). Formulation of betacyclodextrin based nanosponges of itraconazole. *Journal of Inclusion Phenomena and Macrocyclic Chemistry* 57 (1–4): 89–94. https://doi.org/10.1007/s10847-006-9216-9.

Swaminathan, S., Pastero, L., Serpe, L. et al. (2010). Cyclodextrin-based nanosponges encapsulating camptothecin: physicochemical characterization, stability and cytotoxicity. *European Journal of Pharmaceutics and Biopharmaceutics* 74 (2): 193–201. https://doi.org/10.1016/j.ejpb.2009.11.003.

Tejashri, G., Amrita, B., and Darshana, J. (2013). Cyclodextrin based nanosponges for pharmaceutical use: a review. *Acta Pharmaceutica* 63 (3): 335–358. https://doi.org/10.2478/acph-2013-0021.

Ting, Y., Jiang, Y., Ho, C., and Huang, Q. (2014). Common delivery systems for enhancing *in vivo* bioavailability and biological efficacy of nutraceuticals. *Journal of Functional Foods* 7: 112–128. https://doi.org/10.1016/j.jff.2013.12.010.

Trotta, F., Cavalli, R., Martina, K. et al. (2011). Cyclodextrin nanosponges as effective gas carriers. *Journal of Inclusion Phenomena and Macrocyclic Chemistry* 71 (1–2): 189–194. https://doi.org/10.1007/s10847-011-9926-5.

Trotta, F., Dianzani, C., Caldera, F. et al. (2014). The application of nanosponges to cancer drug delivery. *Expert Opinion on Drug Delivery* 11 (6): 931–941. https://doi.org/10.1517/17425247.2014.911729.

Uchegbu, I. and Vyas, S. (1998). Non-ionic surfactant based vesicles (Niosomes) in drug delivery. *International Journal of Pharmaceutics* 172 (1–2): 33–70. https://doi.org/10.1016/s0378-5173(98)00169-0.

Vadlapudi, A. and Mitra, A. (2013). Nanomicelles: an emerging platform for drug delivery to the eye. *Therapeutic Delivery* 4 (1): 1–3. https://doi.org/10.4155/tde.12.122.

Yun, J.W., Kim, S.H., You, J.R. et al. (2015). Comparative toxicity of silicon dioxide, silver and iron oxide nanoparticles after repeated oral administration to rats. *Journal of Applied Toxicology* 35 (6): 681–693. https://doi.org/10.1002/jat.3125.

Zaki, N.M. (2014). Progress and problems in nutraceuticals delivery. *Journal of Bioequivalence and Bioavailability* 6 (3): 75–77. https://doi.org/10.4172/jbb.10000183.

14

Current Challenges and Future Needs for Nanotoxicity and Nanosafety Assessment

Ravish J. Patel[1], *Amit Alexander*[2], *Anu Puri*[3], *and Bappaditya Chatterjee*[4]

[1] *Ramanbhai Patel College of Pharmacy, Charotar University of Science and Technology (CHARUSAT), CHARUSAT campus, Changa, India*
[2] *National Institute of Pharmaceutical Education and Research (NIPER GUWAHATI), Department of Pharmaceuticals, Ministry of Chemicals & Fertilizers, Near Dr. Bhupen Hazarika Regional Govt. Film & Television Institute, Sila Village, Nizsundarighopa, Changsari, Kamrup (R), Guwahati, Assam, India*
[3] *RNA Structure and Design Section, RNA Biology Laboratory (RBL), Center for Cancer Research, NCI-Frederick, NIH, Frederick, MD, USA*
[4] *Shobhaben Pratapbhai Patel School of Pharmacy & Technology Management (SPPSPTM), NMIMS University, Mumbai, MH, India*

14.1 Introduction

In the present era, nanotechnology has emerged as a widespread field in several segments owing to its exceptional and advantageous physical and chemical properties (Salamanca-Buentello et al. 2005). A few of the vital applications of nanotechnology in several areas are indicated in Table 14.1. The optical, electrical, and mechanical properties of a material may be improved by bringing the materials to the nanoscale level (Yang et al. 2015). Enhancement in the surface area to volume proportion will substantiate the surface-associated properties of a material. The higher specific surface area and tiny size of nanomaterials deliver the distinctive features (Kumar and Kumar 2014). Biological systems also have unexpected biological responses on interaction with such nanomaterials (Zhang et al. 2012).

Moreover, different biokinetic behavior and capabilities of nanomaterials enable them to reach deeper in the body compared with their higher size counterparts. On the flip side, enhancement in occupational exposure is observed with an increase in the use and production of nanomaterials. This is an alarming situation for the surrounding environment and disturbances to the ecosystem as well (Gangwal et al. 2011; Mihalache et al. 2017). Such critical observations guided and forced the scientists to systematically examine and comprehend the likely adversative effects of engineered nanomaterials on the well-being and the surrounding environment. This will pave the pathway for exploration of the challenges to evaluate nanomaterials' toxicity.

In the last few years, various reports were published on the assessment of toxicity for the nanomaterials. However, there are many challenges associated with the task for investigation of nanoparticle's interactions with the biological systems. Improper understanding of experimental methods and the lack of precise characterization methods involved in the toxicological assessment of nanomaterials are the probable reasons (Saifi et al. 2018). Although, there is a considerable risk for human health because of the toxicity of the nanotechnology-based engineered materials (Table 14.2) on administration/exposure by various routes of drug delivery (Figure 14.1) (Sharma et al. 2012; Warheit and Sayes 2015).

Nanotechnology in Medicine: Toxicity and Safety, First Edition. Edited by Mahendra Rai, Mrunali Patel, and Rashmin Patel.

Table 14.1 Various applications of nanomaterials in a different segment.

Segment	Applications	References
Health sector	Nano filter, nano-coatings, antimicrobial activities, cosmetic products, nanocarrier for vaccines, therapeutic targets in chemotherapy, cancer nanophotothermolysis, bio-nanosensors	Nassiri Koopaei and Abdollahi (2016), Carvalho et al. (2016) and Parhizkar et al. (2017)
Nanomedicine	*In vivo* imaging via fluorescence, bioimaging via fluorescence, *in vitro* system of diagnostics, delivery of drugs, gene therapy, drug release and gene delivery, tissue engineering, therapy based on photodynamics, photothermal-based controlled drug delivery, cancer management and treatment, bioimaging, Magnetic Resonance Imaging, magnetically directed controlled drug delivery, biosensing via magnetism, vaccines, and gene material, regenerative medicine	Patel and Patel (2013), Vinod et al. (2019), Alexander et al. (2019), Agrawal et al. (2020) and Patel and Parikh (2020)
Security and defense	Robotics, nanocomposites, night vision camera, energy devices, sensors and electronics, modern weapon systems, nano-coatings, smart materials, fuel additives	Frech (2005), Burmaoglu and Sarıtas (2017) and Petkov et al. (2018)
Food and agriculture	Nano fertilizers, nano fungicides, nano pesticides, carbon nanotube, nano delivery vehicles in food, bio-actives, pharma foods, engineered nanomaterials, nanoporous membrane and materials, functional foods, nutraceuticals and their system, storage and packaging of food	He and Hwang (2016), Singh (2016), Yu et al. (2018), Jeevahan and Chandrasekaran (2019) and He et al. (2019a)
Environment and energy	Solar energy, H_2 generation, treatment of wastewater, nano-sized filters and/or membranes, organic/inorganic pollutant adsorption and degradation, lithium-ion batteries, energy storage	Kumar et al. (2011), Lian et al. (2017), Ahmadi et al. (2019) and Vázquez-Núñez et al. (2020)
Building	Cement composite, pigments in paint, photocatalyst, thin film on glass windows, adsorbent, different membranes, fire retardant, sensors for improving mechanical properties, insulation	Sobolev et al. (2008)
Automobile sector	Nano-coatings, composite fillers, nanomaterials, paints, smart lights, catalyst as additives, fuel cells, nano-based lubricants	Shafique and Luo (2019) and Mathew et al. (2019)
Textile sector	Nano-coating, smart fibers, wrinkle-freeness, stain repellence, softness, military applications, high absorbency, and breathability	Yetisen et al. (2016), Afroj et al. (2019), He et al. (2019b) and Lu et al. (2020)
Electronics and devices	Light-emitting diodes, nano-transistor, spintronics, nano-memory device, opto-magnetic, nanogenerator, electrochromic device, OLEDs	Mologni et al. (2008) and Yuan et al. (2019)
Sport sector	Carbon nanotubes containing sports items, ball coatings, nanofibers	Gong (2013) and Harifi and Montazer (2017)

Source: Based on Salamanca-Buentello et al. (2005).

Table 14.2 Probable risks associated with nanoparticles in the human body.

List of nanomaterials	Possible toxicity
Carbon-containing nanomaterials	Enhanced cytotoxicity, pulmonary inflammation, and toxicity, inhibition of DNA enzymes, granulomas, and fibrosis
Gold, silver, and carbon-containing nanomaterials	Toxicity to the liver and respiratory system, disturbing the central nervous system
Nanomaterials containing copper oxide	Damage to cell and DNA, toxicity for aquatic organisms, impairment/defeat the immune system
Nanomaterials containing titanium dioxide	Production of reactive oxygen species, neurotoxicity, genotoxicity, cell damage, metabolic change, reproductive toxicity, skin penetration
Nanomaterials containing cadmium	Toxicity to lungs and liver, potential nephrotoxic agent, fetus malformation, damage to cell and DNA, enhanced cytotoxicity, hampered growth
Quantum dots	Fetus malformation and hampered growth, lung infection and inflammation, decrease in sperm count and quality, cellular damage
Nanomaterials containing silicon dioxide	Tuberculosis, lung cancer, chronic obstructive pulmonary disease (COPD), cell necrosis, membrane damage and lipid peroxidation, dysfunction of mitochondria
Nanomaterials containing zinc oxide	Toxicity to the reproductive system of earthworms, severe liver damage, hepatic oxidative stress

Sources: Adapted from Sharma et al. (2012) and Warheit and Sayes (2015).

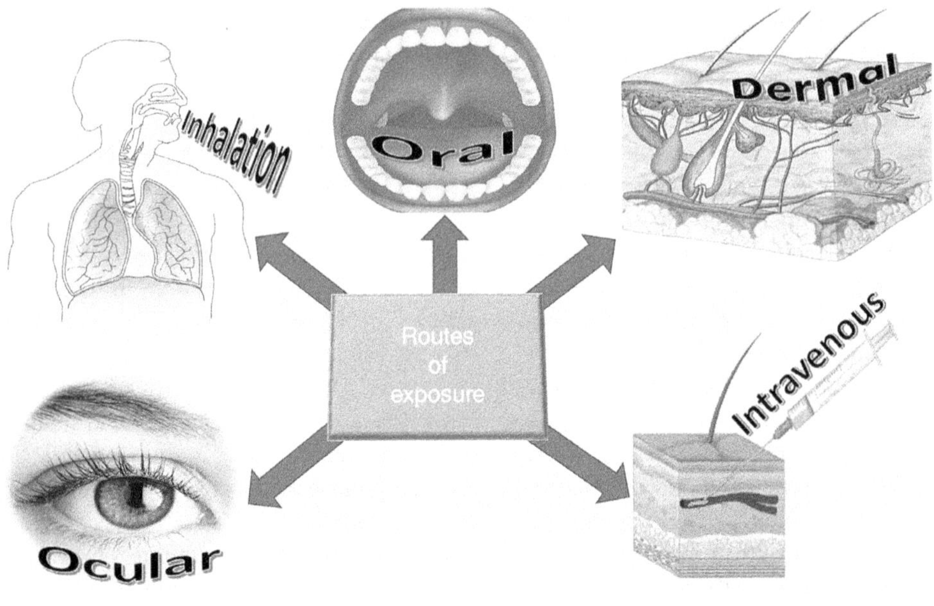

Figure 14.1 Nanomaterial's exposure pathways. *Source:* From Sharma et al. (2012) and Warheit and Sayes (2015).

The global market estimate for nanomaterials is at 11 million tons at a €20 billion market value. The current direct employment is estimated at 300,000–400,000 in Europe in the nanomaterial sector (European-Commission 2020). The potential health risks to labors and customers are remarkable for nanotechnology. They may ascend from contact to nanomaterials either at work (occupational exposure) or through customer products (Roblegga et al. 2009). Little information related to the field of nanotechnology is available on the epidemiological effects of nanomaterials. In the past few years, hundreds of tons of nanomaterials have entered into the environment without having adequate knowledge about their potential responses toward biological systems (Drobne 2007). There is enhanced use of engineered nanoparticles in a wide range of consumer and industrial products. There are many studies on their safety. These studies have shown that specific nanoparticles are cytotoxic. They may cause oxidative stress and damage to the DNA, raising the concerns for human health. Exposure to some types of particles can cause severe health effects. Therefore, it is essential to understand their action or toxicity mechanisms and investigate their possible health risks. Major questions that need to be answered concerning toxicity assessment are as follows: what is the toxicity mechanism of nanoparticles; how do interactions occur between the reactive surface of nanoparticles and cells; and what are the differences between particle size versus particle composition in terms of toxicity? There have been no accurate answers to all these questions; however, investigations are continuing. Nanotechnology products need to be regulated since more nanoparticles and nanoparticle comprising products are developed and carried into commercial use over time. Furthermore, the dose of nanomaterials for toxicological investigation is very complex compared with chemical toxicants. Characterization of the composition and purity of the material is a pillar for verifying the dose or concentration for chemical toxicants (Drobne 2007).

Nanotechnology has its footprints in almost every sector. It is a vital emerging area due to frequent usage (Table 14.1). Still, researchers need to give special attention to nanotechnology products' side effects or toxicities such as carbon nanotubes and nanoparticles. Care to be taken during the formulation of nanomaterials for getting advantageous effects. Else, it may lead to some toxicological reactions. In assessing nanomaterials' toxicity, the primary problem or concern is that there are currently limited analytical techniques. Multiple techniques are available for macroparticles; however, it is essential to develop new and sensitive nanoparticle toxicity assessment methods.

Another point that needs attention is the correlation of data obtained from *in vivo* and *in vitro* experiments reported in the literature. There is a necessity for the database for scientists' experiences on nanomaterial assay, interactions, and nanomaterial toxicity assessment protocols. This database provides information such as the product name, price, quantity, and many other factors that affect nanomaterials' toxicity. Other challenges include the right choice of ingredients, dose, cell line, an animal model, and toxicity assessment protocol suitability; hence, nanomaterial toxicity assessment needs much more attention.

14.2 Nanomaterials: Risk Assessment

Risk analysis, including risk assessment, management, and communication, is the most appropriate for nanomaterials' management. Risk assessment is a policy assisting means for lawmakers. The potential health risks identification is a precondition for a proper evaluation of nanomaterials' benefit and safety. Few nanomaterials may affect biological behaviors at the cellular levels, such as viability and subcellular levels, the generation of reactive oxygen species levels, etc. The human and environmental health risks of nanomaterials are determined through hazard identification and exposure assessment in various materials, life-cycle stages from manufacturing to conveyance for use, discarding, and even reprocessing. Most nanomaterials turn into different shapes by agglomeration, chemical reaction, dissolution, or other interactions with adjacent proximity materials during their life cycle. Therefore, evaluating and addressing the risk becomes more challenging over time (Maynard 2016).

Risk assessment comprises Hazard Identification, Dose–response, Exposure Assessment, and Risk Characterization. The Scientific Committee on Emerging and Newly Identified Health Risks (SCENIHR) concluded that nanomaterials might have various toxicological properties compared with the substances in bulk form. Therefore, their risks need to be evaluated on a case-by-case basis (SCENIHR 2009). Current risk evaluation methodologies need to be modified for the risk assessment of nanotechnological products.

As is known, the first step of any risk assessment process is hazard identification; however, it is a rather challenging step for nanoparticles since the possible toxicity mechanisms brought by exposure of particles are very complex due to the route of exposure, the number of nanoparticles, host response susceptibility, and specific physicochemical properties of each particle. There are different toxicity mechanisms due to different exposures. For example, redox conditions differ from tissue to tissue, with high oxidization at the skin and high reduction in the intestine. This is an essential influence on oxidative stress-driven mechanisms. Thus, a robust and appropriate risk assessment must focus on all of the body systems.

Dose–response assessment is the fundamental concept in characterizing the extent to which different types of particles pose human health risks. The biological effects of several particle types were studied in animals and humans. While some particles are found to be harmless at the doses which humans are potentially exposed to, such as TiO_2, others induce diseases such as fibrosis and cancer (e.g. silica and asbestos). Toxic effects with nanoparticles are related to dose and physicochemical properties (size, shape, chemical composition, surface reactivity, surface charge, solubility/bio-durability, etc.). For dose–response assessment, nanoparticles' characterization should be made correctly and precisely by methods and instrumentation such as Transmission Electron Microscopy and Scanning Electron Microscopy. The correct dose–response relationship should be made for the correct risk analysis and correct dose metric determination of the dose–response curve's shape. In conventional toxicology, the dose metric is generally based on the concentration of the substance; however, in the case of nanomaterials, the exact relationship is not known.

Exposure assessment is determined by the exposure setting such as occupational, environmental, consumer, the route of exposure, the level, duration, and the frequency of exposure. Exposure assessment needs the identification and quantification of exposure scenarios of interest. Therefore, the current approach in the new EU regulatory framework, REACH (Registration, Evaluation, Authorization, and Restriction of Chemicals), suggests a tiered approach for exposure assessment of nanomaterials. Currently, there is a lack of data on nanomaterial exposure in both occupational and consumer contexts. The tiered approach is based on a decision tree structure and consists of a conservative first-tier followed by a more realistic second tier. While a first-tier approach is unacceptable due to being so conservative, a second-tier approach is needed. This would involve actual exposure instead of unrealistic exposure measurements and/or an improved exposure model (ConsExpo 2020). There are different approaches proposed for risk assessment; however, there is no quantitative risk assessment. Traditional risk assessment uses toxicological data from animal and human experiments to obtain the Derived-No-Effect-Level to compare exposure levels among different exposure scenarios for risk assessment.

In the risk characterization, each human population's exposure likely to be uncovered is compared with the appropriate Derived-No-Effect-Level. Humans' risk can be controlled if the estimated exposure levels do not surpass the appropriate Derived-No-Effect-Level. There is no widely accepted risk assessment strategy due to a lack of robust data for the first three steps of risk assessment (ECHA 2016). Regulatory amendments for risk management are necessary. As explained above, regulations and legislation are based on scientific evidence.

14.3 The Hurdles in Toxicity Evaluation of Nanomaterials

It is now theoretically correlated and technically established the functional dependency and properties of nanomaterials are due to combined effects of their shape, size, surface/volume ratio, surface area, solubility, chemical composition, and others. Hence, the study's appropriate design would be incorporated to address the applications of authenticated protocols for nanomaterials' characterization and to study the effects of nanomaterials on living organisms and the surrounding environment. On the same side, it is also significant to highlight the surface properties of nanomaterials affecting their biological behavior. nanomaterials' associated risk/toxicity can be measured by detailed analysis and understanding of observations obtained after nanomaterial's characterization. Otherwise, it is difficult to correlate the specific property of the nanomaterials or even the NM itself to the possible risk/toxic effects. For example, improper observed data and effects could be the outcome of impurities and additional components (Dhawan and Sharma 2010). Hence, critical physicochemical characterization paves the meaningful pathways for the assessment of the biological behavior of nanomaterials.

14.3.1 Physicochemical Properties of Nanomaterials' Characterization

The physicochemical properties characterization of nanomaterials includes a range of parameters such as the size, morphology, surface porosity, roughness, shape, surface area, chemical composition, solubility, analysis of purity, surface chemistry, crystallinity, and reactivity. Alteration in physical and chemical properties may result in variations in the size, elemental composition, or surface properties for nanomaterials.

14.3.1.1 Size of Nanomaterials

Properties of nanomaterials such as solubility, transparency, absorption or emission wavelength, conductivity, melting point, color, and catalytic behavior are changed by varying the particle size of nanomaterials and are also based on the material present in the precursor solutions. nanomaterials possess exceptional physical and chemical properties owing to their size, which affects the mobility and conveyance behavior of nanomaterials.

14.3.1.2 Composition Effects

Difference in the physical and chemical behaviors of the material can be observed by changing the particle compositions.

14.3.1.3 Surface-Related Effects

The lesser the spherical particles' diameter, the higher the surface area-to-volume ratio and the specific surface area. This is going together with conductivity, chemical reactivity, dispersity, catalytic behavior, and optical properties. Therefore, more attention has been required for the surface of nanomaterials rather than its core material. When empty/uncovered nanomaterials interact with a heterogeneous environment, the strong or weak interaction forces enable smaller structures such as atoms, molecules, or macromolecules to attach themselves to the surface of the nanomaterials. Molecules such as proteins and polymers interact with the nanomaterials' surface layer and form an nanomaterials–protein corona in the biological environment. It has also been revealed that the nanomaterials, along with the corona, define the particle-plus-corona composite (Lynch and Dawson 2008; Elsaesser and Howard 2012). This creates a necessity to understand the behavior of nanomaterials along with the environment of biological interactions.

14.3.1.4 Agglomeration

The surface properties and bioavailability of nanomaterials are affected by the agglomeration of particles.

14.3.1.5 Solubility

Reports on the formation of insoluble ion forms of nanomaterials may produce toxicity for the cells, e.g. ZnO, CuO.

14.3.1.6 Surface Charge and Dispersity

Particle solubility in suspension may be affected by the surface charge of the nanomaterials. The tendency of nanomaterials to form agglomerates is dependent on its dispersity.

14.3.1.7 Dose Metric

The exposure metric for nanomaterials has been expressed, based on the surface area, mass, or number. The issue of the proper system of measurement for NPs in workplaces is still controversial. As mentioned, a pulmonary response is well correlated with the surface area concentration irrespective of particle size. However, this may not be true for all types of particles and might also be a function of the agglomeration (NIOSH 2013).

In brief, in-depth knowledge of the nanomaterials' characteristics to be tested is the prerequisite to assess thoroughly the risk/toxicity connected with the nanomaterials. It is advisable to carry out risk/toxicity evaluation before undertaking detailed characterization considering the novel characteristics of nanomaterials.

14.3.2 *In vitro*, *In vivo*, and *In silico* Approaches for the Assessment of Toxicity for the Nanomaterials

An in-depth understanding of the behavior and interactive actions of nanomaterials will determine the path tracked by nano-sized materials inside the organisms, their tenacity, and their assimilation pathways. There are several pathways to trace the route of the nanomaterials and their favorable parameters inside an organism's body. Commonly, this analysis can be classified beneath *in vitro*, *in vivo*, and *in silico* evaluation (Table 14.3).

Toxicity assessment using the *in vitro* techniques is considered the cost-effective, a broad range of accessible, most reliable, wider applicable, and more ethical due to fewer animal studies. However, it is critical to assess the physiological outcomes and also to predict the results of xenobiotics. *In vitro* technique is considered the broadly accepted method due to its speedy results, inclusive detection, and various assays used to examine cell toxicity. The assays are differed by their detection and cell death mechanism (Fotakis and Timbrell 2006; Khalili Fard et al. 2015).

The *in vivo* methods are generally favored and preferred for the primary standards for toxicity assessment. The *in vivo* studies comprise administering the nanomaterials into the body of animals under the test and, subsequently, recording the signs of development over the applications of different available techniques. The *in vitro* results for assessing toxicity are varied from the same obtained from *in vivo* techniques because of the inclusion of several critical factors that cannot be a part of the *in vitro* evaluation. Quite often, an *in vivo* assessment may get affected by the influence of cell–cell, hormonal changes, and cell–matrix interactions. Considerable precautions are to be taken during the *in vivo* studies as most of the time; they are interwoven with several encounters. The actual exposure of the nanomaterials to the body decides the dose for the *in vivo* study. Dose determination for the

Table 14.3 *In vitro*, *in vivo*, and *in silico* approaches for the assessment of toxicity for the nanomaterials.

Approach	Details of the assessment	Instrument/Equipment	References
In vitro technique	• Selecting the appropriate cell lines • Cytotoxicity assays • Cell viability study • Cell stress study • Disadvantages: 1) Unknown physiological pathways 2) Secondary inferences of nanomaterials is not seen	• Optical spectroscopy • Scanning electron microscopy • Transmission electron microscopy • Dynamic light scattering technique	Jones and Grainger (2009), Hillegass et al. (2010); Cohen et al. (2015) and Ajdary et al. (2018)
In vivo technique	• Mice, zebrafish, rodents, and nonhuman primates are selected as a model for living animals • Various organs may get affected due to the diverse intracellular behavior of nanomaterials • Studies significantly depend on the shape, size, surface coating, and surface charge of the nanomaterials • Disadvantages: 1) The prolonged period of evaluation 2) Nonethical nature	• Optical spectroscopy • Electron microscopy techniques like SEM and TEM • Atomic force microscopy study • Magnetic resonance imaging	Zhu et al. (2007), Aillon et al. (2009), Ajdary et al. (2018) and Landvik et al. (2018)
In silico technique	• Nano bio-interfaces demonstrating models are selected • Association between physicochemical properties and nanotoxicity can be evaluated by computational simulation • Risk/Toxicity assessment and hazard control of the nanomaterials • Development of the quantitative structure–activity relationship models and high-throughput screening methods • *In vitro* and *in vivo* studies provide the unfailing experimental toxicity outcomes and can be used to generate the data sets	• Theoretical calculations • Computational simulations • Generate reliable data sets for comparative studies	Oksel et al. (2017), Shityakov et al. (2017), Landvik et al. (2018) and Basei et al. (2019)

in vivo study faces a technical challenge due to the smaller size of nanomaterials and its response unpredictability in the biological system. The vehicle used for the *in vivo* experimental study should be nonreactive and uniformly disperse the nanomaterials to avoid any changes in the nanomaterial dose.

Owing to the larger surface area, agglomeration is the major drawback associated with the nanomaterials. Quite often, agglomeration and nonuniform dispersion may culminate in inappropriate bio-distribution and undesirable outcomes/results. Protein corona formation may be one of the outcomes of the interaction of nanomaterials with protein inside the body. This results in modification in the nanomaterials' properties, their interactive actions, and biological distribution. Furthermore, the structure of protein experiences variations in arrangement and leads to altered bio-purposes and the transformed path of the signal. Hence, interferences of nanomaterials with another substrate are a prerequisite for assessing the toxicity of nanomaterials in a biological system (Guadagnini et al. 2015).

Considering the ethical standards, time requirement, and dependable results, scientists were forced to search for an alternate way to assess the toxicity of the materials. The innovative *in silico* analysis tool is for assessment of toxicity when compared with the general studies. The method for the *in silico* analysis is grounded on the principle of theoretical modeling and simulation of outcomes for several physical and chemical properties of molecules (Figure 14.2) (Raies and Bajic 2016). *In silico* studies, the prediction of toxicity of a material is based on the available toxicity data and their interpolation through multiple mathematical models. Still, the input of the additional experimental verification step is required to verify the toxicological effects. Also, the data gap is one reason for the limited exploration of the web-based tools for the quantitative risk assessment of nanomaterials (Valerio 2009; Fourches et al. 2011).

14.4 Nanosafety Assessment Tools

Using this tool aims to deliver a means for developing a safety sorting for engineered nanomaterials (ENM) built on the in-depth understanding of their interactive actions at molecular, cellular, and organism levels in living organisms. Many critical and essential functions take place at the nanoscale in the living organisms. Numerous systems and processes are controlled by natural nanomaterials, such as proteins and other molecules, in the human body. Nanosafety strategy (Figure 14.3) and nanosafety assessment tools (Table 14.4) are possibly explored to identify and elaborate ENM characteristics that establish their potential of hazard to the biological system.

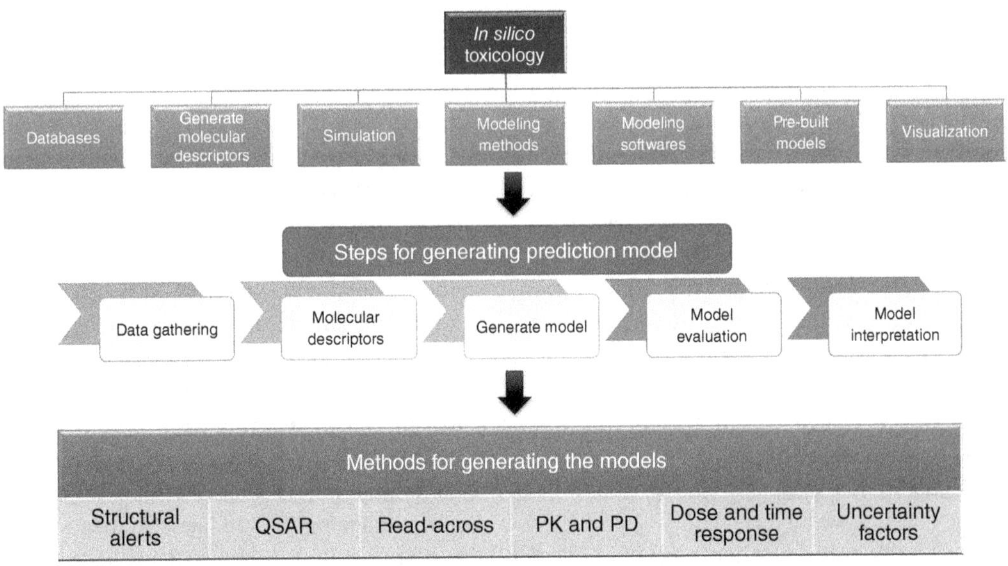

Figure 14.2 *In silico* analysis: a computational method for the evaluation of the chemical risk. *Source:* Modified from Raies and Bajic (2016).

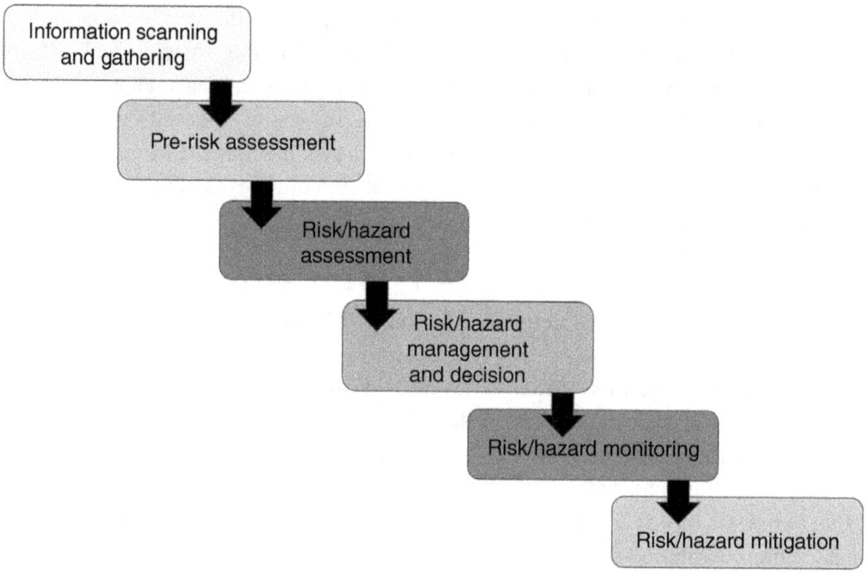

Figure 14.3 Nanosafety strategy.

This includes the capability of ENM to diminish the vital cellular roles by inducing damage/changes at the cellular, tissue, or organism levels after interaction with cellular structures. ENM-induced alterations may mediate these adverse effects in gene expression and translation.

PhysiCell (http://physicell.mathcancer.org) and NanoDoc (http://nanodoc.org) are the *in silico* simulators. These simulators reproduce accurate nanomaterials' movement and their interactive actions within the tumor's environment and with the other nanomaterials. Synthesized nanomaterial based on the utmost talented nanomaterials' designs will be evaluated *in vitro* and *in vivo* on colon and breast cancer (mouse cancer xenografts and microfluidic testbeds featuring cancer microenvironments) stem cells (Ghaffarizadeh et al. 2018).

14.5 Conclusion and Perspectives

Nanotechnology has a wider range of applications. It has a vital role in current advances in drug delivery, environmental clean-up methods, medical diagnosis and treatment, and production of the nanoelectronics. Ongoing research must be critically designed to cater to the need of risk assessment surrounding the manufacture, use, and disposal of nanomaterials as the use of nanomaterials and their potential benefits are increasing rapidly. Many early studies reveal the complication in nanomaterial's risk/toxicity assessment by lack of nanomaterials' characterization, no clear dose metric, and limited exposure information. There is a need to continue the efforts made to address data gaps for current research in risk/toxicity/safety assessment for nanomaterials. There are visible data gaps in exposure assessment of nanomaterials, especially related to the occupational exposures. Very little information and shreds of evidence are available to expose nanomaterials containing different products to end-users through their use, transport, and disposal. The lack of a dose for evaluating the toxic effects of nanomaterials is another large gap of data. Continued coordination, management, and synchronization of the research in the risk/toxicity assessment areas are required to enlighten nanomaterials' future risk. Nanosafety will assist in the

Table 14.4 Objective/application of nanosafety assessment tools.

Name of nanosafety assessment tool	Objective/Application	Source
Nanosolutions	• Determination of the ENM biological identity • Development of the computer program • Prediction of the properties of ENM, their ability to cause health or environmental hazards	http://nanosolutionsfp7.com/
GUIDEnano	• Development of a web-based guidance tool • Help the nano-enabled product design • Apply the most appropriate risk assessment and mitigation strategy for a specific product.	https://www.guidenano.eu/
The Sustainable Nanotechnologies (SUN) project	• Collect the knowledge on environmental and health risks of nanomaterials • Guiding nanomanufacturing by avoiding liabilities	http://www.sun-fp7.eu/
NanoMILE	• Establishing a fundamental understanding of the mechanisms of nanomaterial interactions with living systems and the environment across the entire life cycle of nanomaterials • Developing a framework for ENM classification according to their biological or environmental impacts	http://nanomile.eu-vri.eu/
eNanoMapper	• Common components and application programming interfaces • Address data and models' interoperability challenges	http://enanomapper.net/
ToxBank	• Supporting integrated data analysis • Servicing of alternative testing methods in toxicology	http://www.toxbank.net/
Gov4Nano	• Implementing a functional Nano Risk Governance Model (NRGM) • Addresses the requirements of the transdisciplinary field and innovation in the field of nanotechnology	https://www.gov4nano.eu/
ACEnano	• To introduce adaptability, confidence, and clarity into risk evaluation of nanomaterials • Development of a robust and widely implementable approach for nanomaterials physical and chemical characterization • A reliable framework for grouping of nanomaterials based on exposure or hazard description	http://acenano-project.eu/
NaNoREG	• Nanomaterials safety assessment framework	https://ec.europa.eu/jrc/en/publication/eur-scientific-and-technical-research-reports/nanoreg-framework-safety-assessment-nanomaterials
NanoDefine	• Support the governance challenges • Implementation of the nano legislation	http://www.nanodefine.eu/

determination of appropriate risk management of nanomaterials.

Significant progress has been observed in the development of methods and facts about the nanomaterial's safety. Yet, commercial forms of nanomaterial demand constant venture and focus in the areas of research and innovation. This requires assessing the health and environmental impressions of nano-products under real conditions of exposure such as addressing the real-time exposure of the nanomaterials beyond inhalation at workplace, a critical step toward the commercial acceptance and assessing the product safety throughout its life cycle.

The National Nanotechnology Initiative, a United States government research and development initiative, is involved in the nanotechnology-related initiatives on evolving knowledge and approaches about environmental health and safety of novel nanomaterials Acceptance of the existing conventional systems to address the nanoparticulate aspects of materials by the nanosafety research and policies will set the path for the further future research and investigations in the field of advanced materials and nanotechnologies. Development of nanotoxicology is considered to be innovative, exclusive, and replicable along with the rapid development of nanoscience field. Extensive number of published literatures is available for the so-called first-generation nanomaterials and nanotechnologies. New systems of measurement are emerging with the development of new two-dimensional nanomaterials such as graphene. Even, new initiatives are changing their focus from first-generation nanomaterials to advanced nanomaterials and newer nanotechnologies. Improved focus is required for next-generation nanomaterials and producing technologies for promoting the designing of safe novel nanomaterials and nanotechnologies. Model developed for prediction of biological behavior of one type of nanomaterial based on its physical and chemical properties is not transferable to other type of nanomaterial as well as safer designing of nanomaterials are still few unanswered critical questions even after many years of research and development in the field of nanomaterials and nanotechnologies.

The current resources and organization in the field of nanosafety are in critical stage. They require adaptation of newer generation nanotechnologies to refine the procedures, advancement of knowledge and acceleration in the commercial acceptance and adoption. More consistent and reproducible techniques of measurement along with robust standards, ease in access to database created from publicly funded projects, improved focus on more genuine and less assessed features of safety in the life cycle of product, and focus on progression to evolve and implement newer methodologies for regulatory and acceptance by the market are the critical challenges need to be addressed for the future of nanosafety.

References

Afroj, S., Karim, N., Wang, Z. et al. (2019). Engineering graphene flakes for wearable textile sensors via highly scalable and ultrafast yarn dyeing technique. *ACS Nano* 13 (4): 3847–3857.

Agrawal, M., Saraf, S., Saraf, S. et al. (2020). Recent strategies and advances in the fabrication of nano lipid carriers and their application towards brain targeting. *Journal of Controlled Release* 321: 372–415.

Ahmadi, M.H., Ghazvini, M., Alhuyi Nazari, M. et al. (2019). Renewable energy harvesting with the application of nanotechnology: a review. *International Journal of Energy Research* 43 (4): 1387–1410.

Aillon, K.L., Xie, Y., El-Gendy, N. et al. (2009). Effects of nanomaterial physicochemical properties on in vivo toxicity. *Advanced Drug Delivery Reviews* 61 (6): 457–466.

Ajdary, M., Moosavi, M.A., Rahmati, M. et al. (2018). Health concerns of various nanoparticles: a review of their in vitro and in vivo toxicity. *Nanomaterials (Basel)* 8 (9): 634.

Alexander, A., Saraf, S., Saraf, S. et al. (2019). Amalgamation of stem cells with nanotechnology: a unique therapeutic approach. *Current Stem Cell Research and Therapy* 14 (2): 83–92.

Basei, G., Hristozov, D., Lamon, L. et al. (2019). Making use of available and emerging data to predict the hazards of engineered nanomaterials by means of in silico tools: a critical review. *NanoImpact* 13: 76–99.

Burmaoglu, S. and Sarıtas, O. (2017). Changing characteristics of warfare and the future of Military R&D. *Technological Forecasting and Social Change* 116: 151–161.

Carvalho, I.T., Estevinho, B.N., and Santos, L. (2016). Application of microencapsulated essential oils in cosmetic and personal healthcare products – a review. *International Journal of Cosmetic Science* 38 (2): 109–119.

Cohen, J.M., DeLoid, G.M., and Demokritou, P. (2015). A critical review of in vitro dosimetry for engineered nanomaterials. *Nanomedicine (Lond)* 10 (19): 3015–3032.

ConsExpo (2020). *'ConsExpo web tool and user information.* National Institute for Public Health and the Environment. https://www.rivm.nl/en/consexpo (accessed 1 September 2020).

Dhawan, A. and Sharma, V. (2010). Toxicity assessment of nanomaterials: methods and challenges. *Analytical and Bioanalytical Chemistry* 398 (2): 589–605.

Drobne, D. (2007). Nanotoxicology for safe and sustainable nanotechnology. *Archives of Industrial Hygiene and Toxicology* 58 (4): 471–478.

ECHA (2016). *Guidance on Information Requirements and Chemical Safety Assessment Part E: Risk Characterisation.* European Chemicals Agency https://echa.europa.eu/documents/10162/13632/information_requirements_part_e_en.pdf/1da6cadd-895a-46f0-884b-00307c0438fd (accessed 1 September 2020).

Elsaesser, A. and Howard, C.V. (2012). Toxicology of nanoparticles. *Advanced Drug Delivery Reviews* 64 (2): 129–137.

European-Commission (2020). 'Nanomaterials'. https://ec.europa.eu/growth/sectors/chemicals/reach/nanomaterials/ (accessed 23 August 2020).

Fotakis, G. and Timbrell, J.A. (2006). In vitro cytotoxicity assays: comparison of LDH, neutral red, MTT and protein assay in hepatoma cell lines following exposure to cadmium chloride. *Toxicology Letters* 160 (2): 171–177.

Fourches, D., Pu, D., and Tropsha, A. (2011). Exploring quantitative nanostructure-activity relationships (QNAR) modeling as a tool for predicting biological effects of manufactured nanoparticles. *Combinatorial Chemistry and High Throughput Screening* 14 (3): 217–225.

Frech, C.B. (2005). Our Molecular Future: How Nanotechnology, Robotics, Genetics, and Artificial Intelligence Will Transform Our World (Douglas Mulhall). *Journal of Chemical Education* 82 (5): 685.

Gangwal, S., Brown, J.S., Wang, A. et al. (2011). Informing selection of nanomaterial concentrations for ToxCast *in vitro* testing based on occupational exposure potential. *Environmental Health Perspectives* 119 (11): 1539–1546.

Ghaffarizadeh, A., Heiland, R., Friedman, S.H. et al. (2018). PhysiCell: an open source physics-based cell simulator for 3-D multicellular systems. *PLOS Computational Biology* 14 (2): e1005991.

Gong, Z. (2013). Nanotechnology application in sports. *Advanced Materials Research* 662: 186–189.

Guadagnini, R., Halamoda Kenzaoui, B., Walker, L. et al. (2015). Toxicity screenings of nanomaterials: challenges due to interference with assay processes and components of classic in vitro tests. *Nanotoxicology* 9 (sup1): 13–24.

Harifi, T. and Montazer, M. (2017). Application of nanotechnology in sports clothing and flooring for enhanced sport activities, performance, efficiency and comfort: a review. *Journal of Industrial Textiles* 46 (5): 1147–1169.

He, X. and Hwang, H.-M. (2016). Nanotechnology in food science: Functionality, applicability, and safety assessment. *Journal of Food and Drug Analysis* 24 (4): 671–681.

He, X., Deng, H., and Hwang, H.-m. (2019a). The current application of nanotechnology in food and agriculture. *Journal of Food and Drug Analysis* 27 (1): 1–21.

He, X., Liu, Q., Wang, J., and Chen, H. (2019b). Wearable gas/strain sensors based on reduced graphene oxide/linen fabrics. *Frontiers of Materials Science* 13 (3): 305–313.

Hillegass, J.M., Shukla, A., Lathrop, S.A. et al. (2010). Assessing nanotoxicity in cells in vitro. *WIREs Nanomedicine and Nanobiotechnology* 2 (3): 219–231.

Jeevahan, J. and Chandrasekaran, M. (2019). Nanoedible films for food packaging: a review. *Journal of Materials Science* 54 (19): 12290–12318.

Jones, C.F. and Grainger, D.W. (2009). In vitro assessments of nanomaterial toxicity. *Advanced Drug Delivery Reviews* 61 (6): 438–456.

Khalili Fard, J., Jafari, S., and Eghbal, M.A. (2015). A review of molecular mechanisms involved in toxicity of nanoparticles. *Advanced Pharmaceutical Bulletin* 5 (4): 447–454.

Kumar, N. and Kumar, R. (2014). Chapter 2 - Nano-based Drug Delivery and Diagnostic Systems. In: *Nanotechnology and Nanomaterials in the Treatment of Life-threatening Diseases* (eds. N. Kumar and R. Kumar), 53–107. William Andrew Publishing.

Kumar, V., Govind, A., and Nagarajan, R. (2011). Optical and photocatalytic properties of heavily F--Doped SnO2 nanocrystals by a novel single-source precursor approach. *Inorganic Chemistry* 50 (12): 5637–5645.

Landvik, N.E., Skaug, V., Mohr, B. et al. (2018). Criteria for grouping of manufactured nanomaterials to facilitate hazard and risk assessment, a systematic review of expert opinions. *Regulatory Toxicology and Pharmacology* 95: 270–279.

Lian, X., Li, Y., Tong, X. et al. (2017). Synthesis of Ce-doped SnO2 nanoparticles and their acetone gas sensing properties. *Applied Surface Science* 407: 447–455.

Lu, Y., Sun, H., Cheng, J. et al. (2020). High performance flexible wearable strain sensor based on rGO and AgNWs decorated PBT melt-blown non-woven Fabrics. *Sensors and Actuators A: Physical* 315: 112174.

Lynch, I. and Dawson, K.A. (2008). Protein-nanoparticle interactions. *Nano Today* 3 (1): 40–47.

Mathew, J., Joy, J., and George, S.C. (2019). Potential applications of nanotechnology in transportation: a review. *Journal of King Saud University - Science* 31 (4): 586–594.

Maynard, A.D. (2016). Chapter 1 - The Challenge of Nanomaterial Risk Assessment. In: *Assessing Nanoparticle Risks to Human Health*, 2ee (ed. G. Ramachandran), 1–20. William Andrew Publishing.

Mihalache, R., Verbeek, J., Graczyk, H. et al. (2017). Occupational exposure limits for manufactured nanomaterials, a systematic review. *Nanotoxicology* 11 (1): 7–19.

Mologni, J.F., Goto, F.K., Neto, D.G. et al. (2008). *Potential Nanotechnology Applications on Automotive Electrical / Electronic Architecture*, SAE Technical Paper, 2008 SAE Brasil Congress and Exhibit, 2008-36-0045, 9.

Nassiri Koopaei, N. and Abdollahi, M. (2016). Opportunities and obstacles to the development of nanopharmaceuticals for human use. *DARU Journal of Pharmaceutical Sciences* 24 (1): 23.

NIOSH (2013). *Occupational Exposure to Carbon Nanotubes and Nanofibers*, CURRENT INTELLIGENCE BULLETIN 65. Cincinnati, OH: Department of Health and Human Services, Centers for Disease Control and Prevention, National Institute for

Occupational Safety and Health Publication No. 2013–145.

Oksel, C., Ma, C.Y., Liu, J.J. et al. (2017). Literature Review of (Q)SAR Modelling of Nanomaterial Toxicity. In: *Modelling the Toxicity of Nanoparticles* (eds. L. Tran, M.A. Bañares and R. Rallo), 103–142. Springer International Publishing.

Parhizkar, M., Mahalingam, S., Homer-Vanniasinkam, S., and Edirisinghe, M. (2017). Latest developments in innovative manufacturing to combine nanotechnology with healthcare. *Nanomedicine* 13 (1): 5–8.

Patel, R.J. and Parikh, R.H. (2020). Intranasal delivery of topiramate nanoemulsion: Pharmacodynamic, pharmacokinetic and brain uptake studies. *International Journal of Pharmaceutics* 585: 119486.

Patel, R.J. and Patel, Z.P. (2013). Formulation Optimization and Evaluation of Nanostructured Lipid Carriers Containing Valsartan. *International Journal of Pharmaceutical Sciences and Nanotechnology* 6: 2077–2086.

Petkov, P., Tsiulyanu, D., Popov, C., and Kulisch, W. (2018). *Advanced Nanotechnologies for Detection and Defence against CBRN Agents*, 1e. Springer Netherlands.

Raies, A.B. and Bajic, V.B. (2016). In silico toxicology: computational methods for the prediction of chemical toxicity. *WIREs Computational Molecular Science* 6 (2): 147–172.

Roblegga, E., Sinner, F., and Zimmer, A. (2009). Health risks of nanotechnology. *EURO-NanoTox-Letters* 1: 1–18.

Saifi, M.A., Khurana, A., and Godugu, C. (2018). Chapter 17 - Nanotoxicology: Toxicity and Risk Assessment of Nanomaterials*Equal contribution. In: *Nanomaterials in Chromatography* (ed. C.M. Hussain), 437–465. Elsevier.

Salamanca-Buentello, F., Persad, D., Court, E. et al. (2005). Nanotechnology and the developing world. *PLoS Medicine* 2: e97.

SCENIHR (2009). 'Opinion on: Risk Assessment of Products of nanotechnologies'. Scientific Committee on Emerging and Newly Identified Health Risks. https://ec.europa.eu/health/archive/ph_risk/committees/04_scenihr/docs/scenihr_o_023.pdf (accessed 1 September 2020).

Shafique, M. and Luo, X. (2019). Nanotechnology in transportation vehicles: an overview of its applications, environmental, health and safety concerns. *Materials (Basel, Switzerland)* 12 (15): 2493.

Sharma, V., Kumar, A., and Dhawan, A. (2012). Nanomaterials: exposure, effects and toxicity assessment. *Proceedings of the National Academy of Sciences, India Section B: Biological Sciences* 82 (1): 3–11.

Shityakov, S., Roewer, N., Broscheit, J.-A., and Förster, C. (2017). In silico models for nanotoxicity evaluation and prediction at the blood-brain barrier level: a mini-review. *Computational Toxicology* 2: 20–27.

Singh, H. (2016). 'Nanotechnology applications in functional foods, opportunities and challenges. *Preventive nutrition and food science* 21 (1): 1–8.

Sobolev, K., Flores Vivian, I., Hermosillo, R., and Torres-Martínez, L. (2008). *Nanomaterials and Nanotechnology for High-Performance Cement Composites*, vol. 254. American Concrete Institute.

Valerio, L.G. Jr. (2009). In silico toxicology for the pharmaceutical sciences. *Toxicology and Applied Pharmacology* 241 (3): 356–370.

Vázquez-Núñez, E., Molina-Guerrero, C.E., Peña-Castro, J.M. et al. (2020). Use of nanotechnology for the bioremediation of contaminants: a review. *Processes* 8 (7): 826.

Vinod, K., Ashish, K.C., Prashant, K., and Saurabh, S. (2019). Nanotechnology: nanomedicine, nanotoxicity and future challenges. *Nanoscience and Nanotechnology-Asia* 9 (1): 64–78.

Warheit, D.B. and Sayes, C.M. (2015). Chapter 1.2 - Routes of Exposure to Nanoparticles: Hazard Tests Related to Portal Entries. In: *Nanoengineering* (ed. P.I. Dolez), 41–54. Elsevier.

Yang, S., Wang, C., Sahin, H. et al. (2015). Tuning the optical, magnetic, and electrical properties of ReSe2 by nanoscale strain engineering. *Nano Letters* 15 (3): 1660–1666.

Yetisen, A.K., Qu, H., Manbachi, A. et al. (2016). Nanotechnology in textiles. *ACS Nano* 10 (3): 3042–3068.

Yu, H., Park, J.-Y., Kwon, C.W. et al. (2018). 'An overview of nanotechnology in food science: preparative methods. *Practical Applications, and Safety', Journal of Chemistry* 2018: 5427978.

Yuan, H., Lei, T., Qin, Y., and Yang, R. (2019). Flexible electronic skins based on piezoelectric nanogenerators and piezotronics. *Nano Energy* 59: 84–90.

Zhang, X.-Q., Xu, X., Bertrand, N. et al. (2012). Interactions of nanomaterials and biological systems: Implications to personalized nanomedicine. *Advanced Drug Delivery Reviews* 64 (13): 1363–1384.

Zhu, X., Zhu, L., Li, Y. et al. (2007). Developmental toxicity in zebrafish (Danio rerio) embryos after exposure to manufactured nanomaterials: buckminsterfullerene aggregates (nC60) and fullerol. *Environmental Toxicology and Chemistry* 26 (5): 976–979.

Part IV

Nanotechnology, Ethics, and Regulatory Framework

15

Safety Regulations for the Use of Nanotechnological Products for Biomedical Applications

A Systematic Literature Review

Raquel Von Hohendorff[1], *Wilson Engelmann*[1,2], *and Daniele Weber S. Leal*[3]

[1] *Graduate Program in Public Law, University of Vale do Rio dos Sinos – UNISINOS, São Leopoldo, Brazil*
[2] *Corporate Law and Business, University of Vale do Rio dos Sinos – UNISINOS, São Leopoldo, Brazil*
[3] *CAPES fellow at the Graduate Program in Public Law, University of Vale do Rio dos Sinos – UNISINOS, São Leopoldo, Brazil*

15.1 Introduction

At the beginning of the twentieth century, many bore witness to the creation of the plane and other science theories that revolutionized the world. Now, after so many years later, this new technology is similarly transforming the society and science, showing promises as a solution to more urgent problems across the globe, and in a remarkably wide range of fields, such as environmental sciences. Nanotechnology causes a great impact on biology and medicine, and is of the ideal size to efficiently interact with biological structures. In so doing, they become useful for *in vivo* and *in vitro* biomedical and medical research. These methods can change the method we study and treat some of the most debilitating illnesses, like cardiovascular and Alzheimer's disease (Mirkin 2016).

The most commercialized products enabled for nanotechnology by nano-sectors in 2014 were nano-electronics, magnetic and photonic crystals (including semiconductors, mobile electronics, and displays, packaging, thermal management, batteries, supercapacitors), nanostructured materials and manufacturing (including the production and use of catalysts, coatings, fiber reinforcements/plastic nanotubes, printing and *roll-to-roll* manufacturing), energy and the environment (including improvements in the oil industry and environmental cleaning, fuel cells, solar cells, filtration, sensors and water treatment, and purification), and health applications (including diagnostic and monitoring sensors, cosmetics, food products, and packaging, personal care products, sunscreens, packaging, surgical tools, implantable medical devices, contrast agents, and drug delivery systems) (Roco 2016). A search of the STATNANO databank (https://product.statnano.com) on 19 September 2020 revealed that 1067 products containing nanoparticles had been registered for medical purposes. They were produced by 367 companies in 40 different countries (STATNANO 2020).

Other areas such as electricity and energy, biomedical materials, and food and agriculture have recognized the potential opportunities that nanotechnology can provide for product development.

Nevertheless, it is the field of medicine that has published the most about nanotechnology and presented the most significant discoveries. Mirkin (2016) has been researching spherical nucleic acids (SNAs), and these SNAs are

Nanotechnology in Medicine: Toxicity and Safety, First Edition. Edited by Mahendra Rai, Mrunali Patel, and Rashmin Patel.

synthesized by immobilizing oligonucleotides (functionalized with final alkylthiol groups) on the surface of spherical gold nanoparticles in a highly oriented manner, because nanostructures have shown extraordinary promise in biology and medicine. Typically, nanotechnology has generated several different products, such as numerous medical diagnostic. Nath (2016) and his team have also been using nanoparticles for biomedical purposes, focusing their attention on nanomaterials with multifunctional magnetic, especially Au-Fe_3O_4. Their multi-functionality can be used since therapy for diagnosis, and an important role in drug delivery, and the reason is that these particles can be functionalized with bioactive molecules.

Even so, it must be remembered that it took a long time to understand the properties of small molecules, and study the safety and toxicity (Philbert 2016). Thus, as the fields that use nanotechnology develop and expand, the Science System needs to be careful and cautions about the implementation and structure of this fruitful type of innovative materials. Risks always exist. The System of Law will also need to keep a close eye on the innovative advancements in nanotechnology.

In this context, the main questions that we intend to answer in this study are: (i) have the biomedical applications of nano-products been the subject of research and scientific publications? (ii) what are the main risks of these products? and (iii) are there safety regulations that can be applied to these products?

Specific nanotechnology research[1] has helped to synthesize the evidence available in the literature about interventions and can help professionals and researchers in their daily work. Here, we will show the current knowledge concerning the use of nanotechnological products in biomedicine. This will be followed by a systematic review of scientific perspectives concerning the biomedical applications of nano-products, and their risks and regulations. In closing, there will be a brief discussion about regulated self-regulation and safety by design as possible strategies for uncertainties.

The theoretical framework underlying this chapter is based on Luhmann's Systems Theory, which has created the setting for a new *scientific style* that better comprehends the current complex societies in which we live and that are at the center of contemporary discussions about the meanings of Law and Society (Rocha 2013). It is a systemic-functionalist perspective that aims to establish the connection between the problem and the solution to be forged, particularly through the observation of normative landmarks that can handle the challenges brought by nanotechnology.

This method is conducive to the development of strategies that tend to transform from interdisciplinary to transdisciplinary, since the intention is to deal with actions from different levels of reality, using diverse semantics. Two paths should be openly traveled so that the Law can take into account the challenges that follow the advancement of nanotechnologies.

The first is the active exploration of other fields of knowledge that may help it understand the complex realities that nanotechnology may engender. The second route is to let incoming ideas from other areas of knowledge join those of the Law.

15.2 The State-of-the-art of Biomedical Applications of Nano-Products

One of the greatest potentials of nanotechnology as a pharmaceutical ally lies in antitumor therapies. Nano-oncology includes diagnostic and therapeutic applications. For example, two of the first products available for nano-oncological treatment were doxorubicin preparations loaded into liposomes and a nanoparticle formulation of paclitaxel (Álvarez Díaz 2016). It is probably one of the areas that have shown the greatest development in diagnostic and therapeutic matters.

Clinical diagnoses and medical therapy for malignant and nonmalignant diseases have appeared on recent achievements in this field, with the potential applications. In a recently published study, Lu et al. (2017), gold nanoparticles have been reported to have unique physical and chemical properties, and explored the pharmacological functions as a possibility of gold nanoparticles in mouse melanoma cells. The study concluded that gold nanoparticles measuring 3–5 nm are potential antitumor agents since they were selectively cytotoxic, inhibiting the motility of melanoma cells. In addition, the suppression of the migration of melanoma cells with gold nanoparticles has been acting in the control of these cells. However, further *in vivo* studies in animal models are needed to assess the safety for normal tissues and to investigate the *in vivo* efficacy in reducing tumor burden and metastases (Lu et al. 2017).

PTX is a known antitumor medication, often used in breast treatment, ovary, among others. The low aqueous solubility limits its use intravenously, and commercialized alternatives have been related with a big level of toxicity. A recent study analyzed the influence of the lipid composition of three different lipid nano-systems – solid lipid nanoparticles, nanostructured lipid carriers (NLCs), and PTX encapsulation in nanoemulsion. The incorporation of PTX into lipid nanotransporters has been considered an interesting nontoxic alternative, but the drug load is generally low. Suggesting that the use of NLCs as vehicles for PTX may be a promising strategy in the treatment of breast cancer, its biological response and results demonstrated that PTX was incorporated into NLCs, in correct physical and chemical characteristics for intravenous administration (Marcial et al. 2017).

The cell-membrane-coated nanoparticulate drug delivery system has been widely studied for prolonged *in vivo* circulation, reticuloendothelial system evasion, active tumor selection, and cancer vaccination, as presented by Hu and Gu (2016), reviewing the good technique of using cell-membrane-coated nanoparticles on cancer treatment. However, the resulting cell-membrane-coated nanoparticles need to match patient cell types to minimize immunogenicity.

Using the most updated techniques in colloid nanoscale chemistry, the formulator has greater size control and can generate monodispersed suspensions. A stable colloid supports the self-assembly of a series of components that can protect the active fraction (medication) from regions outside the target, improving effectiveness, as well as preventing the toxicity system. This mechanism has been used with success in the manufacture of a series of nanomedicines for cancer, such as Abraxane and Danorubicin, which are used to treat Kaposi's sarcoma and other types of cancer (Mu et al. 2014; Ojima et al. 2016).

More recently, in Brazil, two promising new drugs have been developed at Universidade Estadual de Campinas (UNICAMP), Brazil, to fight tumors. Although they have been publicized, they are still in the testing phase. Researchers from the Institute of Biology and the Institute of Chemistry at UNICAMP have developed a nanopharmaceutical that has already been tested on animals with highly promising results. This drug has shown the potential to reduce urinary bladder tumors. The technology, which is available for licensing, registered a very promising index: 80% of the analyzed cases demonstrated decreased tumor activity. Besides showing the capacity to promote regression and inhibit tumor progression in 80% of bladder cancer cases, it has exhibited very low toxicity at therapeutic doses. Due to their large surface area and small size, nanopharmaceuticals act more efficiently and penetrate tumors more deeply. As a result of their high effectiveness, they are administered in lower concentrations than chemotherapeutic drugs, eliminating or minimizing their toxic effects. This technology, which has been registered under the trademark *OncoTherad*, is in an advanced stage of development. It is protected by a patent application in Brazil and abroad, and this widens the possibilities regarding the exploration of the product in other markets by the interested company (Octaviano 2017).

Another novel drug in chemotherapy has also been developed at UNICAMP, a new technology that reduces the side effects of chemotherapy. It is the result of Leandro C. Fonseca's Master of Chemistry, and in his thesis, he explains that the drugs traditionally used in chemotherapy are not soluble in water. As blood contains 92% water, a large amount of medication is needed to reach the compromised cells. However, with the new technology, smaller amounts of the drug are transported to the diseased cells with the help of silica. To illustrate, one can think of a silica nanoparticle as a car that more efficiently transports the drug to the cell, without emitting waste. This is because the nanocarrier is soluble in blood and has a hydrophobic interior, where the drug is contained. The effect is high chemotherapy drug retention. In this way, smaller amounts of the drug are used because the substance reaches its target in appropriate amounts for treatment.[2]

Nanotechnology has been used by scientists in CNPEM – National Center for Research in Energy and Material – to develop an antibiotic for the selective elimination of superbugs without major side effects on human cells. In preliminary tests, the nano-antibiotic showed efficiency when facing resistant forms of the *Escherichia coli* bacteria, transporting a large dose of the active ingredient to the target. The method combined tiny silver particles, covered with silica and antibiotic ampicillin molecules to try to overcome the growing resistance of bacteria to conventional drugs. The surface of the nanoparticle was decorated with certain chemical groups that served to direct it to the selected target. The solutions adopted by the researchers to produce this nanoparticle with an antibiotic function and low toxicity may be strategic for the development of new therapies, not only against bacteria but also viruses and tumors (MCTIC 2017).

Another major medical field in which nanotechnological novelties have been emerging is in the area of combating the free radicals produced by spinal cord injuries, strokes, and cardiac arrests. These injuries can be caused by impact, blood vessel rupture, and tissue necrosis. When arteries constrict or obstruct within the adjacent tissues of an organ, hypoxia occurs. This is a pathological process linked to a lack of oxygen that blocks the final reaction in the respiratory chain. In turn, these free radicals destroy cell membranes and start a sequence of reactions that harms, destroying cells and tissues, making the clinical scenario much more complicated (Scientist 2018).

An innovative therapeutic complex based on synthesized nanoparticle antioxidants will help create an effective rehabilitation system, presents the international team of scientists from Russian and US universities, that managed to find a way to solve the problem of the pathological formation of free radicals in health cases such as spinal injury, stroke, and cardiac arrest. The successful rodent test showed that the substance effectively eliminated free radicals, reduced swelling as well as grades of edema, and rehabilitated the subjects more quickly after spinal cord injury, stroke, or cardiac arrest (Scientist 2018).

In veterinary medicine, nanotechnology has also made its debut and has been touted as the technology that will bring more lasting results and abate the use of other veterinary drugs. In this field, small-size nanocapsules that contain active ingredients can be used to penetrate the deeper layers of animal fur and skin. Prolonged release of the asset allows for longer-lasting results and fewer applications.[3]

Extended asset release was one of the aims behind the development of a new drug for the treatment of leishmaniasis. The Institute of Technological Research at the University of São Paulo (USP) has patented new medication against this endemic disease that is caused by protozoa and transmitted by the bite of certain mosquito species. New nanotechnological drug administration, in the form of nanocarriers, serves to transport the drug. Since the nanocarriers open slowly, the drug sustains its effect over a longer time, supported by the development of less invasive and more efficient topical therapy. This type of advancement also reduces the costs of the national public health-care system. Recently, the original formula of the drug is about to undergo preclinical testing. Existing treatment for leishmaniasis is extremely painful and patients depend on hospital admission to receive doses of the medication and control of the side effects, which are intense and can even lead to death. The new nanomedication is topical and the patient can apply the ointment or cream by themselves (Nanotecnologia 2018).

Not all news is good news, however. New risks have also been emerging. While analyzing how porous nanoparticles loaded with drugs interacted with tumor cells, the chemist Oswaldo Alves observed something unexpected. In addition to the drug (camptothecin, a powerful, highly toxic antitumor agent), the nanostructures transported molecules from the cell culture medium into the cells. The medium is the mixture of vitamins, proteins, and salt minerals that keeps cells alive. The discovery suggested that different from what was previously believed, nanoparticles were not that efficient as vehicles for drug transport. Additionally, it meant that even harmful substances might be taken inside cells. Instead of high-precision pharmaceutical adjuvants, the nanoparticles might become Trojan horses (Andrade 2013).

Researchers from the *Supramolecular Nanomaterials and Interface Laboratory* at the *École Polytechnique Fédérale* in Lausanne, in collaboration with scientists from Milan, Turin, Leiden, and Oregon, studied new purposes for therapeutic nanotechnology and developed three additives for simple and cheap vaccines to circumvent problems maintaining vaccines in room temperature environments. This is a huge challenge when administering vaccines in setting outside offices or hospitals. By using small amounts of nanoparticles, or polymer approved by the FDA (polyethylene glycol), or larger amounts of sucrose, the team was able to stabilize the vaccines at room temperature environment for several weeks or, in some cases, months. What is even more interesting is that, after these results, the researchers managed to stabilize a vaccine against the Chikungunya virus for 10 days, even though it was still in the experimental phase (École Polytechnique Fédérale de Lausanne 2016).

Recent influenza vaccines offered protection against circulating of specific influenza viruses (type A), so a flu vaccine would end these limitations of vaccines related to this type. Deng et al. (2018) discussed their methodology for generating two-layer protein nanoparticles for a universal influenza vaccine. In mice, this vaccine induced immunity with robust durability, and completely protected them against the challenges of influenza A viruses, demonstrating that applications with success of nanotechnology brings big promises for the evolution on new generations of vaccines that fight influenza. Moreover, the abiotic nature of protein nanoparticles also increases the storage capacity of vaccines outside refrigerated conditions which is a correct property when administering vaccines in locations outside of a hospital and office environment (Deng et al. 2018).

Restorative medicine is another field that has shown great interest in nanotechnology, with the development of prostheses using materials that generate smaller reactions. In the 1960s and 1970s, the first generation of materials to be used inside the human body was produced and, at that time, they were designed to be biologically inert. In the 1980s, the search for the second generation of biomaterials led to a change: a new preference for re-absorbable or bioactive material. The most current idea is that these properties can be combined to generate materials that induce a healing modality similar to regeneration (i.e. minimizing immune responses and inhibiting infections). Advances in tissue engineering combine cells and their biomolecules in artificial constructions to compensate for bodily functions that have been partially or wholly lost due to disease or accidents (Álvarez Díaz 2016).

Current trends point to a biomimetic strategy; that is, the attempt to mimic what occurs in biological processes. The three basic elements of this strategy are bioactive signaling molecules, cell-based therapies, biomaterials and smart implants (Carmona Ribeiro 2010). The pursuit of biomaterials and *intelligent* implants seeks to design materials that immediately react to modify in the environment to tease specific cellular responses at molecular levels (Schenke-Layland et al. 2009). Another line of research has striven to improve nonabsorbable materials to manipulate their interactions with biological elements at the nanometer scale, to improve the function and duration of transplanted organs.

Recently, an article published by Kumar et al. (2017), related to promises of approaches on neuroscience, summarized basic concepts of neuroscience, as well as the recent way of nanotechnology in the study of neuronal function. The authors described the future applications of this emerging technology, and addressed several concerns about the important role of nanomaterials in neuroscience.

Examples of nanotechnology in neuroscience, like a potential diagnostic and therapeutic application, includes administration of drugs on the central nervous system, using nanomaterials on neuroprotection, nanomaterial-based approaches to neural regeneration, neurosurgery, and for neuroimaging. The main focus of this article was to analyze the implications of these recent discoveries and suggest guidelines for future research and the development of nanomaterials for the advancement of neuroscience. Nano-neuroscience is an emerging field that can greatly impact the understanding of neural circuits and neurological treatment (Kumar et al. 2017).

Given all this information, it is clear that nanotechnology is very much present in the realm of biomedical products and applications. Given this reality, the following is the result of the search we performed for articles on the topic published in the last 10 years.

15.3 The Scientific Perspective on Nanoproducts for Biomedical Applications, Risks, and Regulations

This systematic review was carried using peer-reviewed journals available in the Portal Periódicos Capes databases. Searches included research published in the last 10 years, from all areas of knowledge, using the keywords *nanotechnologies and safety regulations,* and *biomedical applications and risks.* A total of 190 articles were collected and, because of the multidisciplinary nature of the topic at hand, there was no filtering of study areas.

Regarding the practical criteria for the selection and analysis of bibliography – also called inclusion and exclusion criteria by Fink (2014) – we established a period of 10 years to include reports published between 2010 and 2020. Moreover, articles in English were analyzed and checked for relevant search term(s) in their title, abstract, and/or keywords. Titles, abstracts, and keywords were read to identify whether the articles met the established criteria and whether they were related to the theme of our research. Based on the defined criteria, 189 articles were initially obtained. The titles and abstracts of the selected texts were read. After this reading, 30 articles were selected that deal with biomedical risks and applications and 23 that address regulations (3 of which are adequate for both categories).

Figure 15.1 illustrates the distribution of selected articles that deal with risks published within the time frame established in the methodology, and Figure 15.2 shows the studies on the regulation of biomedical products and nanotechnology published between 2010 and 2020.

The abstracts of the selected articles were read in full and their main information is summarized in Table 15.1 (biomedical applications and risks) and Table 15.2 (biomedical regulations and applications).

Table 15.2 summarizes the studies found in our search of the Portal de Periódicos Capes database that mention the regulation of biomedical products and nanotechnology.

Upon analyzing the selected articles, there seemed to be no complete definition of the risks of nanoparticles or any specific regulation on the topic. In the meantime, the systematic review of the literature revealed a series of gaps regarding how to act safely with nano-products – mainly due to the lack of regulation about biomedical products with nanotechnology. Some studies indicated that few risks have been identified by Science about these materials. Nevertheless, it is known that especially because of the different behavior of materials at the nanoscale – due to the diverse properties of matter at this scale – there are numerous possible risks. Therefore, an analysis of the life cycle of products is essential for understanding the real risks, particularly those of nano-products for biomedical applications since they cause even greater concerns.

Although today having benefit of nanotechnologies as general thought – because of considerable advantages they bring to our daily lives – we face potential risks in human health and the environment that should not be underestimated. Due to the particle size, these materials may be organized by physical laws that are quite distinct from the science we are accustomed, starting with opportunities for the nanoparticles to show a higher degree of toxicity than larger particles. That is the "why" of necessity to assess the risks that arise from the manipulation, development, and application of these new technologies. As well as evaluations of the impacts on human and environmental health, also toxicity studies and the establishment of appropriate methods for toxicity testing are also necessary (Hohendorff and Engelmann 2014).

The nanoworld is immense realm that we are just beginning to explore and, under the best of circumstances, we may come across some of the chemical, physical, as well biological rules we know. However, what can happen to materials at the macro scale may not be of any relevance in this new domain. Such exploration requires mostly interdisciplinary background. Paradoxically, nanotechnology can be bane or a boon at the same time. *Nanosafety* is the way (Gatti and Montanari 2015).

Therefore, using nanotechnologies for Health System applications as the object of research, some have sought a basis on which to structure new methods of governance (as opposed to traditional coercive) (Ansell and Baur 2018). New modes that may present great solution for highly uncertain and contested risks, such as those related to new technologies. Nanoworld provide a valuable testing ground and better models of governance, because well-established ideas about appropriate courses of action do not yet exist. The conventional commanding regulation method seems to be insufficient and inappropriate (Stokes 2013). Here, creative spaces of innovative natures are inserted into regulation, seeking to systematize information about risks, risk management, and methodologies for the development of a production cycle for certain nanoparticle-based products, without endangering human life and health or the environment

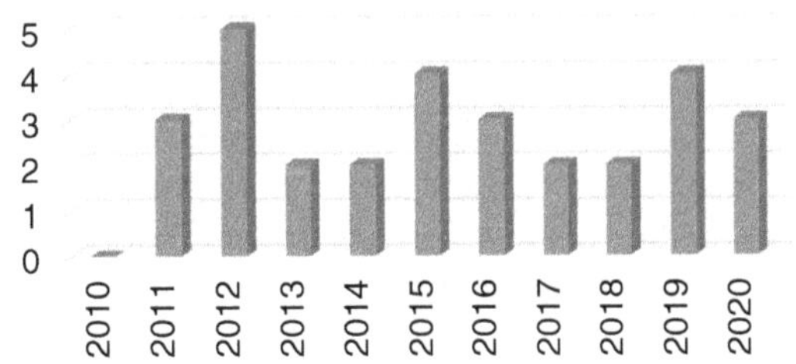

Figure 15.1 The number of studies on the regulation of biomedical nano-products published between 2010 and 2020 on the Portal de Periódicos Capes database (per year).

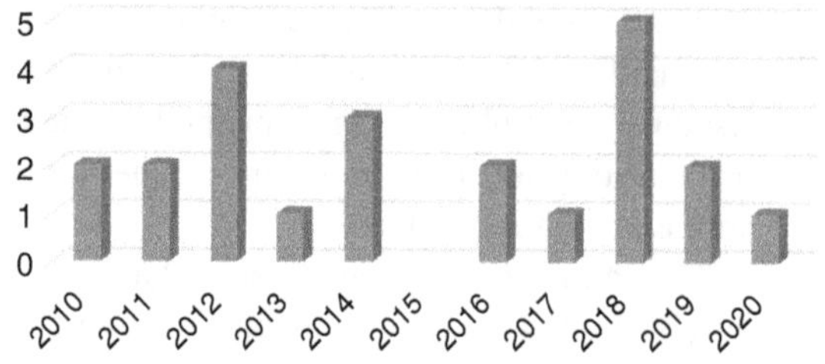

Figure 15.2 The number of studies on the regulation of biomedical nano-products published between 2010 and 2020 on the Portal de Periódicos Capes database (per year).

(Ramanathan 2019). The aim is to make information available to all actors involved with nanotechnologies, without creating obstacles or complexities for scientific, industrial, and economic progress (Malloy 2011), as summarized in Table 15.3.

Table 15.3 shows the challenges that conventional regulation presents, especially due to the lack of clear and sufficient information about risks, limited methodologies, inadequate techniques for toxicological studies, and the rapid pace of the scientific and technological development of nanotechnologies. In the Responsible Research and Innovation (RRI) scenario, new opportunities open for the creation of flexible, provisional, and real models that may help to balance the challenges summarized by Malloy (2011).

15.4 Using RRI as a Methodological Pathway Toward Communication Between the Science System and the Law System

These terms are more integrative, as they cover issues of engineering ethics, participation, technology assessment, early governance, and scientific ethics. The terms development, research, and responsible innovation have been increasingly adopted in current years. They also add reflexivity to the development and design of technology. In this sense, responsible development and innovation can be a new umbrella phrase for new designs that can be characterized by bridging gaps among the practices of innovation, engineering ethics, technology assessment, governance research, and social sciences; more direct involvement in ethical and social issues during the innovation process through integrating approaches to development and innovation; and for presenting new forms of innovation processes and technological governance, according to reflections about responsibility, and transparent distribution of responsibility among actors; and supported constructive paths for the coevolution of technology and regulatory frameworks in society (Grunwald 2011). In *The Vision of Responsible Innovation*, Von Schomberg (2013) mentions that the RRI compose one transparent and interactive process, through which social and innovative actors become sensitive to mutual needs, with one understanding of acceptable ethics, sustainability, conveniences of the social process of innovation, and marketable products. All of this creates true inclusion in the science advances in our world.

Also in 2013, the study of European Commission related to RRI stated an interesting concept of it: *Responsible Research and Innovation* refers to the ways of proceeding in research and innovation that allow those who start and those who are involved in some stage of the processes of research and innovation: to effectively evaluate results and options in terms of moral values (including, among others, welfare, justice, equality, privacy, autonomy, safety, security, sustainability, accountability, democracy, and efficiency); to obtain knowledge relevant to the consequences of the results of their actions and the range of options that they open; as well to use these fists considerations as functional requirements for

Table 15.1 The studies mentioning the risks of biomedical nano-products published between 2011 and 2020 on the Portal de Periódicos Capes database.

Author/year	Study title	Summary of key findings
Roco et al. (2011)	Innovative and Responsible Governance of Nanotechnology for Societal Development.	The governance of nanotechnology development for societal benefit is a challenge with many facets ranging from fostering research and innovation to addressing ethical concerns and long-term human development aspects. In this sense, the US nanotechnology governance approach has aimed to be "transformational, responsible, and inclusive, and [to] allow visionary development."
Kong et al. (2011)	Experimental considerations on the cytotoxicity of nanoparticles.	The unintended adverse effect of nanoparticle exposure is a growing concern both academically and socially. In this context, the toxicity of nanoparticles has been extensively studied; however, several challenges are encountered due to the lack of standardized protocols. In order to improve the experimental conditions of nanoparticle toxicity studies, serious consideration is critical to obtain reliable and realistic data. The cell type must be selected considering the introduction route and target organ of the nanoparticle. In addition, the nanoparticle dose must reflect the realistic concentration of nanoparticles and must be loaded as a well-dispersed form to observe the accurate size- and shape-dependent effect. In deciding the cytotoxicity assay method, it is important to choose the appropriate method that could measure the toxicity of interest without the false-negative or -positive misinterpretation of the toxicity result.
Timmermans et al. (2011)	Ethics and Nanopharmacy: Value Sensitive Design of New Drugs.	Although applications are being developed and have reached the market, nanopharmacy to date is generally still conceived as an emerging technology. It is still debated whether its features give rise to new ethical issues or that issues associated with nanopharma are merely an extension of existing issues in the underlying fields. This article informs that, regardless of the alleged newness of the ethical issues involved, developments occasioned by technological advances affect the roles played by stakeholders in the field of nanopharmacy to such an extent that this calls for a different approach to responsible innovation in this field.
Ahmad et al. (2012)	Environmental applications and potential health implications of quantum dots.	Quantum dots (QDs) are routinely employed for bioimaging applications and detection of pathogens and toxins. Their use as surrogates to study the fate and transport of nonfluorescent nanoparticles is limited due to high cost, detection of limit issues, and lack of sufficient data related to health effects. This review summarizes the studies conducted with QDs with a focus on environmental applications and provides toxicity data important to human health.

(Continued)

Table 15.1 (Continued)

Author/year	Study title	Summary of key findings
Lu et al. (2012)	Ethics in Nanotechnology: What's Being Done? What's Missing?	While numerous technological advancements in nanotechnology have been achieved over the past decade, research into the broader societal issues of nanotechnology is still in its early phases. Based on the data from the Web of Science database, this study points the main path analysis, cluster analysis, and text mining tools to explore the main research fronts and hierarchical structure of these societal issues. In that sense, the research studies fell into four categories: "General Toxicity and EHS (Environment, Health and Safety)," "Medicine and Cytotoxicity," "Assessment and Regulation," and "Environment and Ecotoxicity." These research studies have disclosed much information about the potential effect of nanotechnology on public health and the environment. Relatively speaking, the studies on the assessment, regulation, preventive solutions, and environmental protection are just emerging.
Sharma et al. (2012)	Nanomaterials: Exposure, Effects, and Toxicity Assessment.	The increasing use of engineered nanomaterials (ENMs) in consumer and industrial products has aroused global concern regarding their fate in biological systems and resulting in a demand for parallel risk assessment. A variety of ENMs with different chemical compositions, synthesized through different methods, have been shown to be genotoxic and cytotoxic in different organ specific cell lines (*in vitro*) and mouse. However, there is a dilemma in the selection and validation of the test methods for the ENMs' characterization, dose selection, cytotoxicity, and genotoxicity assessment, because of the altered behavior of ENMs as compared to the chemicals. A multidisciplinary team effort from material scientists, molecular biologists, toxicologists, and physicists is necessary as it will facilitate the interlinking of different facets of nanotoxicology, thus aiding in the understanding of cellular responses to nanomaterials exposure and mechanisms involved.
Hall et al. (2012)	A Portrait of Nanomedicine and Its Bioethical Implications.	Study about the current and future potential of nanomedicine, and its ethical considerations within the comprehensive framework of the four dimensions of medical ethics: Beneficence, Non-Maleficence, Respect, and Justice. While these ethical challenges are not unique to nanomedicine, some require additional consideration, given the envisioned pervasive impact of nanomedicine on society.
Hogle (2021)	Concepts of Risk in Nanomedicine Research.	This article discusses framings of risk in debates over medical innovations such as nanomedicine, and draws attention to organizational and institutional forms of risk which are less visible in bioethical policy debates. While significant, possibly unique risks may exist in specific nano-based products, risk may also arise from the very processes and procedures that develop, test, and oversee any novel technology. This supports recommendations to coordinate efforts through an interagency Working Group and a Secretary-level Advisory Committee to provide flexibility and sensitivity to emerging issues of concern.

Table 15.1 (Continued)

Author/year	Study title	Summary of key findings
Donaldson et al. (2013)	Nanoparticles and the cardiovascular system: a critical review.	Epidemiological, panel, and controlled human chamber studies clearly demonstrate that exposure to combustion-derived nanoparticles (CDNPs) is associated with multiple adverse cardiovascular effects in both healthy individuals and those with pre-existing cardiovascular disease. There is currently little information available on the impacts of manufactured nanoparticles in humans, but experimental studies demonstrate similarities to the detrimental cardiovascular actions of CDNPs. This review describes the evidence for these cardiovascular effects and attempts to resolve the paradox between the adverse effects of the unintentional exposure of CDNPs and the intentional delivery of manufactured NPs for medical purposes.
Sahoo (2013)	Would You Mind, If We Record This? Perceptions on Regulation and Responsibility among Indian Nanoscientists.	The question of regulation and responsibility in the interaction between science, technology, and society is one of the most pressing issues. The paper makes an attempt to contribute significantly to an increased body of knowledge regarding how scientists think and talk about science. Little has been documented about perceptions of nanotechnology regulation and responsibility in developing countries. Given the importance of perceptions in the genetically modified foods debate, the way nanotechnology is perceived holds serious repercussions for the framing of its ethical, legal, and social implications. Through a field-survey that records the opinions of leading Indian nanoscientists, the paper examines scientists' perceptions about nanotech-regulation. Such discussion is imperative to address technological risks and uncertainties.
Lohani et al. (2014)	Nanotechnology-Based Cosmeceuticals.	This review outlines the different nanoparticles used in various classes of cosmeceuticals, nanotechnology-based cosmeceutical products present in the market, and the potential risk caused by nanoparticles on exposure and recent regulatory steps taken to overcome them.
Martirosyan and Schneider (2014)	Engineered nanomaterials in food: implications for food safety and consumer health.	In this study, it was summarized the existing data on the (potential) use of engineered nanomaterials (ENMs) in the food industry, information on the toxicity profiles of the commonly applied ENMs, such as metal (oxide) nanoparticles (NPs), address the potential food safety implications and health hazards connected with the consumption of nanofood. A number of health complications connected with the human exposure to ENMs are discussed, demonstrating that there is a real basis for the arisen concern not only connected with the gut health, but also with the potency to lead to systemic toxicity. The toxicological nature of hazard, exposure levels, and risk to consumers from nanotechnology-derived food are on the earliest stage of investigation and this review also highlights the major gaps that need further research and regulation.

(Continued)

Table 15.1 (Continued)

Author/year	Study title	Summary of key findings
Sajid et al. (2015)	Impact of nanoparticles on human and environment: review of toxicity factors, exposures, control strategies, and future prospects.	Because of their small size and other unique characteristics, nanoparticles have ability to harm human and wildlife by interacting through various mechanisms. This study reviews the characteristics of nanoparticles which form the basis of their toxicity. Also reviews possible routes of exposure of nanoparticles to human body and strategies which can help to control human exposures to toxic nanoparticles.
Stark et al. (2015)	Industrial applications of nanoparticles.	The article highlights established industrial applications of nanoparticles and then moves to rapidly emerging applications in the chemical industry and discusses future research directions. Contributions from large companies, academia, and high-tech start-ups are used to elucidate where academic nanoparticle research has revolutionized industry practice. A nanomaterial-focused analysis discusses new trends, such as particles with an identity, and the influence of modern instrument advances in the development of novel industrial products.
Martin et al. (2015)	Edible Bio-Based Nanostructures: Delivery, Absorption, and Potential Toxicity.	This study will provide an insight into the recent innovations and challenges in the field of delivery via gastrointestinal (GI) tract using bio-based nanostructures. Also, an overview of the approaches followed to ensure an effective deliver (e.g. avoiding physiological barriers) and to enhance stability and absorptive intestinal uptake of bioactive compounds will be provided. Information about nanostructures' potential toxicity and a concise description of the *in vitro* and *in vivo* toxicity studies will also be given.
PourGashtasbi (2015)	Nanotoxicology and challenges of translation.	This paper focuses on the processes of translation in nanomedical research practices and contributes to a context- and object-centered research agenda in Science and Technology Studies. This study addresses how nano-specific issues in medical research are exacerbated by uncertainty and unpredictability. Scientists are involved in highly uncertain processes, which require a contingent and experimental approach to nano-objects in everyday laboratory practices. Consequently, the dealings with nanomedical materials evoke a reformulation of numerous traditional forms of toxicological knowledge and knowledge practices, and challenge the self-concept of toxicology as a testing discipline.
Singh et al. (2016)	Nanotechnology: a future tool to improve quality and safety in meat industry.	Keeping the pace with other industries, the meat industry has adopted the nanotechnology in a range of applications to improve the quality and safety of products. The potential applications include the improvement in the tastes, texture, flavor, production of low fat and salt products, enhanced nutrient absorption, improved packaging techniques, and better pathogen detection system. However, some safety issues need to be addressed before taking a ride on the technology at the full throttle – discussion present in this paper.

Table 15.1 (Continued)

Author/year	Study title	Summary of key findings
Spruit et al. (2016)	Informed Consent in Asymmetrical Relationships: an Investigation into Relational Factors that Influence Room for Reflection.	Although there is a possibility to learn from experiences with informed consent in treatment and research contexts, it should be aware that informed consent traditionally pertains to certain features of the relationships between doctors and patients and researchers and research participants, rather than those between producers and consumers and employers and employees, which are more prominent in the case of engineered nanomaterials. To better understand these differences, it was identified three major relational factors that influence whether valid informed consent is obtainable, namely dependency, personal proximity, and existence of shared interests. Therefore, it was shown that each type of relationship offers different opportunities for reflection and therefore poses distinct challenges for obtaining valid informed consent.
Giannakou et al. (2016)[a]	A comparison of immunotoxic effects of nanomedicinal products with regulatory immunotoxicity testing requirements.	This study compares the current regulatory immunotoxicity testing requirements with the accumulating knowledge on immunotoxic effects of nanomedicinal products (NMPs) in order to identify potential gaps in the safety assessment. This comparison showed that immunotoxic effects, such as complement activation-related pseudoallergy, myelosuppression, inflammasome activation, and hypersensitivity, are not readily detected by using current testing guidelines. Immunotoxicity of NMPs would be more accurately evaluated by an expanded testing strategy that is equipped to stratify applicable testing for the various types of NMPs.
Pillai and Bezbaruah (2017)	Perceptions and attitude effects on nanotechnology acceptance: an exploratory framework.	This paper proposes an exploratory nanotechnology perception-attitude-acceptance framework (Nano-PAAF) to build a systematic understanding of the phenomenon. The framework proposes that perceptions of risks and benefits of nanotechnology are influenced by cognitive, affective, and sociocultural factors. The sociodemographic factors of consumers and contextual factors mitigate the influence of cognitive, affective, and sociocultural factors on the perception of risks and benefits. The perceived risks and benefits in turn influence people's attitude toward nanotechnology, which then influences acceptance of nanotechnology products.
Boyes et al. (2017)	A comprehensive framework for evaluating the environmental health and safety implications of engineered nanomaterials.	This review provides a framework to evaluate the health and safety implications of engineered nanomaterials (ENM) releases into the environment, including purposeful releases such as for antimicrobial sprays or nano-enabled pesticides, and inadvertent releases as a consequence of other intended applications. Emphasis is placed on characterizing ENM as they occur in environmental or biological matrices.

(Continued)

Table 15.1 (Continued)

Author/year	Study title	Summary of key findings
Malsch et al. (2018)[a]	Decision Support for International Agreements Regulating Nanomaterials.	In the project on Sustainable Nanotechnologies (SUN, www.sun-fp7.eu), a Decision Support System (SUNDS) has been developed, primarily for confidential use by risk and sustainability managers inside a company or consortium. In this study, it was formulated a scenario concerning a potential role for an open access decision support system in negotiations on international agreements regulating trade in nanomaterials. The scenario includes design rules for decision support systems as well as procedures for use of such a system in stakeholder dialogue and policy-making on governance of these and other emerging technologies.
Kaphle et al. (2018)[a]	Nanomaterials for agriculture, food, and environment: applications, toxicity, and regulation.	In this paper, it was reviewed metal nanoparticles, nanometal oxides, carbon nanotubes, liposomes, and dendrimers. It was presented the application of these nanomaterials in agriculture, food, and environment for plant protection; disease treatment; packing materials; development of new tastes, textures, and sensations; pathogen detection; and delivery systems. It was discussed risk assessment of nanomaterials and toxicological impacts of nanomaterials on agriculture, food, and environment. Also, it was presented regulatory guidelines for the safer use of nanomaterials.
Jampilek et al. (2019)	Potential of Nanomaterial Applications in Dietary Supplements and Foods for Special Medical Purposes.	As nanotechnology provides "a new dimension" accompanied with new or modified properties conferred to many current materials, it is widely used for the production of a new generation of drug formulations, and it is also used in the food industry and even in various types of nutritional supplements. These nanoformulations of supplements are being prepared especially with the purpose to improve bioavailability, protect active ingredients against degradation, or reduce side effects. This contribution comprehensively summarizes the current state of the research focused on nanoformulated human and veterinary dietary supplements, nutraceuticals, and functional foods for special medical purposes, their particular applications in various food products and drinks as well as the most important related guidelines, regulations, and directives.
Riediker et al. (2019)	Particle toxicology and health – where are we?	Over the past decades, it was acquired considerable experience in how particle properties and particle-bio interactions are linked to human health. This insight is useful for improved risk management in the case of unwanted health effects but also for developing novel medical therapies. The concepts that help to understand particles' and fibers' risks include the fate of particles in the body; exposure, dosimetry, and dose-metrics and the 5 Bs: bioavailability, biopersistence, bioprocessing, biomodification, and bioclearance of (nano)particles. This includes the role of the biomolecule corona, immunity, and systemic responses, nonspecific effects in the lungs and other body parts, particle effects and the developing body, and the link from the natural environment to human health.

Table 15.1 (Continued)

Author/year	Study title	Summary of key findings
Das et al. (2019)	The Sustainability Challenge of Food and Environmental Nanotechnology: Current Status and Imminent Perceptions.	For nanomaterial applications in food processing, such as antimicrobials on food contact surfaces along with the improvement of biosensors, electrospun nanofibers are the most intensively studied ones. The current review aimed to provide an update regarding the sustainability of current nanotechnology applications in food science technology, environment, and public health together with a risk assessment and toxicity evaluation.
Ranjan et al. (2019)[a]	Toxicity and regulations of food nanomaterials.	The presence of nanomaterials in food induces potential risks of toxicity because nanoparticles can easily cross the barriers of human anatomy, through respiratory, dermal, and gastrointestinal routes. This review describes nanomaterial–cell interactions that induce toxicological responses. This study discusses the toxicity of food nanomaterials; mechanisms of oxidative stress, genotoxicity and carcinogenicity, and safety regulations.
Baranowska-Wójcik et al. (2020)	Effects of Titanium Dioxide Nanoparticles Exposure on Human Health – a Review.	The possible risk to health after consuming food containing nanoparticles has been poorly explored but it is supposed that the toxicity of nanoparticles depends on their size, morphology, rate of migration, and amount consumed. Scientific databases inform that titanium dioxide nanoparticles-TiO_2 NPs can induce inflammation due to oxidative stress. They can also have a genotoxic effect leading to, among others, apoptosis or chromosomal instability. In this sense, this paper gives a review of previous studies concerning the effects of exposure to TiO_2 NPs on a living organism (human, animal).
Pramanik et al. (2020)	Application of Nanotechnology in Agriculture.	Nanotechnology (NT) has been identified as a potential technology for reviving the agriculture and food industry and can improve livelihood of poor. In agricultural sector in particular, nanotechnology plays an important role in crop production, food processing and packaging, food security and water purification, environmental remediation, crop improvement, and plant protection. Nanotechnology has the potential of precise delivery of agrochemicals for improving disease resistance, plant growth, and nutrient use. The use of nanomaterials is quite new in agriculture and it requires additional research. Social and ethical repercussions of nanotechnology uses in agriculture have to be considered. Before commercialization and field application, toxicity of nanomaterials has to be evaluated.
Zielińska et al. (2020)	Nanotoxicology and Nanosafety: Safety-by-Design and Testing at a Glance.	This review offers a systematic discussion about nanotoxicology and nanosafety associated with nanomaterials during manufacture and further biomedical applications. A special focus is given to inorganic nanoparticles such as titanium dioxide nanoparticles (TiO_2NPs) and silver nanoparticles (AgNPs). *In vitro* and *in vivo* case studies for the selected nanoparticles are discussed. The final part of this work describes the significance of nano-security for both risk assessment and environmental nanosafety. "Safety-by-Design" is defined as a starting point consisting on the implementation of the principles of drug discovery and development.

Table 15.2 Studies mentioning the regulation of biomedical nano-products published between 2010 and 2020 on the Portal de Periódicos Capes database.

Author/year	Study title	Summary of key findings
Groso et al. (2010)	Management of nanomaterials safety in research environment.	In this work, the authors presented a practical, "user-friendly" procedure for a university-wide safety and health management of nanomaterials, developed as a multi-stakeholder effort (government, accident insurance, researchers, and experts for occupational safety and health). The process starts using a schematic decision tree that allows classifying the nano laboratory into three hazard classes similar to a control banding approach (from Nano 3 – highest hazard to Nano1 – lowest hazard). Classifying laboratories into risk classes would require considering actual or potential exposure to the nanomaterial as well as statistical data on health effects of exposure. Due to the fact that these data (as well as exposure limits for each individual material) are not available, risk classes could not be determined. For each hazard level, we then provide a list of required risk mitigation measures (technical, organizational, and personal).
Quintanilla-Carvajal et al. (2010)	Nanoencapsulation: A New Trend in Food Engineering Processing.	In this paper, the different techniques that have been developed for the production of nanocapsules are discussed, and examples of their application are provided including regulatory aspects on products of nanotechnology. Also, it is illustrated a proposal of classification and characterization of the different structural arrangements of core–shell materials in nanoencapsulate composites found in the literature.
Leach et al. (2011)	The European nanometrology landscape.	The final section describes biological nanometrology, which is perhaps the most interdisciplinary applications area, and presents unique challenges. Within each area, a review is provided of current status, the capabilities and limitations of current techniques and instruments, and future directions being driven by emerging industrial measurement requirements. Issues of traceability, standardization, national and international programmes, regulation and skills development will be discussed in a future paper.
Birnbaum et al. (2011)	Environmental health science for regulatory decision-making.	This paper will discuss the role of the National Institute of Environmental Health Sciences (NIEHS) in supporting the scientific underpinnings of regulatory decision-making by EPA and other agencies through NIEHS's commitment to conducting and funding outstanding research in environmental health sciences. Herein, we present and describe the NIEHS, its mission, and its research program; discuss some basic ideas in environmental health; and then outline some of the major new concepts in the field and explain how they may affect regulatory science.
Lu et al. (2012)[a]	Ethics in Nanotechnology: What's Being Done? What's Missing?	These research studies have disclosed much information about the potential effect of nanotechnology on public health and the environment. Relatively speaking, the studies on the assessment, regulation, preventive solutions, and environmental protection are just emerging. This finding indicates that an abundance of effort should be conducted on these emerging themes to maximize the benefits of nanotechnology while minimizing its potential harm. The implications for various parties in this domain are also presented.

Table 15.2 (Continued)

Author/year	Study title	Summary of key findings
D'Silva et al. (2012)	A game with rules in the making – how the high probability of waiting games in nanomedicine is being mitigated through distributed regulation and responsible innovation.	Nanotechnology in healthcare (often labeled as nanomedicine) presents a situation where waiting games are a strong possibility; however, waiting games have been avoided (so far) through the unlocking of the dominant regulation regime. This paper describes how an innovation impasse in nanomedicine could have ensued if the traditional wait and watch strategies of legislators and regulators had been followed.
Dalton-Brown (2012)	Global Ethics and Nanotechnology: A Comparison of the Nanoethics Environments of the EU and China.	The following article offers a brief overview of current nanotechnology policy, regulation and ethics in Europe, and The People's Republic of China with the intent of noting (dis) similarities in approach, before focusing on the involvement of the public in science and technology policy (i.e. participatory Technology Assessment).
Gouvea et al. (2012)	Emerging Technologies and Ethics: A Race-to-the-Bottom or the Top?	In summary, a race-to-the-top supports invention and commercialization of emerging technologies. A critical finding as it suggests that issues such as corruption and regulation could be critical in limiting the extraction of social and economic benefit from emerging technologies (like nanotechnologies).
Ruggiu (2013)	Temporal Perspectives of the Nanotechnological Challenge to Regulation: How Human Rights Can Contribute to the Present and Future of Nanotechnologies.	This is particularly apparent if we consider the viewpoint of the regulator who faces two different ranges of problems with regard to the market and research and is expected, simultaneously, to manage two different types of regulatory instruments, pursuing at the same time the same goals. Instead, by favoring only the future scenarios, the regulator runs the risk of abruptly shifting from more flexible and elastic tools to forms of hard legislation, wasting, thus, the resources of the soft-regulation, in particular those of self-regulation. By referring primarily to the European context, human rights, also thanks to their normative structure of principles, can help in strengthening both the legislation needed for regulating the present market and the soft instruments needed for steering research and for fostering the stakeholders' participation without sacrificing the coherence of the regulatory response.
Guo et al. (2014)	Development of Taiwan's strategies for regulating nanotechnology-based pharmaceuticals harmonized with international considerations.	This article reviews the trends of regulating nanotechnology-based pharmaceuticals in the international community and proposes strategies for Taiwan's regulation harmonized with international considerations. The draft regulatory measures include a chemistry, manufacturing, and controls (CMC) review checklist and guidance for CMC review of liposomal products. These have been submitted for discussion among an expert committee, with membership comprised of multidisciplinary academia, research institutions, the pharmaceutical industry, and regulators, and are currently approaching final consensus.

(*Continued*)

Table 15.2 (Continued)

Author/year	Study title	Summary of key findings
Ranjan et al. (2014)	Nanoscience and nanotechnologies in food industries: opportunities and research trends.	This review is focused on application of nanotechnology in food industries. It also gives insight on commercial products in market with usage of nanomaterials, current research, and future aspects in these areas. Currently, they are being incorporated into commercial products at a faster rate than the development of knowledge and regulations to mitigate potential health and environmental impacts associated with their manufacturing, application, and disposal. As nanomaterials are finding new application every day, care should be taken about their potential toxic effects.
Mandel and Marchant (2014)	The living regulatory challenges of synthetic biology.	This Article presents one of the first assessments of the regulatory and oversight challenges produced by the beneficial application of synthetic biology, for energy, environmental, medical, and other purposes. Due to the uncertainty present at this early stage of synthetic biology development, and the practical political context, it is unlikely that the significant statutory and regulatory gaps identified herein could be cured directly. This Article recommends instead a selection of "soft law" alternatives that could more quickly provide flexible and adaptive measures to help fill regulatory gaps in a manner that allows this promising technology to develop as rapidly as possible, while still adequately guarding against risks to human health and the environment.
León-Silva et al. (2016)	Silver Nanoparticles (AgNP) in the Environment: a Review of Potential Risks on Human and Environmental Health.	A grading system is proposed to rank and evaluate toxicological properties of AgNP, which can be useful in supplying assistance on the classification of the priorities and concerns in the regulatory and standardization policies of the occupational health and safety issues on nanomaterials.
Giannakou et al. (2016)[a]	A comparison of immunotoxic effects of nanomedicinal products with regulatory immunotoxicity testing requirements.	The aim of this study was therefore, to compare the current regulatory immunotoxicity testing requirements with the accumulating knowledge on immunotoxic effects of NMPs in order to identify potential gaps in the safety assessment. This comparison showed that immunotoxic effects, such as complement activation-related pseudoallergy, myelosuppression, inflammasome activation, and hypersensitivity, are not readily detected by using current testing guidelines. Immunotoxicity of NMPs would be more accurately evaluated by an expanded testing strategy that is equipped to stratify applicable testing for the various types of NMPs.
Evans et al. (2017)	Ethics, Nanobiosensors, and Elite Sport: The Need for a New Governance Framework.	The authors concluded that although the future potential use of Nanobiosensors in sports analysis offers many potential benefits, there is also a fear that it could be abused at a sporting system level. Hence, it is essential for sporting bodies to consider the development of a robust ethically informed governance framework in advance of their proliferated use.

Table 15.2 (Continued)

Author/year	Study title	Summary of key findings
Romero-Franco et al. (2018)	Assessment of information availability for environmental impact assessment of engineered nanomaterials.	Environmental impact assessments of engineered nanomaterials can be hampered by the lack of data/information and thus delay the development of effective regulatory policies. To address this issue, a systematic approach (termed here "IANano") was developed and demonstrated for assessing information availability for environmental impact assessment (EIA) of engineered nanomaterials. In IANano, following the typical EIA process, the required information elements for exposure and hazard potential assessments are classified based on major categories, subcategories, and attributes. Scores for the different information attributes are then assigned, based on a selected scoring scale and weights, and aggregated up to the level of exposure and hazard potential information (EPI and hazard potential information [HPI], respectively), considering both available and unavailable information, via the Dempster–Shafer algorithm.
Garden and Winickoff (2018)	Gene editing for advanced therapies: Governance, policy, and society.	Experts noted that the trajectory of gene editing in research and development and the uptake of future therapies in the clinical setting remain unclear due to uncertainties in the scientific, regulatory, and economic landscapes. Many policy issues are also raised in the context of other emerging technologies. Governance must cope with a moving technical frontier and some level of uncertainty around risks and benefits.
Kaphle et al. (2018)[a]	Nanomaterials for agriculture, food, and environment: applications, toxicity, and regulation.	The authors presented the application of these nanomaterials in agriculture, food, and environment for plant protection; disease treatment; packing materials; development of new tastes, textures, and sensations; pathogen detection; and delivery systems. We discuss risk assessment of nanomaterials and toxicological impacts of nanomaterials on agriculture, food, and environment. We then provide regulatory guidelines for the safer use of nanomaterials.
Malsch et al. (2018)[a]	Decision Support for International Agreements Regulating Nanomaterials.	In this article, the authors formulated a scenario concerning a potential role for an open access decision support system in negotiations on international agreements regulating trade in nanomaterials. The scenario includes design rules for decision support systems as well as procedures for use of such a system in stakeholder dialogue and policy-making on governance of these and other emerging technologies. This article incorporates analysis of results of stakeholder engagement on nanomaterials as well as literature and internet sources suggested by these stakeholders.
Lai et al. (2018)	Regulation of engineered nanomaterials: current challenges, insights, and future directions.	To address the fast-growing nanotechnology market with limited resources, four ENMs (nanoparticles of Ag, TiO_2, ZnO, and Fe_2O_3) have been prioritized for research. Compulsory reporting schemes (registration and labeling) for commercial products containing ENMs should be adopted. Moreover, to accommodate their potential risks in time, an integrative use of quantitative structure–activity relationship and adverse outcome pathway (QSAR-AOP), together with qualitative alternatives to conventional risk assessment are proposed as tools for decision-making of regulators.

(Continued)

Table 15.2 (Continued)

Author/year	Study title	Summary of key findings
Ranjan et al. (2019)[a]	Toxicity and regulations of food nanomaterials.	This review describes nanomaterial–cell interactions that induce toxicological responses. We discuss the toxicity of food nanomaterials; mechanisms of oxidative stress, genotoxicity and carcinogenicity; and safety regulations.
Zhang and Wang (2019)	Chinese Public and Nanoresearchers' Perceptions of Benefits and Risks of Nanotechnology.	Public and experts' perceptions of benefits and risks of nanotechnology constitute an important element of nanoethics studies. On the one hand, compared with traditional ethics, nanoethics is a future-oriented ethics. The construction of ethical norms requires public participation. On the other hand, nanotechnology is characterized by uncertainty.
Nelson et al. (2020)	Emerging Standards and Analytical Science for Nanoenabled Medical Products.	In this review, the focus is on providing the reader with an introduction to the landscape of commercially available nanotechnology-enabled medical products as well as an overview of the international documentary standards and reference materials that support and facilitate efficient regulatory evaluation and reliable manufacturing of this diverse group of medical products.

[a] Articles that apply to both researched categories.

Table 15.3 Overview of problems and consequences of the regulation (or lack thereof) of nanoparticles.

Problem characteristics	Consequences
There are no available methodologies or data regarding the use, risks, or exposure to nanoparticles.	With missing methodology and data, conventional direct regulation is not feasible.
Government agencies have limited technical capacity, knowledge, and resources.	Governing mechanisms should rely on the capacity, knowledge, and resources of third-party businesses and organizations.
The beneficent development and implantation of nanotechnology is growing at a rapid pace, together with potential risks.	The balanced implantation of governing mechanisms should occur at a compatible pace.

Source: Adapted from Malloy (2011).

the design and development of new research, products, and services (Blind et al. 2013).

Based on these requirements, in this European scenario, the following topics should reflect the standard of products, research, and innovation: orientation toward the needs of society which includes guidance for contributing to the achievement of sustainable development goals (consisting of economic, social, and environmental aspects) and collaboration to achieve normative goals such as the *equality of men and women* or an improvement in the *quality of life*, which are also the main European objectives expressed in the Treaty on European Union, and ethical acceptance, which includes compliance with both the European Union Charter on fundamental rights and product safety and acceptable product risk (Blind et al. 2013).

Different definitions of RRI have been showed by different social actors and government and nongovernment initiatives. Although they are not all fully convergent, without a doubt, some elements enable essential features to be identified. In this way, Forsberg and Wickson (2015) present four characteristics that permeate the diverse attempts to conceptualize RRI, such as the presentation of a specific focus on meeting social needs and challenges, a search process that involves and answers the parties involved, and the collective effort to anticipate potential problems, identify alternatives, and reflect on underlying values. Additionally, relevant actors are willing to work and adapt per content. These authors also mentioned the questions that should be made about how these characteristics may be achieved in practice. These questions need to remain open as they address the motivations of the different actors related and the way to identify and measure progress toward a new paradigm of innovation (Forsberg and Wickson 2015).

The expansion of responsible RRIs leads to the emergence of some requirements that must be observed. Four integrated dimensions have initially been suggested: (i) anticipation, which is amplified through imagining the future and understanding how current promising dynamics shape this future; (ii) reflectiveness, which is a consequence of learning the first, second, and third orders; (iii) inclusion, which is the involvement of a variety of parties such as users and NGOs, in the initial development of science and technology; and (iv) a capacity for mutual response, which aims to answer to emerging knowledge, perspectives, points of view and standards (Stilgoe et al. 2013; RRI tools 2017b).

Anticipation involves systematic thinking to increase resilience while revealing new opportunities for innovation and the development of socially robust risk research agendas (Stilgoe et al. 2013). On this way, anticipation leads researchers and organizations to ask "*What if*? to consider every contingency regarding what is studied and recognized, what is plausible, and what is possible. Systematic thinking is needed on Anticipation, to increase resilience while revealing new opportunities for innovation and the development of socially robust risk research agendas" (Stilgoe et al. 2013). Regarding reflectiveness, responsibility requires reflection on the part of institutions and partners, but this is not directly defined. At institutional-level practice, reflectiveness means holding a mirror to action, commitments, and assumptions, while being aware of the limits of knowledge and of the fact that the framing of a particular issue may not be universally feasible. Bora (2012) mentions that reflectiveness could be understood as raising awareness of the operation itself; an awareness still connected to a kind of radical decentralization of concepts, originating from the theory of society that stimulates legal learning processes in regulatory fields. Thus, for the legal system to understand the risks, it is essential to base decision-making assumptions on reflectiveness. Only then will it be possible to include risks as aspects of observed decisions. These institutions not only have a duty to reflect on their value systems but also to help develop a capacity within the practice of science and innovation (Stilgoe et al. 2013).

The curtailment of specialist authority and the formulation of top-down policies have been associated with an increase in the insertion of new voices in the governance of science and innovation as part of a search for legitimation, as for inclusion. Over the past two decades, particularly in northern Europe, new deliberative forums have been established for issues involving science and innovation, in addition to engagement with stakeholders to include members of the general public (Stilgoe et al. 2013). Inclusion processes inevitably require the consideration of issues of power. The agencies that commission such exercises, facilitators, and public participants may have different expectations about the instrumental, substantive, or normative benefits of the dialogue (Stirling 2008).

Stilgoe et al. (2013) have stated that innovation systems must be treated as moldable so that they are as sensitive as possible since responsiveness involves responding to new knowledge, perspectives, points of view, and standards as they materialize. There has been significant criticism regarding the impact of public involvement and the limited capacity to empower agency on social and in technological choice and modulate innovation trajectories.

Innovation and Responsible research is a dynamic and different process through all characters involved in the practice of these activities (i.e. researchers, politicians, industries, citizens, and educators) become mutually responsible for each other and share responsibility for the results and processes of RRI. Some requirements of the process or dimensions integrated into the process are (i) diversity and inclusion: involves in advance high types of actors and audiences at research and innovation practices, deliberations to increase benefits. This strengthens democracy and expands the sources of knowledge, disciplines, and perspectives; (ii) anticipation and reflectivity: predict general reflects on underlying assumptions, values, and purposes to better understand how research and innovation shape the future. This produces valuable information and increases our ability to decide on what we know (here we can include the issue of systemic-constructivist methodology and the use of bases that are not part of traditional law. This enables the connection between and application of other regulatory institutes, and the interconnection with other systems, such as the inter-systemic communication already suggested for the systems of Law and Science); (iii) openness and transparency: means balanced communication of conclusions, implications, meaningful methods, results, to allow public scrutiny and dialogue. This benefits the visibility and understanding of research and innovation; and (iv) responsibility and adaptation to change: being able to modify modes of thinking and behavior, and comprehensive organizational structures, in response to changing circumstances, knowledge, and perspectives (RRI tools 2017a).

These characteristics of RRI, which qualitatively involve the actions of organizations, generate unprecedented temporal possibilities, and concern for an uncertain future. Because of this scenario, "[. . .] responsible innovation means taking care of the future through the collective management of science and innovation in the present" (RRI tools 2017b).

Other European Commission document (European Commission 2017) addresses the four characteristics of RRI and mentions that the principle of asks researchers to insert others perspectives in research and innovation processes and study about various possibilities for developing socially robust agendas for risk research and risk management. The principle of reflectivity asks researchers and innovators to think about their ethics and political or social assumptions and contemplate their roles and responsibilities in research and innovation, as well as in public dialogue. The inclusion calls for researchers to involve several stakeholders (such as users and NGOs) in the process of expanding and diversifying knowledge sources and perspectives. Regarding the last characteristic of responsibility, this implies research and innovation declaring themselves responsible and having the capacity to change their direction or form when it becomes evident that current developments do not correspond to the needs of society or are ethically challenged. RRI must be a fundamental part of the research and innovation process and should be implemented at an early stage and at different levels, such as the individual projects of scientists, research activities, and acts of innovation, as well as in research and development programs. At the institutional level, RRI should be understood as a stakeholder strategy to foster mutual reception and anticipate research and innovation results that support the great challenges of our time. Challenges for which they share responsibility. For modern innovations, responsibility for the consequences of implementation is primarily related to the properties and characteristics of products or technology and less so to the privileged owners and creators of the technology (Von Schomberg 2013). Thus, RRI can help to ensure that the positive and negative impacts of new technologies are better governed and exploited at stages well before products enter the market. Here, two interrelated dimensions can be identified: the product dimension and the process dimension (Von Schomberg 2013)

Normative anchoring points must reflect in the product dimension. They should have: (i) ethical acceptability: in the context of the European Union, this refers to mandatory compliance with the fundamental values of the European Union Charter concerning fundamental rights. It also refers to product *safety* in terms of acceptable risks. Continuous risk assessments are part of the procedure to gauge the safety issues of acceptable products. However, the question of security must be taken from a broader perspective. The United Kingdom's largest public foundation for basic innovative research, the Engineering and Physical Science and Research Council, has asked candidates to report the major problems and potential risks related with research proposals in the field of nanosciences; (ii) sustainability: this contributes to the objective of sustainable development in the European Union. The European Union follows the 1997 *definition* of sustainable development, which consists of economic, social, and environmental dimensions in mutual interdependence; and (iii) social desirability: this principle captures the relevant and more specific normative anchor points of the Treaty on European Union, such as *quality of life*, and *equality between men and women*. In the development and evaluation of products, these points will lead beyond the simple profitability of the market, although the latter can be a precondition for the viability of products in the competitive economies of the market (Von Schomberg 2013). Thus, regarding a product's dimension, they must be evaluated and designed based on their normative anchoring points: high levels of protection for the environment, human health, sustainability, and social desirability.

Regarding the details of the process, challenge is to achieve an integrated management of the innovation process, with a more responsive and adaptive way. Whereby technical innovators become sensitive to the needs of society and social actors become co-responsible for the process of innovation through constructive contributions to the definition of socially desirable product, a multidisciplinary approach with the involvement of several actors and other interested parties should lead to an inclusive innovation process (Von Schomberg 2013). Product and process dimensions are naturally interrelated. Implementation is underpinned by five mechanisms: innovative governance, stakeholder involvement and public engagement, technological assessment and prospection, the application of the precautionary principle, and the conception of normative/ethical principles for technology (Von Schomberg 2013).

The RRI concept gives new responsibilities, not only to scientists, universities, innovators, and organizations but also to policymakers and research financiers. Responsible innovation evokes a responsibility of collective care which entails on rethinking what one hopes to accomplish with innovation. Then, one must consider how goals can become viable in uncertainty face. By recognizing innovation power to build our collective future, RRI challenges us, above all, to ask ourselves what kind of future we want innovation to bring to the world (Owen et al. 2012). The Law must be made to participate more actively in discussions about new creative uses for the regulation of nanotechnologies. Today, this only occurs with the hard sciences, without the due contribution of the social sciences (soft). Shulz (2017) touches on the difficult communication between the different fields of science and, specifically in the case of nanotechnologies, states that: "gaps persist between the discourses of different sciences (and between different audiences) facing common issues. Mutual distrust between scientists in the exact and social fields is also frequent."

15.5 Final Considerations and New Propositions

Today's society can be described as a producer of risks arising from the multiple possibilities that have developed since the advent of revolutionary milestones and modified the environment, especially through technological performance. Nanotechnologies are an excellent example of acts of innovation that are developing rapidly with no specific regulatory framework. Additionally, insufficient attention has been paid to risk management and the concerns surrounding the consequences of nano-production in the present and the future of the Planet.

We are in times of intense technological and scientific growth, but these have also generated a series of important questions about the future of technology: Are there any negative environmental or health impacts? Who controls the use of nanotechnology? What will perish in society when nanotechnology becomes popular? How do we deal with the responsibility? Who will be the beneficiary of technology or the casualty of damage? What are the ethical problems? (European Commission. 2013).

Presently, the knowledge that the System of Science holds contains very little about the possible reactions of nanoparticles in the environment. Therefore, questions regarding the behavior of nanomaterials continue to exist, especially since the toxicological tests that have been performed on nano-products have, almost always, had the disadvantage of being carried out in controlled environments (*in vitro*). This generates data that are far different from what may happen in natural environments, because of the numerous biotic and abiotic factors. Moreover, among the existing uncertainties is the concern about the possible health impacts on those who will eventually handle nanoparticles. More questions and insecurities add themselves to this risk scenario: how do we safely handle, transport, store, and dispose of nanomaterials? For those who will receive drugs made with nanoparticles, how do we ensure their protection so that they just enjoy the benefits of nano-drugs?

Despite the advantages that nanotechnology offers, many workers are not aware that they work with it and scientists still know little about harmful side effects, since they remain unclear. Experts have advised care and caution and that we think, first of all, about the safety and health of humans and the environment. While nanotechnologies offer possibilities never before imagined, new strategies for managing risks and expectations must be simultaneously developed. Simply making a law or establishing regulations or standards of good practice along the traditional lines will not be enough. Complex innovations require complex analyses to determine the best protection for the ecosystem. Therefore, to provide a safer and more sustainable environment for the development of nanotechnologies for biomedical applications, building a production environment linked to ethical vectors and governed by precaution is the minimum that we can do. Since the law does not regulate such activities or other developments of this new technology, other areas and systems have assumed this role, through the most diverse instruments and oriented by decision-making.

Policymakers face the challenge of making decisions about new publicly funded research priorities and regulations. Responding society's concerns, it is crucially important initiating a dialogue on the risks and benefits and of nanotechnology, including the governmental aspects, legal, social, ethical, and relating large parts of the public using informed judgment. Any action should reflect current global scenario, as socioeconomic scenario and supporting the impetus to (i) manage scientific efforts to deal better with unknown and complex, systemic challenges and complement this with traditional knowledge professional, lay and local; (ii) reorder the rank of economic and human and natural capital, financial capital into social; and (iii) develop more adaptable and resilient governing systems to deal with multiple threats and systemic surprises (Hansen et al. 2013). This is the crossroads at which the Law and the other areas of knowledge involved with the scientific Nanotechnological Revolution have found themselves. That is, at a point where there is a responsibility to find safe and responsible alternatives to deal with the present and future of the life of all living beings on the face of the Earth, through the healthy use of scientific contributions produced by human intelligence.

The challenge of nanotechnologies is not to maintain and win a slice of an emerging market. Rather, more simply, the question is how can we as a society reap the maximum possible benefits at the nanoscale and, at the same time, avoid, limit, or at least be prepared to repair the damage associated with the human health and environmental risks that this new technology may create. Although it is not possible to predict all potential outcomes associating with nanoworld, important is decisions concerning the evolution and development of the nanoworld to mainly prepare us to reduce risks and protect biotic and abiotic beings.

Notes

1 To answer research questions, a systematic literature review was carried out. This is a technique focused on scientific evidence and predetermined criteria, consistent on explicit application and systematic search methods, strict methodology, and a clear presentation of results.

2 Moreover, the researcher highlighted another benefit: nanoparticles are pegylated. That is, they contain polyethylene glycol, which are hair-like strands in their surroundings that are able to deflect white cells; the cells that identify foreign bodies in the blood. "The use of this polymer allows the nanoparticles to bypass these cells, increasing the probability of not being detected and circulating in the blood for a longer time, thus optimizing the treatment', he explains. The new technique, however, is still taking a long time to reach patients. For cancer treatments, at least 8 to 10 years of laboratory tests are required. *In vitro* studies are still underway, with animal models, because there are different types of cancer and toxicological tests are essential. [. . .]. Licensing is in charge of Unicamp's innovation agency Inova." The university filed a patent application at INPI – the National Institute of Industrial Property – and is seeking companies interested in forming partnerships for research and future productions. NEW technology reduces the side effects of chemotherapy (LQES – Laboratório de Química do Estado Sólido 2017).

3 "Nanotechnology has brought to tutors and pets a novelty for the hygiene, hydration, and nutrition of animals skin and hair. It ensures a greater cuteness and a better smell and shine to our puppies and kittens. Nanotechnology is nothing more than small particles that contain some medications, also called active ingredients, which are so small that they are able to enter the deepest layers of the skin, enhancing and prolonging the effects of the products. And of course, for you to become an expert in the subject and in caring for your pet, DrogaVET Manipulated has 3 main assets, with very different names. **Nanocoating**: Used for hygiene and cleaning. This asset forms a protective barrier against dirt, preventing bad odors; **Nanohydrate**: Through a moisturizing action, this asset acts mainly on the skin barrier, which is nothing more than a protective layer of the skin that works as a

defense mechanism of the organism. It is especially recommended for animals with atopic dermatitis to help reduce itching; **Nanomelaleuca**: This asset has moisturizing, anti-inflammatory, and regenerating properties. That is, it works by healing your Pet. It also acts as an antifungal agent to treat mycoses, superficial candidiasis, and other fungal and bacterial infections. Did you know that nanotechnology could help your pet that much? So, get in touch with a trusted veterinarian and ask them about Manipulação Veterinária and also find out more about how these tiny particles can help your pet keep their shiny fur, young skin and shiny tails"(emphasis added) (Nanotechnology: pelos brilhantes, pele jovem e rabinhos reluzentes 2017).

References

Ahmad, F., Pandey, A.K., Herzog, A.B. et al. (2012). Environmental applications and potential health implications of quantum dots. *Journal of Nanoparticle Research* 14: 1038. https://doi.org/10.1007/s11051-012-1038-7. (accessed 10 April 2021).

Álvarez Díaz, J.A. (2016). Aspectos éticos de la nanotecnología en la atención a la salud. Coyoacán: Ed. Universidade Autônoma Metropolitana. Serie académicos, n. 128.

Andrade R.O. (2013). Uma carona indesejada: nanopartículas podem transportar moléculas intrusas para o interior das células. Revista Pesquisa FAPESP, São Paulo, October. https://revistapesquisa.fapesp.br/uma-carona-indesejada/ (accessed 16 September 16 2020).

Ansell, C. and Baur, P. (2018). Explaining trends in risk governance: how problem definitions underpin risk regimes. *Risk, Hazards & Crisis in Public Policy* 9 (4): 397–430.

Baranowska-Wójcik, E., Szwajgier, D., Oleszczuk, P. et al. (2020). Effects of titanium dioxide nanoparticles exposure on human health-a review. *Biological Trace Element Research* 193 (1): 118–129. https://doi.org/10.1007/s12011-019-01706-6 (accessed 10 April 2021).

Birnbaum, L., Jung, P., and Newton, S. (2011). Environmental health science for regulatory decisionmaking. *Duke Environmental Law and Policy Forum* 21 (2): 259–293. https://www.researchgate.net/publication/241802682_Environmental_Health_Science_for_Regulatory_Decisionmaking (accessed 11 April 2021).

Blind, K., Guske, A.L., and Riera, C.M. (2013). *Options for strengthening responsible research and innovation: report of the expert group on the state of art in Europe on responsible research and innovation.* European Commission. http://ec.europa.eu/research/swafs/pdf/pub_public_engagement/options-for-strengthening_en.pdf (accessed 16 September 2020).

Bora, A. (2012). Capacidade de lidar com o futuro e responsabilidade por inovações: para o trato social com a temporalidade complexa. In: SCHWARTZ G (Org.). Juridicização das esferas sociais e fragmentação do direito na sociedade contemporânea. Porto Alegre: Livraria do Advogado.

Boyes, W., Thornton, B.L.M., Al-Abed, S.R. et al. (2017). A comprehensive framework for evaluating the environmental health and safety implications of engineered nanomaterials. *Critical Reviews in Toxicology* 47 (9): 771–814. https://doi.org/10.1080/10408444.2017.1328400 (accessed 10 April 2021).

Carmona-Ribeiro, A.M. (2010). Biomimetic nanoparticles: preparation, characterization and biomedical applications. International Journal of Nanomedicine, Auckland 5. https://www.ncbi.nlm.nih.gov/pmc/articles/PMC2865020/ (accessed 16 September 2020).

D'Silva, J., Robinson, D.K.R., and Shelley-Egan, C. (2012). A game with rules in the making – how the high probability of waiting games in nanomedicine is being mitigated through distributed regulation and responsible innovation. *Technology Analysis & Strategic Management* 24 (6): 583–602. https://doi.org/10.1080/09537325.2012.693671 (accessed 11 April 2021).

Dalton-brown, S. (2012). Global ethics and nanotechnology: a comparison of the nanoethics environments of the EU and China. *NanoEthics* 6: 137–150. https://doi.org/10.1007/s11569-012-0146-2 (accessed 11 April 2021).

Das, G., Patra, J.K., and Paramithiotis, S. (2019). The sustainability challenge of food and environmental nanotechnology: current status and imminent perceptions. *International Journal of Environmental Research and Public Health* 16 (23): 4848. https://doi.org/10.3390/ijerph16234848 (accessed 10 April 2021).

Deng, L., Mohan, T., Chang, T.Z. et al. (2018). Double-layered protein nanoparticles induce broad protection against divergent influenza A viruses. *Nature Communications* 9: 359. https://www.nature.com/articles/s41467-017-02725-4 (accessed 16 September 2020).

Donaldson, K., Duffin, R., Langrish, J.P. et al. (2013). Nanoparticles and the cardiovascular system: a critical review. *Nanomedicine (London)* 8 (3): 403–423. https://doi.org/10.2217/nnm.13.16 (accessed 10 April 2021).

École Polytechnique Fédérale de Lausanne (2016). A nanotechnology method for storing vaccines at room temperature. LQES - Laboratório de Química do Estado Sólido. http://lqes.iqm.unicamp.br/canal_cientifico/lqes_news/lqes_news_cit/lqes_news_2016/lqes_news_novidades_2174.html (accessed 16 September 2020).

European Commission (2013). Options for strengthening responsible research and innovation: report of the expert group on the state of art in Europe on responsible research and innovation. Chair: Jeroen van den Hoven Rapporteur: Klaus Jacob Members: Linda Nielsen, Françoise Roure, Laima Rudze, Jack Stilgoe Contributors: Knut Blind, Anna-Lena Guske, Carlos Martinez Riera. Brussels. http://ec.europa.eu/research/swafs/pdf/pub_public_engagement/options-for-strengthening_en.pdf (accessed 16 September 2020).

European Commission (2017). Science for Environment Policy. Assessing the environmental safety of manufactured nanomaterials: in-depth report 14. Bristol, August. http://ec.europa.eu/environment/integration/research/newsalert/pdf/assessing_environmental_safety_nanomaterials_IR14_en.pdf (accessed 16 September 2020).

Evans, R., Mcnamee, M., and Guy, O. (2017). Ethics, Nanobiosensors and elite sport: the need for a new governance framework. *Science and Engineering Ethics* 23: 1487–1505. https://doi.org/10.1007/s11948-016-9855-1 (accessed 11 April 2021.

Fink, A.G. (2014). *Conducting Research Literature Reviews: From the Internet to Paper*. Thousand Oaks, CA: Sage 257pp.

Forsberg, E.M. and Wickson, F. (2015). Standardising responsibility? The significance of interstitial spaces. *Science and Engineering Ethics*, [S.l.] 21: 5. https://link.springer.com/article/10.1007/s11948-014-9602-4 (accessed 16 September 2020).

Garden, H. and Winickoff, D. (2018). Gene editing for advanced therapies: governance, policy and society. In: *OECD Science, Technology and Industry Working Papers*, n. 2018/12. Paris: OECD Publishing https://doi.org/10.1787/8d39d84e-en (accessed 11 April 2021).

Gatti, A.M. and Montanari, S. (2015). *Case Studies in Nanotoxicology and Particle Toxicology*. Cambridge: Academic Press.

Giannakou, C., Park, M., de Jong, W. et al. (2016). A comparison of immunotoxic effects of nanomedicinal products with regulatory immunotoxicity testing requirements. *International Journal of Nanomedicine* 11: 2935–2952. https://doi.org/10.2147/IJN.S102385 (accessed 10 April 2021).

Gouvea, R., Linton, J.D., Montoya, M. et al. (2012). Emerging Technologies and Ethics: A race-to-the-bottom or the top? *Journal of Business Ethics* 109 (4): 553–567. https://doi.

org/10.1007/s10551-012-1430-3 (accessed 11 April 2021).

Groso, A., Petri-Fink, A., Magrez, A. et al. (2010). Management of nanomaterials safety in research environment. *Particle and Fibre Toxicology* 7 (40) https://doi.org/10.1186/1743-8977-7-40 (accessed 11 April 2021).

Grunwald, A. (2011). Responsible innovation: bringing together technology assessment, applied ethics, and STS research. *Enterprise and Work Innovation Studies*: 7. https://run.unl.pt/bitstream/10362/7944/1/Grunwald9-31.pdf (accessed 16 September 2020).

Guo, J.-W., Lee, Y., Huang, H.W. et al. (2014). Development of Taiwan's strategies for regulating nanotechnology-based pharmaceuticals harmonized with international considerations. *International Journal of Nanomedicine* 9: 4773–4783. https://doi.org/10.2147/IJN.S68134 (accessed 11 April 2021).

Hall, R., Sun, T., and Ferrari, M. (2012). A portrait of nanomedicine and its bioethical implications. *The Journal of Law, Medicine & Ethics* 40 (4): 763–779. https://doi:10.1111/j.1748-720X.2012.00705.x (accessed 9 April 2021).

Hansen, S.F., Maynard, A., Baun, A. et al. (2013). Nanotechnology - early lessons from early warnings. In: *European Environment Agency. Late Lessons from Early Warnings: Science, Precaution, Innovation*. Luxemburgo: Publications Office of the European Union. https://www.eea.europa.eu/publications/late-lessons2 (accessed 16 September 2020).

Hogle, L. (2021). Concepts of risk in nanomedicine research. *The Journal of Law, Medicine & Ethics* 40 (4): 809–822. https://doi.org/10.1111/j.1748-720X.2012.00709.x (accessed 10 April 2021).

Hohendorff, R.V. and Engelmann, W. (2014). *Nanotecnologias aplicadas aos agroquímicos no Brasil: a gestão do risco a partir do diálogo entre as fontes do direito*. Curitiba: Juruá Editor.

Hu, Q. and Gu, Z. (2016). Cell membrane-mediated anticancer drug delivery. In: *Nanotechnology: Delivering on the Promise*, ACS Symposium Series, v. 1224). Chapter 10, vol. 2 (eds. H.N. Cheng, L. Doemeny, C.L. Geraci and D.G. Schmidt), 197–211. Washington: American Chemical Society.

Jampilek, J., Kos, J., and Kralova, K. (2019). Potential of nanomaterial applications in dietary supplements and foods for special medical purposes. *Nanomaterials (Basel, Switzerland)* 9 (2): 296. https://doi.org/10.3390/nano9020296. (accessed 10 April 2021).

Kaphle, A., Navya, P.N., Umapathi, A. et al. (2018). Nanomaterials for agriculture, food and environment: applications, toxicity and regulation. *Environmental Chemistry Letters* 16: 43–58. https://doi.org/10.1007/s10311-017-0662-y (accessed 11 April 2021).

Kong, B., Seog, J.H., Graham, L.M. et al. (2011). Experimental considerations on the cytotoxicity of nanoparticles. *Nanomedicine (London)* 6 (5): 929–941. Disponível em: https://doi.org/10.2217/nnm.11.77. (accessed 10 April 2021).

Kumar, A., Tan, A., Wong, J. et al. (2017). Nanotechnology for neuroscience: promising approaches for diagnostics, therapeutics and brain activity mapping. *Advanced Functional Materials, Weinheim* 27: 39. http://onlinelibrary.wiley.com/doi/10.1002/adfm.201700489/full (accessed 16 September 2020).

Lai, R.W.S., Yeung, K.W.Y., Yung, M.M.N. et al. (2018). Regulation of engineered nanomaterials: current challenges, insights and future directions. *Environmental Science and Pollution Research* 25: 3060–3077. https://doi.org/10.1007/s11356-017-9489-0 (accessed 11 April 2021).

Leach, R.K., Boyd, R., Burke, T. et al. (2011). The European nanometrology landscape. *Nanotechnology* 11 (22). https://doi.org/10.1088/0957-4484/22/6/062001 (accessed 11 April 2021).

León-Silva, S., Fernández-Luqueño, F., and López-Valdez, F. (2016). Silver Nanoparticles (AgNP) in the environment: a review of potential risks on human and environmental health. *Water, Air, and Soil Pollution* 227 (9): 1–20. https://doi.org/10.1007/s11270-016-3022-9 (accessed 11 April 2021).

Lohani, A., Verma, A. Himanshi, J. et al. (2014). Nanotechnology-Based Cosmeceuticals, International Scholarly Research Notices, vol. 2014. https://doi.org/10.1155/2014/843687 (accessed 11 April 2021).

LQES - Laboratório de Química do Estado Sólido (2017). Nota do Scientific Editor: os resultados deste projeto estão associados ao Instituto Nacional de C,T&I em Materiais Complexos Funcionais (INOMAT) e Laboratório de Síntese de Nanoestruturas e Interação com Biossistema (NanoBioss/SisNano), financiados pelo CNPQ, Fapesp e MCTIC e sediados no Instituto de Química da Unicamp. http://lqes.iqm.unicamp.br/canal_cientifico/lqes_news/lqes_news_cit/lqes_news_2017/lqes_news_novidades_2362.html (accessed 16 September 2020).

Lu, L., Lin, B., Liu, J. et al. (2012). Ethics in Nanotechnology: What's being done? What's missing? *Journal of Business Ethics* 109 (4): 583–598. https://doi.org/10.1007/s10551-012-1432-1 (accessed 10 April 2021).

Lu, P.-H., Li, H.-J., Chang, H.-H. et al. (2017). Gold nanoparticles induce cell death and suppress migration of melanoma cells. *Journal of Nanoparticle Research, Netherlands*: 19, October. https://link.springer.com/article/10.1007/s11051-017-4036-y (accessed 16 September 2020).

Malloy, T.F. (2011). Nanotechnology regulation: a study in claims making. *ACS Nano* 5 (1): 5–12.

Malsch, I., Mullins, M., Semenzin, E. et al. (2018). Decision support for international agreements regulating nanomaterials. *NanoEthics* 12 (1): 39–54. https://doi.org/10.1007/s11569-018-0312-2 (accessed 10 April 2021).

Mandel, G. and Marchant, G.E. (2014). The living regulatory challenges of synthetic biology. *Iowa Law Review* 100 (1): 155–201. http://dx.doi.org/10.2139/ssrn.2410179 (accessed 11 April 2021).

Marcial, S.P.S., Carneiro, G., and Leite, E.A. (2017). Lipid-based nanoparticles as drug delivery systemfor paclitaxel in breast cancer treatment. *Journal of Nanoparticle Research, Netherlands* 19 https://link.springer.com/article/10.1007/s11051-017-4042-0 (accessed 16 September 2020).

Martin, J.T., Ramos, O.L., Pinheiro, A.C. et al. (2015). Edible bio-based nanostructures: delivery, absorption and potential toxicity. *Food Engineering Reviews* 7: 491–513. https://doi.org/10.1007/s12393-015-9116-0 (accessed 9 April 2021).

Martirosyan, A. and Schneider, Y.J. (2014). Engineered nanomaterials in food: implications for food safety and consumer health. *International Journal of Environmental Research and Public Health* 11 (6): 5720–5750. https://doi.org/10.3390/ijerph110605720 (accessed 10 April 2021).

MCTIC (2017). Antibiótico nanotecnológico desenvolvido pelo CNPEM pode eliminar superbactérias. Sala de Imprensa, Brasília, DF, November. http://www.mctic.gov.br/mctic/opencms/salaImprensa/noticias/arquivos/2017/11/Antibiotico_nanotecnologico_desenvolvido_pelo_CNPEM_pode_eliminar_superbacterias.html (accessed 16 September 2020).

Mirkin, C.A. (2016). Nanotechnology at northwestern university: delivering on the promise. In: *Nanotechnology: delivering on the promise*, ACS Symposium Series, v. 1220, vol. 1 (eds. H.N. Cheng, L. Doemeny, C.L. Geraci, et al.). American Chemical Society.

Mu, X., Qiao, J., Qi, L. et al. (2014). Poly (2-vinyl-4,4-dimethylazlactone) – functionalized magnetic nanoparticles as carriers for enzyme immobilization and its application. *ACS Applied Materials & Interfaces, Washington* 6 (23) https://www.ncbi.nlm.nih.gov/pubmed/25360545 (accessed 16 September 2020).

Nanotechnology: pelos brilhantes, pele jovem e rabinhos reluzentes (2017). DrogaVET, São Paulo. https://www.drogavet.com.br/geral/nanotecnologia-pelos-pele-saudaveis/ (accessed 16 September 2020).

Nanotecnologia viabiliza novo medicamento contra leishmaniose (2018). Diário da Saúde. http://www.diariodasaude.com.br/news.php?article=nanotecnologia-viabiliza-novo-medicamento-contra-leishmaniose&id=12569&nl=nlds (accessed 16 September 2020).

Nath, M. (2016). Multifunctional magnetic nanomaterials for diverse applications. In: *Nanotechnology: Delivering on the Promise*, vol. 2 (eds. H.N. Cheng, L. Doemeny, C.L. Geraci, et al.). American Chemical Society ACS Symposium Series, v. 1224.

Nelson, B.C., Minelli, C., Doak, S.H. et al. (2020). Emerging standards and analytical science for nanoenabled medical products. *Annual Review of Analytical Chemistry (Palo Alto, California)* 13 (1): 431–452. https://doi.org/10.1146/annurev-anchem-091619-102216 (accessed 11 April 2021).

Octaviano, C. (2017). "Nanofármaco mostra potencial para redução de tumor na bexiga: Inova busca empresa para licenciamento da tecnologia desenvolvida na Unicamp". Jornal da Unicamp, Campinas. http://www.unicamp.br/unicamp/ju/noticias/2017/10/06/nanofarmaco-mostra-potencial-para-reducao-de-tumor-na-bexiga (accessed 16 September 2020).

Ojima, Y., Lichtenthal, B., Lee, S. et al. (2016). Taxane anticancer agents: a patent perspective. *Expert Opinion on Therapeutic Patents, London* 26 (1) https://www.ncbi.nlm.nih.gov/pubmed/26651178 (accessed 16 September 2020).

Owen, R., Macnaghten, P., and Stilgoe, J. (2012). Responsible research and innovation: from science in society to science for society, with society. *Science and Public Policy, London* 39 (6) https://academic.oup.com/spp/article/39/6/751/1620724 (accessed 16 September 2020).

Philbert, M.A. (2016). Nanomaterials: promise in balance with safety. In: *Nanotechnology: delivering on the promise*, vol. 1 (eds. H.N. Cheng, L. Doemeny, C.L. Geraci, et al.). Washington: American Chemical Society ACS Symposium Series, v. 1220.

Pillai, G.R. and Bezbaruah, A.N. (2017). Perceptions and attitude effects on nanotechnology acceptance: an exploratory framework. *Journal of Nanoparticle Research* 19 (41). https://doi.org/10.1007/s11051-016-3733-2 (accessed 9 April 2021).

Pourgashtasbi, G. (2015). Nanotoxicology and challenges of translation. *Nanomedicine (London, England)* 10 (20): 3121–3129. https://doi:10.2217/nnm.15.131 (accessed 11 April 2021).

Pramanik, P., Krishnan, P., Maity, A. et al. (2020). Application of nanotechnology in agriculture. *Environmental Nanotechnology* 4: 317–348. https://doi.org/10.1007/978-3-030-26668-4_9 (accessed 10 april 2021).

Quintanilla-Carvajal, M.X., Camacho-Díaz, B.H., Meraz-Torres, L.S. et al. (2010). Nanoencapsulation: a new trend in food engineering processing. *Food Engineering Reviews* 2: 39–50. https://doi.org/10.1007/s12393-009-9012-6 (accessed 9 April 2021).

Ramanathan, A. (2019). Toxicity of nanoparticles – challenges and opportunities. *Applied Microscopy* 49: 1–11.

Ranjan, S., Dasgupta, N., Chakraborty, A.R. et al. (2014). Nanoscience and nanotechnologies in food industries: opportunities and research trends. *Journal of Nanoparticle Research* 16 (2464). https://doi.org/10.1007/s11051-014-2464-5 (accessed 11 April 2021).

Ranjan, S., Dasgupta, N., Singh, S. et al. (2019). Toxicity and regulations of food nanomaterials. *Environmental Chemistry Letters* 17: 929–944. https://doi.org/10.1007/s10311-018-00851-z (accessed 9 April 2021).

Riediker, M., Zink, D., Kreyling, W. et al. (2019). Particle toxicology and health - where are we? *Particle and Fibre Toxicology* 16 (1). https://doi.org/10.1186/s12989-019-0302-8 (accessed 10 April 2021.

Rocha, L.S. (2013). Da epistemologia jurídica normativista ao construtivismo sistêmico II. In: *Introdução à teoria do sistema autopoiético do direito*. 2. ed. rev. e ampl (eds. L.S. Rocha, G. Schwartz and J. Clam). Porto Alegre: Livraria do Advogado.

Roco, M.C. (2016). Building foundational knowledge and infrastructure for nanotechnology: 2000-2030. In: *Nanotechnology: delivering on the promise*, ACS Symposium Series, v. 1220, vol. 1 (eds. H.N. Cheng, L. Doemeny, C.L. Geraci, et al.). Washington: American Chemical Society.

Roco, M.C., Harthorn, B., Guston, D. et al. (2011). Innovative and responsible governance of nanotechnology for societal development. In: *Nanotechnology Research Directions for Societal Needs in 2020*, Science Policy Reports, vol. 1. Dordrecht: Springer. https://doi.org/10.1007/978-94-007-1168-6_14 (accessed 9 April 2021).

Romero-Franco, M., Bilal, M., Godwin, H.A. et al. (2018). Assessment of information availability for environmental impact assessment of engineered nanomaterials. *Journal of Nanoparticle Research* 20 (324) https://doi.org/10.1007/s11051-018-4402-4 (accessed 11 April 2021).

RRI tools (2017a). About RRI. [S.l.], 2017. https://www.rri-tools.eu/about-rri (accessed 16 September 2020).

RRI tools (2017b). The project. RRI Tools: building a better relationship between science and society. [S.l.]. http://www.rri-tools.eu/project-description (accessed 16 September 2020).

Ruggiu, D. (2013). Temporal perspectives of the nanotechnological challenge to regulation: how human rights can contribute to the present and future of nanotechnologies. *NanoEthics* 7: 201–215. https://doi.org/10.1007/s11569-013-0184-4 (accessed 11 April 2021).

Sahoo, S. (2013). Would you mind, if we record this? Perceptions on regulation and responsibility among Indian nanoscientists. *NanoEthics* 7: 231–249. https://doi.org/10.1007/s11569-013-0182-6 (accessed 10 April 2021).

Sajid, M., Ilyas, M., Basheer, C. et al. (2015). Impact of nanoparticles on human and environment: review of toxicity factors, exposures, control strategies, and future prospects. *Environmental Science and Pollution Research* 22: 4122–4143. https://doi.org/10.1007/s11356-014-3994-1 (accessed 10 April 2021).

Schenke-layland, K., Rofail, F., Heydarkhan, S. et al. (2009). The use of three-dimensional nanostructures to instruct cells to produce extracellular matrix for regenerative medicine strategies. *Biomaterials* 30 (27) https://www.ncbi.nlm.nih.gov/pubmed/19524289. (accessed 16 September 2020).

Scientists synthesize nanoparticle-antioxidants to treat strokes and spinal cord injuries (2018). Nanotechnology Now: Your gateway to Everything Nanotech. [S.l.]. http://www.nanotech-now.com/products/newsdigest/?view=newsletter688-January192018.htm#story54929 (accessed 16 September 2020).

Sharma, V., Kumar, A., and Dhawan, A. (2012). Nanomaterials: exposure, effects and toxicity assessment. *Proceedings of the National Academy of Sciences, India Section B: Biological Sciences* 82 (1): 3–11. https://doi.org/10.1007/s40011-012-0072-7 (accessed 10 April 2021).

Shulz, P. (2017). "Ciência aberta para quem? Nanotecnologia, sociedade e meio ambiente". Jornal da Unicamp, Campinas, 27 November. http://www.unicamp.br/unicamp/ju/artigos/peter-schulz/ciencia-aberta-para-quem-nanotecnologia-sociedade-e-meio-ambiente (accessed 16 September 2020).

Singh, P.K., Jairath, G., and Ahlawat, S.S. (2016). Nanotechnology: a future tool to improve quality and safety in meat industry. *Journal of Food Science and Technology* 53 (4): 1739–1749. https://doi.org/10.1007/s13197-015-2090-y (accessed 10 April 2021).

Spruit, S.L., Van de Poel, I., and Doorn, N. (2016). Informed consent in asymmetrical relationships: an investigation into relational factors that influence room for reflection. *NanoEthics* 10: 123–138. https://doi.org/10.1007/s11569-016-0262-5 (accessed 10 April 2021).

Stark, W.J., Stoessel, P.R., Wohlleben, W. et al. (2015). Industrial applications of nanoparticles. *Chemical Society Reviews* 44 (16): 5793–5805. https://doi.org/10.1039/c4cs00362d. (accessed 11 April 2021).

STATNANO (2020). Nanotechnology Products Database – NPD. [S.l.]. http://product.statnano.com/ (accessed 19 September 2020).

Stilgoe, J., Owen, R., and Macnaghten, P. (2013). Developing a framework for responsible innovation. *Research Policy, Amsterdam* 42 (9) http://www.sciencedirect.com/science/article/pii/S0048733313000930 (accessed 16 September 2020).

Stirling, A. (2008). “Opening up” and “closing down”: power, participation, and plu- ralism in the social appraisal of technology. *Science Technology & Human Values, New York* 33 (2) http://journals.sagepub.com/doi/pdf/10.1177/0162243907311265 (accessed 16 September 2020).

Stokes, E. (2013). Demand for command: responding to technological risks and scientific uncertainties. *Medical Law Review* 21 (Winter): 11–38.

Timmermans, J., Zhao, Y., and Van Den Hoven, J. (2011). Ethics and nanopharmacy: value sensitive design of new drugs. *NanoEthics* 5: 269–283. https://doi.org/10.1007/s11569-011-0135-x. (accessed 10 April 2021).

Von Schomberg, R. (2013). A vision of responsible innovation. In: *Responsible Innovation: Managing the Responsible Emergence of Science and Innovation in Society* (eds. R. OWEN, J. BESSANT and M. HEINTZ). Wiley: Nova Jersey.

Zhang, J. and Wang, G. (2019). Chinese public and nanoresearchers’ perceptions of benefits and risks of nanotechnology. *NanoEthics* 13 (11): 155–171. https://doi.org/10.1007/s11569-019-00354-3 (accessed 11 April 2021).

Zielińska, A., Costa, B., Ferreira, B.C. et al. (2020). Nanotoxicology and nanosafety: safety-by-design and testing at a glance. *International Journal of Environmental Research and Public Health* 17 (13): 4657. https://doi:10.3390/ijerph17134657 (accessed 9 April 2021).

16

Nanoethics and Nanotechnology

Iti Chauhan[1], Madhu Verma[1], and Mohd Yasir[2]

[1] *Department of Pharmaceutics, I.T.S College of Pharmacy, Muradnagar, Ghaziabad, UP, India*
[2] *Department of Pharmacy, College of Health Sciences, Arsi University, Asella, Oromia Region, Ethiopia*

16.1 Introduction

In recent years, nanotechnology is creating a buzz in the field of science and technology. Before moving on to nanoethics, it will be better to go through in brief/historical development, the contributions of various scientists in this field. The idea of nanotechnology was primarily discussed by Richard Phillips Feynman, an American Physicist in the talk "There's Plenty of Room at the Bottom" at a Physical Society meeting at the California Institute of Technology on 29 December 1959 (Toumey 2009). A decade later, in 1974, Professor Norio Taniguchi of Tokyo Science University used the term nanotechnology while exploring ultraprecision machining. He mentioned nanotechnology as the separation, consolidation, and deformation of materials by one atom or one molecule (Botticello 2020). In 1986, K Eric Drexler in his book Engines of Creation used the term nanotechnology to conceptualize Feynman's vision (Sampath and Perumalraj 2006).

Various scientific and government organizations have explained nanotechnology differently (Table 16.1).

16.1.1 Types of Nanotechnology

The various forms of nanotechnology are categorized according to how they proceed (top-down or bottom-up) and the medium in which they work (dry or wet) (Iberdrola 2019). Figure 16.1 shows the types of nanotechnology.

16.1.2 Applications of Nanotechnology

Nanotechnology is considered as the futuristic industrial revolution which exploits extraordinary properties of ordinary materials to compose high-precision products. The universe of nanotechnology is expanding rapidly in multiple sectors viz. information, electronics, pharmaceuticals, cosmetics, biotechnology, defense and security, energy, agriculture, optical engineering, food processing, textiles, devices and sensors, environmental science, etc. Figure 16.2 and Table 16.2 represent a glimpse of applications of nanotechnology in various fields.

Despite spread to such a vast stretch, Nanotech is also gaining considerable attention in mainstream society – with special reference in

Nanotechnology in Medicine: Toxicity and Safety, First Edition. Edited by Mahendra Rai, Mrunali Patel, and Rashmin Patel.

Table 16.1 Widely accepted definitions of nanotechnology.

S.no	Organization	Definition	References
1	National Nanotechnology Initiative (NNI), US	Study and application of extremely small things which can be used across all the other science fields, such as chemistry, biology, physics, materials science, and engineering	United States National Nanotechnology Initiative (n.d.-a)
2	British Standards Institution (BSI), UK	The design, characterization, production and application of structures, devices and systems by controlling shape and size at the nanoscale.	Scientific Committee on Emerging and Newly Identified Health Risks (2006)
3	National Aeronautics and Space Administration (NASA), US	Creation of functional materials, devices, and systems through control and manipulation of matter on the nanometer length scale (1–100 nm). The technology is able to exploit novel phenomenon and material properties (physical, chemical, biological, mechanical, or electrical) at this scale.	ASK OCE, APPEL Knowledge Services (2006)
4	Royal Netherlands Academy of Arts and Sciences	Ability to work at the scale of atoms, molecules, and supramolecular, individually targetable structures (from 1 to 100 nm) so as to create complex functional structures with a fundamentally new molecular organization	Wood et al. (2007)
5	Environmental Protection Agency (EPA), US	The creation and use of structures, devices, and systems that have novel properties and functions because of their small size	Khare et al. (2014)
6.	National Institute of Health (NIH) U.S. Department of Health and Human Services	Understanding and control of matter at dimensions of roughly 1–100 nm, a scale at which unique properties of materials emerge that can be used to develop novel technologies and products.	National Institutes of Health (2016)

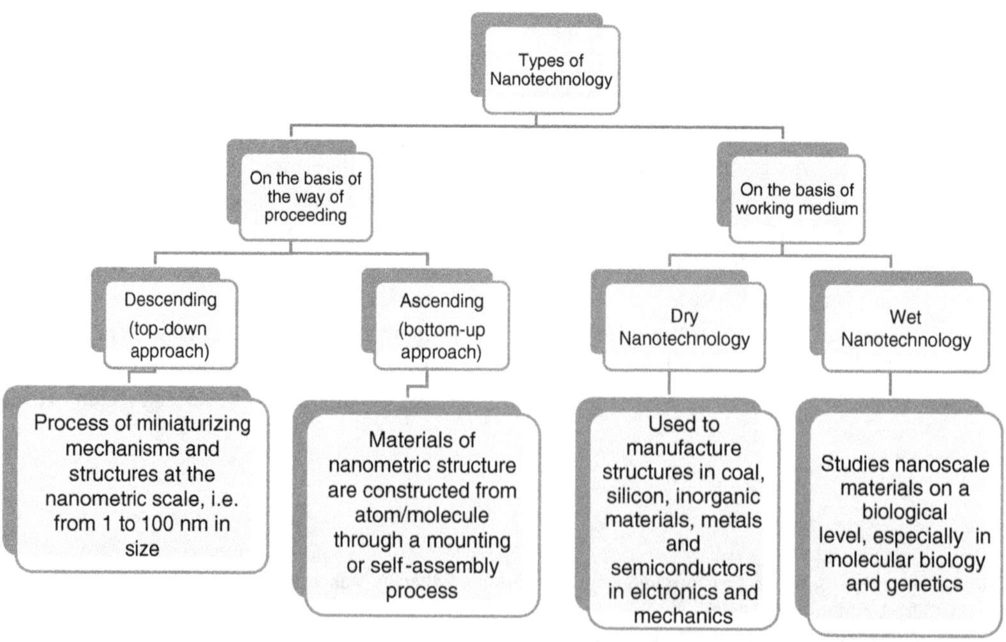

Figure 16.1 Types of nanotechnology.

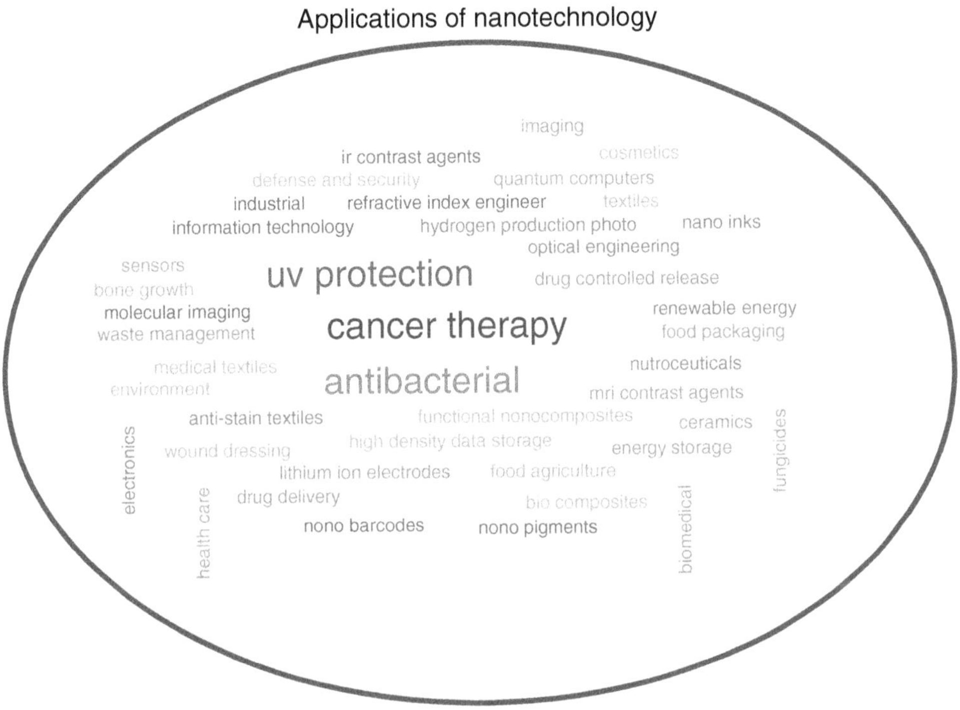

Figure 16.2 Applications of nanotechnology.

science-fiction films and TV shows like Terminator series, G.I. Joe: The Rise of Cobra, Avengers: Endgame, Transcendence, Red Dwarf, Stargate SG-1, and Revolution. Computer games like the Crysis series, Marvel: Ultimate Alliance 2, Ratchet & Clank series, and SpaceChem also feature nanotechnology in many aspects. Nanobot, Nano-mist, nanomites, and nanomachines are referenced in the Japanese anime and comic series Black cat, Sky girls, Battle Angel Alita, Mai-Otome, Kiddy Grade, and Project ARMS. Governments and organizations are investing huge amounts for research and development in nanotechnology (Keiper 2003). Despite big investments, there exist a dubious feeling about the status of impact and influence of nanotechnology. The scientific community is still facing questions about the responsibilities and independence of nanoscience (Lin and Allhoff 2007). Undoubtedly, nanotechnology is reflecting revolutionary changes in human lives but also show its unfavorable impressions on humankind, environment, nanomedicines, privacy, security, and economy around the globe. These unfitting ventures of nanotechnology have raised various concerns regarding the responsible and safe use of nanomaterials and nanoscience. These developments have given birth to a new discipline "Nanoethics" to uncover the moral and ethical questions raised due to expansion of technology.

The chapter aims to highlight aforementioned social and ethical facets of nanotechnology. It addresses the ethical matters in the field of nanotechnology focusing on global equity, privacy, security, safety, medicine, human enhancement, and environment. The chapter also discusses the regulatory guidelines by several regulatory agencies with respect to nanotechnology.

16.2 Nanoethics

The revolutionary applications of nanotechnology are considered as a pandora box of prospective ethical concerns that runs from nanoweapons to privacy violations to

Table 16.2 Applications of nanotechnology.

S. no.	Applications of nanotechnology	Examples
1.	Drug discovery and drug delivery	Nanosphere, Solution arrays, Nanobiosensors, Nanoneedles, Nanocapsule, Dendrimers, Vaccine delivery, Gold nanoparticles, Solid lipid nanoparticles, Nanostructured lipid carriers
2.	Tissue engineering	Bone and neural tissue engineering, Dental applications, Graphene nanoribbons.
3.	Biomedical imaging and diagnostics	Nuclear magnetic resonance, Lab-on-a-chip, Semiconductor quantum dots, Nanoscale encapsulated fluorophores, Nanowires
4.	Electronics and IT applications	Quantum computers, Nano transistors, Nano diodes, Ultrahigh definition displays and televisions Semiconductor nanomembranes in smartphone and e-reader displays Flash memory chips for smart phones and thumb drives Conductive inks for printed electronics Data memory (Magnetic random access memory, GMR-HD), Optical switches
5.	Textiles	Zinc oxide nanoparticles for UV protection and antibacterial finish Nano finishing for water repellence, flame retardant, antistatic properties, wrinkle resistance, self-cleaning properties Graphene-coated "e-textile" that detect noxious gases Smart clothes integrated with sensors
6.	Environmental remediation	Nanomembranes for water purification, desalination Industry pollutant scavenger Nanotechnology-based air filters Pollution monitoring sensors Self-assembled monolayers on mesoporous supports, Dendrimers and carbon nanotubes for toxic remediation Efficient and cheaper nano catalyst Nanotech-based soil set to avoid soil erosion
7.	Agriculture	Nano pesticides, Nanoherbicides, Nanogenetic manipulation of crops Controlled release of nano fertilizer Nano biosensors, Nano biofarming Control of plant diseases (Silver, silica, aluminosilicate nanomaterial)
8.	Automotive industry	Polymer nanocomposites structural parts Nano varnish, Carbon nanotube sensors, Wear-resistant tyres, Anti-fogging nanocoatings, Polymeric nano composites, Nanograined engine/body, Nanomaterials in fuel cells/battery, Self-cleaning nano films, Glazing nanopolymer, Nanoinclusions in lubricants, Nanoadditives in fuel, Catalyst nanomaterials
9.	Energy storage and conversion	Hydrogen production photocatalysts Photovoltaic cells, Thermophotovoltaics, Lithium ion battery electrodes Fuel cell catalysts, Dye-sensitized solar cells Nanotechnology in wind turbines Ultracapacitors, Flywheels, Piezoelectrics

Table 16.2 (Continued)

S. no.	Applications of nanotechnology	Examples
10.	Food processing and packaging	Nano additives to improve nutritional value Nanoencasulated flavor enhancer Nanotubes and nanoparticles as gelation and viscosifying agent Nanoemulsions for better dispersion of nutrients Nano biosensors for pathogen detection Biodegradable sensors for temperature and moisture monitoring Nano clays and nano films to prevent spoilage
11.	Construction	Nano titanium dioxide coated self-cleaning glass Nano fibril wood Nano structure modified steel with ultrahigh strength Nano silica fume concrete Nanotech in fire protection Nanotech-enhanced smart paints
12.	Health monitoring	Battery-free printed graphene sensors for clinical monitoring, measuring glucose levels, stamina, fitness, bacteria in the mouth Smart contact lens

Sources: Based on Cox (2019), Singh et al. (2017), Surendiran et al. (2009) and United States National Nanotechnology Initiative (n.d.-b).

undesirable impact on health and the environment. The onus is on industry, academia, and government regulatory bodies to highlight the potential risks of nanotechnology. The biggest challenge is to tame the technological changes and to assess associated risks. To acquire globally accepted status, every novel technology should possess some basic ethical attributes viz. safe, beneficial, fair, and respect autonomy. The novel debate on expected risks has opened the gateway to a whole new discipline, i.e. Nanoethics (Lu et al. 2012). To bear the responsibility and to maintain a balance between risks and benefits associated with any new technology, ethical considerations should be taken care of. It can be either reactive or proactive. Reactive ethics is mainly correcting the problems as they arise whereas proactive is thinking about probable issues and correcting them before. New technology creates new situations for which there are no policies (Bucaro 2016). It demonstrates "development first and restriction afterward" patterns like nuclear and cloning technology (Feynman 1960). Nanotechnology is the first technology in human history of which the ethical issues are considered and studied in the process of technology development depicting proactive ethics. Proactive ethics defines the direction, sets rules of accountability, and builds an environment to identify and assess the questions raised (Feynman 1960; Bucaro 2016).

Nanotechnology is a common term which has exhibited its influence in almost every industrial sector namely the information and communications sector, including electronic and optoelectronic fields, food technology, energy technology, and the medical products sector, including many different facets of pharmaceuticals, drug and gene delivery applications, diagnostics, and medical technology (United States National Nanotechnology Initiative n.d.-b). Nanotechnology products may also offer novel challenges for the reduction of environmental pollution. However, when a phenomenon takes place at the nanoscale, it behaves quite differently than those occurring at larger dimensions. The researchers aim to exploit this attribute for the benefit of mankind. These products on the other side expose the humans and environment to new health threats with unexplained mechanisms of interference with the physiology of living organisms (Khan 2015).

These speculations are based on the fact that nanoparticles generated in nanotechnology processes are either released into the environment (intentionally or unintentionally) or actually delivered directly to individuals during the utilization of nanotechnology-based products. Those individuals will receive special concerns whose job profile compels them to be in sustained contact with nanoparticles. During evolution, the human beings have evolved defense mechanisms against environmental agents, commonly encountered, within which size is a significant factor. The exposure to nanoparticles having unrevealed properties may pose a threat to the normal defense mechanisms related to immune and inflammatory systems. There is all probability of influence of nanoparticles on the environment pertaining to its dispersion and existence. As per the standard operating process of risk management, whenever a new risk is identified, a comprehensive investigation should be done to identify the nature of the risk (Lauterwasser 2005; Scientific Committee on Emerging and Newly Identified Health Risks 2006; Lin and Allhoff 2008b).

Some of the key areas – such as security, global inequalities, social justice, research funding, weapons, and privacy, are social implications of nanotechnology. If the risks and challenges associated with these fields are not met timely, they can have a devastating impact on the lives of people. With each new advancement, nanotech may become a potential abuser for society. Threats to individual privacy and private health data theft are the serious concerns raised by nanodevices. It is imperative to foresee the long-term effect of nanotech in shaping and transforming society. We must educate all nanotechnology stakeholders about the short-term and long-term benefits to promote sustainable, ethical, and economic nanotechnological growth. Nanoethics has evolved as a domain concerned with various ethical issues of this particular nanoscale technology (Khan 2015; Gökçay and Arda 2015). From a search in the Web of Science (WoS) using "ethics and nanoparticles" and "ethics and nanotechnology" as keywords, about 400 publication works found to be indexed, of which only 110 works were mentioned in ethics-related editions, which highlights a new forum of debate (Souto et al. 2020).

Generally, before nanomaterials are processed and used in multiple consumer goods, risk factors and public issues associated with nanotechnology should be greatly minimized. There arises a need of drafting some definite guidelines to ensure the safe utilization of nanoparticles. However, some regulatory agencies have stepped into this field to end uncertainties (Allhoff and Lin 2006).

16.2.1 Regulatory Agencies

U.S Environmental Protection Agency (EPA) developed a significant new use rule (SNUR) to ensure that nanoscale materials receive an adequate regulatory examination. SNUR would require manufacturers and importers to submit a Significant New Use Notice (SNUN) to EPA at least 90 days prior to the start of any processing activity of new nanoscale materials based on chemical substances listed in the TSCA (Toxic Substances Control Act).

EPA will evaluate the new application of nanomaterials based on the provided information such as chemical identification, chemical uses, physical and chemical properties, production volume, fate as well as toxicity data. This information will help the agency to deter activities that could be detrimental to human health or the environment (Hamburg 2012).

Environment Agency of Canada has issued regulations that ensure that any engineering processes of nanomaterials or import into Canada is safe for human health and the environment. Regulatory frameworks have been designed to collect information and assess the risk associated with nanomaterials. The Canadian regulation determines whether nanoscale materials are considered new substances (Health Canada 2011).

European Union has introduced a new Classification, Labeling, and Packaging (CLP) regulation. This regulation states that if the physical state of a substance is changed due to nano scaling, an evaluation should be conducted to determine whether the hazard classification of nanoscale material will be changed or not (European commission 2015).

U.S. National Institute for Occupational Safety and Health published provisional guidelines for working with the nanomaterials in a report titled "Managing the Health and Safety Concerns Associated with Engineered Nanomaterials" (National Institute for Occupational Safety and Health [NIOSH] 2012).

U.S. Food and Drug Administration has given FDA regulation of Nanotechnology to regulate the nano-based products. To facilitate the responsible production of FDA-regulated products that may include nanomaterials or otherwise require the application of nanotechnology, the FDA is implementing a regulatory policy for nanotechnology. FDA has given a strategic plan to assess the efficacy, quality, and safety of products (United States Food and Drug Administration 2018).

16.2.1.1 Nanotechnology and State Regulation (India)

The Ministry of Science and Technology, Government of India is the nodal ministry for the promotion of R&D in the area of nanotechnology. It administers its functions through three departments: The Department of Science and Technology (DST), the Department of Biotechnology (DBT), and the Department of Scientific and Industrial Research (DSIR). Other than the programs within the Ministry of Science and Technology, the Ministry of Information and Communication Technology and the Ministry of Defense have also shown interest in promoting nanotechnology R&D pertaining to their areas of activity.

The Drugs and Cosmetics Act, 1940 is an Indian parliament act to regulate the "import, manufacture, distribution and sale of drugs and cosmetics." The act also provides for the establishment of the Drugs Technical Advisory Board that would function as a general advisory body to the central and state governments on "technical matters arising out of the administration of this Act." The licensing authority for new drugs has been divided between the Central Drugs Standards Control Organization (CDSCO) and the state drug controllers. The Ministry of Health is involved in the regulation of nanotechnology applications through its Directorate General of Health Services, under which the CDSCO is situated. The Ministry of Environment and Forests (MoEF) deals with environmental impacts or hazards that may emanate from new applications of existing or new chemicals and substances. The Central Pollution Control Board (CPCB) discharges most of the functions relating to the prevention and control of pollution, including through hazardous materials. The State Pollution Control Boards (SPCBs) are the state-level authorities under the 1986 Environment Protection Act (EPA) (Government of India 2003; Srivastava and Chowdhury 2011).

Indian Council of Medical Research (ICMR) is the primary scientific body funded and supported by Department of Health Research, Ministry of Health and Family Welfare involved in the development, coordination, evaluation, and promotion of biomedical research in India. Currently, there are 31 intramural research-oriented institutes/medical research centers of the ICMR (Indian Council of Medical Research n.d.). However, nanotechnology research is primarily promoted through extramural fellowships. The ICMR has recognized the emerging importance of nanotechnology in the development of medicines and devices and has funded

154 nanomedicine-based projects (Bhatia et al. 2018). The Council for Science and Industrial Research (CSIR) is analogous to the ICMR and focuses its efforts into scientific research and industrial growth for economic, environmental, and societal benefits. Under the aegis of CSIR, Central Drug Research Institute (CDRI) is a premier institute for modernized drug research and development.

The Indian Institute of Toxicology Research (IITR) is a CSIR laboratory engaged in specialized research, innovation, and risk assessment for nanotechnology's effective regulation. Other center of excellence like the National Institute of Pharmaceutical Education and Research (NIPER) and Central Food Technology Research Institute (CFTRI), also imparting educational expertise in toxicology, can help to fill the knowledge void in health applications about nanotechnology-related threats (Srivastava and Chowdhury 2011).

16.2.2 Distributive Justice

Distributive justice is one of the simplest social concepts about the allocation of resources. The whole of mankind foresees nanotechnology as the epicenter of development and innovation owing to its potential impact on our daily life. But a part of the world is also worried that technology will increase the disparity between developed and emergent countries. While taking into account the ever-increasing discrepancy between developed and developing nations, will the rich–poor gap will be managed or the potential nanotech benefits will be distributed and shared? Nano-divide or nano-gap is the term used for this ethical concern since 2001. As per Cozzens and Wetmore, the term Nano-divide can be explained in two ways (Schroeder et al. 2016):

1) **Nano-innovation divide** – It is the inequality between high-, middle-, and low-income economies based on the development and retention of special knowledge and innovation. Developing countries are dominating nanotechnology-based research, innovations, and business while poor countries are still focusing on poverty, clean air, potable water, better education, and health services. The top innovating companies are investing heavily in nanotech practices and attracting significant developments while low income and politically unstable nations are not in a position to catch the new technological pace (Heller and Peterson 2014; Gonzalez 2015).
2) **Nano-orientation divide** – This term highlights the inequity between nanotech-focused fields and fields targeting fundamental human needs. It assumes that the nanotechnology industry will be more business-oriented expecting high financial returns rather than serving humanity (Maclurcan 2009; Schroeder et al. 2016).

The nano innovation is found to be pervasive only where nano knowledge is generated and conserved. This divide can be represented as information-rich and information-poor nations. As per Thai and Australian informants, nanotechnology research and innovation is confined to global wealthy leaders like the United States, Japan, Germany, Sweden, the United Kingdom, Australia, Taiwan, Hong Kong, and France. This nano-divide has created a chasm between leaders and trailers. However, several Asian economies have experienced fast growth over the last 30 years; in particular, South Korea and Taiwan have looked to high-tech development to become global leaders. Nanotechnology developments have gained enough attention in treating lifestyle-related diseases and cancer that is common with people in advanced countries. However, in developing nations, these diseases are more prevalent. The concern arises whether the poor and underdeveloped nations will be provided access to these high-cost advanced healthcare services based on advances in nanotechnology. It is the need of the hour to thoroughly evaluate the gap in the environment of innovation and equal global sharing (Maclurcan 2009; Flament 2013; Schroeder et al. 2016).

In developed nations, employers do not warn their staff about the nano-products being used in manufacturing procedures, for example through inhaling nanoparticles, posing a health danger to such employees. Besides this, the shift of poor workers from developing countries to rich nations with superior infrastructure and transparent work situations raises important concerns about how equal nanotechnology progress has been and what can be done to regain the balance (Nickerson 2013).

16.2.2.1 Bridging the Gap

The revolutionary technology has other facets also. On one hand, if nanotechnology is presumed to build a gap and divide the world into "Haves" and "Have not," it also has the potential to fill the void between rich and developing nations. Multiple international and humanitarian groups around the globe have been working to diminish the gap. Nanotechnology and its anticipated market advantages unlock the opportunities for social and economic growth, while for poor nations, the technology may offer benefits that would help to knock out poverty. Nanoscience can bring significant transformations in the lives of the poor populations of the world. There are plenty of nanotech applications that could answer a majority of the recurrent difficulties confronted by developing nations. Nano innovation can change the third world for disease eradication, quality medicines, potable water, abundant food, cheap energy sources, better agricultural tools, and faster and affordable shipping (Flament 2013; Heller and Peterson 2014).

Efficient nanosystems would likely allow cheaper and more productive manufacturing processes that could reduce the price of all products. Besides supporting the economy of rich and poor nations, it is the necessity of the moment that nanotech should also serve humanity. Government and nongovernment organizations should take concrete and early steps to multiply the chances of the positive universal impact of nanotechnology. If opened up to the developed world, nanotechnology may have deeply beneficial effects on the progressive development of those countries (Sachs 2003; Schroeder et al. 2016).

In fact, not just the advanced technology countries like China and India have demonstrated their potential to become pioneers in global nanotechnology space, but even small nations like Brazil, Mexico, and Malaysia are eager to invest in nanotech research and win an entry ticket to rich economies (Flament 2013). China is the only southern nation holding its position actively in nanotech and putting a hefty amount in nano-investigations and inventions. Countries such as Costa Rica, Nigeria, and Thailand are now involved in nano-innovation, with the Government of Thailand spending 300 million baht (approx USD 10 million) in 2010 for nanotechnology products. "Nanodialogues," conducted in Zimbabwe in 2006, aimed to encourage bottom-up nanotech initiatives by collaborating local groups and scientists from both the South and the North to determine the suitability of nanotechnologies for the community (Maclurcan 2010).

Brazil has played a pivotal role in Latin America since a national plan for nanotechnology was launched in 2000. Brazilian Ministry of Science, Technology, and Innovation hold account of over 3000 nanotech scholars working till 2010. Brazil, along with South Africa and India, has played an influential role in the Nanotechnology Initiative for India, Brazil, and South Africa (IBSA) which seeks to promote economical and societal partnerships among these developing countries. The South African nanotech policy focuses on investment in food, water, health, and energy. Other African nations like Zimbabwe, Kenya, and Nigeria are also showing interest in nanoscience to bridge to nano-divide (Flament 2013).

However, the journey to reap the real fruit of nanotechnology by poor countries is still long. In both the advanced and underdeveloped economies, deliberate attempts should be made to make certain that the nanotechnology techniques diffuse as broadly as possible. Without stringent policies, it would be uncertain to assist the poor societies with nanotech innovations without any further delay. Such conscious initiatives may include expanded involvement in promoting global nanotechnology research, knowledge diffusion, and exempting poor countries from specific patents rules (Heller and Peterson 2014).

16.2.3 Nanoweapons

Nanotechnology is playing a noteworthy role in the modern battlefield by driving innovations in security, nano-fabric for armor, weapons, and military personnel health. Even though nanotechnological innovations are serving as a potential "game-changer" in various military applications, these developments may also lead to new kinds of accidental wars and abuses. These military advancements escalate various societal and ethical concerns (Brehm 2017).

The incessant designing and manufacturing of nano weaponry carries a risk of massive suicidal destruction. Worldwide, many emerging economies are investing heavily in nanotechnology-based weaponry. Conventional weapons are being refined and this modernization may trap the human race in the web of nano-engineered arms. Unlike nuclear weapons, nanoweapons can create more havoc and security issues worldwide. Nanotechnology can prove to be a destructive technology that can lead humanity to face biological and chemical terrorism and ultimately a risk of life extinction. Novel bioweapons can be built to selectively target and affect only a certain geographical area or a genetically distinct group of people (Van de Poel 2008). The integrated applications based on information and nanotechnology facilitate universal, pervasive, and covert surveillance and stealth ways of data gathering which may raise human and public security concerns. Another key concern is the probable access of nanoweapons and devices to nonmilitary individuals or groups because of simple admittance to sheer knowledge. These military developments would be hard to detect and do not require special facilities and rare crude material (Hoven and Vermaas 2007).

The development and utilization of nanotechnology in warfare do not highlight its potential impacts on military personnel. These nanomaterials used in military uniforms or battle suits might address serious health hazards to the soldiers that are wearing the material. Topical exposure to nanoparticles is quite possible because of worn-out uniforms. The uptake of these undesirable particles into the soldiers' skin could have profound biological effects. Extensive studies and workplace monitoring are necessary to identify the essentials of risk assessment and management framework. Nano-engineered material may readily enter the brain as a result of small size and cause varying degrees of brain damage (Brehm 2017). These nanoscale particles may also reach the systemic circulation of military personnel by inhalation from fuel or scattering of the spectrum of particles from bombing through the environment. Lungs, nasal cavity, skin, gastrointestinal tract, and eyes are the key routes of concern being investigated for the identification of hazards in military personnel (Nanoscience Conference 2020).

Secret military advancements make it hard to raise ethical concerns relating to the use of nanotechnology. Despite this secrecy fact, it is pivotal to specify ethical concerns related to the harmful aspects of nanotechnology. Military applications like biological weapons, autonomous weapons if turned hostile may lead to undesirable consequences and can put societal ethics at stake.

16.2.4 Ethics in Nanomedicine

Nanotechnology has produced revolutionary medical results and its influence on society is substantial. Medical science has taken a giant step in the nanoscale domain and the practice is known as nanomedicine. Nanomedicine is characterized as the exhaustive monitoring, diagnosis, treatment, repair, defense, and enhancement of human biological systems, working from the molecular scale, using nano-engineered devices and nanostructures (Berger 2008).

Most of the recent technological applications of nanomedicine are novel nano-targeting therapeutic systems and enhancement of bioavailability, approximately accounting for 78% global sales and 58% of patent filing. Innovative modalities like nanostructure scaffolds for tissue replacement, nanodevices, remote nanoprobe regulation, integrated sensory electronic system, nanopores, nanobubbles, gene therapy, nanofiber brain implants, nanosomes, and respirocytes will eventually be integrated (Bawa and Johnson 2007; Surendiran et al. 2009).

Nanomedicine could significantly award a quality life to patients, provide early diagnosis of medical issues, reduce the treatment duration, minimize healthcare costs, and cite better clinical outcomes. While the nanomedicine research and innovation are in a hustle, the ethical, educational policies and laws are still struggling to match up. It is indispensable to proactively discuss social, ethical, and regulatory facets of nanomedicine to avoid any unintended harm to humans and the environment (Gökçay and Arda 2015). It will also alleviate any societal opposition close to that experience for other emerging technologies. People are still dubious about the toxicological risks accompanying nanotechnological techniques and products. There also arise key questions about the management of nano waste and ecological pollution from nanomedical device manufacturing. Healthcare practitioners are entering a new era of nanotherapies and nano-based diagnostic interventions accompanying a different set of ethical questions regarding the benefits and risks (Bawa and Johnson 2007; Gökçay and Arda 2015). Depending on the size, morphology, shape, and concentration, nanoparticles may interact with cellular processes and compromise cell function, resulting in a potential threat of cytotoxicity and/or genotoxicity (Ebbesen et al. 2006; Van de Poel 2008). Although most complex living organisms have sophisticated immune systems, the impact of nanoparticles still needs to be assessed at a preclinical level. For a nanoformulation to reach clinical trials, its quality, efficacy, and safety must be ensured (Allhoff 2007).

16.2.4.1 Ethics in Research and Development

The area has attracted ethical dialogues from academic scholars, researchers, and various regulatory bodies before extensive development in technical fields. Before utilizing nanomedicine for diagnostic, preventive, and therapeutic purposes, there must be rigorous preclinical and clinical testing. On a prima facie basis, scientists have initiated investigating the pharmacological effects, immunological properties, toxicity, and exposure pathways of various nanomaterials (Resnik and Tinkle 2007). Various government and regulatory bodies are focusing on creating and implementing exploratory programs to pin down the risks associated. National Cancer Institute in association with the National Institute of Standards and Technology (NIST) and the U.S. Food and Drug Administration (FDA) has established a Nanotechnology Characterization Laboratory (NCL) in 2004 to perform preclinical characterization of therapeutics, imaging agents, and diagnostics used in cancer. One of the objectives of the NCL is to perform *in-vitro* and *in-vivo* characterization of nano-engineered material for safety, efficacy, and toxicity. To train and educate students, researchers, and scientists from various disciplines, the laboratory is aimed to communicate related findings and developments through journal articles, conferences, public meetings, the media, and news releases (Nanotechnology Characterization Laboratory n.d.).

United States environmental protection agency recommends four basic steps to evaluate the nature and probability of adverse health effects in humans due to exposure to chemicals and metals. The risk assessment involves hazard identification, dose-response assessment, exposure assessment, and risk characterization (United States Environmental Protection Agency 2016).

The available literature on hazards unique to these nano-engineered materials is still nebulous. The safety and adoption of this emerging technology is a matter of concern due to a lack of reported cases and documentation of health hazards from nanomaterials (Yokel and MacPhail 2011). It has been studied that a nanoparticle may show characteristics different from its normal material. Size and shape changes can have significant and unexpected effects on nanomaterials' physical and chemical properties and hence can be more reactive and toxic (Canadian Centre for Occupational Health and Safety 2018).

Since nanomaterials rely heavily on their microenvironment, they can undergo variations in size or shape inside any organism. A particle of larger dimension could disintegrate into particles of 1 nm, or 1 nm could agglomerate to form larger particles. Continuous expansion of nanotechnology has increased human exposure to nanosize particles through dermal uptake, ingestion, inhalation, and injection. Nanosize particles of natural as well as anthropogenic sources may produce deleterious effects like toxicity, oxidative stress, cellular dysfunction, and increased inflammatory activity. Several epidemiological studies have documented evidence of the adverse impact of nanosize particles on the cardiovascular and respiratory system resulting in morbidity and mortality (Oberdörster et al. 2005).

When inhaled, the small size of nano-engineered materials promotes cellular uptake and easy translocation to bone marrow, lymph nodes, spleen, and heart. Key findings have shown that engineered nanomaterials can easily cross the cell membrane, mobilize along axons, dendrites, and induce toxicity. Titanium dioxide nanoparticles (used in sunscreen products) have been found to cause liver damage by inducing DNA and chromosomal changes. Inhaled particles may impair the ability of macrophages and hence affect innate and adaptive immune responses. The distribution of various types of nanoparticles calls for exhaustive studies that have not been carried out yet. More extensive *in vitro* and *in vivo* studies are required to fill the literature voids. Scientists are also investigating alternative methods like high-throughput systems, *in silico* models, *in vitro* modeling, and quantum mechanics-based computational modeling to perform toxicity screening of nano-engineered material (Kumar et al. 2014). The actual challenge associated with Nanomedicine research is risk identification, risk management, and communication. The current scenario requires the formulation and implementation of ethical guidelines and regulations to achieve the maximum benefit of technology without harming the human race (Costa et al. 2011). Concerns were raised for protecting the rights of patients treated with nanomedicine or devices and there must be an act against the violation of rights. Patients should be given complete information about the effectiveness, safety, and probable effects of the newly used nanosubstances. At the same time, it is also necessary to ensure the protection of medical practitioners practicing nanomedicine technology (Yasri and Wiwanitkit 2017).

16.2.4.2 Health and Safety Issues

Health is one of the major ethical issues and involves the possible effects of nanomaterials and nanodevices on human safety and health. Nanomaterials provide highly effective surface area as compared with non-nano materials and hence they are better absorbed through the various routes of administration like inhalation, dermal exposure, as well as ingestion (Van de Poel 2008). It is already reported by several scientists that the nanoparticulate system can easily penetrate the cell via several mechanisms like membrane penetration, endocytosis, transmembrane channels, etc., resulting in positive as well as negative effects on the organism body. Positive effects include treatment of diseases like cancer, heart attacks, and Alzheimer's disease, and adverse effects give birth to deadly diseases (Ebbesen et al. 2006). Since both nanotechnology and its related ethical issues are rising rapidly, they raise several questions related to human safety as well as health. These questions are, what are the possible health and safety threats with nanotechnology applications, and who is accountable if anything happens? What are the rights of people harmed by the products and processes of nanotechnology? How can we guard society from the negative consequences of nanotechnology and nanomaterials/devices? (Sparrow 2009).

16.2.4.3 Toxicity Associated with Nanoparticulate System

Nanoparticles have been prepared from metals, nonmetals, polymeric materials, and bioceramics. Human is exposed to various nanoscale materials in daily life since childhood, and the newly emerging field of nanotechnology has become a new threat to human life. Due to the fine size of nanoparticles, they can easily enter the human body and can cross the various biological barriers, and can reach the most sensitive organs. Nanoparticles of size less than 10 nm are gas-like and readily penetrate human tissues and interfere with the biochemical environment of a cell. Experimentation studies on animals and human have shown that after inhalation or oral exposure, nanoparticles assess distribution to the kidney, heart, spleen, liver, and brain in addition to the lungs and gastrointestinal tract (Vishwakarma et al. 2010). The half-life of nanoparticles is estimated to be 700 days in the human lungs which is harmful to the respiratory system. Radziun et al. have reported that aluminum oxide NPs showed no substantial toxicity on the viability of mammalian cells at a concentration of 400 μg/ml (Radziun et al. 2011). Regarding toxic effects, Chen et al. have reported that aluminum oxide NPs disturb the cell viability, alter mitochondrial function, increase oxidative stress, and also alter tight junction protein expression of the blood–brain barrier (BBB) (Chen et al. 2008). Similarly, another study has reported a dose-dependent (25–40 μg/ml) cytotoxicity as the effect of aluminum oxide NPs (160 nm) on human mesenchymal stem cells. Besides this, dose-dependent genotoxicity was observed due to aluminum oxide NPs. Gold NPs are relatively safe because of their inert core and nontoxic nature. Cytotoxicity associated with gold NPs depends on the dose, side chain (cationic), and the stabilizer used. The variation in toxicity for different cell lines has been observed in human lung and liver cancer cell lines (Patra et al. 2007). Copper oxide NPs are used in semiconductors, antimicrobial reagents, heat transfer fluids, and intrauterine contraceptive devices. Copper oxide NPs (50 nm) have the potential to cause genotoxicity and cytotoxicity along with disrupting cell membrane and prompting oxidative stress (Ahamed et al. 2010). Historically, silver has long been known as an antibacterial substance. Silver NPs are used in the form of wound dressings, coating of surgical instruments, and prostheses. In comparison to others, these NPs have shown more toxicity in terms of cell viability, generation of reactive oxygen species (ROS), and lactate dehydrogenase (LDH) leakage (Patra et al. 2007).

16.2.5 Privacy and Confidentiality

The key issues to be discussed when using nanotechnology for health purposes are data ownership and confidentiality (Yasri and Wiwanitkit 2017). Nanotechnology-based modern implantable/swallowed diagnostic devices may facilitate the medical surveillance of individuals at the human cellular level which can be securely transferred to a scientific database system for review and afterward tracked using diagnostic software. The potential of nanotechnology to miniaturize implants and their administration in human beings is one of the primary concerns. It is laudable that nano implants/chips can furnish various advantages like dispensing a tailored amount of drug, monitoring cellular activities, assisting Alzheimer's patients, or sensing biochemical changes within organs or tissues (Bawa and Johnson 2007; Berger 2008). But maintaining that these benefits do not come at the cost of the security and privacy of the public is quite important. The unfair exploitation of technology to prepare implant devices for controlling, surveillance, and tracking may foster nano-panopticism (Gutierrez 2004).

The recent development in technology has made real-time remote monitoring possible by simply implanting radiofrequency identification (RFID) chips in human skin. However, in 2007, the American Medical Association's Council on Ethical and Judicial Relations (CEJA) released a study arguing that embedded RFID chips could violate privacy since there is no guarantee that microchip information can be adequately secured. While implants could assist in tracking patients' medical records in medical emergencies, it may also allow governments and surveillance agencies to track individuals (Halan 2019).

Another disturbing ethical question is about the management and storage of this enormous health information. Can the advanced nanotech-based healthcare system handle, accumulate, process, and review large-scale data? The real challenge is of an extensive data repository, its storage, and mining of data to analyze related health issues. Also, it will be of paramount importance to maintain anonymity and confidentiality while nanodevices are working on the molecular communication concepts. Serious ethical issues may be posed in context with securing a person's personal information and health interests by a system lacking appropriate management of information. Ethical responsibility calls for preventive and security measures to tackle potential data thefts and hacking of data (Pramanik et al. 2020).

16.2.6 Human Enhancement

The noun "enhancement" comes from the verb "enhance," meaning "to increase or improve." The concept of human enhancement can be defined as modifications and refinement of the natural abilities of the human species to achieve a high-performance body and mind. Human enhancement has a decent history; nearly two decades ago, biomedical interventions were in the center of development for the improvement and functioning of human health. With time and advancements in technology, the range of these practices has now increased and there are many kinds of genetic, biomedical, and pharmaceutical, and nanotechnological interventions aimed to improve human capabilities and wellbeing (Almeida and Diogo 2019).

In today's world, we have seen many possible scenarios of human enhancement through steroids to become stronger and plastic surgery to be more attractive. The exploitation of drugs to augment creativity and attentiveness, the perception has become a common practice. There are many possible scenarios like the power to see in the dark, non-visible spectrums of light, insertion of nanocomputers in bodies for better and enhanced functioning of human organs which today we have seen in science fiction movies will be possible through nanotechnology. Since nanotechnology basically manipulates at the molecular and atomic level and human beings are made of the same building blocks, it will be easy for the scientist to get the desired result without getting noticed. The enhancement will look more natural rather than a technological tool. Eventually, nanotechnology would allow us to examine and heal any physical condition in the body. This would mean that nanotechnologies would be able to repair someone who is damaged or diseased back to full health; an aged body and brain could be restored to a youthful state. Ultimately, the outcome could be the end of pain, sickness, disease, or aging. These kinds of innovations should be welcomed by modern society and liberals because they are just the extension of modern medicine. Just take a simple example in today's time when the world is facing the pandemic of Coronavirus and many patients are dying because of lack of ventilators; would it not made a difference if we have artificial red blood cells "respiratory" which can directly be embedded in the blood and patient lives could have been saved (Jotterand 2008; Lin and Allhoff 2008a; Université de Genève 2019)?

With all the positive sides of nanotechnology, it comes with many controversial and debatable parts as well. Some have raised concerns that human enhancements may lead to unnecessary pressures. Would it be possible to compete for a job if everyone else was enhanced mentally, to compete in athletics if everyone else was strength-enhanced, or compete for a spouse if you were not physically enhanced? Others fear that enhancements enabled by nanotechnology will worsen existing economic and social inequalities, especially when being costly the advancements are restricted to the elite class of people. Rich people can easily exploit the technology to boost their physical strength, intelligence quotient, or any particular trait and hence may become superior and evolved. As a result, the disparity would expand dramatically between those who can meet the expenses of enhancements and those who do not have resources. This individual modification will hamper social relations and may create a different class of un-enhanced people. There is a further threat that those that control all the resources, the enhanced class, would feel increasingly disconnected from the underclass, and as a result, there may be an enhancement divide. There can be a possible scenario that if the parents are given the choice to choose their child's traits like color of hair, eyes, intelligence level, etc., the diversity of species can be lost forever. The raised concerns are all valid, but they are not substantial enough to forbid enhancement technologies. Nanoethics need to meet the ethical challenges posed by human improvement technology. It is important to frame regulations and restore the technology uses for human interest (Heller and Peterson 2006; Jotterand 2008; Lin 2009).

16.2.7 Nanopollution

Even though the nano blast will bring an abundance of positive changes, there is a great discussion concerning what extent industrial and commercial use of nanomaterials will influence living organisms and ecosystems (Soutter 2012; Chauhan n.d.). The revelation of carbon nanotubes and graphene, and our expanding comprehension of the properties of materials on the nanoscale, is prompting a blast in the commercialization of next-generation advancements. The health effects of nanopollution are yet to be completely perceived, making nanopollution one more manmade environmental impact with dubious impacts in the long term (Soutter 2012; Chauhan n.d.).

Nanopollution is a conventional name for all nanosized waste generated by nanoscopic devices or during the production process of nanomaterials. It mentions all the naturally occurring nano contaminants and the introduction of nanoengineered materials produced by human activity into the environment (Berger 2016). This sort of waste might be exceptionally hazardous because of its size. It can exist all around and may effectively penetrate animal and plant cells, even through the skin causing unknown effects. Most human-made nanoparticles do not occur in the environment naturally, so living beings might not have suitable inbuilt mechanisms to manage nanowaste. Nanopollution causes instability, several disorders, harm, or discomfort to the ecosystem whether they are naturally occurring nanoparticles or nano-engineered nanoparticles (NENPs) (Maity and Pramanik 2013). Moreover, it is simultaneously giving rise to a persistent form of pollution which is too small to detect or contain easily (Chauhan n.d.). The potential risk of nanopollutants is segregated in the following categories.

16.2.7.1 Health Issues

Universities and research organizations around the globe are undergoing extensive research programs to explore the effect of nanoparticles on humans. Specialists have expressed that individuals are already exposed to significant levels of nanopollution in urban zones because of the fumes from vehicles and manganese oxide from construction sites, and researchers are working hard to determine and characterize the hazardousness of these pollutants. Nanoparticles are small to such an extent that they can easily penetrate living cells. The human body is programmed to recognize unfamiliar agents and generate phagocytes to destroy the foreign agents. If the body's phagocytes are constantly engaged in digesting nanoparticles, the cells cannot perform other functions efficiently inside the body (Maity and Pramanik 2013).

- An investigation led by analysts at Johns Hopkins Hospital earlier in 1941 revealed that nanoscopic particles, when inhaled, could enter the brain via the olfactory tract, as was confirmed by experimentation with chimps and rhesus monkeys (Soutter 2012).
- Another research by toxicologist Eva Oberdorster from Southern Methodist University involved exposing captive large-mouth bass to different degrees of carbon60, or Buckminster fullerene. Two days later, she discovered an immune response in the livers of the fish population. This proved that the nanoparticles were capable to bypass the fishes' immune systems, thereby raising the concerns that these nanoparticles could cause some harm to the living systems before identification (Soutter 2012).
- Researchers at Trinity College Dublin have associated rheumatoid arthritis and the initiation of autoimmune diseases to nanoparticle exposure. These particles in human cells contributed to the modification of the arginine (amino acid) into citrulline, which can burgeon autoimmune conditions viz. rheumatoid arthritis (Soutter 2012).
- In 2004, scientists at NASA exhibited that inhalation of carbon nanotubes reaches lungs. They are more hazardous than equivalent quantities of carbon black and quartz. The cytotoxicity of carbon nanotubes was found similar to chrysotile asbestos (Maity and Pramanik 2013).
- To provide strength to the above evidence, at a meet of the Society of Toxicology, scientists from Tottori University displayed the first series of images to highlight the entry of carbon nanotubes in the systemic circulation within a minute of exposure to the lungs. The negatively charged carbon nanotubes interact with complement system components, red blood cells, and proteins in the blood, leading to further intricacies.
- Nanoparticles from automobile discharge are the upmost devastating constituent of atmospheric pollution which plays a key role in inducing cholesterol plaque buildup in the arteries which ultimately lead to heart attack and stroke. Toxic metallic nanoparticles enter the tissues of heart and cause serious damages to these vital powerhouses of human body.
- Odor-free garments having silver nanoparticles are newly launched items by the sports clothing industries. These silver particles, when compared with bleach, are two times more toxic to bacteria. These could cause complications at waste water treatment setups, which are based on bacterial beds to purify water.
- "Buckyballs" – microscopic football-shaped cages of carbon – can be taken up by simple organisms, which can further accumulate in toxic concentrations in the food chain (Maity and Pramanik 2013).

Although there are numerous unanswered questions, research like this is gradually throwing more light on the hazards of nanopollution, which will help in creating more awareness and hopefully lead to safer workplaces and a cleaner environment.

16.2.7.2 Measuring Nanomaterials' Concentration in the Environment

Another main field of ongoing research is the determination of the nanopollutants' concentration in a given environment. The minute size of nanoparticles makes it extremely difficult to extract and measure, so novel and innovative methods are needed before we even fully understand the magnitude of the problem. A modern, cost-effective invented technique raises expectations of carbon nanotube concentration measurement in plants using microwave heating. Carbon nanotubes are an unavoidable element in all carbon-based nanomaterials and therefore require close supervision. This method is capable of both qualitative and quantitative estimation of nanoparticles. This method has the potential to become a testing standard for the U.S. Environmental Protection Agency (Maity and Pramanik 2013).

16.2.7.3 Environmental Issues

There is no doubt that the use of nanotechnological materials and devices will impact the environment. The burning topic of debate is concerning how much extent industrial and commercial use of nanomaterials will influence living beings and biological systems. Nanotechnology's environmental impact can be divided into two viewpoints: the potential for nanotechnological innovations to help improve the environment, and perhaps the novel sort of contamination that nanotechnological materials might cause if delivered into the environment (Chauhan n.d.). Several instances of negative impact have been cited by researchers:

- Many plant species can absorb various nanoparticles in a variable amount if present in abundance. Nanosilver is of great concern as this silver has the ability to be transmitted to animals that feed on or are fed these plants.
- In several pieces of research, scientists have also reported a substantial decline in biomass in many plant species in reaction to numerous nanoparticles suggesting that it has the ability to modify community composition.
- In few experiments, soil microbes display a decline in microbial biomass and surface soil activity that can slow down the role of microbes in decomposition but a rise in nitrous oxide production, which are both significant greenhouse gas and actually the most active stratospheric ozone-depleting material.
- The growing usage of nanomaterials deliberately contributes the nanoparticles into the waste streams, viz. wastewater treatment facilities leading to its recirculation and accumulation in biological systems.
- The main toxicological issue is that some of the processed and engineered nanomaterials are redox-active, while others are transferred via cell membrane and in particular, through mitochondria.
- The nanomaterials have a wide surface area per volume unit that gives rise to new electronic and photochemical properties compared with traditional chemicals. These unique properties, which make nanomaterials more useful, can also allow certain nanomaterials to pose hazards to humans and the environment under some circumstances (Maity and Pramanik 2013).

Consequently, the toxicity of nanoparticles that comes into the spotlight raises questions about the sustainability and future of nano-products.

16.2.7.4 Social Issues

These issues highlight the effects of nanotechnological gadgets on politics and human interactions on people working in multinational companies. Besides the aforementioned considerations, nanotechnology has wider societal influence and poses broader social challenges. Social scientists have proposed that the societal challenges of nanotechnology should be considered and analyzed not solely as downstream threats or impacts. Instead, the challenges should be incorporated into upstream research and decision-making to ensure technology development that meets social objectives. Social scientists and civil society organizations suggest that technology assessment and governance should also include public participation (Chauhan n.d.).

In 2003, more than 800 nano-related patents were awarded, rising to about 19000 worldwide by 2012. Corporations are now seeking large-scale patents on discoveries and inventions on the nanoscale. For example, two companies, Nippon Denki Kabushiki-gaisha (NEC) and International Business Machines (IBM), own one of the major key elements of nanotechnology, the basic patents on carbon nanotubes. Carbon nanotubes have a variety of applications and have likely to become essential to several industries from electronics and computers, to improved components for drug delivery and diagnostics. Carbon nanotubes have proved to become the most marketed item with the ability to substitute major conventional raw materials. Nanotechnologies can provide new opportunities for the millions of people in developing countries lacking access to primary services, such as safe water, reliable energy, healthcare, and education. The 2004 United Nations Task Force on Science, Technology, and Innovation highlighted that nanotechnology can boost production using little labor, land, or maintenance, high output, low cost, and minimal requirements for materials and energy (Chauhan n.d.).

However, the major concerns are that the benefits of nanotechnology will be unevenly distributed, and will only reach affluent nations. Longer-term issues revolve around the impact that whether these could lead to either a post-scarcity economy or exaggerate the wealth gap between developed and developing nations. Social scientists and social sciences workers recommended knowing and understanding the social and ethical issues and explaining to the public. It is also recommended that the public should participate in the decision made by the scientist as well as the government related to nanotechnology (Wamocha et al. 2013). A lot of research has been done in the field of nanotechnology and several publications, as well as patents, have been generated and published. But still, no perfect rules and regulations are there in respect of manufacturing and marketing of nano-based products. It is the responsibility of the authorities, industries, and universities to look at the safety of all participants as well users of the nano-based product because absence of rules and regulation can affect the scientists, engineers, students, technicians, and workers in a distant time (Schulte and Salamanca-Buentello 2007).

16.2.7.5 Speculative Issues

The exceptional dangers pertaining to the theoretical vision of molecular nanotechnologies, such as "Grey goo"(Fries 2018). Sci-fi novels can act as a helpful option in such scenarios since they depend on "projected time" rather than on "occurring time" (Dupuy 2007). However, authors have a future vision and they hope that their readers will respond to their dreams which will trigger societal changes. The examples of grey goo in context to nanotechnology, their diversity and uncontrollable outburst to global "ecophagy" [a term coined by nano scientist Robert Freitas (2000)], to depict circumstances when entire environments are consumed by nanorobots have gone awry.

The depiction of effects perceived by our five senses raises surprise, amazement, bewilderment, fear, and horror. Blue goo suggests "police" nanorobots to defeat grey goo. Green goo talk about the power to utilize DNA as a computer code and related nanotechnology with life sciences, thus directing toward the possibility to hybrid genetically modified organisms (GMOs) which multiply unpredictably. Khaki or red goo represents nanoweapons that destroy everything in the enemy's territory. From Drexler's view point, "grey goo" is the outcome of a bottom-up manufacturing approach and human error to program nanorobots to replicate themselves exponentially (Fries 2018).

This concept has highlighted that such types of accidents are not affordable and exemplify the possible dangers of nanotechnology. Ethically grey goo is debatable and a compelling reason for future foresightedness and should not be seen a story to fun (Fries 2018).

16.2.8 Educational Issues

The advancement of nanotechnology offered remarkable benefits to human life and the environment and leading to huge educational consequences. Since the last two decades, research-based innovations, development, manufacturing of nano-based products have grown but educational progress is still in slow motion or we can say that skill training is lacking for the individual working in the industry to gull fill need of consumers as well industry. Therefore, a compulsory parallel training program is mandatory in the terms of ethical and social implications of technology (Asmatulu et al. 2012).

Multidisciplinary and interdisciplinary training program on advancement of nanotechnology and its ethics should be conducted for students and scientists together with engineers, technologists, medical practitioners, workers, social scientists, and humanists in every field to ensure the proper knowledge. Offering new class sessions, conferences, workshops, seminars, invited talk, guest lecturers, etc., in the field of nanotechnology and nanoethics can assist in getting the goals (Allhoff 2007). In order to reach the needed goal/focus in nanotechnology and nanoethics, new courses can be added for the undergraduate and graduate students or biologists, chemists, physicists, or engineers who want to pursue their carrier in this. To disseminate the knowledge based on nanotechnology, media, television, online news, leaflet, etc., can be exploited along with conferences, workshops, seminars, invited talk, and guest lectures. Study material should be made available on website and update should be done on regular basis (Asmatulu et al. 2012). It is the responsibility of the government agencies, such as the Environmental Protection Agency (EPA), the Food and Drug Administration (FDA), the National Institutes of Health (NIH), the Department of Health and Human Services (DHHS), and the National Science Foundation (NSF), to provide grants and financial support to regularly participate in education and ethics related to nanotechnology. Private companies (e.g. IBM, Intel, General Electric, and General Motors) should also invest as their participation will increase awareness.

16.2.9 Biological Issues

Biotechnology is the innovative field of biology associated with the manifestation of a living body or its part to improve the living life. Bioethics relates to ethical challenges during the research and development of biotechnological and medical nano-products (Asmatulu et al. 2012). The basic principles of bioethics are:

1) Confidentiality: things should be kept confidential.
2) Beneficence: development and advance in technology should not give any harm to the population.
3) Justice: future benefits and related risk of improved or newly developed technique and being transparent to all parties.
4) Autonomy: basically means the freedom of every human being to control their life and fundamental rights.

Ethical behavior demands us to understand the consequences of our decisions and actions and to make wise choices (Wamocha et al. 2013). Recently, several nano-based materials like Au, Ag, Pt, TiO_2, Fe_3O_4, CNTs, and many others have been developed. There are so many benefits of these nanomaterials, but on the other, their toxic effects cannot be ignored. Therefore, it is important to highlight the issues related to ethics, environment, health, and medicines during the research and development, and make it understandable for the general public and users (Asmatulu et al. 2010).

Besides the above-stated issues, nanotechnology poses many other legal and regulatory challenges. There are several agencies such as EPA, FDA, National Science Foundation (NSF), and NIH that hold their own regulation to control and provide standard operating procedures (SOPs) for the use of nanomaterial to the human being (Gökçay and Arda 2015).

16.3 Conclusion

Nanotechnology is rapidly growing, evolving, and influencing academic, science, and industrial domains. Advances and innovations in this domain are offering immense possibilities in several industries. Nanotechnology is already creating ripples in electronics, information technology, energy resources, textile, healthcare, food, and agriculture. It is said as the next big force and game changer in industrial revolution. However, like every other technology, nanotech also has its own bright and dark spots. Certain nanotech applications are capable to raise significant social, legal, economical, and ethical concerns. A number of environmental, health, safety, and privacy issues have come to surface with nanoscience. This demands a paradigm shift of our focus to the ethical concerns for better governance of technological developments. Nanoethics is an interdisciplinary platform with a vision to define the ethical and societal implications of nanotechnology. As a discipline, nanoethics seeks to investigate the potential pitfalls and beneficial impact of nanotechnology applications. This area also furnishes action lines to train, prohibit, and restraint the use of various toxic nano-substances. Government organizations, decision-makers, industry, and academia people should frame and implement regulations to guard the environment and public health. These initiatives can assist to achieve maximum benefits and minimize the uncertainties and risks. The scientific community should focus to call off the fears associated with unforeseen environmental and health risks, nanoweapons, and privacy invasion. Effective strategies should be designed to make sure the fair distribution of costs and profits of nanotechnology so that the last-mile delivery of innovations could be possible. Nanoethics also creates awareness regarding nanotechnology implications in medicine. It encourages the medical community to understand the ethical complexities related to research, production, regulation, clinical practice, and toxicity of nanomedicine. Nanotechnology has proved its potential in various ways; however, it is also necessary to fix its rising ethical concerns to reap its actual fruit.

References

Ahamed, M., Siddiqui, M.A., Akhtar, M.J. et al. (2010). Genotoxic potential of copper oxide nanoparticles in human lung epithelial cells. *Biochemical and Biophysical Research Communication* 396 (2): 578–583.

Allhoff, F. (2007). On the autonomy and justification of Nanoethics. *Nanoethics* 1: 185–210.

Allhoff, F. and Lin, P. (2006). What's so special about Nanotechnology and Nanoethics? *International Journal of Applied Philosophy* 20: 179–190.

Almeida, M. and Diogo, R. (2019). Human enhancement: Genetic engineering and evolution. *Evolution, Medicine, and Public Health* 1: 183–189.

ASK OCE, APPEL Knowledge Services (2006). It's a small. *Small World: NASA Nanotechnology* 1 (8) https://appel.nasa.gov/2010/02/26/ao_1-8_f_small-html/ (accessed 26 September 2020).

Asmatulu, R., Khan, W.S., Asmatulu, E., and Ceylan, M. (2010). 'Biotechnology and bioethics in engineering education'. *ASEE Midwest Conference*, Lawrence, Kansas, 10pp.

Asmatulu, R., Asmatulu, E., and Zhang, B. (2012). Recent progress in Nanoethics and its possible effects on engineering education. *International Journal of Mechanical Engineering Education* 40: 1–10.

Bawa, R. and Johnson, S. (2007). The ethical dimensions of Nanomedicine. *Medical Clinics of North America* 91: 881–887.

Berger, M. (2008). 'Ethical aspects of nanotechnology in medicine'. https://www.nanowerk.com/spotlight/spotid=3938.php (accessed 27 September 2020).

Berger, M. (2016). 'Identifying nanopollution'. https://www.nanowerk.com/spotlight/spotid=42220.php (accessed 27 September 2020).

Bhatia, P., Vasaikar, S., and Wali, A. (2018). A landscape of nanomedicine innovations in India. *Nanotechnology Reviews* 7 (2): 131–148.

Botticello, C. (2020). 'What is Nanotechnology? A closer look at nanotechnology'. https://medium.com/@caseybotticello/what-is-nanotechnology-e2faf828542f (accessed 26 September 2020).

Brehm, M. (2017). 'Nanoweapons, Discussion paper for the convention on certain conventional weapons (CCW).' http://www.article36.org/weapons/weapons-review/discussion-paper-nanoweapons/ (accessed 26 September 2020).

Bucaro, F. (2016). 'Is your ethics training proactive or reactive?'. https://www.frankbucaro.com/is-your-ethics-training-proactive-or-reactive/ (accessed 27 September 2020).

Canadian Centre for Occupational Health and Safety (2018). 'Nanotechnology-general'. https://www.ccohs.ca/oshanswers/chemicals/nanotechnology.html (accessed 27 September 2020).

Chauhan, G.S. (n.d.). 'Nanotechnology and future impacts of Nanopollution'. https://madhavuniversity.edu.in/nanotechnology.html (accessed 27 September 2020).

Chen, L., Yokel, R.A., Hennig, B., and Toborek, M. (2008). Manufactured aluminum oxide nanoparticles decrease expression of tight junction proteins in brain vasculature. *Journal of Neuroimmuno Pharmacology* 3 (4): 286–295.

Costa, S.H., Sethe, S., Pego, A.P., and Olsson, I.A. (2011). Scientists' perception of ethical issues in nanomedicine: a case study. *Nanomedicine* 6 (4): 681–691.

Cox, L. (2019). 'Small but mighty: 6 applications of Nanotechnology'. https://disruptionhub.com/6-applications-of-nanotechnology/ (accessed 27 September 2020).

Dupuy, J.P. (2007). Complexity and uncertainty: a prudential approach to nanotechnology. In: *Nanoethics: the Ethical and Social Implications of Nanotechnology* (eds. F. Allhoff, P. Lin, J.H. Moor, et al.), 119–131. Hoboken, NJ: Wiley.

Ebbesen, M., Andersen, S., and Besenbacher, F. (2006). Ethics in Nanotechnology: starting from scratch? *Bulletin of Science, Technology & Society* 26: 451–462.

European Commission (2015). 'Classification and labelling (CLP/GHS)'. https://ec.europa.eu/growth/sectors/chemicals/classification-labelling_en (accessed 24 September 2020).

Feynman, R. (1960). There's plenty of room at the bottom. *Engineering and Science* 23: 22–36.

Flament, G. (2013). 'Closing the gap: the impact of Nanotechnologies on the global divide, Nanotechnology industries association'. https://nanotechia.org/sites/default/files/files/20131126_NIA_Closing_the_gap.pdf (accessed 26 September 2020).

Freitas, R. (2000). *Some limits to Global ecophagy by biovorous Nanoreplicators, with public Policy recommendations*. Foresight Institute. https://foresight.org/nano/Ecophagy.html (accessed 27 September 2020).

Fries, M. (2018). 'Nanotechnology and the Grey Goo Scenario: narratives of Doom?.' http://journals.openedition.org/ilcea/4687 (accessed 27 September 2020).

Gökçay, B. and Arda, B. (2015). Nanotechnology, nanomedicine; ethical aspects. *Revista Română de Bioetică* 13 (3). http://www.bioetica.ro/index.php/arhiva-bioetica/article/view/829/pdf.

Gonzalez, A. (2015). '5 ways to close the global innovation divide'. https://www.weforum.org/agenda/2015/09/5-ways-to-close-the-global-innovation-divide/ (accessed 26 September 2020).

Government of India (2003). 'The Drugs and Cosmetics Act 1940'. https://cdsco.gov.in/

opencms/export/sites/CDSCO_WEB/Pdfdocuments/acts_rules/2016DrugsandCosmeticsAct1940Rules1945.pdf (accessed 24 September 2020).

Gutierrez, E. (2004). *Privacy Implications of Nanotechnology*. Electronic Privacy Information Center. http://www.epic.org/privacy/nano/ (accessed 27 September 2020).

Halan, D. (2019). 'Microchip implants: Are you ready for them?.' https://www.electronicsforu.com/technology-trends/tech-focus/healthcare-microchip-implants (accessed 27 September 2020).

Hamburg, M.A. (2012). Science and regulation. FDA's approach to regulation of products of nanotechnology. *Science* 336 (6079): 299–300.

Health Canada (2011). 'Policy Statement on Health Canada's Working Definition for nanomaterials'. http://www.hc-sc.gc.ca/sr-sr/pubs/nano/pol-eng.php (accessed 24 September 2020).

Heller, J. and Peterson, C. (2006). 'Human enhancement and nanotechnology: a foresight nanotech institute policy issues brief, Foresight Nanotech Institute'. http://www.foresight.org/policy/brief2.html (accessed 28 September 2020).

Heller, J. and Peterson, C. (2014). 'Nanotechnology, poverty, and disparity'. https://foresight.org/policy/brief6.php (accessed 26 September 2020).

Hoven, J.V.D. and Vermaas, P.E. (2007). Nano-Technology and privacy: on continuous surveillance outside the panopticon. *Journal of Medicine and Philosophy* 32 (3): 283–297.

Iberdrola (2019). Nanotechnology: a small solution to big problems. https://www.iberdrola.com/innovation/nanotechnology-applications (accessed 25 September 2020).

Indian Council of Medical Research (n.d.). 'Institutes'. https://main.icmr.nic.in/institutes (accessed 29 Septempter 2020).

Jotterand, F. (2008). Beyond therapy and enhancement: the alteration of human nature. *Nanoethics* 2: 15–23.

Keiper, A. (2003). The nanotechnology revolution, The new atlantis. https://www.thenewatlantis.com/publications/the-nanotechnology-revolution (accessed 25 September 2020).

Khan, A. (2015). Ethical and social implications of nanotechnology. *QScience Proceedings (Engineering Leaders Conference 2014)* 2015: 57.

Khare, S., Williams, K., and Gokulan, K. (2014). Nanotechnology. In: *Encyclopedia of Food Microbiology* (eds. C.A. Batt and M.L. Tortorello), 893–900. USA: Academic Press.

Kumar, A., Kumar, P., Anandan, A. et al. (2014). Engineered Nanomaterials: knowledge gaps in fate, exposure, toxicity, and future directions. *Journal of Nanomaterials* 2014: 130198. 16 pages.

Lauterwasser, C. (2005). 'Small sizes that matter: opportunities and risks of Nanotechnologies report in co-operation with the OECD International Futures Programme'. http://www.oecd.org/chemicalsafety/nanosafety/44108334.pdf (accessed 27 September 2020).

Lin, P. (2009). 'Ethics of human enhancement: 25 questions & answers'. https://www.nanotech-now.com/columns/?article=351 (accessed 27 September 2020).

Lin, P. and Allhoff, F. (2007). Nanoscience and nanoethics: defining the disciplines. In: *Nanoethics: The Ethical and Social Implications of Nanotechnology* (eds. F. Allhoff, P. Lin, J.H. Moor, et al.), 3–14. Hoboken, NJ: Wiley.

Lin, P. and Allhoff, F. (2008a). Untangling the debate: the Ethics of human enhancement. *Nanoethics* 2: 251–264.

Lin, P. and Allhoff, F. (2008b). 'Introduction: nanotechnology, society, and ethics'. https://digitalcommons.calpoly.edu/cgi/viewcontent.cgi?article=1006&context=phil_fac (accessed 27 September 2020).

Lu, L.Y.Y., Lin, B.J.Y., Liu, J.S., and Yu, C.Y. (2012). Ethics in Nanotechnology: What's being done? What's missing? *Journal of Business Ethics* 109: 583–598.

Maclurcan, D.C. (2009). Southern roles in global Nanotechnology innovation: perspectives from Thailand and Australia. *Nanoethics* 3: 137–156.

Maclurcan, D. (2010). 'A more equitable approach to nano-innovation is needed'. https://www.scidev.net/global/innovation/opinion/a-more-equitable-approach-to-nano-innovation-is needed.html (accessed 26 September 2020).

Maity, A. and Pramanik, P. (2013). Nanopollution a growing issue. *National Environmental Science Academy* 16 (7): 1–4.

Nanoscience Conference (2020). Nanoweapons. https://nanoscience.conferenceseries.com/events-list/nano-weapons (accessed 27 September 2020).

Nanotechnology Characterization Laboratory (n.d.). 'Mission and objectives'. https://ncl.cancer.gov/about-ncl/mission-objectives (accessed 27 September 2020).

National Institute for Occupational safety and health (NIOSH) (2012). Department of Health and Human Services, Centers for Disease Control and Prevention. 'General Safe Practices for Working with Engineered Nanomaterials in Research Laboratories', no. 2012–147.

National Institutes of Health (2016). Nanotechnology at NIH. https://www.nih.gov/research-training/nanotechnologynih#:~:text=Nanotechnology%20is%20defined%20as%20the,develop%20novel%20technologies%20and%20products (accessed 26 September 2020).

Nickerson, J. (2013). 'Innovation's global nanotechnology divide causes controversy'. https://www.ft.com/content/1b78b9aa-2d0f-11e3-8281-00144feab7de (accessed 26 September 2020).

Oberdörster, G., Oberdörster, E., and Oberdörster, J. (2005). Nanotoxicology: an emerging discipline evolving from studies of ultrafine particles. *Environmental Health Perspectives* 113 (7): 823–839.

Patra, H.K., Banerjee, S., Chaudhuri, U. et al. (2007). Cell selective response to gold nanoparticles. *Nanomedicine: Nanotechnology, Biology and Medicine* 3 (2): 111–119.

Pramanik, P.K.D., Solanki, A., Debnath, A. et al. (2020). Advancing Modern Healthcare with Nanotechnology, Nanobiosensors, and Internet of Nano Things: taxonomies, applications, architecture, and challenges. 8: 65230–65266.

Radziun, E., Dudkiewicz Wilczyńska, J., Książek, I. et al. (2011). Assessment of the cytotoxicity of aluminium oxide nanoparticles on selected mammalian cells. *Toxicology in vitro* 25 (8): 1694–1700.

Resnik, D.B. and Tinkle, S.S. (2007). Ethics in Nanomedicine. *Nanomedicine (London)* 2 (3): 345–350.

Sachs, J.D. (2003). The Global Innovation divide. In: *Innovation Policy and the Economy* (eds. A.B. Jaffe, J. Lerner and S. Stern), 132–140. MIT Press.

Sampath, V.R. and Perumalraj, R. (2006). 'Nano Technology – enabling technologies for innovative industries'. https://www.fibre2fashion.com/industry-article/994/nano-technology-enabling-technologies-for-innovative-industries (accessed 25 September 2020).

Schroeder, D., Dalton-Brown, S., Schrempf, B., and Kaplan, D. (2016). Responsible inclusive innovation and the Nano-Divide. *Nanoethics* 10: 177–188.

Schulte, P.A. and Salamanca-Buentello, F. (2007). Ethical and scintific issues of Nanotechnology in the workplace. *Environmental Health Perspectives* 115 (1): 5–12.

Scientific Committee on Emerging and Newly Identified Health Risks (2006). Nanotechnologies. https://ec.europa.eu/health/scientific_committees/opinions_layman/en/nanotechnologies/l-3/1-introduction.htm (accessed 25 September 2020).

Singh, T., Shukla, S., Kumar, P. et al. (2017). Application of Nanotechnology in food science: perception and overview. *Frontiers in Microbiology* 8: 1501.

Souto, E.B., Silva, G.F., Dias-Ferreira, J. et al. (2020). Nanopharmaceutics: part I—clinical trials legislation and good manufacturing practices (GMP) of nanotherapeutics in the EU. *Pharmaceutics* 12: 146.

Soutter, W. (2012). 'Nanopollution: hype or health risk?', AZoNano. https://www.azonano.com/article.aspx?ArticleID=3138 (accessed 27 September 2020).

Sparrow, R. (2009). The social impacts of Nanotechnology, an ethical and political analysis. *Bioethical Inquiry* 6: 13–23.

Srivastava, N. and Chowdhury, N. (2011). Nanotechnology and state regulation (India). In: *Nanotechnology and Global Sustainability* (eds. D. Maclurcan and N. Radywyl), 252–254. CRC Press, Taylor and Francis.

Surendiran, A., Sandhiya, S., Pradhan, S.C., and Adithan, C. (2009). Novel applications of nanotechnology in medicine. *Indian Journal of Medical Research* 130: 689–701.

Toumey, C. (2009). 'Feynman and nanotechnology – anniversary reflections'. https://www.nanowerk.com/spotlight/spotid=13169.php (accessed 26 September 2020).

United States Environmental Protection Agency (2016). 'Human health and risk assessment'. https://www.epa.gov/risk/human-health-risk-assessment (accessed 27 September 2020).

United States Food and Drug Administration (2018).'FDA's Approach to regulation of Nanotechnology products'. https://www.fda.gov/science-research/nanotechnology-programs-fda/fdas-approach-regulation-nanotechnology-products (accessed 24 September 2020).

United States National Nanotechnology Initiative (n.d.-a). 'What is nanotechnology?' https://www.nano.gov/nanotech-101/what/definition (accessed 25 September 2020)

United States National Nanotechnology Initiative (n.d.-b). 'Benefits and application'. https://www.nano.gov/you/nanotechnology-benefits (accessed 27 September 2020).

Université de Genève (2019). 'Human enhancement: Is it good for society?', ScienceDaily. https://www.sciencedaily.com/releases/2019/02/190211114300.htm (accessed 27 September 2020).

Van de Poel, I. (2008). How should we do Nanoethics? A network approach for discerning ethical issues in Nanotechnology. *Nanoethics* 2: 25–38.

Vishwakarma, V., Samal, S.S., and Manoharan, N. (2010). Safety and risk associated with nanoparticles – a review. *Journal of Minerals and Materials Characterization and Engineering* 9 (5): 455.

Wamocha, H.L., Misak, H.E., Song, Z. et al. (2013). Cytotoxicity of release products from magnetic Nanocomposites in targeted drug delivery. *Journal of Biomaterials Applications* 11: 661–667.

Wood, S., Jones, R., and Geldart, A. (2007). Nanotechnology: a report for the ESRC, ISBN978-0-86226-255-6. http://umassk12.net/nano/2008summer/Gibson/ESRC_Nanotechnology.pdf (accessed 26 September 2020).

Yasri, S. and Wiwanitkit, V. (2017). Important ethical issues for nanomedicine. *Journal of Research in Medical Sciences* 22: 138.

Yokel, R.A. and MacPhail, R.C. (2011). Engineered nanomaterials: exposures, hazards, and risk prevention. *Journal of Occupational Medicine and Toxicology* 6: 7.

17

Current Regulatory Framework in Nanotechnology and Medicine

Nitin Dubey[1] *and Nidhi Dubey*[2]

[1] *IPS Academy College of Pharmacy, Indore, MP, India*
[2] *School of Pharmacy, Devi Ahilya Vishwavidyalaya, Indore, MP, India*

17.1 Introduction

The term "Nano" was derived from the Greek word "nanos," means "dwarf." "Think big" do not attract global scientists anymore; the community has started thinking really small. "Think small" and "small is different" are the new mantra for the twenty-first century. Big outcomes are oozing out of a significantly small world. Physicist Richard Feynman explored the possibility to examine and control matter at the nanoscale in his lecture "There's plenty of room at the bottom" bringing it to practical conceptualization (Feynman 1960), Japanese researcher Taniguchi, used the term "nanotechnology" in 1974 (Taniguchi 1974). Sophisticated scanning tunneling microscopy (STM) technique and the atomic force microscopy (AFM) technique helped to manipulate atoms or molecules over surfaces, design nanostructures, or to understand a surface at atomic resolution by capturing its image (Xie and Régnier 2012; Huan-Cheng et al. 2018).

The world is currently focusing on six key enabling technologies (KETs) for rapid industrial growth and innovations, namely advancement in materials, its manufacturing and processing, nano and microelectronics, biotechnology, photonics, and nanotechnology. The 1990s have witnessed the strategic development of nanotechnology by national initiatives from the United States, Japan, and China. The European Union along with various international and regional bodies carried forward the drive in the decade of 2000 (National Science and Technology Council 2000; Eurlex 2011; National Science and Technology Council 2016; Huan-Cheng et al. 2018). According to a current industrial report, nanotechnology is a rapidly evolving technique; the current global nanotechnology market size is approximately \$50 billion as of 2018 and is expected to grow to approximately \$350 billion by 2025. This sector has observed a substantial investment of \$67.5 billion in R&D by public-funded sources (McWilliams 2016; NSTC 2016; Grand View Research 2017). As per the observations available with service launched with StatNano as indicated in Table 17.1, around 43 types of nanomaterials were used to formulate around 3135 products spanning approximately 198 697 patents and 1 680 392 relevant research articles (StatNano 2018). It is interesting to note that the nanoparticles category alone has approximately 2 437 products with 45 626 patents and 401 014 articles. The year 2017 witnessed 11 042 and 4 019 patents published by USPTO and EPO, respectively. The United States Strategy report "Nanotechnology Research

Nanotechnology in Medicine: Toxicity and Safety, First Edition. Edited by Mahendra Rai, Mrunali Patel, and Rashmin Patel.

Directions for Societal Needs in 2020" provides a useful insight into the requirements for safe and sustainable development of nanotechnology (McCray 2006). However, these data are not limited to the medicinal category. Nanotechnology is rapidly transforming itself as "horizontal technology" to provide new means of general methodology for product development and innovation (Roco et al. 2011).

The understanding of processes and phenomena and the application of science and technology to organisms, organic and inorganic materials, as well as parts, products, and models thereof, at the nanometer-scale (but not exclusively below 100 nm) in one or more dimensions, with the onset of size-dependent phenomena usually enables novel applications. These applications utilize the properties of nanoscale materials that differ from the properties of individual atoms, molecules, and bulk matter for the production of knowledge, goods, and services, like improved materials, devices, and systems that exploit these new properties (OECD 2018b). It includes approaches to fabricate materials at 1–100 nm dimensions by understanding and controlling relevant properties of the material (Japan Science and Technology Agency 2018).

As per 2011/696/EU, nanomaterials are defined as: "A natural, incidental or manufactured material containing particles, in an unbound state or as an aggregate or as an agglomerate and where, for 50% or more of the particles in the number size distribution, one or more external dimensions is in the size range 1–100 nm" (Eurlex 2011).

17.1.1 Classification of Nanomaterials

As per the ISO/TS 80004-1, the nanomaterials are classified into two general groups (ISO/TS 80004-12:2016):

1) *Nano-objects:* These are substances with at least one nanoscale dimension. Nano-objects may be further classified as one-dimensional, two-dimensional, and three-dimensional products. It includes a wide variety of one-dimensional products such as borophene, graphene, graphyne, langmuir blodget, nanocoating, nanofilm, nanolayer, nanoplate (nanosheet), nanoribbon, quantum well, and self-assembled layers. Two-dimensional products include nanofibers, nanorods, nanotubes, nanowire, and quantum wires. Three-dimensional products include dendrimer, diamondoid, fullerene, nanocapsule, nanoclusture, nanohorn, nanoliposome, nanomicelle, nanoion, nanoparticles, nanopore, nanoshell, nanosphere, nanostar, and quantum dot.
2) *Nanostructured materials:* These are not nanoscale structures. They may be morphological or microscopically nanoscale structures or contain at least one nano-object in their composition. It includes aerogel, nanocolloid, nanocomposite, nanocrystalline materials, nanaoelectromechanic system, nanofluid, nanofoam, nanoporus materials, porous silicon, qubit, and supermolecule.

17.1.2 Nanoparticles

The approach of defining nanoparticle involves two fundamentals – particle dimension and dimension-dependent properties. Currently, all those products are termed "nanotechnology product" or "product involving application of nanotechnology," which contains or are manufactured using nanoscale materials or exhibit related dimension-dependent property and such particles are known as "nanoparticles" (ISO/TS 80004-12:2016). In case the particles were created intentionally, the same applies in defining "engineered nanoparticles" (ASTM E2535–07: 2018; ISO/TR 12885:2018). If these nanoparticles are fixed in a medium to abolish their potential to exposure or prevent their chance to become airborne, then they are termed as "bound engineered nanoparticles." Nanoparticles having a high potential of exposure in absentia of such a medium are known as "unbound engineered nanoparticles" (UNP) (ASTM E2535–07 2018; Trotta and Mele 2019).

Table 17.1 Nanotechnology Products Database (NPD) indicating status of nanotechnology products of all categories as available with Statnano.com (up to June 2018).

No.	Countries	Products	Types	Companies	Patents EPO in 2018 (Rank)	Patents USPTO in 2018 (Rank)	Number of technology standard* cumulative (Rank)
1.	USA	2975	372	596	856 (1)	4343 (1)	98 (3)
2.	Japan	407	68	56	318 (3)	640 (3)	83 (5)
3.	Taiwan	186	64	51	20 (18)	427 (5)	76 (6)
4.	India	134	45	45	21 (16)	63 (14)	21 (16)
5.	Germany	692	226	200	355 (2)	309 (6)	47 (11)
6.	France	101	35	27	288 (4)	236 (7)	65 (7)
7.	South Korea	382	100	73	251 (5)	887 (2)	47 (12)
8.	UK	389	112	95	114 (6)	133 (9)	141 (1)
9.	Switzerland	456	46	26	103 (7)	85 (12)	11 (25)
10.	China	849	189	199	90 (8)	520 (4)	216 (1)
11.	Netherlands	68	25	21	89 (9)	101 (11)	43 (13)
12.	Entire World	8674	1104	2181	2908	8493	263

Source: Data from StatNano 2019.

17.2 Quality Attributes and Regulatory Concerns of Nanomaterials

Nanomaterial may be classified based on dimensionality, shape, morphology, chemistry, composition, functionality, complexity, generation, and biointerface. They may be having synthetic or biological origin. The inherent risk associated with nanomaterials is greater when compared with its bulkier counterparts because the properties of the same chemical molecule may differ at nanoscale. It is mainly due to the quantum effects and increased relative surface area of nanoparticles which in turn affect the chemical reactivity, strength, and electrochemical properties. Quantum effect may be defined as an observed physical effect due to the quantum nature of particles, interactions, and secondary effects in quasi-particles (elementary excitation) in a physical system (Japan Science and Technology Agency 2018). These quantum effects may affect the electrical, magnetic, and optical properties of materials. Mechanical strength also improves. It clearly indicates a major change in its surface or interface-related properties. Higher particulate surface area plays an important role in its therapeutic and toxicological properties. It is an established fact that with a decrease in size, greater proportion of atoms is found on the surface leading to highly improved chemical reactivity. Further, random movement of such small particles (Brownian motion) makes it extremely difficult to control them. Identification of risk associated is critical to design a strategy for any hazard assessment and prevention (Taniguchi 1974; Chang 2018; Trotta and Mele 2019). Following parameters require careful considerations to understand the safety, efficacy, and hazard assessment of the nanomaterials in medicine.

17.2.1 Physicochemical Characterization

The study of material structure and its complexity along with physicochemical stability helps to understand changes in its property due to its nanostructure as shown in

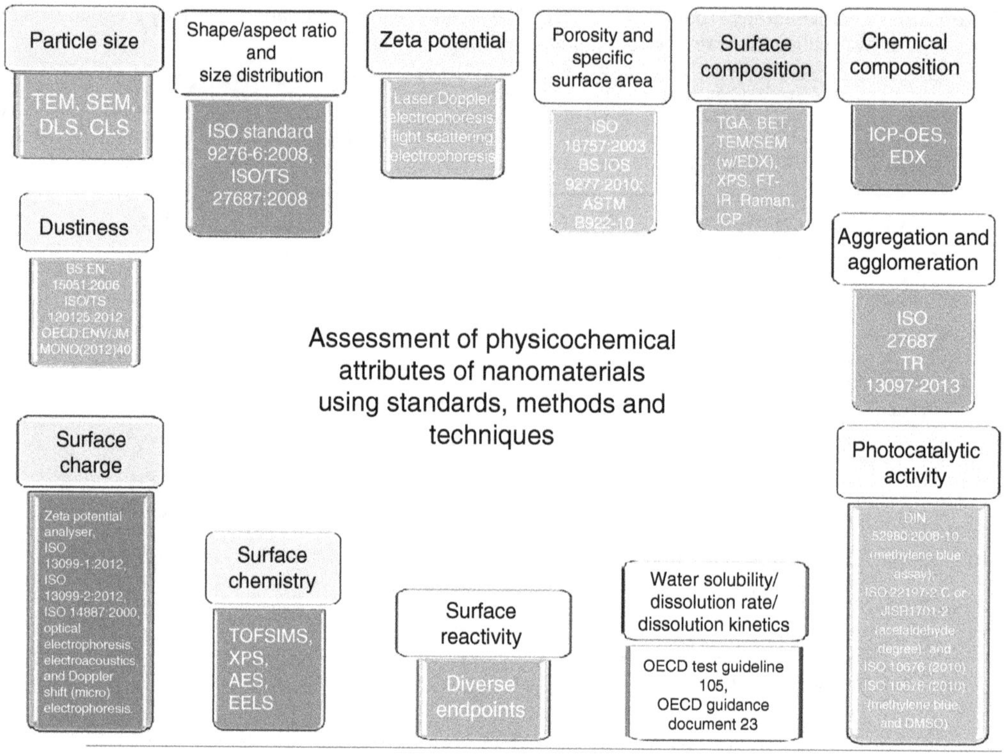

Figure 17.1 Assessment of physicochemical attributes of nanomaterials using standards, methods, and techniques.

Figure 17.1. The quality parameters of the nanomaterial should include the general structure, crystalline nature, crystal form; surface characteristics such as charge, chemical reactivity, hydrophilicity, hydrophobicity, area; surface peculiarities such as roughness or smoothness, porosity, coating properties, shape and morphology along with aspect ratio, chemical composition, assay, the ratio of free to bound drug in the solution or carrier, particle concentration, average particle size, particle size distribution along with modalities or polydispersity; physical parameters indicating stability such as agglomeration, precipitation, separation, state and degree of dispersion, diffusivity; and chemical parameters indicating stability, impurity, endotoxin levels, sterility, and *in vitro* release. The analytical methods to identify and control the above parameters should be described with relevant methodology, sampling frequency, and number along with tolerance limits and acceptance criteria. The establishment of a robust manufacturing procedure should ensure the reproducible synthesis of representative nanomaterial.

17.2.2 Nonclinical Studies for Nanomaterials

It should follow ICH or other relevant guidelines. Increased safety concerns require evaluations of its interaction with biological barriers such as placenta and blood–brain barriers. Biodistribution studies involving radio or fluorescent material labeling may be conducted to evaluate the same. The specific route of administration may require specific testing owing to specific risk considerations. Penetration of nanomaterial may vary based on its interaction with diseased, damaged, or intact skin. It may alter the interaction property of the skin with sunlight. Their use in dermal or topical products may exhibit higher penetration, different sensitization potential, or deposition possibilities compared with their bulkier counterpart. Their subcutaneous administration may increase the allergenic response to other allergens. Variation in hemocompatibility or tissue distribution may alter the property of intravenous products. Variation in distribution or deposition in respiratory tissue may change its toxicity profile compared with larger particles, in the case of inhalation products. The bioavailability of active ingredients may improve in the case of oral products. However, insoluble nanomaterials may have different accumulation potential. Thus, when nanomaterial carriers are used for API delivery, their inherent biological activity and its impact on overall product quality should be investigated. Such impact may include different immunogenicity, disposition, or enhance permeation retention.

17.2.3 Clinical Studies for Nanomaterials

When compared with any existing reference product, product containing nanomaterials should undergo extensive pharmacokinetic and pharmacodynamic studies capable of demonstrating clinical equivalence or difference based on absorption, release, availability at the site of action, drug disposition, exposure–response relationship, and bioequivalence studies. The objective of such studies may include demonstration of non-inferiority or superior performance over the reference product. Such studies may involve biomarkers to identify pharmacokinetic endpoints to justify the conclusions. Due to the complexity and diversity of nanomaterials, product-specific guidelines for specific parenteral products are available (USFDA 2014a). Biosimilar product development to a reference biological product containing nanomaterials requires to comply with specific biosimilar requirements (www.biac.org). Nanomaterials are cleared primarily by phagocytes; the study of cytotoxic potential of nanomaterials is required due to their large accumulation tendency inside these cells. Similarly, nanomaterials may affect enzymes, cytochrome P450 enzyme systems, or GIT transporters impacting *in vitro* clearance system of the body. Adequate sensitive bioanalytical methods should be developed to estimate total, free, and nanomaterial-bound drugs along with major active metabolites in biological fluid. The effect of plasma protein binding and biological fluid such as blood and plasma over nanomaterials should be understood. Environmental assessment of the nanomaterial may be required to study its exposure, fate, and other impacts on human environment including terrestrial and aquatic organisms. "Safety by design" is the current major concept to ensure safety. Currently, the observational science of toxicity studies should be replaced with a predictive model of toxicology with the aid of computational modeling along with a paradigm shift in the design of validated *in vitro* and *in vivo* assay models using lower model organisms (Table 17.2).

17.2.4 Identification of Hazardous Material

The product should be carefully examined to identify the presence of any harmful, toxic, very toxic, irritant, or corrosive substance. Appropriate labeling of products must be done by including name, address, telephone numbers, etc., details of suppliers, quantity, product identifier, hazard identifier pictogram or statement, and supplementary information on physical and health properties. Appropriate packaging should be able to prevent any leakage or damage during usage and shipment. Development of material safety data sheet (MSDS) as per ISO/TR13329:2012 should include details of health risks, toxicological information, guidance related to spillage management, first-aid requirements, and fire hazards prevention. It should be noted that most of the world/regions have accepted the "Globally Harmonized System of Classification

Table 17.2 *In vitro* systems for nanotoxicology.

Sr. no.	*In vitro* test	Purpose/Use
1.	MTT (3-(4,5-dimethylthiazol-2-yl)-2,5-diphenyltetrazolium bromide) assay (Mossman, 1983), or variations of this assay (e.g. MTS, XTT, WST-1, etc.)	For determination of mitochondrial function by measuring the activity of mitochondrial enzymes
2.	Lactate dehydrogenase (LDH) assay	For determination of cell viability. Reduced cell viability increases the leakiness of the plasma membrane leading to release of the LDH enzyme
3.	Trypan Blue exclusion	Intact cell membrane prevents dye uptake
4.	Fluorescent dye propidium iodide	Intact cell membrane prevents dye uptake
5.	Assessment of cellular ATP content	Luminescence to assess the ATP content of cell
6.	Fluorescent dye 2,7-dichlorofluorescin (DCFH)	Measurement of reactive oxygen species by fluorescent product using fluorimetry
7.	The plasmid assay	Unwinding and linearization of a coiled bacterial DNA plasmid indicate free radical exposure
8.	*o*-Phthalaldehyde (OPT) method	Changes in the GSH:GSSG ratio can be used as an indicator of oxidative stress
9.	Thiobarbituric acid reactive substances (TBARS) assay	Measurement of lipid peroxidation
10.	Trolox equivalent antioxidant capacity assay (TEAC)	Measurement of lipid peroxidation
11.	Enzyme-linked immunosorbent assay (ELISA)	Determination of cytokine protein content

and Labelling of Chemicals" (GHS) developed by the mandate of The United Nations Conference on Environment and Development (UNCED). So, MSDS and other labels are also prepared as per the format recommended in the GHS. COSHH (The Control of Substances Hazardous to Health Regulations 2002) requires a risk assessment approach to be in place if a material is classified as very toxic, harmful toxic, irritant, or corrosive and/or possess health risk and/or has an approved workplace exposure limit (WEL) and/or present as dust at 8 hour time-weighted average values as respirable fraction (4 mg/m^3) or inhalable fraction (10 mg/m^3). Risk management of nanoparticles should involve a combination of means such as partial or total enclosure of process with minimum exposure, appropriate ventilation, regular cleaning of surfaces, prohibition of ingesting food or drink at workplace, and adequate protective gears. Such materials should have documented monitoring and control measures for routine as well as emergency situations. Appropriate cleaning and maintenance of machineries and workplaces should be done periodically with proper protective gear and adequate training.

17.2.5 Exposure Risk to Humans and Environment During Nanomaterial Manufacturing, Storage, or Disposal

The effect of nanomaterials through the inhalation route after exposure at the workplace should be studied for the initiation or progression of toxicological symptoms. Appropriate nasal lavage, urine, or blood biomarkers to enhance its predictability should be identified. Pulmonary, neurological, cardiovascular, immunological, or reproductive responses may also help to establish the consequences of exposure. Various *in vitro* tests such as cell transformation, genotoxicity, and fibrogenicity may also help to predict the exposure. Exposure measurement and dose–response data may help in understanding nanoparticle disposition and associated health risk in bench workers. The effectiveness of protective clothing and preventive measures implemented during manufacturing to minimize exposure should be established. The handling of dry powders requires special precautions. All physical and chemical properties that may affect the dustiness, explosibility, flammability, or other hazards should be investigated thoroughly. Detailed study of such factors and their possible impact may help to develop newer predictive tools. Realistic exposure prediction tools should be developed, which are capable of discriminating engineered nanomaterial from other available nanomaterials developed during the manufacturing process as fumes or dust. Similarly, appropriate analytical methods should be selective enough to identify metallic nanomaterials in the presence of their metallic oxides. Predictive tools to understand the possible transformation of nanomaterials in finished pharmaceuticals due to chemical factors such as oxidation, reduction, or physical factors such as absorption on the surface, agglomeration, solubilization, and deposition should be developed.

Nanomaterials may easily cross biological barriers to get translocated, leading to higher exposure potential compared with large particles. Their diffusivity along the biological system should be investigated. Employers must ensure a safe working environment, reporting of health issues to regulatory agencies, proper storage and transport facilities along with competent supervision. Proper disposal and waste treatment procedure to protect any environmental hazards. In case of a situation indicating health, safety, or welfare hazards, the regulatory authorities are empowered to exercise one or more actions such as investigation, issuance of warning/suspension/withdrawal/recall notices or forfeiture, and destruction of material. Employers must look into various components of the health and safety of employees as indicated in Figure 17.2. Manufacturing spaces should be analyzed periodically to identify any emission during manufacturing activities. Various techniques such as dark-field hyperspectral imaging microscopy, scanning electron microscopy, and transmission electron microscopy may be used. These techniques are currently being used for the measurement of carbon nanotubes. Other techniques or sensor-based devices may also assist in direct reading or time-dependent sampling of the area.

The National Institute for Occupational Safety and Health (NIOSH) of the U.S. Department of Health and Human Services, Centers for Disease Control and Prevention has issued various guidelines to prevent health hazards during nanomaterial manufacturing, storage, or disposal (COSHH 2002). A toxicological study report was published by NIOSH in 2003 for the assessment of nanotube cytotoxicity followed by a comprehensive guidance report to minimize the health risks from exposure of nanomaterials in 2005 and 2016 (COSHH 2002; CDC 2018).

ISO developed the first technical report (TR 12885) on the safety of nanomaterials under the leadership of NIOSH in 2008. General guidance over the handling of nanomaterials in a research laboratory was published in 2012 (NIOSH 2009). A description of composition and charge of the material, morphology, or complexation of the material along with diagram should be available for regulatory submissions. Complex structures with multiple layers or compartments, coatings, ligands, and core-shell requires further elaboration. Such details improve as drug advances through clinical trial stages. Such descriptions should be revised and updated to improve our understanding. The role of a specific component or ingredient should be clearly described. It includes descriptions such as carrier, active ingredient, solubilizing agent, acidifier, disintegrant, etc. Nanomaterials are also used as excipients in free, aggregated, or agglomerated state to improve processability or product performance. It should have careful consideration in terms of its human exposure experience related to the workplace and patient safety. Adequate safety evaluations should be provided for marketing approval in such cases to assess its contribution to the product toxicology in the presence of API. Stability data and shelf-life testing design should include the possibility of nanomaterial interaction with the container closure system. Other stability issues that may affect nanomaterial properties may include surface charges, aggregation, change in particle size distribution, colloidal stability, morphology, change in crystal structure, possible drug leakage from nanomaterial carrier, degradation, or dissolution rate of API. Studies should also evaluate any possible interaction with primary packaging material or delivery devices. Post-approval changes to the approved marketed product containing nanomaterials should evaluate and classify the impact of change. Any major or moderate change should follow appropriate *in vivo* bioequivalence studies, ADME profiling, and physicochemical comparisons with product before and after the change.

17.3 Quality Assessment of Nanomedicines

Nanomedicines may be defined as the medical applications of nanotechnology based products. The different factors affecting the quality assessment of nanomedicines are described as under:

1) Choice of the route of administration may have a significant impact on the overall efficacy and safety of the drug. The physical form of the drug may play a critical role in achieving the desired therapeutic effect.
2) Specific production methodology may have its own merits to affect the stability of the finished formulation. If physicochemical attributes of drugs are understood completely, it may be correlated with pharmacokinetic and pharmacodynamic

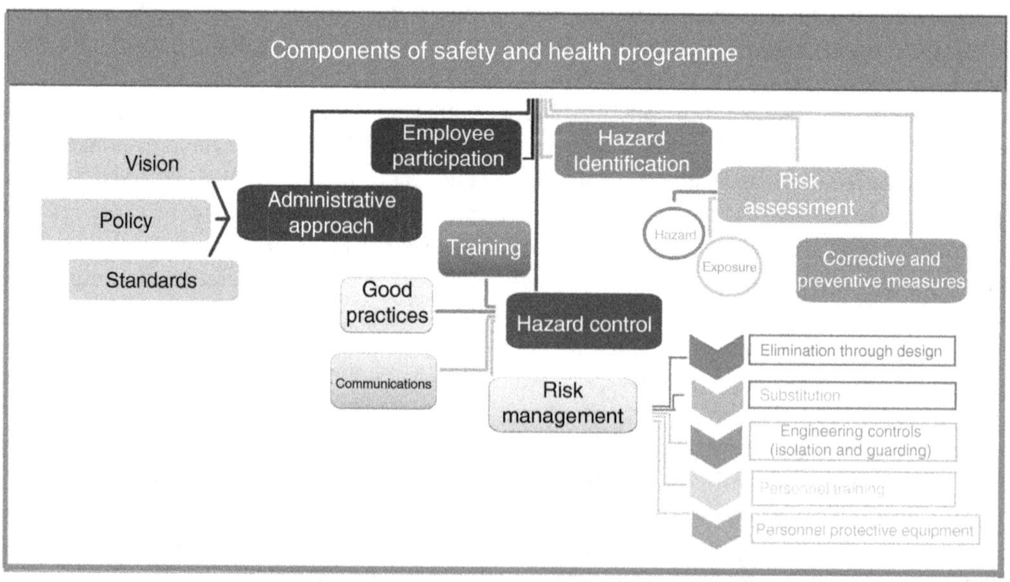

Figure 17.2 Components of safety and health programme.

parameters of choice. Appropriate analytical methodologies should be devised to identify such physicochemical indicators that may confirm the desired physicochemical form of the finished formulation. The effect of changes in the manufacturing process and their impact on the quality of the finished product should be well documented. The manufacturing process should be carefully controlled by identifying critical processing parameters. Appropriate in-process controls and in-process quality control checks should ensure the desired quality of the finished product.

3) The finished product containing OTC drugs should follow general quality, safety, and efficacy-related requirements. The responsibility lies with the manufacturer to confirm related legal requirements.
4) It is customary to develop the product according to the principles of cGMP to ensure that the finished product complies with all quality requirements despite structural complexity, unique product processing requirements, peculiarity of analytical method, and properties of nanomaterials used. The interaction mechanism between the nanomaterials and biological systems requires careful considerations. A nanomaterial may interact actively or passively with a specific site based on charge distribution or specificity of the receptor. The following factors affect the interaction between the nanomaterials and the biological system:
   A) Intrinsic factors: Age, sex, disease of the patient.
   B) Extrinsic factors: Presence of other drugs in the body.
   C) Effect of intrinsic and extrinsic factors over drug response.
   D) Role of enzymes.
   E) Role of transporters involved in the disposition of drugs.
   F) Size and charge of the nanomaterials.
   G) Property affecting receptor binding.
   H) Property affecting protein binding.
   I) Immunogenic potential of nanomaterials.
   J) Properties of nanomaterial–protein complex.
   K) Hydrophilic or hydrophobic nature.
   L) Effect of phagocyte system.

Endocytosis may help in the uptake of nanomaterial–protein complex in targeted tissue while phagocyte macrophage system may eliminate it predominantly in some organs such as liver and spleen. The effect of various attributes of nanomaterials over the overall identity, strength, purity, safety, and efficacy of the finished formulation must be investigated carefully. "Critical quality attributes" (CQAs) of the nanomaterial affecting the product should be identified and documented. CQAs will help to control the formulation or manufacturing process based on risk assessment and analytical checks. Parameters describing the intermediate stages of the products should also be identified and controlled (Stone et al. 2009; NIOSH 2012; Steinhäuser and Sayre 2017; Boverhof et al. 2015; Kim and Yu 2016; NIOSH 2016; Friedrichs 2018; Jantunen et al. 2018; Rasmussen et al. 2018; Urbán et al. 2018; Lee and Choi 2019; OECD 2020; SAFENANO 2020; NanoSafetyCluster 2018).

17.4 Current Regulatory Framework over Nanomaterials

High international scientific focus on various aspects of nanotechnology has compelled to develop standards for better risk assessment and management for deliverable regulatory compliance (National Research Council, USA 2012). Figure 17.3 indicates general overview of regulatory control over nanotechnology through regulations or applicable standards in various domains. Diversified definitions based on size and properties have delayed the development of the regulatory framework. The analytical tools responsible for dimension measurement still require validated practical guidelines. Modern measurement tools do not effectively cover current

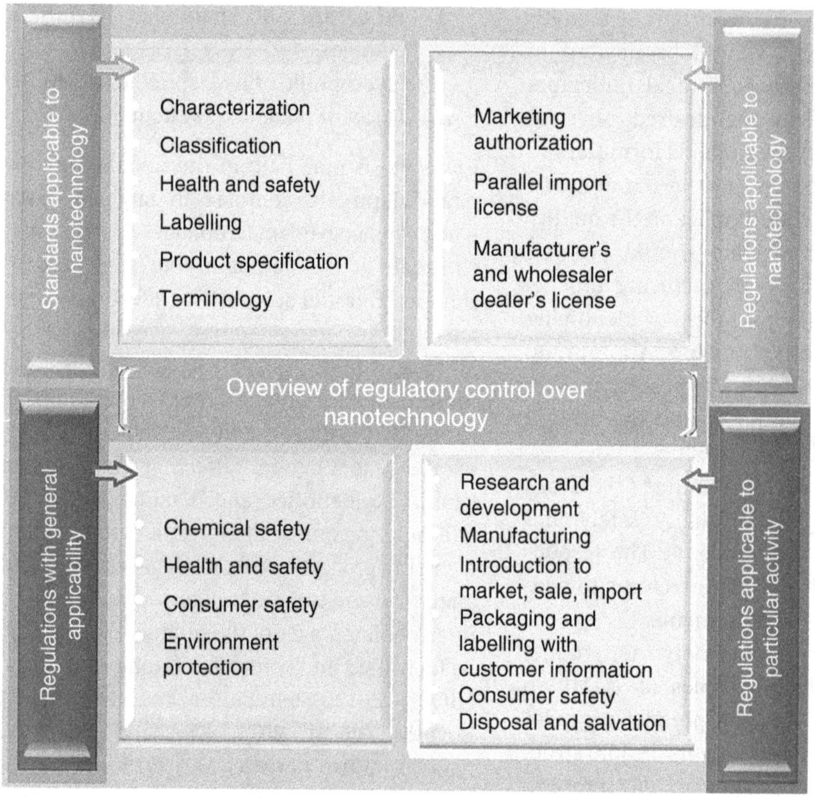

Figure 17.3 Overview of regulatory control over nanotechnology.

requirements due to complicated dimensional criteria (Division of Occupational Health and Safety, USA 2014). The biggest question unanswered was equivocally related to the status of nanomaterials, should it be regulated differently than "chemical" or "drug"? It took a while to understand and establish its uniqueness, and the lack of knowledge deepened the dilemma, but finally, it is realized and strategies have ensured the health and safety concerns are adequately addressed. Multiple pieces of legislation and different approaches applicable over nanomaterials indicate a long way to go (OECD 2013).

National standards bodies of the countries like the United States, EU, Japan, China, Korea, etc., and many organizations are playing an important role in the process. However, ISO, ASTM, European Committee for Standardization (CEN), British Standards Institution (BSI), and OECD Working Party on Manufactured Nanomaterials (WPMN) may be considered as leading organizations in the field (Figure 17.4). The details of the different measures taken by earlier stated countries and organizations are as under.

17.4.1 USA

The National Science and Technology Council (NSTC) establishes clear national goals for Federal science and technology investments to develop research and development strategies across federal agencies in the United States (Rauscher et al. 2017). The Office of Science and Technology Policy (OSTP) was established by the National Science and Technology Policy, Organization, and Priorities Act of 1976 to assist the president of USA regarding policy decisions of various technologies. National

Figure 17.4 Various regulatory agencies providing guidance on regulation of nanomaterials. *Sources:* BIS; U.S. Department of Commerce; U.S. Food and Drug Administration; US Department of Defense; NASA; National Institute of Standards and Technology; U.S. Consumer Product Safety Commission; National Institutes of Health; U.S. Department of Health and Human Services; United States Department of Energy; National Science Foundation; United States Department of Commerce; United States Department of Agriculture; United States Department of Labor; United States Department of Health and Human Services; United States Environmental Protection Agency; Organisation for Economic Co-Operation and Development (OECD); International Organization for Standardization (ISO); European Committee for Standardization; European Union; ASTM International; AFNOR; European Commission; British Standards Institution (BSI); Australian Government Department of Health; Asia Nano Forum (ANF); Health Canada; European Chemicals Agency; National Institute of Advanced Industrial Science and Technology; European Patent Organisation; Ministry of Health and Family Welfare; Pharmaceuticals and Medical Devices Agency; China National Standards; Ministry of Food and Drug Safety; State Administration for Market Regulation.

Nanotechnology Coordination Office (NNCO) provides necessary support to the Nanoscale Science, Engineering, and Technology (NSET) Subcommittee for National Nanotechnology Initiative (NNI). It is a research and development (R&D) initiative of the US government involving the nanotechnology-related activities of 20 Federal department and agency units (Patravale et al. 2012). Since 2001, research, development, and commercialization of nanotechnology under NNI has attracted more than $25 billion fund from US agencies. NSET is an interagency body which hosts representatives of 20 federal agencies to coordinate various activities for ambitious $1.4 billion NNI. NNI coordinates various research and regulatory roles and responsibilities in the field of nanotechnology. Center for Nanoscale Science and Technology (CNST) hosts a state-of-the-art advance instrumentation facility under office of the director (DOC), National Institute of Standards and Technology (NIST). Nanotechnology Environmental and Health Implications Working Group is a Subcommittee of NSET dedicated to environmental and workplace safety. Department of Energy Order 456.1A, "The Safe Handling of Unbound Engineered Nanoparticles" (DOE O 456.1A), requires the development of standards to protect health and the environment under the Occupational Safety and Health Act (OSHA) of 1970 (EPA 2017). Voluntary standards, technical specifications, technical recommendations, national standards, or industry code of practices are various means through which nanomaterials are regulated. A wide array of acts and rules directly or indirectly controls the use of nanomaterials throughout the globe covering research and development, manufacturing, packaging, labeling, import, product safety, occupational health, and environmental safety including waste disposal (Table 17.3).The EPA regulates nanomaterials under the Toxic Substances Control Act (TSCA) (NANO 2020). Once approved, the material is listed in the TSCA Inventory of Chemical Substances. U.S. EPA Federal Insecticide, Fungicide, and Rodenticide Act (FIFRA) regulates a nanomaterial, whenever there are bacterial involvement in a product (OSHA 1970). US EPA has issued a direct final rule promulgating "significant new use rules" (SNUR) under TSCA 5(a)(1)(B) for 145 substances subject (under 40 CFR part 721, subpart E) of premanufacture notices (PMN) applicable to manufacture, import, or process. EPA requires to be notified to characterize "potential health and/or environmental effects" of chemical substance at least 90 days before commencing significant new activity effective October 2018 (TSCA 1976). Once notified, manufacturing or processing may be initiated only after due assessment and permission by EPA after completion of review. Clean Air Act or Clean Water Act may also apply to nanoproduct during manufacturing or use. US FDA had issued a draft guidance document for the industry on drug products, including biological products containing nanomaterials in December 2017. The document does not provide any regulatory definition of nanomaterials, nanotechnology, or nanoscale. However, it includes the products which have materials with at least one external dimension between 1 and 100 nm. It also includes products that exhibit size-dependent physical, chemical, or therapeutic attributes, even if their dimensions are up to 1000 nm or 1 μm. This is noteworthy that naturally occurring nanosized biological material such as protein, vaccine, cell, etc., are not considered as nanomaterials. Deliberated manipulation of size or dimension-dependent properties is critical to such classification. As per American National Standards Institute's Nanotechnology Standards Panel, nanotechnology deals with imaging, measuring, modeling, and manipulating matter of 1–100 nm dimensions (EPA 2018). National Nanomanufacturing Network, USA has developed an information platform "Internano" for the nanomanufacturing community (InterNano 2019). It defines nanomanufacturing as the "controllable manipulation of materials structures, components, devices, and systems at the nanoscale (1 to 100 nanometers) in one, two, and three

Table 17.3 General rules and regulation applicable in various regions having direct or indirect control over the products containing nanomaterials in the United States and European Union.

No	Particulars	Remarks	Country
1.	OSHA Act Section 5(a)(1)	(29 U.S.C. 654) Workplace free from recognized hazards which are likely to cause death or serious physical harm to his employees	USA
2.	OSHA Act 1904	Requires reporting occupational injuries and illness in case of employing more than 10 employees	USA
3.	OSHA Act 1910.132	Requires personal protective equipment, general requirements	USA
4.	OSHA Act 1910.133	Requires eye and face protection	USA
5.	OSHA Act 1910.134	Requires respiratory protection	USA
6.	OSHA Act 1910.138	Requires hand protection	USA
7.	OSHA Act 1910.1200	Requires creation of safety data sheets and communication regarding hazards with employees	USA
8.	OSHA Act 1910.1450	Occupational exposure to hazardous chemicals in laboratories	USA
9.	OSHA Act 1910.1027	Substance specific standards, such as cadmium	USA
10.	TSCA Section 4	Testing of chemicals wherever exposures related concern or risks	USA
11.	TSCA Section 5	Notification of "new chemical substances" for even existing chemicals manufactured at the nanoscale	USA
12.	TSCA Section 5	Significant new use rules (SNURs) identifies a "significant new use" capable of exposure or release	USA
13.	Biocidal Products Regulation (EU) No 528/2012	Requires specific assessment, procedural, and labeling requirements for nanomaterials	EU
14.	Waste Electrical and Electronic Equipment (WEEE) Directive No 2012/19 s	Requires evaluation of the risks and treatment methodology of nanomaterials	EU
15.	Restriction of Hazardous Substances (RoHS) Directive No 2011/65	Examination of inclusion of nanomaterials in products	EU
16.	Food Labelling Regulation (EC) No 1169/2011	Labeling of ingredients containing nanomaterials is required	EU
17.	The Chemicals (Hazard Information and Packaging for Supply) Regulations 2009 (EC) No 1272/2008	Chemical hazard information and packaging for supply (CHIP) and classification, labeling, and packaging of substances and mixtures (CLP)	EU
18.	Plastic Food Contact Material Regulation (EC) No 10/2011	Prohibitions and restrictions for nanomaterials in food contact	EU
19.	REACH (EC) No 1907/2006	Regulation for the registration, evaluation, authorization, and restriction of chemicals (REACH) Annex II, section 9.1 requires development of safety data sheet for solid materials with granulometry and specific surface area details	EU
20.	COSHH (2002)	The Control of Substances Hazardous to Health Regulations (COSHH)	EU

(Continued)

Table 17.3 (Continued)

No	Particulars	Remarks	Country
21.	Dangerous Substances and Explosive Atmospheres Regulations 2002	Identification and control of dangerous substances with fire and explosion risk in the workplace	EU
22.	Health and Safety at Work etc. Act 1974	Identification and control of dangerous substances with health and safety risk in the workplace	EU
23.	Workplace (Health, Safety and Welfare) Regulations 1992	Identification and control of dangerous substances with health and safety risk in the workplace	EU
24.	The Reporting of Injuries, Diseases and Dangerous Occurrences Regulations 2013	Requires reporting occupational injuries and illness of employees	EU
25.	Management of Health and Safety at Work Regulations 1999	Requires risk assessment for health and safety of the employees in the workplace	EU
26.	UK Directive 2001/95/EC	The General Product Safety Regulations 2005 requires only safe products to be placed in markets	EU
27.	Consumer Protection Act 1987	Consumers injured by defective products may sue the manufacturers for damages	EU
28.	The Environmental Permitting (England and Wales) Regulations 2010, Schedule 21, 22	Requires to identify and control pollutants	Britain
29.	The Pollution Prevention and Control (Industrial Emissions) Regulations (Northern Ireland) 2013	Requires to identify and control pollutants	Ireland
30.	The Pollution Prevention and Control (Scotland) Regulations 2012	Requires to identify and control pollutants	Scotland
31.	The Waste Management Licensing Regulations (Northern Ireland) 2003 (as amended) The Waste Management Licensing Regulations (Scotland) 2011	Waste management	Scotland
32.	The Environmental Damage (Prevention and Remediation) Regulations 2009	Environmental protection	Scotland
33.	Regulation (EC) No 1223/2009	Along with The Cosmetic Products Enforcement Regulations 2013 controls the cosmetics for safety and labeling requirements	EU
34.	The General Food Regulations 2004	Along with Regulation (EC) No 258/97, The Novel Foods and Novel Food Ingredients Regulations 1997, and Food Labeling Regulations 1996 controls the food materials	EU
35.	Regulation (EC) No 1829/2003	For control of on genetically modified food and feed	EU
36.	Regulation (EC) No 1333/2008	For control of food additive	EU
37.	The Human Medicines Regulations 2012	For control of manufacturing, import, and wholesale of medicinal products	EU

Table 17.4 Various international standards given by NCL in the field of nanotechnology assisting cancer research.

No.	NCL method	Title	Remarks
1.	ITA-9	Phagocytosis Assay	Evaluation of nanoparticle internalization
2.	ITA-14	Analysis of Nanoparticle Effects on Maturation of Monocyte Derived Dendritic Cells *in vitro*	Use of dendritic cells (DC) as antigen-presenting cells
3.	ITA-8	Chemotaxis Assay	Measures of the chemo attractant capacity
4.	ITA-6	Leukocyte Proliferation Assay	Measurement of leukocyte proliferation
5.	ITA-12	Coagulation Assays	Effect on plasma coagulation
6.	ITA-5.2	Quantitative Analysis of Complement Activation	Complement activation by enzyme-immuno assay
7.	ITA-2	Analysis of Platelet Aggregation	Anticoagulant property evaluation

dimensions for large-scale reproducibility of value-added components and devices". It collaborated with ISO for the development of protocols. Nanotechnology Characterization Laboratory (NCL) has also developed various assay procedures and methodologies in the field of cancer research. NCL of the National Cancer Institute (NCI) is part of National Institutes of Health (NIH), Department of Health and Human Services (HHS), USA. Some of the protocols are tabulated in Table 17.4.

17.4.1.1 ANSI

The American National Standards Institute has formed the Nanotechnology Standards Panel (ANSI-NSP) to develop the standards in the area of materials properties, characterization procedures, measurement, and testing of nanomaterials in 2004. ANSI incorporated by reference (ANSI IBR) portal developed by ISO hosts various standards cited by US code of federal regulation (CFR) in "read only" format. The portal was developed by ISO and other standards developing organizations (SDOs).

17.4.1.2 ASTM

ASTM International, formerly known as American Society for Testing and Materials, has also issued some standard methods, such as ASTM E2859-11 (2017), as a standard guide for size measurement of nanoparticles using atomic force microscopy. Other standards issued by ASTM are summarized in Table 17.5. The United States has initiated significant progress with the implementation of the "21st Century Nanotechnology Research and Development Act" in 2003, for the appropriations of nanoscience, nanoengineering, and nanotechnology research.

17.4.2 European Union

The European Commission's document published in 2004 "Toward a European Strategy for Nanotechnology" was the first official vision document in the field of nanotechnology (European Commission 2004). European Commission (EC) proposed (2011/696/EU) a definition for the term nanomaterial in 2011 covering natural, incidental, and manufactured materials. The same was reviewed in 2015. The ISO definition (ISO/TS 27687:2008; ISO/TS 80004-1:2010) has broader context and can be used in all possible area; however, the EC finds it to be "less quantitative" and do not align with its definition for regulatory purpose. The definition was reviewed in 2015 and published as "JRC science for policy report – Towards a review of the EC Recommendation for a definition of the term nanomaterial." It defines "Nanomaterial' as a natural, incidental or manufactured material containing particles,

Table 17.5 Various international standards given by ASTM in the field of nanotechnology.

No.	Standards	Description
1.	ASTM E2834-12(2018)	Standard Guide for Measurement of Particle Size Distribution of Nanomaterials in Suspension by Nanoparticle Tracking Analysis (NTA)
2.	ASTM E2859-11(2017)	Standard Guide for Size Measurement of Nanoparticles using Atomic Force Microscopy
3.	ASTM F1877-16	Standard Practice for Characterization of Particles
4.	ASTM E2834-12(2018)	Standard Guide for Measurement of Particle Size Distribution of Nanomaterials in Suspension by Nanoparticle Tracking Analysis (NTA)
5.	ASTM E2490-09(2015)	Standard Guide for Measurement of Particle Size Distribution of Nanomaterials in Suspension by Photon Correlation Spectroscopy (PCS)
6.	ASTM E2909-13	Standard Guide for Investigation/Study/Assay Tab-Delimited Format for Nanotechnologies (ISA-TAB-Nano): Standard File Format for the Submission and Exchange of Data on Nanomaterials and Characterizations
7.	ASTM E2535-07(2018)	Standard Guide for Handling Unbound Engineered Nanoscale Particles in Occupational Settings
8.	ASTM E2526-08(2013)	Standard Test Method for Evaluation of Cytotoxicity of Nanoparticulate Materials in Porcine Kidney Cells and Human Hepatocarcinoma Cells
9.	ASTM E2524-08(2013)	Standard Test Method for Analysis of Hemolytic Properties of Nanoparticles

in an unbound state or as an aggregate or as an agglomerate and where, for 50% or more of the particles in the number size distribution, one or more external dimensions is in the size range 1–100 nm." The size distribution threshold may be replaced between 1 and 50%, if required. It is interesting to note that the EU was the first regulatory body to formally announce health and safety-related provisions for nano-specific products. The EU has announced its eighth framework programme in 2014 known as "Horizon 2020" (H2020) for research and technology development from 2014 to 2020 in the field of nanotechnology with a budget of approximately €80 billion (European Commission 2004).

British Standards Institute (BSI) is the recognized national standards body of the United Kingdom. It has established national committee NTI/1 "Nanotechnologies" (Nanotechnologies Standardization Committee) in 2004. It has published voluntary PAS 137:2013 guidance to develop standards. It is working with ISO TC 229 for the development of various standards. The European Union controls the chemicals with two approaches; first one controls the "placing of chemicals" in the market, while the second deals with health, consumer, occupational safety, and environmental protection-related aspects. It regulates nanomaterial under the definition of chemical "substance" by the European Chemical Agency (ECHA) through REACH and CLP regulations. The Regulation (EC) No. 1907/2006 on the registration, evaluation, authorization, and restriction of chemicals is commonly known as "REACH" (Rauscher et al. 2015). "REACH" ensures human health and environment protection in the EU. It is an umbrella regulatory framework inside the European Union. It requires registration of chemicals with more than one ton per year production quantity, followed by evaluation and authorization. Substance with unacceptable risk to health or environment may be restricted and listed in Annex XVII. Annex 24 identifies and classifies "substance of very high concern"

(SVHC), which cannot be placed in EU markets without authorization. These substances are carcinogens, mutagens, and reproductive toxins in nature. They may be persistent, bioaccumulative, and toxic (PBT) or very persistent, very bioaccumulative (VPVB). These chemicals are listed in three inventories as per the regulations, namely the European Inventory of Existing Commercial Chemical Substances (EINECS), European List of Notified Chemical Substances (ELINCS), and No-longer Polymers (NLP) list. The regulation (EC) No. 1272/2008 on Classification, Labeling, and Packaging of Substances and Mixtures, commonly known as CLP Regulation, is synchronized with GHS requirements. It requires the classification and labeling of new chemicals within one month of "placing" the substance on the market for the first time. It is also known as C&L notification. CLP article 9 requires the evaluation of hazard information for substances and mixtures. The European Union Medical Device Regulation (2017/745) released in 2017 is the only regulation having nano-specific provisions related to approval, labeling, and safety of health products (ECHA 2020a; Europa 2020). Other regulations may also apply to different stages of the development of products (EU 2017/745). It may include the Biocidal Products Regulation (BPR, Regulation (EU) 528/2012), the Prior Informed Consent Regulation (PIC, Regulation (EU) 649/2012), the Chemical Agents Directive (Directive 8/24/EC), and the Carcinogens and Mutagens Directive (Directive 2004/37/EC) (GoNano 2018). Some EU states have required products containing nanomaterials to be notified in national product register or "nano-registers." Swedish Chemicals Agency "Kemi" requires compulsory notification of nanomaterials according to Swedish chemical agency regulations (KIFS 2017:7) chapter 3. French regulatory authority requires an annual declaration of nanoparticle substances as defined in Articles L. 523-1 to L. 523-8 of the French Environmental Code and Article L. 5161-1 of the French Public Health Code effective from 2013 (EINECS 2020; France 2020). The Danish Environmental Protection Agency also requires notification of the same (Statutory Order no. 644, 2014) (DEPA 2014; ECHA 2020a). Health, Food chain safety and Environment agency of Belgium requires nanomaterials to be notified in "BE nanoregistry" as directed by Royal Decree in 2014 (ECHA 2020b; Federal Public Service 2020). The objective of such registries is to understand the characteristics, potential ecotoxicity, and impact of the nano-product over the market. Europe had initiated its seventh framework programme (FP7) in 2014 with a focus on the toxicological aspects of nanomaterials after the successful completion of sixth framework programme (FP6). The European NanoSafety Cluster was a part of FP7 to address comprehensively the issues related with safety of nanotechnology synergistically with FP6 and extending to H2020. Projects such as "NanoReg2" with the Safe Innovation Approach (SIA) are serving as a proactive system for safety risk evaluation. Various NanoSafety academic research centers such as "NanoSafety-Austria/EURO-NanoTox," the "Danish NanoSafety Centre," "Finnish NanoSafety Research Centre," and "Namur NanoSafety Centre" are also involved in research-related activities. Occupational safety and health-related academic group includes nanoSafety expert platforms such as "KIR nano" of Dutch government, "BioNanoNet" of Austria, and "SAFENANO." The academic-industry collaboration includes "EHS-Advance" of Spain, "Nanocentre," "NanoHouse," "INERIS," "CRANN," and "FENAC." The EU has constituted "The Advisory Group for Nanotechnologies, Advanced Materials, Biotechnology and Advanced Manufacturing and Processing" (NMBP) for guidance and policymaking in the field. Currently, research and development focus on environmental, health, and safety issues (EHS) and ethical, legal, and social aspects (ELSA) of nanotechnology. Recently, H2020 project "Gov4nano" was initiated in January 2019 to develop a "nano risk governance model" (NRGM) with an integrative approach to efficiently manage scientific, regulatory, and market-related aspects of risk governance of nanotechnology. It has 32 partners including institutions of the EU, Korea, and South Africa. The objectives of the projects include the development of a nanosafety governance portal, machine readable database, harmonized guidance, establishment of a Nanotechnology Risk Governance Council (NRGC) catering to the needs of chemicals, biocides, cosmetics, medicine, and food sectors across regulatory sectors of Europe and beyond.

17.4.2.1 CEN

European CEN has established a technical committee CEN/TC 352–Nanotechnologies–under Association Française de Normalisation (AFNOR), France and Czech office for standards, metrology and testing (UNMZ) secretariats to cover standardization activity in the field of nanotechnologies. It coordinates internationally with bodies like ISO and IEC and locally between CEN/TC 137 (assessment of workplace exposure to chemical and biological agents) and CEN/TC 195 (air filters for general air cleaning) responsible for execution of Mandate M/461. The mandate M/461 requests CEN, European Committee for Electrotechnical Standardization (CENELEC), and European Telecommunications Standards Institute (ETSI) to develop the standardization deliverables in the field of nanotechnologies and nanomaterials. The mandate M/461 focuses on four areas of development as per annex-1 and annex-II:

- Nanomaterial characterization strategies, toxicity, and ecotoxicity testing at raw material and finished product form of manufacturing.
- Sampling and measurement of workplace, environment, and consumer exposure.
- Methodologies to simulate exposure of nanomaterial.
- Health, safety, and environment.

It has three working groups known as CEN/TC352/WG1, CEN/TC352/WG2, and CEN/TC352/WG1. These working groups are tasked with measurement, characterization, and performance evaluation; commercial and other stakeholder aspects and health, safety, and environmental aspects, respectively. These working groups have published one, one, and four standards, respectively (WHO 2012).

17.4.3 Taiwan

Taiwan established a national nanotechnological programme in 2003.The Taiwan Nano Standard Technical Council (TNSC) was founded in 2009 at the Bureau of Standards, Metrology, and Inspection, Ministry of Economic Affairs, Taiwan. Approved chemicals are added to Taiwan's chemical substance inventory (TCSI). Taiwan's Ministry of Health and Welfare (MOHW) regulates pharmaceutical products under Pharmaceutical affairs act, 1970 amended in 2018. Some of the International standards and documents given by TNSC (Taiwan) in the field of nanotechnology is mentioned in Table 17.6.

17.4.4 Iran

Iran Nanotechnology Initiative Council (INIC) was established in 2003. Iran Nanotechnology Standardization Department was established in 2006. It has more than 80 international standards with 4 standards published with ISO/TC 229. It has published policy document "Nanotechnology in Iran 2025: National Nanotechnology Development Program" in 2017 (Iran Regulations 2020).

17.4.5 Canada

Policy statement on Health Canada's working definition considers nanomaterial to be manufactured at or within the nanoscale (1–100 nm inclusive), or have internal or surface structures in the nanoscale (www.canada.ca).

Health product and food branch (HPFB), Canada is a participant of TC 229. It established The RCC Nanotechnology Initiative under Canada-US Regulatory Cooperation Council (RCC) in 2011. It regulates nanomaterials

Table 17.6 Some of the international standards and documents given by TNSC (Taiwan) in the field of nanotechnology.

No.	Particulars	Title
1.	IEC/TS 62844	Guidelines for quality and risk assessment for nano-enabled electrotechnical products in 2006
2.	IEC 62607-4-4	Nanomanufacturing – Key control characteristics – Part 4-4 in 2006
3.	TNS M017001-2012	Nano-product – surface particle size measurement – Scanning Electron Microscope in 2012
4.	TNS A001001	Nanotechnology – the terminology and definition in 2012
6.	TNS003-2013	Nanotechnology – the terminology and definition in 2013
7.	Guidelines for Applying for Packing of Nano-Contained Food Containers	Published in 2017
8.	Nano-Product Certification System of Nanomark	Published in 2003
9.	TNS007(TWA4)	Nanotechnology (material specifications). Standardized the guideline of nano-objects in 2005
10.	TN-052-2013	V&V Standard on nano antibacterial products in 2003

under Canadian Environmental Protection Act, 1999, the Feeds Act, the Fertilizers Act, the Pest Control Products Act, and the Food and Drugs Act. It has defined the term nanomaterial in Policy Statement on Health Canada's Working Definition for Nanomaterial under Canadian Environmental Protection Act, 1999 (CEPA 1999), effective 2011. In 2005, a "section 71" notice collected the commercial status of approximately 206 nanoscale materials existing in Canada. The inventory of chemical substances is maintained in Canada Domestic Substances List (DSL) and Non-domestic Substance List (NDSL). Currently, all those nanomaterials which are not listed in domestic substance list (DSL) should follow "New Substances Notification Regulations" (NSNR) under CEPA. Canada has revised Drug Submission Application Form for Human, Veterinary, Disinfectant Drugs, and Clinical Trial Application/Attestation (HC/SC 3011) to include section 59, which allows identification of nanomaterials among API and excipients listed in section 56 or 57 (Government of Canada 2020a). NSNR requires following OECD documents to be considered for generating any data for new nanomaterial (Government of Canada 2020b).

- OECD Series on the Safety of Manufactured Nanomaterials. No. 36, Guidance on Sample Preparation and Dosimetry for the Safety Testing of Manufactured Nanomaterials.
- OECD Series on the Safety of Manufactured Nanomaterials. No. 33, Important Issues on Risk Assessment of Manufactured Nanomaterials.
- OECD Series on the Safety of Manufactured Nanomaterials. No. 15, Preliminary Review of OECD Test Guidelines for their Applicability to Manufactured Nanomaterials.

17.4.6 Australia

Nanomaterials in Australia are governed by seven different bodies (Australian Academy of Science 2020). The therapeutic goods administration (TGA) regulates nanoparticles in therapeutic goods and medical devices as per Therapeutic Goods Act 1989. Worksafe Australia is responsible for workers' protection at workplace. The National Industrial Chemicals Notification and Assessment Scheme (NICNAS) is responsible for the regulation of industrial nanomaterials, while The Australian Pesticides and Veterinary Medicines Authority (APVMA) regulates it use in agriculture sector. Food Standards Australia New Zealand (FSANZ) regulates nanotechnologies in foods articles. Department of the Environment, Australia regulates environmental concerns related to nanomaterials. The Australian Competition and Consumer Commission (ACCC) regulates all other consumer products. The National Nanotechnology Strategy (NNS) is aimed for safer nanomaterials paving the way for the formation of TGA Nanotechnology Focus Group. International collaboration includes OECD-WPMN and OECD-WMN. It has adapted TC 299 terminologies. Approved products are listed in Australian Register of Therapeutic Goods (ARTG) (TGA 2016; Solaiman et al. 2017). The nanoinformatics 2030 roadmap combines the vision about future research efforts and interconnecting issues related to regulatory issues and nanomaterial risk management by the United States and EU (Department of Chemical and Product Safety, Germany 2020). It includes generation and funneling of data to a structured database for predictive capability using various descriptor computational models for environment, health, and safety risk assessment and hazards prevention. It proposes to use computational modeling and informatics to combine data from medicine, physicochemical characterization, chemistry, biology, laboratory data, materials science, human toxicology, and ecotoxicology. It proposes FAIR (Findability, Accessibility, Interoperability and Reusability) data principles for optimal data usage based on algorithms, tools, and workflows that operate on it. The Nano Particle Ontology (NPO) defines a NM (NPO 199) as equivalent to a chemical substance that has as a constituent an engineered NM, nanoparticle formulation, nano-object, nanoparticle, or nanostructured material. Nano-informatics data and infrastructure is supposed to have a great impact on the advancement, risk analysis, and safety assessment of NM from a regulatory perspective.

17.4.7 Japan

Japan regulates nanotechnology through its six departments namely the Ministry of Education, Culture, Sports, Science and Technology (MEXT), the Ministry of the Environment, the Ministry of Internal Affairs and Communications, The Ministry of Economy, Trade and Industry (METI), The Ministry of Health, Labour and Welfare, and the Ministry of Agriculture, Forestry and Fisheries. Along with many universities involved in research and development of nanotechnology and related standards, The National Institute of Advanced Industrial Science and Technology (AIST) and Center for Research and Development Strategy (CRDS) also play a crucial role. The fifth science and technology plan having a vision up to 2020 is expected to bring substantial changes in policies and prioritization in the field of nanotechnology (Japan 2018a). Pharmaceutical affairs Act regulating the pharmaceuticals was revised in 2014 (Japan 2018b). Presently "Law for Ensuring Quality, Efficacy, and Safety of Drugs and Medical Devices" (commonly called the Drugs and Medical Devices Law), The Evaluation of Chemical Substances and Regulation of Their Manufacture, etc. (Revised Act No. 117 of 1973, effective from 2018) also known as Japan Chemical Substance Control Law (Japan CSCL) along with other relevant laws regulate nanopharmaceuticals in Japan (Japan 2018c). Once approved, the nanomaterial is updated in the inventory of existing and new chemical substances (Japan ENCS) (Japan 2018d).

17.4.8 India

India regulates nanotechnology through three departments namely the Department of Science and Technology (DST), the Department of Scientific and Industrial Research (DSIR), and the Department of Biotechnology (DBT). Ministry of Health and Family Welfare regulates nanopharmaceuticals through Central Drugs Standard Control Organisation (CDSCO) under schedule Y of Drugs and Cosmetics Act,

1940 and Rules, 1945. Marketing approval is granted by Drug Controller General of India (DCGI). Nanomaterials are also regulated by the Food Safety and Standards Act, 2006 (FSSA), Environment Protection Act 1986, the Water (Prevention and Control of Pollution) Act 1974, and the Air (Prevention and Control of Pollution) Act 1981. Department of Biotechnology, Government of India and Indian Society of Nanomedicine (ISNM) have released "Guidelines for Evaluation of Nanopharmaceuticals in India covering quality, safety and efficacy" for the finished pharmaceuticals and API in 2018. It classifies the nanomaterials into four basic categories (Table 17.7). It requires to comply with schedule Y for manufacture, import, stability testing, preclinical studies, and clinical trials related requirement. It requires nanomaterials to be evaluated on a 'case-by-case' basis. However, it suggests to submit a descriptive data along with the application (India 2020; TANIDA 2020).

17.4.9 People's Republic of China

Standardization Administration of the People's Republic of China (SAC) is the central administration body for standardization in China. National Steering Committee for Nanoscience and Nanotechnology (NSCNN) along with SAC and National Nanotechnology Standardization Technical Committee (NSTC) constituted a coordinating committee SAC/TC279 for development of standards. The NSTC-TC also develops the protocols and technical standards. National medical products administration, "NMPA" (formerly, China food and drug administration, CFDA) regulates nanopharmaceuticals in China along with State Administration of Standardization (SAS),

Table 17.7 Classification of nanomaterials as per Indian guideline.

I	**According to degradability of nanomaterial**		
	A. Biodegradable	:	Poly-alkyl-cyanoacrylates, PEG, albumin, PLA, chitosan, gelatin, polycaprolactone, PLGA
	B. Nonbiodegradable	:	Metals such as gold, platinum silver, and titanium oxide, iron oxide
II	**According to nature of nanomaterial**		
	A. Organic nanoparticles	:	Liposome, albumin, polymer–protein, or polymer–drug conjugates
	B. Inorganic nanoparticles	:	Metallic nanoparticles, semiconductor quantum dots, iron oxides
	C. Multicomponent nanoparticles	:	Magnetic liposomes
III	**According to nano form of the ingredient**		
	A. Nanocarriers loaded with API	:	Liposomal amphotericin B, albumin-bound paclitaxel, liposomal doxorubicin
	B. APIs converted to nano form	:	Sirolimus, tacrolimus, fenofibrate, cyclosporin, griseofulvin
IV	**According to the approval status of drug and nanomaterial**		
	A. NCE and new nanocarrier	:	IND submission required
	B. NCE and approved nanocarrier	:	IND submission required
	C. Approved drug and new nanocarrier	:	IND submission required, Safety and efficacy data required
	D. Approved drug and approved nanocarrier	:	ANDA submission required

State Administration of Work Safety (SAWS), and AQSIQ for environmental protection and related control. General Administration of Quality Supervision, Inspection, and Quarantine (AQSIQ) under SAC has issued national standards also known as Guobiao Standards (GB). They fall into three categories such as GB (mandatory), GB/T (voluntary), and GB/Z (technical guide). It has issued many standards such as GB/T 19619-2004, which describes the definitions of nanomaterials and GB/T 37129-2018 for nanomaterial risk evaluation. A keyword search with word "nano" on its portal displays 118 standards. The National High Technology Research and Development Program, known as the "863 program," in 1986 followed by China's National Basic Research Program ("973 Program") in 1997 to the current "MLP 2006-2020" plan ensured sustained development of nanotechnology. Any approved substance is added to standards for the Inventory of Existing Chemical Substances in China (IECSC) (China National Standards 2020a,b).

17.4.10 Republic of Korea

Ministry of Food and Drug Safety (MFDS) of the Republic of Korea regulates pharmaceuticals under "pharmaceutical affairs act" (amended in 2016) through Korea Food and Drug Administration (KFDA). The Korean Toxic Chemicals Control Act (TCCA) was implemented in 1991 by the Ministry of Environment (MOE). It requires new chemicals to be listed in Korea Existing Chemicals List (KECL) as per TCCA and the amended act on registration and evaluation, etc., of chemical substances in South Korea (K-REACH). Revised K-REACH became effective in January 2019. It requires mandatory prenotification or registration prior to manufacture or import of new substances in a quantity 0.1 metric ton or more per year. It has some similarities with US-REACH and EU-REACH. MFDS have also established a toxicity information-providing system called "Tox-info," which is a searchable toxicity, Poisoning and Product Information database (K-REACH 2019).

17.4.11 Russia

First Russian national initiative "Nanotechnology Industry Development Strategy" was implemented in 2007. Federal Law "On the Russian Nanotechnology Corporation" paved the way for the formation of "RUSNANO," which is responsible for implementation of state policy in the field of nanotechnology since 2011. Medicinal products including nanomaterials are regulated by Federal Service for Surveillance in Healthcare (Roszdravnadzor) under "Federal Law on Circulation of Medicines" amended in 2010 (Russian Federation 2020).

17.4.12 WHO

WHO regional office for Europe listed nanotechnology as a potential health and environment challenge as reflected in 2010 Parma Declaration on Environment and Health signed by the 53 Member States (WHO 2012). WHO has 194 member countries. It suggested: The risk governance framework (International Risk Governance Council, IRGC) for risk analysis of nanomaterial. The Guideline Development Group (GDG) has published a guideline on protecting workers from potential risks of manufactured nanomaterials in 2017 (Table 17.8). It suggests GRADE (Grading of Recommendations, Assessment, Development and Evaluations) approach to be followed for qualitative and prognostic studies as well as for getting systematic review conclusions. It also recommends GHS system to be followed for assigning hazard classes to the nanomaterials (WHO 2017). WHO-GDG and OECD are members of the Inter-Organization Programme for the Sound Management of Chemicals (IOMC). IOMC coordinates activities related to the chemicals and nanomaterials among international public organizations and national bodies.

17.4.13 OECD

Organization for Economic Co-operation and Development (OECD) is an intergovernmental organization of 36 member countries.

Table 17.8 Summary of WHO recommendations to protect workers from the potential risks of MNMs.

A. **Assessment of health hazards:**

1) Assigning hazard classes as per GHS system. (SR,M-QE)[a]
2) Updating safety data sheets with MNM-specific hazard information or indicating inadequate testing. (SR,M-QE)[a]
3) Use of provisional classification of nanomaterials of the same group for the respirable fibers and granular biopersistent. (CR,L-QE)[b]

B. **Assessment of worker's exposure at workplace:**

4) Use of recommended methods for assigning OEL in workplace. (CR,L-QE)[b]
5) If exposure exceeds a proposed OEL value than chosen OEL should be at least as protective as indicated for the bulk form of the material.(CR,L-QE)[b]
6) In the absence of specific recommended OELs, a stepwise approach for inhalation exposure should be used. It must assess potential for exposure followed by basic exposure assessment and comprehensive exposure assessment as per OECD or CEN. (CR,M-QE)[c]. However, merit of any method could not be ascertained in case of dermal exposure assessment.

C. **Control exposure to MNMs:**

7) Inhalation exposure should be reduced as much as possible. (SR,M-QE)[a]
8) Highest level of controls should be in place to reduce exposures during cleaning, maintenance, addition, or withdrawal of material in manufacturing operations. A specifically planned approach is required in case of specific toxicology information is not available. (SR,M-QE)[a]
9) The principle of hierarchy of controls should be used for implementation of measures to prevent exposure. The first measure should be able to eliminate the cause of exposure, followed by dependent measures. It includes engineering controls followed by respiratory protection. (SR,M-QE)[a]
10) Dermal exposure should be using occupational hygiene measures including cleaning of surface, use of protective gloves. (CR,L-QE)[b].
11) Control banding should be used when assessment and measurement by a workplace safety expert is not available. However, merit of any method could not be ascertained. (CR,vL-QE)[b]

D. **Health surveillance:**

- No specific recommendation over existing health surveillance programmes due to lack of evidence.

E. **Training and involvement of workers:**

- No specific recommendation over best training programmes due to lack of studies.

[a] SR,M-QE, strong recommendation, moderate-quality evidence.
[b] CR,L-QE, conditional recommendation, low-quality evidence.
[c] CR,vL-QE, conditional recommendation, very low-quality evidence.
[d] CR,M-QE, conditional recommendation, moderate-quality evidence.

Committee for Scientific and Technological Policy (CSTP) (Directorate for Science, Technology and Industry) of OECD formed Working Party on nanotechnology (OECD-WPN) in 2006 which was replaced by Working Party on Biotechnology, Nanotechnology, and Converging Technologies (BNCT) in 2015. BNCT addresses policy issues related to biotechnology, nanotechnology, and their convergence with other technologies. OECD defines nanomaterial as materials engineered at the atomic, molecular, or macromolecular levels, in the length scale of approximately 1–100 nm range and includes the controlled manipulation of nanoscale structures and their integration into larger material components, systems, and architectures (OECD 2018b). OECD's Working Party on Manufactured Nanomaterials (WPMN) was established in 2006 to address human health and environmental safety implications of manufactured nanomaterials. WPMN has pooled number of testing methods which are part of mutual acceptance of data programme among 42 signatory countries.

Table 17.9 Various testing methods and report on safety of manufactured nanomaterial by OECD.

No.	Particulars test guideline	Title
1.	No. 318	Dispersion Stability of Nanomaterials in Simulated Environmental Media
2.	No. 412	28 days (Subacute) Inhalation Toxicity Study
3.	No. 413	90 days (Subchronic) Inhalation Toxicity Study
4.	No. 82	Strategies, Techniques, and Sampling Protocols for Determining the Concentrations of Manufactured Nanomaterials in Air at the Workplace
5.	No. 85	Evaluation of *in vitro* methods for human hazard assessment applied in the OECD Testing Programme for the Safety of Manufactured Nanomaterials
6.	No. 88	Investigating the Different Types of Risk Assessments of Manufactured Nanomaterials
7.	No. 80	Alternative testing strategies in risk assessment of manufactured nanomaterials: current state of knowledge and research needs to advance their use
8.	No. 77	Gold Nanoparticle Occupational Exposure Assessment in A Pilot Scale Facility
9.	No. 74	Exposure Assessment of Nano-Silver (AgNP): Case Study
10.	No. 65	Physical–chemical properties of nanomaterials: Evaluation of methods applied in the OECD-WPMN testing programme
11.	No. 36	Guidance on Sample Preparation and Dosimetry for the Safety Testing of Manufactured Nanomaterials
12.	No. 35	Inhalation Toxicity Testing: Expert Meeting on Potential Revisions to OECD Test Guidelines and Guidance Document

WPMN has four steering groups dedicated to "The publications on the safety of manufactured nanomaterials" which encompasses various test guidelines and reports summarized in Table 17.9. Business at OECD (BIAC) Nanotechnology Expert Group contributes toward safety assessment of nanomaterials. It has published its comments about nanomaterials in waste streams (BIAC 2016).

17.4.14 ISO

ISO is a nongovernment organization with 164 national standards bodies as members. It has established a technical committee "ISO/TC 229" under the secretariat of BSI with specific task of developing standards related to nanotechnology. ISO/TC 229 has five working groups. The first (ISO/TC 229/WG 1), second (ISO/TC 229/WG 2), and third (ISO/TC 229/WG 3) working group are tasked with terminology and nomenclature; Measurement and characterization and Health, Safety and Environmental Aspects of Nanotechnologies, respectively. ISO/TC 229 has published 68 standards (including updates) while 40 standards are under development (ISO/TC 229 2018a). Belgium, Germany, Russia, and South Africa follow ISO definition.

After the inception of ISO/TC 229 and IEC/TC 113, the terminologies of ISO have evolved. It has developed 21 standards mostly in collaboration with ISO. ISO/TS 80004-1:2015 defines nanomaterial, nano-object, or nanostructured material as the material with a measure of 10–9 units ranging from 1 to 100 nm. The lower limit of approximately 1 nm helps to avoid misunderstanding with single or groups of atoms, from being designated as nano-objects or elements of nanostructures. If a nanomaterial is intentionally produced to have selected properties or composition, it is known as "manufactured nanomaterial." "Engineered nanomaterial" is designed for specific purpose or function while "incidental nanomaterial" is unintentionally produced byproduct. The term transitive nanoparticles refers to nanoparticles that exhibit a size-related intensive property that differs significantly from that seen in fine particles or bulk materials (ASTM E2456-2006).

The International Organization for Standardization has issued ISO 22412:2017 (particle size analysis-dynamic light scattering) method to measure particle size distribution in dilute solutions or suspension containing nanoparticles, emulsions, and fine bubbles dispersed in liquid. It uses the principle of dynamic light scattering (DLS) or quasi-elastic light scattering (QELS) or photon correlation spectroscopy (PCS) for the measurement of average hydrodynamic particle size and distribution. The graphical representation and calculation of average particle sizes/diameters and moments from particle size distributions are given as per ISO 9276-1 and ISO 9276-2, respectively. It replaces ISO 22412:2008 and ISO 13321:1996. Many other relevant guidance, technical specification and technical reports issued by ISO are summarized in Tables 17.10 and 17.11. ISO have also formed a working group known as ISO TC-229 (technical committee) to develop standards of occupational safety related to utilization of nanotechnologies in 2006. It has published 67 standards while 41 are under development (ISO/TC 229 2018b).

17.4.15 VAMAS

Versailles Project on Advanced Materials and Standards (VAMAS) was established in 1982 by G7 nations and European members to promote world trade by innovation and use of advance materials. It is instrumental in harmonization of measurement methods using technical working area (TWA) projects. TWA 34 is active with six projects mainly focusing on the development of various methods and standard used in measurement of nanoparticle population area. TWA 33 primarily focuses on measurement methods in the field of polymer nanocomposites (PNC) and their fillers (VAMAS 2018).

Table 17.10 Various international standards and guidance given by ISO in the field of nanotechnology.

No.	International standards/Guidance	Description
1.	ISO 10801:2010	Nanotechnologies – Generation of metal nanoparticles for inhalation toxicity testing using the evaporation/condensation method
2.	ISO 19007:2018	Nanotechnologies – In vitro MTS assay for measuring the cytotoxic effect of nanoparticles
3.	ISO 10808:2010	Nanotechnologies – Characterization of nanoparticles in inhalation exposure chambers for inhalation toxicity testing
4.	ISO/DIS 19749	Nanotechnologies – Measurements of particle size and shape distributions by scanning electron microscopy
5.	ISO 20579-4:2018	Surface chemical analysis
6.	ISO 18337:2015	Surface chemical analysis, Surface characterization
7.	ISO 18747-1:2018	Determination of particle density by sedimentation methods

Table 17.11 Various technical reports and specifications given by ISO in the field of nanotechnology.

No.	Technical report/Specification	Description
1.	ISO/TR 27628:2007	Workplace atmospheres – Ultrafine, nanoparticle, and nano-structured aerosols – Inhalation exposure characterization and assessment
2.	ISO/TR 14187:2011	Surface chemical analysis – Characterization of nanostructured materials
3.	ISO/TR 12885:2018	Nanotechnologies – Health and safety practices in occupational settings
4.	ISO/TR 10993-22:2017	Biological evaluation of medical devices – Part 22: Guidance on nanomaterials
5.	ISO/TR 16197:2014	Nanotechnologies – Compilation and description of toxicological screening methods for manufactured nanomaterials
6.	ISO/TR 19601:2017	Nanotechnologies – Aerosol generation for air exposure studies of nano-objects and their aggregates and agglomerates (NOAA)
7.	ISO/TR 19716:2016	Nanotechnologies – Characterization of cellulose nanocrystals
8.	ISO/TR 13014:2012	Nanotechnologies – Guidance on physicochemical characterization of engineered nanoscale materials for toxicologic assessment
9.	ISO/TR 13121:2011	Nanotechnologies – Nanomaterial risk evaluation
10.	ISO/TS 19590:2017	Nanotechnologies – Size distribution and concentration of inorganic nanoparticles in aqueous media via single particle inductively coupled plasma mass spectrometry
11.	ISO/TS 16550:2014	Nanotechnologies – Determination of silver nanoparticles potency by release of muramic acid from *Staphylococcus aureus*
12.	ISO/TS 14101:2012	Surface characterization of gold nanoparticles for nanomaterial specific toxicity screening: FT-IR method
13.	CEN ISO/TS 17200:2013	Nanotechnology – Nanoparticles in powder form – Characteristics and measurements
14.	ISO/TS 13830:2013	Nanotechnologies – Guidance on voluntary labeling for consumer products containing manufactured nano-objects
15.	CEN/TS 17276:2018	Nanotechnologies – Guidelines for Life Cycle Assessment
16.	CEN/TS 17010:2016	Nanotechnologies – Guidance on measurands for characterizing nano-objects and materials that contain them
17.	EN ISO 10808:2010	Nanotechnologies – Characterization of nanoparticles in inhalation exposure chambers for inhalation toxicity testing
18.	EN ISO 29701:2010	Nanotechnologies – Endotoxin test on nanomaterial samples for *in vitro* systems
19.	PD CEN ISO/TS 80004-1:2014	Nanotechnologies. Vocabulary. Core terms

Table 17.12 Timeline of various initiatives, strategies, milestones, research plans, and reports by countries or organizations.

Year	By	Description	Remarks
2018	USA	Continuing To Protect the Nanotechnology Workforce: NIOSH Nanotechnology Research Plan for 2018–2025	It is a research plan of NIOSH-NTRC (CDC)
2018	USA	"Significant new use rules" (SNUR) under TSCA applicable for 14 nanomaterials	Premanufacture notices (PMN) required. Initially proposed in 2013
2017	USA	Chemical Substances When Manufactured or Processed as Nanoscale Materials; TSCA Reporting and Recordkeeping Requirements	Final reporting and recordkeeping requirements for certain chemical substances when they are manufactured or processed at the nanoscale by EPA (82 FR 22088, 40 CFR 704)
2017	EU & USA	EU US Roadmap Nanoinformatics 2030	World policy roadmap document
2016	USA	Building a Safety Program to Protect the Nanotechnology Workforce: A Guide for Small to Medium Sized Enterprises	Workplace safety issues
2015	USA	NASA Technology Roadmaps: TA 10: Nanotechnology	It is technology area (TA) 10, part of 16 sections of NASA technology requirements and goals plan for next 20 years
2016	USA	Cancer Nanotechnology Plan	Provide the current technical advancements and future trends in treatment and diagnosis of cancer
2016	USA	National Nanotechnology Initiative Strategic Plan	It is national strategic plan Under the 21st Century Nanotechnology Research and Development Act of 2003 DEVELOPED BY nni-NSTC43
2012	USA	A Research Strategy for Environmental, Health, and Safety of Engineered Nanomaterials	General guidelines, Information and resources for health and safety in laboratory
2012	USA	Nanotechnology Signature Initiatives (NSIs) – Nanosensors and the Nanotechnology Knowledge Infrastructure (NKI)	The NNI launched two more NSIs totaling it to five NSIs
2003	USA	21st Century Nanotechnology Research and Development Act	It is an important public law enacted in 2003. 45
2014	USA	Nanotechnology Safety and Health Program	General guideline, Information and resources for health and safety in NIH laboratory
2014	EU	JRC reports "Towards a review of the EC Recommendation for a definition of the term 'nanomaterial'"	Important mile stone for terminologies development

(Continued)

Table 17.12 (Continued)

Year	By	Description	Remarks
2015	Iran	Nanotechnology In Iran 2025: National Nanotechnology Development Program	Vision and policy document
2012	France	French registry "r-nano" launched	Specific registration requirements
2011	USA	The Safe Handling Of Unbound Engineered Nanoparticles	Specific guideline, Information and resources for health and safety in laboratory
2011	Japan	Committee on Safety Management for Nanomaterials established	Focus on risks and management procedures
2011	Korea	First Master Plan on Management of Nanomaterials Safety (2012–2016).	Guidance on Safety Management of nano-based products was issued.
2010	USA	Nanotechnology Safety Act of 2010	Safety of nanotechnology in consumer products
2010	China	National Steering Committee for Nanoscience and Nanotechnology established	NSCNN, China was responsible for strategy and policy making
2009	EU	Regulation (EC) No 1272/2008 on Classification, Labeling and Packaging (CLP) dangerous substances and mixtures	Covers nanomaterials
2008	USA	NNI Strategy for Nanotechnology-Related Environmental, Health, and Safety (EHS) Research	Publication of first official report
2007	OECD	Establishment of The Working Party on Nanotechnology (WPN)	Coordinates international collaborations
2007	Canada	Requirements for nanomaterials under the New Substances Notification Regulations (NSNR)	Release of advisory note
2006	EU	Regulation (EC) No 1907/2006 to cover nanomaterial as "substance"	Regarding Registration, evaluation, authorization, and restriction of chemicals under "REACH"
2006	OECD	Establishment of The Working Party on Manufactured Nanomaterials (WPMN)	Coordinates international collaborations
2006	Australia	Initiated National Industrial Chemicals Notification and Assessment Scheme (NICNAS)	First international government to undertake a voluntary reporting scheme
2005	ISO	Establishment of ISO Technical Committee TC229 (Nanotechnologies)	International group involving UK, US, Japan, and Canada initiated the standards development
2005	Australia	Establishment of National Nanotechnology Strategy Taskforce (NNST)	Strategy development for regulatory frameworks
2005	USA	ASTM Committee E56 on Nanotechnology established	Focus on nanotechnology standard development

Table 17.12 (Continued)

Year	By	Description	Remarks
2004	EU	The EC adopted the Communication "Towards a European Strategy for Nanotechnology"	Leads to institutionalizing European efforts toward an integrated strategy, and funding for nanotechnology R&D
2004	International Body	International Council on Nanotechnology (ICON)	ICON founded for policy/regulation development
2004	USA	USPTO created different class for nanotechnology	Class 977 was created for nanotechnology patents
2004	USA	Creation of workgroup in EPA for environmental impact assessment	Cross Agency Nanotechnology Workgroup to study implication of nanotechnology
2001	Canada	National Institute of Nanotechnology (Canada)	National institute was founded
2000	USA	National Nanotechnology Initiative(NNI), The Initiative and Its Implementation Plan	A $495 million coordinative initiative for R&D and Policy design with defined goals and timeline
1999	USA	**A report** "Nanotechnology Research Directions: Vision for the Next Decade"	It defined the vision leading to the formation of the U.S. National Nanotechnology Initiative
1996	Scientists of USA and UK	Probably the first Nobel Prize in the field of nanotechnology	Nobel in field of chemistry to Robert Curl, Harold Kroto, and Richard Smalley reporting of carbon fullerenes
1991	USA	First US federal nanotechnology program	The NSF initiated and funded the Nanoparticle Synthesis and Processing
1990	Japan	Initiatives for funding nanotechnology	Formal commitment for R&D
1990	China	"Climbing Project on Nanometer Science"	Initiation of a 10-year plan
1986	AFM	Development of new instrument at (IBM Zurich Research Lab) known as The Atomic Force Microscope (AFM)	Important role in surface imaging at molecular level, and materials problems including measurement of infinitesimal forces at nanonewton scale
1981	Scanning Tunneling Microscope	Development of instrument	Important innovation in the field of nanotechnology
1974	"Nanotechnology"	Professor Norio Taniguchi Coined the term	Beginning of the nanotechnology era

17.5 Conclusion and Future Outlook

This compilation offers a birdeye view about the efforts and status of the individual and combined efforts that are made at various parts of the globe (Table 17.12). The status of key enabling technology has given the field necessary push it required to identify and utilize the potential of this magic bullet. The pace at which research and development are progressing in the field of nanotechnology is really admirable. Efforts made in last decades have helped the domain to acquire a status of maturity. It may be attributed primarily to the regulatory proactiveness and efficiency improvements that it offers for existing drugs and delivery systems. International synchronization of efforts have further boosted the pace of the advancement and improved our understanding. The sudden strike of pandemic has stalled the progress of the sector for a while, until it was realized that calamity may actually be the driving force for an industry that requires extensive and prolong testing of new drugs and approaches. It gives a midas touch to improve the efficacy of most of the existing sanitizers, chemicals, vaccines, protective gears, drugs, dosage forms, devices, and numerous other things. Definitely, significant improvements in efficiency of existing drugs and dosage form and possibility of better outcomes using nanotechnology may serve as an important and significant tool for mankind to fight the challenges of the future.

References

(EU) 2017/745 (2017). European regulations on medical devices. European parliament and of the council (5 April 2017).

ASTM E2535–07 (2018). *Standard Guide for Handling Unbound Engineered Nanoscale Particles in Occupational Settings*. ASTM International.

Australian Academy of Science (2020). Nanotechnology regulation in Australia. Australian standards. https://www.science.org.au/curious/technology-future/nanotechnology-regulation-australia (accessed 4 December 2020).

BIAC (2016). Comments of the BIAC Nanotechnology Expert Group on the OECD publication. http://biac.org/wp-content/uploads/2016/07/2016-04-BIAC-Comments-NM-in-Waste-Streams1.pdf (accessed 4 April 2019).

Boverhof, D.R., Bramante, C.M., Butala, J.H. et al. (2015). Comparative assessment of nanomaterial definitions and safety evaluation considerations. *Regulatory Toxicology and Pharmacology* https://doi.org/10.1016/j.yrtph.2015.06.001.

CDC (2018). Nanotechnology. https://www.cdc.gov/niosh/topics/nanotech/pubs.html (accessed 4 April 2019).

ChemSafetyPRO (2020). European Inventory of Existing Commercial Chemical Substances (EINECS). https://www.chemsafetypro.com/Topics/EU/EU_Chemical_Inventory_EINECS_ELINCS_NLP.html (accessed 4 December 2020).

China National Standards (2020a). Chinese standards. http://www.gbstandards.org/index/GB_Search.asp (accessed 18 December 2020).

China National Standards (2020b). China IECSC: inventory of existing chemical substances in China. https://www.chemsafetypro.com/Topics/China/China_IECSC_Inventory_Existing_Substances.html (accessed 18 December 2020).

COSHH (2002). The Control of Substances Hazardous to Health Regulations 2002, USA.

Danish Environmental Protection Agency (DEPA) (2014). Guideline for the Danish inventory of nanoproducts. Danish Standards. https://www2.mst.dk/Udgiv/publications/2014/08/978-87-93178-89-2.pdf (accessed 4 December 2020).

Department of Chemical and Product Safety, Germany (2020). EU US Roadmap Nanoinformatics 2030. https://www.nanosafetycluster.eu/wp-content/uploads/NSC%20Outputs/Roadmaps/EU%20US%20Nanoinformatics%20Roadmap%202030.pdf?_t=1542815805 (accessed 12 December 2020).

Division of Occupational Health and Safety (2014). Nanotechnology Safety and Health Program: Office of Research Services,

National Institutes of Health, Technical Assistance Branch, USA.

ECHA (2020a). Understanding REACH. https://echa.europa.eu/regulations/reach/understanding-reach (accessed 22 December 2020)

ECHA (2020b). Legislation. https://echa.europa.eu/legislation (accessed 4 December 2020).

Environmental Protection Agency (EPA) (2017). Nanoscale Materials; TSCA Reporting and Recordkeeping Requirements, 82 FR 22088. Office of the Federal Register, National Archives and Records Administration, USA

EPA (2018). USA agriculture. www.epa.gov/agriculture (accessed 4 December 2020).

EURLEX (2011). 2011/696/EU, Commission Recommendation of 18 October 2011 on the definition of nanomaterial Text with EEA relevance, OJ L 275, 20.10.2011, p. 38–40. http://data.europa.eu/eli/reco/2011/696/oj (accessed 4 December 2020)

Europa (2020). What is Horizon 2020? https://ec.europa.eu/programmes/horizon2020/en/what-horizon-2020 (accessed 22 December 2020).

European Commission (2004). Toward a European Strategy for Nanotechnology, Belgium (communication).

Federal Public Service (2020). https://www.health.belgium.be/sites/default/files/uploads/fields/fpshealth_theme_file/1._be_nanoregister_setting_the_scene.pdf (accessed 4 December 2020).

Feynman, R.P. (1960). There's plenty of room at the bottom. *California Institute of Technology Journal of Engineering and Science* 4 (2): 23–36.

France (2020). Declaration. French Declaration for nanomaterial. https://www.r-nano.fr/?locale=en (accessed 4 December 2020).

Friedrichs, S. (2018). *Report on Statistics and Indicators of Biotechnology and Nanotechnology*, OECD Science, Technology and Industry Working Papers, vol. 2018/06. OECD https://doi.org/10.1787/3c70afa7-en.

GoNano (2018). Risk governance and research & innovation priorities in nanotechnologies. http://gonano-project.eu/. http://gonano-project.eu/wp-content/uploads/2018/08/5.1.pdf (accessed 4 December 2020).

Government of Canada (2020a). Policy statement on Health Canada's working definition for nanomaterial. https://www.canada.ca/en/health-canada/services/science-research/reports-publications/nanomaterial/policy-statement-health-canada-working-definition.html (accessed 4 December 2020).

Government of Canada (2020b). Nanotechnology-based health products and food. https://www.canada.ca/en/health-canada/services/drugs-health-products/nanotechnology-based-health-products-food.html (accessed 4 December 2020).

Grand View Research (2017). Nanomedicine Market Size Worth $350.8 Billion By 2025 Press Release (April 2017).

Huan-Cheng, C., Wesley, W.H., and Meng-Chih, S. (2018). Introduction to nanotechnology. In: *Fluorescent Nanodiamonds*. Wiley https://doi.org/10.1002/9781119477099.ch1.

India (2020). Indian guideline. http://www.dbtindia.nic.in/wp-content/uploads/Modified-Guidelines-for-Evaluation-of-Nanopharmaceuticals-in-India-converted-2.pdf (accessed 18 December 2020).

Internano (2019). Resources for nanomanufacturing 2018. http://internano.org/ (accessed 22 March 2019).

Iran Regulations (2020). Laws and regulations of Iran. http://dotic.ir/news/621 (accessed 4 December 2020).

ISO/TC 229 (2018a). Technical Committees on Nanotechnology. https://www.iso.org/committee/381983.html (accessed 22 December 2020).

ISO/TC 229 (2018b). Technical Committees on Nanotechnology. https://www.iso.org/committee/381983/x/catalogue/ (accessed 22 December 2020).

ISO/TR 12885:2018 (2018). *Nanotechnologies: Health and Safety Practices in Occupational Settings*. International Organization for Standardization.

ISO/TS 27687:2008 (2008). *Nanotechnologies: Terminology and Definitions for Nano-Objects–Nanoparticle, Nanofibre and Nanoplate*. NSAI.

ISO/TS 80004-1:2010 (2010). *Nanotechnologies: Vocabulary–Part 1: Core Terms*. International Organization for Standardization.

ISO/TS 80004-12:2016 (2016). *Nanotechnologies: Vocabulary–Part 12: Quantum Phenomena in Nanotechnology*. International Organization for Standardization.

Jantunen, A.K., Gottardo, S., Rasmussen, K., and Crutzen, H.P. (2018). An inventory of ready-to-use and publicly available tools for the safety assessment of nanomaterials. *NanoImpact* 12: 18–28. https://doi.org/10.1016/j.impact.2018.08.007.

Japan (2018a). https://www8.cao.go.jp/cstp/english/index.html (accessed 20 December 2020).

Japan (2018b). http://www.jpma.or.jp/english/parj/pdf/2015.pdf (accessed 18 December 2020).

Japan (2018c). https://www.meti.go.jp/policy/chemical_management/english/cscl/files/laws/laws_cscl.pdf (accessed 12 December 2020).

Japan (2018d). Japanese ENCS: list of existing and new chemical substances. https://www.chemsafetypro.com/Topics/Japan/Japan_ENCS_Inventory_of_Existing_and_New_Chemical_Substances.html (accessed 14 December 2020).

Japan Science and Technology Agency (2018). CRDS-FY2017-XR-02 Nanotechnology and Materials R&D in Japan (2018): An Overview and Analysis. Nanotechnology/Materials Unit, Center for Research and Development Strategy, Japan Science and Technology Agency, Japan.

Kim, J. and Yu, I.J. (2016). National survey of workplaces handling and manufacturing nanomaterials, exposure to and health effects of nanomaterials, and evaluation of nanomaterial safety data sheets. *BioMed Research International* https://doi.org/10.1155/2016/8389129.

K-Reach Database (2019). https://en.k-reach.com/ (accessed 18 December 2020).

Lee, D. and Choi, K. (2019). Comparison of regulatory frameworks of environmental risk assessments for human pharmaceuticals in EU, USA, and Canada. *Science of The Total Environment* 671 (24): 1026–1035.

McCray, W.P. (2006). Will small be beautiful? Making policies for our nanotech future. *History and Technology* https://doi.org/10.1080/07341510500103735.

McWilliams, A. (2016). The Maturing Nanotechnology Market: Products and Applications. 4 NAN031G, Global Markets, BBC Research Report, USA, 2016.

NANO (2020). What it is and how it works. www.nano.gov/nanotech-101/what (accessed 4 December 2020).

NanoSafetyCluster (2018). Strategic direction enhancing synergies. https://www.nanosafetycluster.eu/ (accessed 4 December 2020).

National Research Council (2012). *A Research Strategy for Environmental, Health, and Safety Aspects of Engineered Nanomaterials*. Washington, DC: The National Academies Press https://doi.org/10.17226/13347.

National Science and Technology Council (2000). National Nanotechnology Initiative the Initiative and its Implementation Plan. National Science and Technology Council Committee on Technology Subcommittee on Nanoscale Science, Engineering, and Technology. Executive Office of the President of USA.

National Science and Technology Council (2016). National nanotechnology initiative strategic plan, National Science and Technology Council Committee on Technology Subcommittee on Nanoscale Science, Engineering, and Technology. Executive Office of the President of USA.

NIOSH (2009). *Approaches to Safe Nanotechnology: Managing the Health and Safety Concerns Associated with Engineered Nanomaterials*, DHHS (NIOSH) Publication No., vol. 2009-125. Cincinnati, OH: U.S. Department of Health and Human Services,

Centers for Disease Control and Prevention, National Institute for Occupational Safety and Health.

NIOSH (2012). *General Safe Practices for Working with Engineered Nanomaterials in Research Laboratories*, DHHS (NIOSH) Publication No., vol. 2012-147. Cincinnati, OH: U.S. Department of Health and Human Services, Centers for Disease Control and Prevention, National Institute for Occupational Safety and Health.

NIOSH (2016). *Building a Safety Program to Protect the Nanotechnology Workforce: A Guide for Small to Medium-Sized Enterprises*. By Hodson L. and Hull, M, DHHS (NIOSH) Publication No., vol. 2016-102. Cincinnati, OH: U.S. Department of Health and Human Services, Centers for Disease Control and Prevention, National Institute for Occupational Safety and Health.

OECD (2013). *Regulatory Frameworks for Nanotechnology in Foods and Medical Products: Summary Results of a Survey Activity*, OECD Science, Technology and Industry Policy Papers, No., vol. 4. Paris: OECD Publishing http://dx.doi.org/10.1787/5k47w4vsb4s4-en.

OECD (2018a). http://www.oecd.org/sti/emerging-tech/34360743.pdf (accessed 22 December 2020).

OECD (2018b). *Revised Proposal for the Revision of the Statistical Definitions of Biotechnology and Nanotechnology*, OECD Science, Technology and Industry Working Papers, vol. 2018/1. OECD Publishing https://doi.org/10.1787/18151965.

OECD (2020). Publications in the series on the safety of manufactured nanomaterials. http://www.oecd.org/env/ehs/nanosafety/publications-series-safety-manufactured-nanomaterials.htm (accessed 4 December 2020).

OSH Act of 1970, USA.

Patravale, V., Dandekar, P., and Jain, R. (2012). Regulatory aspects of nanoparticulate drug delivery systems. In: *Nanoparticulate Drug Delivery*, 157–186. USA: Woodhead Publishing Ltd https://www.sciencedirect.com/science/article/pii/B978190756898550005X?via%3Dihub.

Rasmussen, K., Rauscher, H., Mech, A. et al. (2018). Physico-chemical properties of manufactured nanomaterials – characterisation and relevant methods. An outlook based on the OECD Testing Programme. *Regulatory Toxicology and Pharmacology* https://doi.org/10.1016/j.yrtph.2017.10.019.

Rauscher, H., Roebben, G., Sanfeliu, A.B. et al. (2015). JRC science for policy report. In: *Towards a review of the EC Recommendation for a definition of the term "nanomaterial" part-3* (eds. H. Rauscher and G. Roebben). Publications Office of the European Union.

Rauscher, H., Rasmussen, K., and Sokull-klüttgen, B. (2017). Regulatory aspects of nanomaterials in the EU. *Chemie ingenieur technik* 89: 224–231.

Roco, M.C., Hersam, M., and Mirkin, C.A. (2011). *WTEC Panel Report on Nanotechnology Research Directions for Societal Needs in 2020 Retrospective and Outlook*. Springer Netherlands.

Russian Federation (2020). Federal law on circulation of medicines. https://advocatetanmoy.com/2020/08/13/russian-federal-law-on-circulation-of-medicines/ (accessed 22 December 2020).

SAFENANO (2020). Current nanotechnology standardisation activities. https://www.safenano.org/knowledgebase/standards/ (accessed 4 December 2020).

Solaiman, S.M., Yamauchi, Y., Kim, J.H. et al. (2017). Nanotechnology and its medical applications: revisiting public policies from a regulatory perspective in Australia. *Nanotechnology Reviews* https://doi.org/10.1515/ntrev-2016-0095.

StatNano (2018). https://statnano.com/nanomaterials (accessed 20 December 2019).

Steinhäuser, K. and Sayre, P. (2017). Reliability of methods and data for regulatory assessment of nanomaterial risks. *NanoImpact* 7: 66–74.

https://www.sciencedirect.com/science/article/pii/S2452074817300472.

Stone, V., Johnston, H., and Schins, R.P.F. (2009). Development of in vitro systems for nanotoxicology: methodological considerations. *Critical Reviews in Toxicology* 39 (7): 613–626.

TANIDA (2020). Taiwanese regulations. http://www.tanida.org.tw/tanida_e.php?mn=tnsc_e&nc=tnscStandard_e (accessed 18 December 2020).

Taniguchi, N. (1974). On the basic concept of nanotechnology. In: *Proceedings of the International Congress on Prod Eng*. Tokyo, Japan: JSPE.

TGA (2016). Regulation of nanomedicines. https://www.tga.gov.au/sites/default/files/tga-presentation-nanoparticle-therapeutics-2016-20-october-2016.pdf (accessed 12 April 2020).

The Control of Substances Hazardous to Health Regulations (2002). Globally harmonized system of classification and labelling of chemicals. http://www.unece.org/fileadmin/DAM/trans/danger/publi/ghs/ghs_rev05/English/ST-SG-AC10-30-Rev5e.pdf (accessed 4 April 2019).

Toxic Substances Control Act (TSCA) of 1976, 15 U.S.C. 2601 et seq. (1976). United States Congress

Trotta, F. and Mele, A. (eds.) (2019). Nanomaterials: classification and properties. In: *Nanosponges: Synthesis and Applications*. Wiley.

Urbán, P., Liptrott, N.J., and Bremer, S. (2018). Overview of the blood compatibility of nanomedicines: a trend analysis of in vitro and in vivo studies. *Wiley Interdisciplinary Reviews: Nanomedicine and Nanobiotechnology* https://doi.org/10.1002/wnan.1546.

USFDA (2014a). US Food and Drug Administration. Guidance for Industry Considering Whether an FDA-Regulated Product Involves the Application of Nanotechnology. Office of the Commissioner, U.S. Department of Health and Human Services Food and Drug Administration, USA

VAMAS (2018). Versailles Project on Advanced Materials and Standards (VAMAS). http://www.vamas.org/ (accessed 22 December 2020).

WHO (2012). *Nanotechnology and Human Health: Scientific Evidence and Risk Governance. Report of the WHO Expert Meeting 10–11 December 2012, Bonn, Germany*. Copenhagen: WHO Regional Office for Europe, 2013.

WHO (2017). *Guidelines on Protecting Workers from Potential Risks of Manufactured Nanomaterials*, Department of Public Health, Environmental and Social Determinants of Health, Cluster of Climate and Other Determinants of Health, 2017, published by. Switzerland: World Health Organization.

Xie, H. and Régnier, S. (2012). 3D Micro/nanomanipulation with force spectroscopy. In: *Encyclopedia of Nanotechnology* (ed. B. Bhushan). Dordrecht: Springer https://doi.org/10.1007/978-90-481-9751-4_203.

Index

Nanotechnology in Medicine: Toxicity and Safety, First Edition. Edited by Mahendra Rai, Mrunali Patel, and Rashmin Patel.

e

Z

www.ingramcontent.com/pod-product-compliance
Lightning Source LLC
LaVergne TN
LVHW062109200926
848484LV00007B/1623